ADVANCED TOPICS IN MULTIVARIATE APPROXIMATION

APPROXIMATIONS AND DECOMPOSITIONS

Editor-in-Chief: CHARLES K. CHUI

Vol. 1: Wavelets: An Elementary Treatment of Theory and Applications
Tom H. Koornwinder, ed.

Vol. 2: Approximate Kalman Filtering
Guanrong Chen, ed.

Vol. 3: Multivariate Approximation: From CAGD to Wavelets
Kurt Jetter and Florencio I. Utreras, eds.

Vol. 4: Advances in Computational Mathematics: New Delhi, India
H. P. Dikshit and C. A. Micchelli, eds.

Vol. 5: Computational Methods and Function Theory
Proceedings of CMFT '94 Conference, Penang, Malaysia
R. M. Ali, St. Ruscheweyh and E. B. Saff, eds.

Vol. 6: Approximation Theory VIII
Approximation and Interpolation – Vol. 1
Wavelets and Multi-level Approximation – Vol. 2
C. K. Chui and L. L. Schumaker, eds.

Vol. 8: Advanced Topics in Multivariate Approximation
F. Fontanella, K. Jetter and P. J. Laurent, eds.

Forthcoming
Vol. 7: Introduction to the Theory of Weighted Polynomial Approximation
H. N. Mhaskar

Series in Approximations and Decompositions – Vol. 8

Proceedings of the International Workshop

ADVANCED TOPICS IN MULTIVARIATE APPROXIMATION

Montecatini Terme, Italy
September 27 – October 3, 1995

edited by

F. Fontanella
Università di Firenze
Italy

K. Jetter
Universität Hohenheim
Germany

P. J. Laurent
Université Joseph Fourier
France

World Scientific
Singapore • New Jersey • London • Hong Kong

Published by

World Scientific Publishing Co. Pte. Ltd.

P O Box 128, Farrer Road, Singapore 912805

USA office: Suite 1B, 1060 Main Street, River Edge, NJ 07661

UK office: 57 Shelton Street, Covent Garden, London WC2H 9HE

Library of Congress Cataloging-in-Publication Data

Advanced topics in multivariate approximation : proceedings of the international workshop : Montecatini Terme, Italy, September 27–October 3, 1995 / edited by F. Fontanella, K. Jetter, P. J. Laurent.

380 p., 21.5 cm. -- (Series in approximations and decompositions : vol. 8)

ISBN 981022852X (alk. paper)

1. Approximation theory -- Congresses. 2. Functions of several real variables -- Congresses. I. Fontanella, F. II. Jetter, K. III. Laurent, Pierre Jean, 1937– . IV. Series.

QA297.5.A28 1996

511'.4--dc20 96-40987

CIP

British Library Cataloguing-in-Publication Data

A catalogue record for this book is available from the British Library.

Printed in Singapore by Uto-Print

Approximations and Decompositions

During the past decade, Approximation Theory has reached out to encompass the approximation-theoretic and computational aspects of several exciting areas in applied mathematics such as wavelets, fractals, neural networks, and computer-aided-geometric design, as well as the modern mathematical development in science and technology. The objective of this book series is to capture this exciting development in the form of monographs, lecture notes, reprint volumes, textbooks, edited review volumes, and conference proceedings.

This is the eighth volume of the series. Based on the invited presentations at the Fifth International Workshop on Multivariate Approximation and Interpolation, this volume consists of 24 carefully reviewed and edited papers on current topics of interest that range from the basic theory to practical algorithms and applications. The series editor would like to congratulate the editors and their assistants for their excellent effort in editing this volume and the authors for their fine contributions.

World Scientific Series in
APPROXIMATIONS AND DECOMPOSITIONS

Editor-in-Chief: CHARLES K. CHUI
Texas A&M University, College Station, Texas

Preface

During the week of September 27 – October 3, 1995, an International Workshop on Multivariate Approximation and Interpolation was held at Montecatini Terme, Italy, organized by F. Fontanella, C. Manni and R. Morandi (Università degli Studi di Firenze), M. Bozzini (Università degli Studi di Milano) and P. Costantini (Università degli Studi di Siena). Previous conferences in this conference series took place in 1983 in Gargnano (Italy), 1986 and 1992 in Santiago (Chile), and 1989 in Duisburg (Germany).

This workshop was attended by 59 mathematicians and engineers from 18 different countries. The program included 38 invited talks dealing with the following topics: subdivision algorithms, constrained and shape preserving approximation, thin plate splines, radial basis functions, treatment of scattered data, rational surfaces and offsets, blossoming, grid generation, surface reconstruction, algebraic curves and surfaces, and neural networks. This book consists of 24 refereed papers based on the invited lectures, and we believe that it provides a good selection of advanced topics of current interest in multivariate approximation.

We acknowledge the helpful advice from the following members of the international organizing committee, C. K. Chui, N. Dyn, C. A. Micchelli and F. I. Utreras, for setting up the scientific perspective of the conference. Our thanks go to the European Commission - DG XII, the Comitato per la Matematica del C.N.R, the Gruppo Nazionale Informatica Matematica, the Università degli Studi di Firenze, the Regione Toscana, the Provincia di Firenze, and the ATP Montecatini/Valdinievole for their support.

We are also grateful for the assistance from the following people: the secretaries of the Dipartimento di Energetica "S. Stecco" and the group in Numerical Analysis, in particular C. Conti, C. Manni, M. L. Sampoli and A. Sestini for running the conference and editing this volume. Last but not least, we would like to express our thanks to the series editor, C. K. Chui, his wife Margaret, and the publisher for their support.

F. Fontanella
K. Jetter
P. J. Laurent

Florence, July 1996

List of Participants

In the following list of participants, the numbers in parenthesis indicate the page on which the author's contribution(s) begin.

Silvia Bertoluzza, *I.A.N.-C.N.R., Via Abbiategrasso 215, 27100 Pavia,* ITALY [aivlis@dragon.ian.pv.cnr.it].

Peter G. Binev, *Institute of Mathematics, Bulgarian Academy of Sciences, Sofia 1090,* BULGARIA [binev@fmi.uni-sofia.bg].

Mira Bozzini (1), *Dipartimento di Matematica, Universitá di Milano, Via C. Saldini 50, 20133 Milano,* ITALY [bozzini@vmimat.mat.unimi.it].

Martin D. Buhmann, *Mathematik Departement, ETH Zentrum, 8092 Zürich,* SWITZERLAND [mdb@sam.math.ethz.ch].

Nicola Castanó, *I.M.A.-C.N.R., Via De Marini 6, 16143 Genova,* ITALY [nicola@ima.ge.cnr.it].

Albert Cohen (33), *Laboratoire d' Analyse Numérique Université Pierre et Marie Curie, 4 Place Jussieu Paris, 75005 Paris,* FRANCE [cohen@ann.jussieu.fr].

Valentina Colla, *Scuola Superiore di Studi Universitari e di Perfezionamento Sant' Anna, Via Carducci 40, 56127 Pisa,* ITALY [colla@shamash.sssup.it].

Costanza Conti, *Dipartimento di Energetica, Universitá di Firenze, Via Lombroso 6/17, 50134 Firenze,* ITALY [conti@ingfi1.ing.unifi.it].

Paolo Costantini, *Dipartimento di Matematica, Universitá di Siena Via del Capitano 15, 53100 Siena,* ITALY [costantini@unisi.it].

Luisa D'Amore, *Dipartimento di Matematica e Appl., Universitá di Napoli Federico II, Via Cintia, 80126 Napoli,* ITALY [damore@ matna2.dma.unina.it].

Stefano De Marchi, *Dipartimento di Matematica Pura e Applicata, Universitá di Padova, Via Belzoni 7, 35131 Padova,* ITALY [stef@ hilbert.math.unipol.it].

Vittoria Demichelis, *Dipartimento di Matematica, Universitá di Torino Via Carlo Alberto 10, 10123 Torino,* ITALY [demichelis@dm.unito.it].

Uwe Depczynski (57), *Inst. für Angewandte Mathematik und Statistik, Universität Hohenheim, 70593 Stuttgart,* GERMANY [depczyns@uni-hohenheim.de].

Nira Dyn (33), *School of Mathematical Sciences, Tel-Aviv University, Ramat Aviv, 69978 Tel Aviv,* ISRAEL [niradyn@math.tau.ac.il].

Hans Georg Feichtinger, *Department of Mathematics, University of Vienna, Strudlhofg. 4, A-1090 Wien,* AUSTRIA [fei@tyche.mat.univie.ac.at].

Yu Yu Feng (71), *Department of Mathematics, University of Science and Technology of China, Hefei, Anhui 230026,* P. R. of CHINA [mailer@ustc.ac.cn].

Ferruccio Fontanella, *Dipartimento di Eergetica, Universitá di Firenze, Via S. Marta 3, 50139 Firenze,* ITALY [fontanella@ingfi1.ing.unifi.it].

Emanuele Galligani, *Dipartimento di Matematica, Universitá di Modena, Via Campi 213/b, 41100 Modena,* ITALY [galligani@dipmat.unimo.it].

Tim N. T. Goodman (125), *Department of Mathematics and Computer Science, University of Dundee, Dundee DD1 4HN Scotland,* UK [tgoodman@mcs.dundee.ac.uk].

George Goodsell, *Department of Applied Mathematics and Theoretical Physics, University of Cambridge, Silver Street, Cambridge CB3 9EW,* UK [G.Goodsell@damtp.cam.ac.uk].

Laura Gori, *Dipartimento di Metodi e Modelli Matematici per le Scienze Applicate, Universitá "La Sapienza", Via A. Scarpa 16, 00161 Roma,* ITALY [symorpol@itcaspur.bitnet].

Hakop Hakopian (97), *Department of Mathematics, Yerevan State University, A. Manoukian, 1 375049, Yerevan,* ARMENIA [gayane@arminco.com].

Armin Iske (87), *SINTEF SI, Postboks 124, Blindern, N-0314 Oslo,* NORWAY [Armin.Iske@si.sintef.no].

Kurt Jetter (57), *Inst. für Angew. Mathematik und Statistik, Universität Hohenheim, 70593 Stuttgart,* GERMANY [kjetter@uni-hohenheim.de].

Rong-Qing Jia (139), *Department of Mathematical Sciences, University of Alberta, Edmonton, T6G 2G1,* CANADA [jia@xihu.math.ualberta.ca].

Panagiotis D. Kaklis (113), *Department of Naval Arch. and Mar. Eng., National Technical University of Athens, 9 Heroon Polytechneiou St., Zografou, Athens 157 73,* GREECE [kaklis@deslab.naval.ntua.gr].

Anna Kamont, *Mathematical Institute of Polish Academy of Sciences, ul. Abrahama 18, 81-825 Sopot,* POLAND [A.Kamont@impan.gda.pl].

Alex K. Kushpel (155), *Department of Mathematics and Computer Sciences, University of Leicester, University Road, Leicester, LE1 7RH,* UK [ak99@mcs.le.ac.uk].

Boris I. Kvasov (181), *Institute of Computational Technologies, Russian Academy of Sciences, Siberian Branch, 6, Lavrentyeva Avenue, Novosibirsk, 630090,* RUSSIA [kvasov@ictnw.scnet.nsk.su], [kvasov@sura1.sut.ac.th]

Pierre-Jean Laurent (227), *LMC-IMAG, Université Joseph Fourier, BP 53X, 38041 Grenoble,* FRANCE [pjl@imag.fr].

Alain Le Méhauté, *Departement ITI ENST-Bretagne, B.P. 832, 29285 Brest cedex,* FRANCE [Alain.LeMehaute@enst-bretagne.fr].

Licia Lenarduzzi (1), *Istituto Applicazioni Matematica Informatica C.N.R., Via Ampere 56, 20131 Milano,* ITALY [licia@hp1.iami.mi.cnr.it].

Burkhard Lenze (197), *Fachbereich Informatik, Fachhochschule Dortmund, Postfach 105018, D-44047, Dortmund* GERMANY [lenze@fh-dortmund.de].

Dany Leviatan, *School of Mathematical Sciences, Tel-Aviv University, Ramat Aviv, 69978 Tel Aviv,* ISRAEL [leviatan@math.tau.ac.il].
Department of Mathematics, University of South–Carolina, Columbia SC 29208, USA [leviatan@math.sc.edu].

David Levin (33, 47), *School of Mathematical Sciences, Tel-Aviv University, Ramat Aviv, 69978 Tel Aviv,* ISRAEL [levin@math.tau. ac.il].

Tom Lyche, *University of Oslo, Institutt for Informatikk, P.O. Box 1080 Blindern, 0316 Oslo,* NORWAY [tom@ifi.uio.no].

Lucia Maddalena, *Centro di Ricerche per il Calcolo Parallelo e i Supercalcolatori, C.N.R., c/0 Dipartimento di Matematica e Appl., Universitá di Napoli Federico II, Via Cintia, 80126 Napoli,* ITALY [lucia@matna2.dma.unina.it].

Carla Manni (211), *Dipartimento di Energetica, Universitá di Firenze, Via Lombroso 6/17, 50134 Firenze,* ITALY [manni@ingfi1.ing.unifi.it].

Marie-Laurence Mazure (227), *LMC-IMAG, Université Joseph Fourier, BP 53X, 38041 Grenoble,* FRANCE [Marie-Laurence.Mazure@imag.fr].

Rossana Morandi (243), *Dipartimento di Matematica, Universitá di Perugia, Via Vanvitelli 1, 06123 Perugia,* ITALY [morandi@gauss.dipmat.unipg.it].

Almerico Murli, *Dipartimento di Matematica e Appl., Universitá di Napoli Federico II, Via Cintia, 80126 Napoli,* ITALY [murli@matna2.dma.unina.it].

Boon Hua Ong (259), *School of Math. and Computer Sciences, Universiti Sains Malaysia, 11800 Penang,* MALAYSIA [bhong@cs.usm.my].

Juan Manuel Peña (17), *Departamento de Matemática Aplicada, Edificio de Matématicas, Universidad de Zaragoza, 50009 Zaragoza,* SPAIN [jmpena@posta.unizar.es].

Helmut Pottmann (275), *Institut für Geometrie, Technische Universität Wien, Wiedner Hauptstrasse 8-10/113, A-1040 Wien,* AUSTRIA [pottmann@geometrie.tuwien.ac.at].

Mike J. D. Powell (303), *DAMTP, University of Cambridge, Silver Street, Cambridge CB3 9EW,* UK [mjdp@amtp.cam.ac.uk].

Luigia Puccio, *Dipartimento di Matematica, Universitá di Messina, Contrada Papardo Salita Sperone 31, 98166 Messina,* ITALY [gina@imeuniv.unime.it].

Christophe Rabut, *Centre de Matematiques, Institut National des Sciences Appliquées, Complexe Scientifique de Rangueil, F-31077 Toulouse Cedex,* FRANCE [rabut@cict.fr].

Paul Sablonnière (323), *Institut National des Sciences Appliquées, 20, avenue des Buttes de Coësmes, 35043 Rennes Cedex,* FRANCE [sablonni@perceval.univ-rennes1.fr].

Maria Lucia Sampoli, *Dipartimento di Matematica, Universitá di Milano, Via Saldini 50, 20133 Milano,* ITALY [sampoli@fonty.ing.unifi.it].

Elisabetta Santi, *Dipartimento di Energetica, Facoltá di Ingegneria, Universitá de L'Aquila, 67040 Monteluco Roio, L'Aquila,* ITALY.

Giovanni Schmid, *Dipartimento di Matematica e Appl., Universitá di Napoli Federico II, Via Cintia, 80126 Napoli,*ITALY [schmid@matna2.dma.unina.it].

Sebastiano Seatzu, *Dipartimento di Matematica, Universitá di Cagliari, Via Merello 92, 09123 Cagliari* ITALY [seatzu@tex.unica.it].

Alessandra Sestini (243), *Dipartimento di Energetica, Universitá di Firenze, Via Lombroso 6/17, 50134 Firenze,* ITALY [sestini@ingfi1.ing.unifi.it].

Joachim Stöckler (359), *Inst. für Angew. Mathematik und Statistik, Universität Hohenheim, 70593 Stuttgart,* GERMANY [stockler@uni-hohenheim.de].

Manfred Tasche (287), *Inst. für Mathematik, Universität zu Lübeck, Wallstr. 40, 23560 Lübeck,* GERMANY [manfred.tasche@informatik.mv-luebeck.de].

Riccardo Ulivelli, *C.C.S., Via del Guarlone 60, 50135 Firenze,* ITALY.

Maria L. Varas, *Departamento de Ingeniería Matemática, Universidad de Chile, Casilla 5272 - Correo 3 Santiago,* CHILE [mlvaras@dim.uchile.cl].

Ren-Hong Wang (355), *Institute of Mathematical Sciences, Dalian University of Technology, Dalian 116024,* P. R. of CHINA

Ermanno Zanazzi, *C.C.S., Via del Guarlone 60, 50135 Firenze,* ITALY.

Luca Zanni, *Dipartimento di Matematica, Universitá di Modena, Via Campi 213/b, 41100 Modena,* ITALY [zanni@dipmat.unimo.it].

Table of Contents

Preface vii

List of Participants ix

Recovering a Function with Discontinuities from Correlated Data
M. Bozzini and L. Lenarduzzi 1

Bases with Convexity Preserving Properties
J. M. Carnicer, M. García-Esnaola and J. M. Peña 17

Stability and Inter-Dependence of Matrix Subdivision Schemes
A. Cohen, N. Dyn and D. Levin 33

Guided Multi-Dimensional Reconstruction from Cross-Sections
D. Cohen-Or and D. Levin 47

Multiscale Decompositions on the Interval Based on DCT-I Transforms
U. Depczynski and K. Jetter 57

On the Dimension of the C^1 Spline Space for the Morgan-Scott Triangulation from the Blossoming Approach
Y. Y. Feng, J. Kozak and M. Zhang 71

Thinning and Approximation of Large Sets of Scattered Data
M. S. Floater and A. Iske 87

Numerical Curves and their Applications to Algebraic Curves
H. Gevorgian, H. Hakopian and A. Sahakian 97

Sectional-Curvature Preserving Skinning Surfaces with a 3D Spine Curve
A. I. Ginnis, P. D. Kaklis and N. C. Gabrielides 113

Pairs of Refinable Bivariate Splines
T. Goodman 125

The Subdivision and Transition Operators Associated with a Refinement Equation
R. Q. Jia 139

Approximation of Smooth Functions by sk-Splines
A. K. Kushpel, J. Levesley and W. A. Light 155

Shape Preserving Spline Approximation via Local Algorithms
B. I. Kvasov 181

Introducing Dilation and Translation for Incomplete Sigma-Pi Neural Networks of Hopfield-Type
B. Lenze 197

Parametric Shape-Preserving Hermite Interpolation by Piecewise Quadratics
C. Manni 211

Marsden Identity, Blossoming and de Boor-Fix Formula
M. L. Mazure and P. J. Laurent 227

Precise Controls in Numerical Grid Generation
R. Morandi and A. Sestini 243

A C^1 Positivity Preserving Scattered Data Interpolation Scheme
B. H. Ong and H. C. Wong 259

Designing Rational Surfaces with Rational Offsets
M. Peternell and H. Pottmann 275

Kernels of Spherical Harmonics and Spherical Frames
D. Potts, G. Steidl and M. Tasche 287

A Review of Algorithms for Thin Plate Spline Interpolation in Two Dimensions
M. J. D. Powell 303

B-Splines on Uniform Meshes of the Plane
P. Sablonnière 323

A Laurent Operator Technique for Multivariate Frames and Wavelet Bases
J. Stöckler 339

On Piecewise Algebraic Curves
R. H. Wang 355

ADVANCED TOPICS IN MULTIVARIATE APPROXIMATION

Recovering a Function with Discontinuities from Correlated Data

M. Bozzini and L. Lenarduzzi

Abstract. Assume that a function is measured on a grid with constant mesh size and that the data are affected with correlated errors. Assume that there are irregularities in the function. There is a need for a fast method which gives a good graphical representation of the function. The aim is obtained by using the discrete wavelet transform and by using a partial spline model with Hermite bases for the parametric part.

§1. Introduction

In this paper we consider a regression model from noisy data, both for dimension one and two, when the regression function allows singularities.

The model presented here makes use of an expansion with wavelets. Because of the localization properties of wavelets, the graphic of the approximating function fits the graphic of the unknown function in a better way. This was already remarked by Antoniadis in [1], when comparing his method with the classical method of smoothing splines.

To apply our method, one needs to know the locations of the singularities and the estimates of the stochastic parameters of the process. For such a purpose one can use the method by Rossini in [8], that provides such unknown elements by the wavelet transform and the estimate of the spectral density of the process. In what follows, we assume that we know the estimate of the spectrum and the locations of the singularities, and that they are coincident with some data locations.

Advanced Topics in Multivariate Approximation
F. Fontanella, K. Jetter and P. J. Laurent (eds.) pp. 1–16.

ISBN 981-02-2852-X

1.1 A Short Review of Wavelets

We recall the properties of orthogonal wavelets and of the multiresolution analysis (MRA) which we shall use for our estimator.

Wavelets are functions generated by translations and dilations of a basis function. With a suitable choice of the function $\psi \in L^2(\mathbb{R})$ the family

$$\{\psi_{j,k}(t) = 2^{-j/2}\psi(2^{-j}t - k) \quad j, k \in \mathbb{Z}\}$$

constitutes an orthogonal basis of $L^2(\mathbb{R})$.

For the construction of such a basis we make use of the orthogonal MRA.

A MRA of $L^2(\mathbb{R})$ consists of a sequence of closed subspaces V_j, $j \in \mathbb{Z}$ of $L^2(\mathbb{R})$

$$\ldots \subset V_j \subset V_{j-1} \subset \ldots$$

such that

$$\cap V_j = \emptyset \qquad \overline{\cup V_j} = L^2(\mathbb{R})$$

and

$$f(x) \in V_j \Leftrightarrow f(2x) \in V_{j-1}.$$

In addition there is a scaling function ϕ such that the translates constitute an orthonormal basis for the space

$$V_0 = \text{clos}_{L_2(\mathbf{R})}\text{span}\{\phi(\cdot - k); k \in \mathbb{Z}\}.$$

An orthogonal basis of V_j is then given by the family

$$\phi_{j,k}(t) = 2^{-j/2}\phi(2^{-j}t - k).$$

A MRA is called regular if, for $\phi \in C^r$:

$$| D^\alpha \phi | < C_m(1 + | x |)^{-m} \qquad \forall m \in \mathbb{N} \qquad \alpha \in [0, r].$$

The MRA leads to the construction of the set of subspaces mutually orthogonal $\{W_j, j \in \mathbb{Z}\}$ of $L^2(\mathbb{R})$ obtained by considering for each V_j its orthogonal complement W_j with respect to V_{j-1}.

The space W_0 has as a basis the translates of the function ψ defined by

$$\psi(x) = \sum_{k \in \mathbb{Z}} (-1)^k c_{-k+1}\phi(2x - k)$$

where the c_k are given by the scaling equation

$$\phi(x) = \sum_{k \in \mathbb{Z}} c_k \phi(2x - k)$$

which derives from the fact that $V_0 \subset V_{-1}$; by such a construction we have that $\oplus W_j = L^2(\mathbb{R})$, and so the family $\{\psi_{j,k} : j, k \in \mathbb{Z}\}$ constitutes an orthogonal basis for $L^2(\mathbb{R})$.

In the bidimensional case we take tensor product functions generated by two one-dimensional orthonormal wavelets:

$$\Psi_{j,h;l,m}(x,y) = \psi_{j,h}(x)\psi_{l,m}(y);$$

see Daubechies [3, p. 313].

In this basis x and y are dilated separately; the MRA is the following one:

$$(V_i \otimes V_j) = (V_{i+1} \otimes V_{j+1}) \oplus (W_{i+1} \otimes W_{j+1}) \bigoplus_{k\geq 2}(W_{i+1} \otimes W_{j+k}) \bigoplus_{k\geq 2}(W_{i+k} \otimes W_{j+1}).$$

§2. One-Dimensional Case

Let be assigned the data

$$\{y_h = v(h\Delta) + \epsilon(h\Delta) \qquad \Delta = (2^n - 1)^{-1} \qquad N = 2^n \qquad h = 1, \ldots, N\}$$

of an ergodic Gaussian stationary process belonging to a Hölder space of order m, where $v(t)$ is the trend function; for ϵ we assume $E(\epsilon(t)) = 0$, $t \in \mathbb{R}$, and the covariance function $\sigma(u) = E(\epsilon(t)\epsilon(t+u))$ with non-zero spectral density

$$s(\omega) = \int_{\mathbf{R}} \sigma(u) e^{-i\omega u} du.$$

Since the function $v(t)$ may not belong to $L^2(\mathbb{R})$, and on the other side our interest is in approximation restricted to the interval $(0,1]$, we consider instead of the function $v(t)$ a function $g(t)$, which belongs to the Besov space $B_2^{m,2}(\mathbb{R})$, which has compact support in $[-A, 1+A]$, with A arbitrary and small, and such that

$$v(t) = g(t) \qquad \forall t \in [0,1].$$

As $g \in H_2^m(\mathbb{R})$, the Sobolev space of order m, and $L^2(\mathbb{R}) = V_{-n} \oplus (\oplus_{l\leq -n} W_l)$, the following expansion holds true:

$$\begin{aligned} g(t) &= \sum_{k\in\mathbb{Z}} (g, \phi_{-n,k})\phi_{-n,k}(t) + \sum_{l\leq -n, k\in\mathbb{Z}} (g, \psi_{l,k})\psi_{l,k}(t) \\ &= P_{-n} g(t) + \sum_{l\leq -n, k\in\mathbb{Z}} (g, \psi_{l,k})\psi_{l,k}(t) \end{aligned}$$

where

$$(g,\phi_{-n,k}) = \int_{-\infty}^{\infty} 2^{n/2}\phi(2^n x - k)g(x)dx, \quad (g,\psi_{-n,k}) = \int_{-\infty}^{\infty} 2^{n/2}\psi(2^n x - k)g(x)dx.$$

In the following we put $(g,\phi_{-n,k}) = g_{-n,k}$ for the scaling coefficients.

We shall use the coiflets, see Daubechies [3].

A coiflet is associated with an r-regular MRA of $L^2(\mathbb{R})$ (see Meyer [7]). Both the scaling function ϕ and the associated orthonormal wavelet have compact support, and they satisfy

$$\int_{-\infty}^{\infty} \phi(t)dt = 1$$

$$\int_{-\infty}^{\infty} t^k\phi(t)dt = 0 \qquad k = 1,\ldots,M-1$$

$$\int_{-\infty}^{\infty} t^k\psi(t)dt = 0 \qquad k = 0,\ldots,M-1$$

where for the current problem we choose $M \geq [m]+1$.

Because of the r-regularity of the MRA we know that

$$\| g - P_{-n}g \|_{L^2} = o(2^{-mn}) \quad \text{as} \quad n \to \infty$$

see Meyer [7]; that is to say

$$E(y_h - P_{-n}g(t_h)) = o(2^{-nm}).$$

Due to the compact support of g and ϕ, one has

$$P_{-n}g(t) = \sum_{k\in\mathbb{Z}} g_{-n,k}\phi_{-n,k}(t) \simeq \sum_{k=1}^{N} g_{-n,k}\phi_{-n,k}(t). \tag{1}$$

So the problem leads to the approximation of the coefficients $\{g_{-n,k}\}_{k=1}^{N}$.

Remark 1. *Because of the presence of the singularities in the function g, we have avoided to consider projections onto some space $V_{-n+j(n)} \subset V_{-n}$, because this would imply the loss of the irregularity points and could result in a decrease of the approximation order as well, which in general is determined by a parameter $m \in (0,1]$.*

2.1 Determination of the Estimator $\{\tilde{g}_{-n,k}\}_{k=1}^{N}$

The model is constructed by periodization of the data on $(0,1]$, and restriction to the interval $[-A, 1+A]$. Moreover, we add the hypothesis of circularity and of positive eigenvalues for $\{\sigma_h\}_1^N$, the discretized covariance.

We write the function $g(t)$ as

$$g(t) = f(t) + \sum_{l=1}^{L} \alpha_l \nu_l(t) = f(t) + p(t)$$

where $f(t)$ is a smooth part belonging to $H_2^2(\mathbb{R})$ and the functions $\{\nu_l(t)\}_{l=1}^{L}$ are chosen as explained below. We indicate with $\{x_s\}_1^p$ the locations of the singularities in $(0,1)$, which are assumed to be coincident with some data locations. Let $B = \{\nu_l\}_1^L$, the set of functions, be assigned as follows.

For the generic locations $x = t_{s-1}$, $x = t_s$, $x = t_{s+1}$ we consider the following cubics:

ζ_1 with support $[t_{s-1}, t_s]$, satisfying $\zeta_1(t_s) = 1$, $\zeta_1(t_{s-1}) = \zeta_1'(t_{s-1}) = \zeta_1'(t_s) = 0$,
ζ_2 with support $[t_s, t_{s+1}]$, satisfying $\zeta_2(t_s) = 1$, $\zeta_2(t_{s+1}) = \zeta_2'(t_s) = \zeta_2'(t_{s+1}) = 0$,
ζ_3 with support $[t_{s-1}, t_s]$, satisfying $\zeta_3'(t_s) = 1$, $\zeta_3(t_{s-1}) = \zeta_3(t_s) = \zeta_3'(t_{s-1}) = 0$,
ζ_4 with support $[t_s, t_{s+1}]$, satisfying $\zeta_4'(t_s) = 1$, $\zeta_4(t_{s+1}) = \zeta_4(t_s) = \zeta_4'(t_{s+1}) = 0$,
and we also put:

$$\zeta_5 = \zeta_5|[t_{s-1}, t_s, t_{s+1}] := \begin{cases} \zeta_1 & \text{on} \quad [t_{s-1}, t_s], \\ \zeta_2 & \text{on} \quad (t_s, t_{s+1}], \end{cases}$$

$\zeta_6 = \zeta_6|[t_{s-1}, t_s, t_{s+1}] := \zeta_3 + \zeta_4$, with support $[t_{s-1}, t_{s+1}]$.

For $\{x_s\}_1^p$ we include in B the elements:

- $\zeta_1|[x_{s-1}, x_s], \zeta_2|[x_s, x_{s+1}], \zeta_3|[x_{s-1}, x_s], \zeta_4|[x_s, x_{s+1}]$, if $x = x_s$ is a jump point of g;
- $\zeta_5|[x_{s-1}, x_s, x_{s+1}], \zeta_3|[x_{s-1}, x_s], \zeta_4|[x_s, x_{s+1}]$, if $x = x_s$ is a jump point of g'.

For $x = 0$ we include in B the elements:

- $\zeta_1|[x_p, 1], \zeta_3|[x_p, 1], \zeta_4|[0, x_1]$, if it is a jump point of g;
- $\zeta_3|[x_p, 1], \zeta_4|[0, x_1]$, if it is a jump point of g';
- $\zeta_6|[x_q, 0, x_1]$, if it is a continuity point of g'.

It follows that because of (1) one has:

$$P_{-n}g(t) = \sum_{k=1}^{N} \{f_{-n,k} + \sum_{l=1}^{L} \alpha_l \nu_{l,-n,k}\} \phi_{-n,k}(t)$$

where the elements to be approximated are the coefficients $\{f_{-n,k}\}_{k=1}^{N}$ and $\{\alpha_l\}_{l=1}^{L}$ on the basis of the data $\{y_h\}_1^N$ of the process.

Therefore we consider the vector of the scaling coefficients of the stochastic process: $Y = \{y_{-n,k}\}_{k=1}^{N}$, then $F = \{f_{-n,k}\}_{k=1}^{N}$ and $P = \{\sum_{l=1}^{L} \alpha_l \nu_{l,k}\}_{k=1}^{N}$.

As the Gaussian process belongs to the Hölder space of order m, by making use of the Taylor expansion of $E(y(t))$, one can prove that

$$E\{y_{-n,k} - y(k\Delta)/2^{n/2}\} = \mathcal{O}(2^{-nm}) \tag{2}$$

and

$$| y_{-n,k} - y(k\Delta)/2^{n/2} | \leq C_1 2^{-mn} \quad \text{with Probability} \geq 1 - \frac{1}{C_1^2}. \tag{3}$$

In fact the process has derivatives in mean up to order $[m]$, and for the derivative $y^{[m]}$ one has:

$$E \mid y^{[m]}(t_1) - y^{[m]}(t_2) \mid \leq C \mid t_1 - t_2 \mid^{m-[m]} \quad \text{with} \quad C > 0, \quad \mid t_1 - t_2 \mid < \delta;$$

since it is Gaussian, one has:

$$E \mid y \mid = \sqrt{\frac{2}{\pi} E(y^2)}$$

Therefore

$$P\{| (y_{-n,k} - y(k\Delta)/2^{n/2}) - E(y_{-n,k} - y(k\Delta)/2^{n/2}) | \\ < C E(y_{-n,k} - y(k\Delta)/2^{n/2})^2\} \geq 1 - \frac{1}{C^2},$$

and on the basis of (2), the formula (3) is proved. ■

In what follows we put $(g, \psi_{j-n,k}) = g_{j,k}, \quad j = 1, \ldots, n$, and analogously for $y_{j,k}$ and $\nu_{r,j,k}$; k runs on the locations left at level j through the downsampling process; for a generic function ξ we put ξ_0 for its mean value.

Let consider the minimization of the following functional with respect to f:

$$Q(f) = (Y - F - P)^T R^{-1} (Y - F - P) + \lambda J(f) = LS + \lambda J(f) \ ; \tag{4}$$

here, $\lambda > 0$, R is the correlation matrix of the noise, and $J(f)$ is the smoothness measure related to the Besov norm of the function $f \in H_2^2$ (the frequency zero term is not penalized):

$$J(f) = \sum_{l=1}^{n} 2^{4(n-l)} \sum_{k} f_{l,k}^2.$$

Existence and unicity of the solution of this semiparametric model is guaranteed by [4].

As R^{-1} is circulant with positive eigenvalues, the term LS can be written as $LS = \xi^T \xi$ with $\xi = C(Y - F - P)$ and C circulant. By considering the wavelet transform, it is $LS = \sum_l \sum_k \xi_{l,k}^2$. By moving from the time to the frequency domain and taking into account that the frequencies of W_{-n+l} are approximately restricted to a band and that the Fourier transformation diagonalizes C, we have for LS:

$$LS = \frac{1}{2\pi}\{\sum_{l=1}^{n}\sum_{\omega_j\in I_l} \mid \hat{f}(\omega_j) + \sum_{r=1}^{L} \alpha_r \hat{\nu}_r(\omega_j) - \hat{y}(\omega_j) \mid^2 \mid \hat{\psi}_{l,1}(\omega_j) \mid^2 (\frac{s^l(\omega_j)}{\sigma(0)})^{-1}\} + \xi_0^2$$

where ω_j runs over discrete frequencies in $I_l : \{-\pi 2^{-l+1} < \omega_j \leq -\pi 2^{-l}\} \cup \{\pi 2^{-l} \leq \omega_j < \pi 2^{-l+1}\}$; $s^l(\omega_j)$ is the component of the periodogram relative to the frequency ω_j on the space W_{-n+l}.

By applying the theorem of the mean and substituting the unknown value $(s^l(\omega^\star))^{-1}$ by the averaged value

$$\bar{s}^l = \frac{\sum_{\omega_j\in I_l}(s^l(\omega_j))^{-1}}{2^{n-l}}$$

the approximation error is of order 2^{-l}. We indicate

$$d_l = (\frac{\bar{s}^l}{\sigma(0)})^{-1} \quad \text{for} \quad l = 1,\dots,n \quad \text{and} \quad d_0 = (\frac{s(0)}{\sigma(0)})^{-1}.$$

The solution of our problem is calculated by solving:

$$\min_{f:J(f)<\infty} \sum_{j,k} d_j(y_{j,k} - p_{j,k} - f_{j,k})^2 + d_0(y_0 - p_0 - f_0)^2 +$$

$$+\lambda \sum_{j,k} 2^{4(n-j)} f_{j,k}^2 = \min LS(f, d_j) + \lambda J(f)$$

which has the solution

$$\tilde{f}_{j,k} = \gamma_{j,k}(\lambda)(y_{j,k} - \sum_{l=1}^{L} \alpha_l \nu_{l,j,k})$$

$$\text{where} \quad \gamma_j(\lambda) = d_j/(d_j + \lambda 2^{4(n-j)}); \quad j = 1,\dots,n$$

$$\tilde{f}_0 = y_0 - \sum_{1}^{L} \alpha_l \nu_{l,0}.$$

For the parametric part, taking into account the results in Speckman [9] for kernel estimators, we minimize:

$$\min LS_{\alpha_l}(\tilde{f}, d_j).$$

To solve the associated linear system, it is suitable to run an orthogonalization of the basis $\{\nu_l\}$ according to the semi-inner product:

$$(\xi, \eta) = \sum_{j,k} (1 - \gamma_j(\lambda))^2 \xi_{j,k} \eta_{j,k}.$$

Let us call $\{\mu_k\}$ the set of functions obtained in this way and let us rewrite the sum $\sum_{l=1}^{L} \alpha_l \nu_{l,j,k}$ as $\sum_{l=1}^{L} \beta_l \nu_{l,j,k}$. We obtain

$$\tilde{\beta}_l = \frac{\sum_{j,k} (1 - \gamma_j(\lambda))^2 y_{j,k} \mu_{l,j,k}}{\sum_{j,k} (1 - \gamma_j(\lambda))^2 (\mu_{l,j,k})^2}.$$

Therefore the estimator $\tilde{g}$ has coefficients

$$\tilde{g}_{j,k} = \gamma_j(\lambda)(y_{j,k} - \sum_{l=1}^{L} \tilde{\beta}_l \mu_{l,j,k}) + \sum_{l=1}^{L} \tilde{\beta}_l \mu_{l,j,k}, \quad j = 1, \ldots, n; \quad \tilde{g}_0 = y_0 \; .$$

By performing the multiresolution backwards we obtain the smoothed scaling coefficients of the function:

$$\tilde{g}_k = \tilde{f}_k + \sum_{l=1}^{L} \tilde{\beta}_l \mu_{l,k}. \quad \blacksquare$$

In [6] the mean squared error was studied:

$$\mathrm{MSE}(\tilde{g}) = E \sum_{h=1}^{2^n} (g_h - \tilde{g}_h)^2 = \mathrm{Bias}^2(\tilde{g}) + \mathrm{Var}(\tilde{g})$$

in the case of uncorrelated noise. The same arguments can be applied in case of correlated noise. Due to lack of space we do not give the details, but simply present the following expressions:

$$\mathrm{Bias}^2(\tilde{g}) \asymp \sum_{j=1}^{n} \frac{(\lambda 2^{4(n-j)})^2}{(d_j + \lambda 2^{4(n-j)})^2} \sum_k f_{j,k}^2$$

$$\mathrm{Var}(\tilde{g}) \asymp \frac{\sigma(0)}{2^n d_0} + \frac{\sigma(0)}{2^n} \sum_{j=1}^{n} \frac{d_j 2^{n-j}}{(d_j + \lambda 2^{(4(n-j))})^2}. \quad \blacksquare$$

Remark 2. *The size of* $\mathrm{MSE}(\tilde{g})$ *is still of order* $(2^n)^{-4/5}$ *subject to* $f \in H_2^2(0,1]$, *but with stronger assumptions one obtains an improved order.*

About the determination of λ we follow the criterium of taking λ so as to minimize the following expression:

$$T(\lambda) = \frac{\sum_{j,k}(y_{j,k} - \tilde{g}_{j,k})^2}{(1 - \sum_j \gamma_j(\lambda)2^{-j} + 2^{-n})^2}.$$

This is obtained from the standard Generalized Cross Validation expression by substituting in the denominator the trace of the matrix A associated to the current semiparametric model (A transforms from $\{y_k\}$ to $\{\tilde{g}_k\}$) with the trace of K associated to the smoothing model without parametric part. This is for computational efficiency; it is justified as the difference between $\mathrm{tr}A/2^n$ and $\mathrm{tr}K/2^n$ is very small.

§3. Bidimensional Case

Let us consider a bidimensional stochastic process $z(x, y)$ from which we have the data:

$$z(k_x \cdot \Delta, k_y \cdot \Delta) = g(k_x \cdot \Delta, k_y \cdot \Delta) + \epsilon(k_x \cdot \Delta, k_y \cdot \Delta); \quad k_x, k_y \in [1, \ldots, 2^n]$$

at the grid points on the domain $D = (0, 1] \times (0, 1]$.

It is assumed that the trend function $g(x, y)$ has jumps or its partial derivative $\frac{\partial g}{\partial x}$ has jumps through known lines $x = x_s, \quad s = 1, \ldots, p$, coincident with data abscissas; so we model $g(x, y)$ as

$$g(x, y) = f(x, y) + \sum_{l=1}^{L} \alpha_l(y)\nu_l(x)$$

The model is constructed by periodization of the data on $(0, 1] \times (0, 1]$ and restriction to the domain $[-A, 1 + A] \times [-A, 1 + A]$.

The choice of the functions $\{\nu_l\}_{l=1}^{L}$ is the same as in subsection 2.1.

For the random part of the process $\epsilon(x, y)$, we assume that $E(\epsilon(x, y)) = 0$ and covariance $\sigma(u, v)$ with spectral density well different from zero; after discretization, let the correlation matrix be R, block circulant structured with positive eigenvalues (a block circulant has circulant blocks with squared and circulant blocks).

We put the following notations for scaling and wavelet coefficients of a generic function ξ, with ψ a coiflet of order M:

$$\xi_{k_x,k_y} = \int_{-\infty}^{\infty} \int_{-\infty}^{\infty} \xi(t, v)\phi_{-n,k_x}(t)\phi_{-n,k_y}(v)dtdv;$$

$$\xi_{h,j,x,y} = \int_{-\infty}^{+\infty} \int_{-\infty}^{+\infty} \xi(t, v)\psi_{h-n,x}(t)\psi_{j-n,y}(v)dtdv$$

$$\xi_{n,j,\cdot,y} = \int_{-\infty}^{+\infty}\int_{-\infty}^{+\infty} \xi(t,v)\psi_{j-n,y}(v)dtdv$$

$$\xi_{h,n,x,\cdot} = \int_{-\infty}^{+\infty}\int_{-\infty}^{+\infty} \xi(t,v)\psi_{h-n,x}(t)dtdv;$$

x, y run over the dyadic locations remaining at levels h, j through the downsampling process; $h, j \in [1,\ldots,n]$. We put $Z = \{z_{k_x,k_y}\}_{k_x,k_y}$, $F = \{f_{k_x,k_y}\}_{k_x,k_y}$ and $P = \{\sum_{l=1}^{L} \alpha_{l,k_y}\nu_{l,k_x}\}_{k_x,k_y}$. We consider the minimization of the following functional, with respect to f, where $\lambda > 0$:

$$Q(f) = (Z - F - P)^T R^{-1}(Z - F - P) + \lambda J(f).$$

The solution of our problem is calculated by solving:

$$\begin{aligned} \min_{f:J(f)<\infty} & \sum_{h,j,x,y} d_{h,j}(z_{h,j,x,y} - p_{h,j,x,y} - f_{h,j,x,y})^2 + \\ & + \sum_{j,y} d_{n,j}(z_{n,j,\cdot,y} - p_{n,j,\cdot,y} - f_{n,j,\cdot,y})^2 + \\ & + \sum_{h,x} d_{h,n}(z_{h,n,x,\cdot} - p_{h,n,x,\cdot} - f_{h,n,x,\cdot})^2 + \\ & + d_{0,0}(z_{0,0} - p_{0,0} - f_{0,0})^2 + \\ +\lambda\{ \sum_{h,j,x,y} 2^{2(2n-h-j)} f^2_{h,j,x,y} & + \sum_{j,y} 2^{2(n-j)} f^2_{n,j,\cdot,y} + \sum_{h,x} 2^{2(n-h)} f^2_{h,n,x,\cdot}\} = \\ = \min LS(f, d_{h,j}) + \lambda J(f); & \quad h \in [1,\ldots,n], \quad j \in [1,\ldots,n] \end{aligned}$$

where $\{d_{h,j}\}$ are obtained from averaging the inverses of the frequencies of the periodogram, normalized with $\sigma(0,0)$, within the band of frequencies:

$$\{-\pi 2^{-h+1} < \omega_x \leq -\pi 2^{-h}\} \cup \{\pi 2^{-h} \leq \omega_x < \pi 2^{-h+1}\} \times$$

$$\times \{-\pi 2^{-j+1} < \omega_y \leq -\pi 2^{-j}\} \cup \{\pi 2^{-j} \leq \omega_y < \pi 2^{-j+1}\}; \quad d_{0,0} = \left(\frac{s(0,0)}{\sigma(0,0)}\right)^{-1}.$$

Existence and unicity of the solution of our semiparametric model is again guaranteed by [4]. The solution is the following one:

$$\tilde{f}_{h,j,x,y} = \gamma_{h,j}(\lambda)(z_{h,j,x,y} - p_{h,j,x,y})$$

$$\tilde{f}_{n,j,\cdot,y} = \gamma_{n,j}(\lambda)(z_{n,j,\cdot,y} - p_{n,j,\cdot,y})$$

$$\tilde{f}_{h,n,x,\cdot} = \gamma_{h,n}(\lambda)(z_{h,n,x,\cdot} - p_{h,n,x,\cdot})$$

$$\tilde{f}_{n,n,\cdot,\cdot} = \gamma_{n,n}(\lambda)(z_{n,n,\cdot,\cdot} - p_{n,n,\cdot,\cdot}) \qquad \text{as} \quad h = 1, \ldots, n \quad j = 1, \ldots, n$$

with

$$\gamma_{h,j}(\lambda) = \gamma_{h,j}(d_{h,j}, \lambda) = \frac{d_{h,j}}{d_{h,j} + \lambda 2^{2(2n-h-j)}}$$

$$\tilde{f}_{0,0} = z_{0,0} - p_{0,0}.$$

For the parametric part, we minimize

$$\min_{\alpha_l} LS(\tilde{f}, d^N_{h,j}).$$

For the estimate of the parametric part we have used new weights $\{d^N_{h,j}\}$ with the feature of negative correlation along the y direction. In fact as already said in the introduction, the wavelet transform has the property of localizing the signal and therefore of reconstructing better the function but at the same time of smoothing less the random component of the process. This fact is more evident in the case that the data are correlated among themselves. By such weights we weight more low frequencies and therefore we obtain an approximating function without wiggles.

We put $\gamma^N_{h,j}(\lambda) = \gamma_{h,j}(d^N_{h,j}, \lambda)$.

We rewrite

$$\sum_{l=1}^{L} \alpha_{l,j,y}\nu_{l,h,x} = \sum_{l=1}^{L} \beta_{l,j,y}\mu_{l,h,x}$$

after orthogonalizing according to the seminner product:

$$(\xi, \eta) = \sum_{h=1,\ldots,n;x} (1 - \gamma^N_{h,j}(\lambda))^2 \xi_{h,x}\eta_{h,x} + (1 - \gamma^N_{n,j}(\lambda))^2 \xi_n \eta_n;$$

it results

$$\tilde{\beta}_{l,j,y} = \frac{\sum_{h,x}(1 - \gamma^N_{h,j}(\lambda))^2 z_{h,j,x,y}\mu_{l,h,x} + (1 - \gamma^N_{n,j}\lambda)^2 z_{n,j,\cdot,y}\mu_{l,n}}{\sum_{h,x}(1 - \gamma^N_{h,j}(\lambda))^2 \mu^2_{l,h,x} + (1 - \gamma^N_{n,j}(\lambda)^2 \mu^2_{l,n,j}} \quad j = 1, \ldots, n.$$

Then we rewrite

$$\sum_{l=1}^{L} \alpha_{l,n,\cdot}\nu_{l,n,\cdot} = \sum_{l=1}^{L} \beta_{l,n,\cdot}\mu_{l,n,\cdot}$$

after orthogonalization with respect to the seminner product:

$$(\xi, \eta) = \sum_{h=1,\ldots,n;x} (1 - \gamma^N_{h,n}(\lambda))^2 \xi_{h,x}\eta_{h,x};$$

it results

$$\tilde{\beta}_{l,n,\cdot} = \frac{\sum_{h,x}(1 - \gamma^N_{h,n}(\lambda))^2 z_{h,n,x,\cdot}\mu_{l,h,x}}{\sum_{h,x}(1 - \gamma^N_{h,n}(\lambda))^2 \mu^2_{l,h,x}}.$$

Therefore the estimator $\tilde{g}$ has coefficients

$$\tilde{g}_{0,0} = z_{0,0}$$

$$\tilde{g}_{h,j,x,y} = \gamma_{h,j}(\lambda)(z_{h,j,x,y} - \sum_{l=1}^{L} \tilde{\beta}_{l,j,y}\mu_{l,h,x}) + \sum_{l=1}^{L} \tilde{\beta}_{l,j,x}\mu_{l,h,x}$$

$$\tilde{g}_{n,j,\cdot,y} = \gamma_{n,j}(\lambda)(z_{n,j,\cdot,y} - \sum_{l=1}^{L} \tilde{\beta}_{l,j,y}\mu_{l,n}) + \sum_{l=1}^{L} \tilde{\beta}_{l,j,y}\mu_{l,n}$$

$$\tilde{g}_{h,n,x,\cdot} = \gamma_{h,n}(\lambda)(z_{h,n,x,\cdot} - \sum_{l=1}^{L} \tilde{\beta}_{l,n}\mu_{l,h,x}) + \sum_{l=1}^{L} \tilde{\beta}_{l,n}\mu_{l,h,x}.$$

About the determination of λ we follow the criterium of taking λ which minimizes the following expression:

$$T(\lambda) = \frac{NT(\lambda)}{(1 - \frac{1}{4^n} trK(\lambda))^2} = \frac{NT(\lambda)}{(1 - \sum_{(h,j)\neq(n,n)} \gamma_{h,j}(\lambda)2^{-h-j} + 4^{-n})} \quad \text{with}$$

$$NT(\lambda) = \sum_{h,x} (z_{h,j,x,y} - \tilde{g}_{h,j,x,y})^2 + \sum_{j,y} (z_{n,j,\cdot,y} - \tilde{g}_{n,j,\cdot,y})^2 + \sum_{h,x} (z_{h,n,x,\cdot} - \tilde{g}_{h,n,x,\cdot})^2.$$

$T(\lambda)$ is obtained from the standard Generalized Cross Validation expression by substituting in the denominator the trace of the matrix A associated to the current semiparametric model (A transforms $\{z_{k_x,k_y}\}$ into $\{\tilde{g}_{k_x,k_y}\}$) with the trace of K associated to the smoothing model without parametric part. In fact the difference between $\text{tr}A/N$ and $\text{tr}K/N$ is very small as already shown in [2].

§4. Numerical Examples

On an experimental basis we have chosen coiflets with $M = 6$.

First we show the results for $d = 1$ for the following test function:

$$g_1(x,y) = \begin{cases} |\cos(x)| & 0 \le x \le 2.75 \\ \sin(2x) & 2.75 < x \le 9.75 \\ x & 9.75 < x \le 15.9375 \end{cases}$$

The noise is structured as a moving average process of the first type $\epsilon_t = e_t + \theta_0 e_{t-1}$, $\theta_0 \in \mathbb{R}$ with $\{e_t\}$ i.i.d. Gaussian variables with $E(e_t) = 0$ and variance $\sigma(0)$.

The approximation error is measured by $r = \sqrt{\frac{\sum(\tilde{g}(j)-g(j))^2}{N}}$, with $N = 2^n$.

Figure 1 and 2 refer to the same data set, with positive correlation: the result in Figure 1 is obtained putting $\{d_k = 1\}_{k=0}^{n}$, that is to say with a model that does not take into account the correlation; on the contrary the result in Figure 2 is obtained with our technique. The true function is given by the dashed line, and the approximation is represented by the solid line.

In Figure 3 we treat the same example, but with negative correlation, that is to say with $\theta_0 < 0$.

Next we show the results for $d = 2$; the noise is structured as a moving average process of the first type $\epsilon_{t,\tau} = e_{t,\tau} + \theta_{10}e_{t-1,\tau} + \theta_{01}e_{t,\tau-1} + \theta_{11}e_{t-1,\tau-1}$, $\theta_{t,\tau} \in \mathbb{R}$ with $\{e_{t,\tau}\}$ i.i.d. Gaussian variables with $E(e_{t,\tau}) = 0$ and variance $\sigma(0,0)$.

The weights $\{d^N_{h,j}\}$ are taken as those of a moving average process of the first type with $\theta_{10} = 0$, $\theta_{01} = -0.85$ and $\theta_{11} = 0$.

To take into account the lack of periodicity in the y direction we have done a one-dimensional smoothing with the model of subsection 2.1 for $y = y_{min}$ and $y = y_{max}$; we have constructed the linear blending interpolation of these two curves and subtracted it from the data and we have treated the new set of data with the method of §3; at the end we add the linear blending interpolation to the smoothed data.

The approximation error is measured by $r = \sqrt{\frac{\sum(\tilde{g}(i,j)-g(i,j))^2}{N}}$, with $N = 2^n \cdot 2^n$.

We consider the following test function:

$$g_2(x,y) = \begin{cases} 1 + \sin(4\pi x)y^2 & 0 \le x \le 0.5 \quad 0 \le y \le 1 \\ 1 & 0.5 < x \le 1 \quad 0 \le y \le 1 \end{cases}$$

where there are discontinuities of the partial derivative across $x = 0.5$ and $x = 1$; the discontinuities decrease as y decreases.

The result is shown in Figure 4.

Next we consider the test function:

$$g_3(x,y) = \begin{cases} (x+y); \quad 0 \le x \le 6.75 + 2\cos(y), \quad 0 \le y \le 16 \\ 3(x+y) - 2 * (6.75 + (2\cos(y) + y)); \\ 6.75 + 2\cos(y) < x \le 16, \quad 0 \le y \le 16 \end{cases}$$

where the discontinuities of the function do not belong to lines of the type $x =$ const, see the result in Figure 5. In fact the method described in the current paper can be adapted to more general cases; in this respect, see the paper by Guicciardi [5].

Remark 3. *The cost of the procedure is of order $N \log N$ in any case, where N is the sample size.*

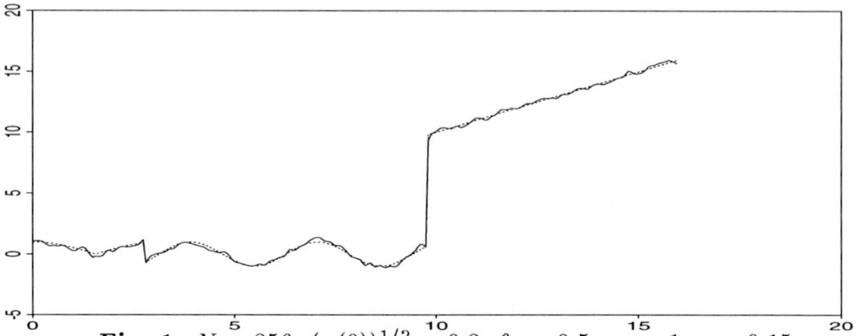

Fig. 1. $N = 256$ $(\sigma(0))^{1/2} = 0.2$ $\theta_0 = 0.5$ $\tau_k = 1$ $r = 0.15$.

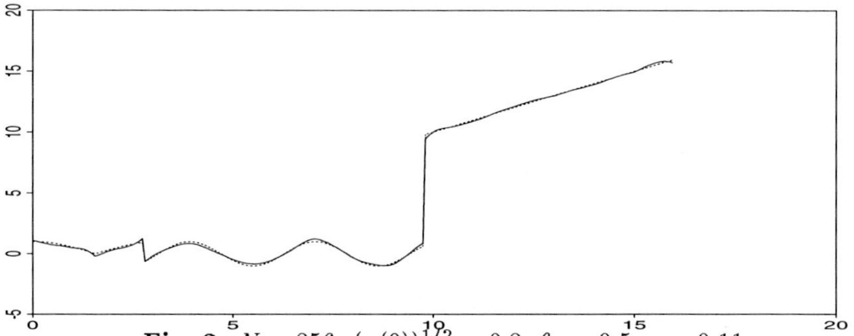

Fig. 2. $N = 256$ $(\sigma(0))^{1/2} = 0.2$ $\theta_0 = 0.5$ $r = 0.11$.

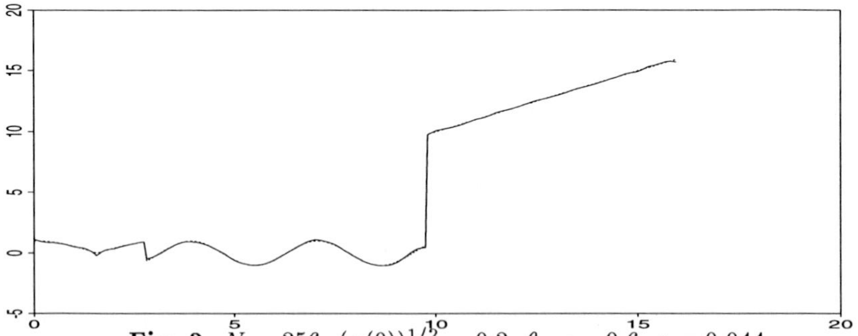

Fig. 3. $N = 256$ $(\sigma(0))^{1/2} = 0.2$ $\theta_0 = -0.6$ $r = 0.044$.

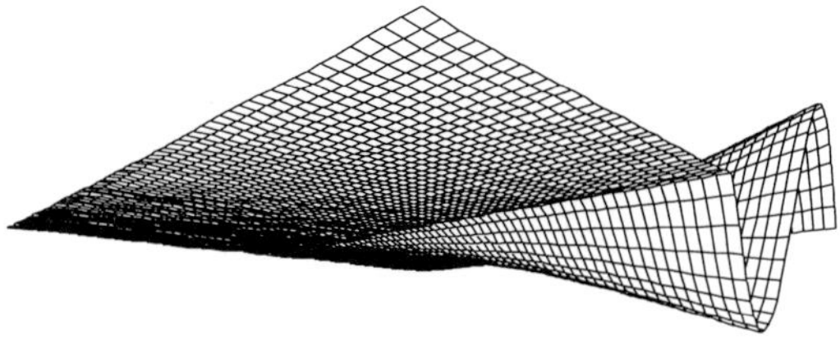

Fig. 4. $N = 64 \times 64$ $(\sigma(0,0))^{1/2} = 0.01$, $\theta_{01} = 0.25$, $\theta_{10} = 0.25$, $\theta_{11} = 0.03$, $r = 0.0035$.

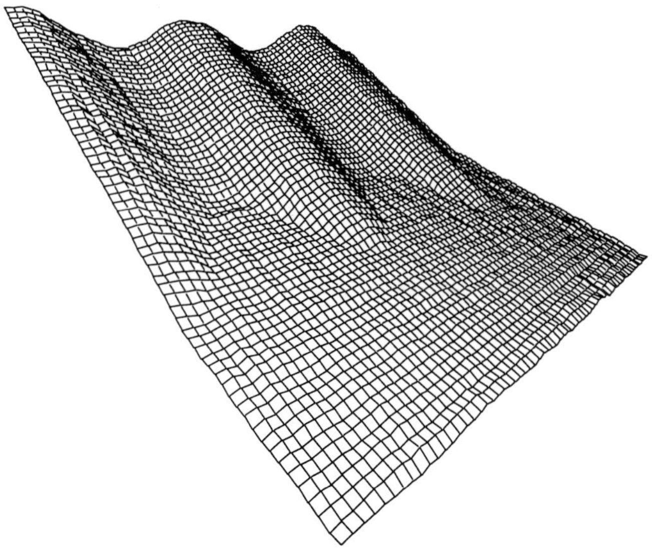

Fig. 5. $N = 64 \times 64$ $(\sigma(0,0))^{1/2} = 0.6$, $\theta_{01} = -0.25$, $\theta_{10} = -0.25$, $\theta_{11} = -0.03$, $r = 0.193$.

References

1. Antoniadis, A., Smoothing noisy data with coiflets, RT 93 IMAG(1993).
2. Bozzini, M. and L. Lenarduzzi, Recovering a function with discontinuities in dimension two, in *Mathematical Methods in Computer Aided Geometric Design III*, M. Dæhlen, T. Lyche and L. L. Schumaker (eds.), Vanderbilt University Press, Nashville, 1995, 41-54.
3. Daubechies, I., *Ten Lectures on Wavelets*, CBMS-NSF **61**, SIAM, Philadelphia, 1992.
4. Girard, D. and P.J. Laurent, Splines and estimation of nonlinear parameters, in *Mathematical Methods in Computer Aided Geometric Design*, T. Lyche and L. L. Schumaker (eds.), Academic Press, New York, 1989, 273-298.
5. Guicciardi, S., Reconstruction of bidimensional functions discontinuous along C^1 curves, using wavelets, IAMI 95.13 (1995).
6. Guicciardi, S. and L. Lenarduzzi, Accuracy of a semiparametric model with coiflets for smoothing noisy data, IAMI 95.14 (1995).
7. Meyer, Y., *Wavelets and Operators*, Cambridge Studies in Advanced Mathematics **37**, Cambridge University Press, 1992.
8. Rossini, M., Irregularity detection from noisy data in one and two dimensions, *Quaderno Istituto Matematico*, Univ. Milano, preprint (1995).
9. Speckman, P., Kernel smoothing in partial spline models, J. R. Statist. Soc. B **3** (1988), 413-436.

M. Bozzini
Dip. di Matematica, Universitá di Milano
Via Saldini 50
20133 Milano, ITALY
bozzini@vmimat.mat.unimi.it

L. Lenarduzzi
I.A.M.I. C.N.R.
via Ampère 56
20131 Milano, ITALY
licia@hp1.iami.mi.cnr.it

Bases with Convexity Preserving Properties

J. M. Carnicer, M. García-Esnaola and J. M. Peña

Abstract. We analyze some shape properties of a control polygon which are preserved by the corresponding curve and explore which characteristics of the basis functions lead to the mentioned shape preserving properties. Rational curves and the relationship with total positivity are also discussed. We survey some recent papers on this subject and provide a unified approach.

§1. Introduction

This survey collects some results on characterizations of convexity preserving representations developed by the authors in some recent papers ([**8, 2, 3, 4**]). Here we present a unified approach and also shed new lights on this subject.

Let $(f_0, \ldots, f_n)$ be a blending system of functions on $[a, b]$, that is, a system of nonnegative functions defined on $[a, b]$ such that $\sum_{i=0}^n f_i(t) = 1$ for all $t \in [a, b]$. Given $P_0, \ldots, P_n \in \mathbb{R}^s$ we may define a curve $\gamma(t) = \sum_{i=0}^n P_i f_i(t)$. The points $P_0, \ldots, P_n$ are called *control points* of γ and the polygon $P_0 \cdots P_n$ is called the *control polygon* of γ with respect to the system $(f_0, \ldots, f_n)$.

Given $u_0, \ldots, u_n$ nonnegative functions defined on $[a, b]$ such that $\sum_{i=0}^n u_i(t) > 0$ for all $t \in [a, b]$ and any positive weights $w_0, \ldots, w_n > 0$, a blending system is given by

$$\left(\frac{w_0 u_0}{\sum_{i=0}^n w_i u_i}, \ldots, \frac{w_n u_n}{\sum_{i=0}^n w_i u_i} \right). \tag{1.1}$$

Let us remark that $\sum_{i=0}^n w_i u_i$ never vanishes because $\sum_{i=0}^n u_i(t) > 0$ for all $t \in [a, b]$ implies that $\sum_{i=0}^n w_i u_i(t) > 0$ for all $t \in [a, b]$ (see Proposition 2.4 of [**4**]). Taking

$$f_j := \frac{w_j u_j}{\sum_{i=0}^n w_i u_i}, \quad j = 0, \ldots, n,$$

Advanced Topics in Multivariate Approximation
F. Fontanella, K. Jetter and P. J. Laurent (eds.) pp. 17–31.

ISBN 981-02-2852-X

and $P_0, \ldots, P_n \in \mathbb{R}^s$, the curve $\gamma(t) = \sum_{i=0}^n P_i f_i(t)$ is called the *rational curve* generated by the system $(u_0, \ldots, u_n)$ with weights $w_0, \ldots, w_n$.

We shall consider bases $(u_0, \ldots, u_n)$ of a given space $\mathcal{U}$ such that the corresponding representations $c_0 u_0 + \cdots + c_n u_n$ of a function $u \in \mathcal{U}$ provide some information about the geometry of a curve associated to the function. On one hand, we shall deal with functional curves, that is, curves regarded as graphs of functions $u : [a, b] \to \mathbb{R}$. We shall study bases $(u_0, \ldots, u_n)$ for which some shape properties of the curve can be derived from the behaviour of the coefficients: this is the functional point of view. On the other hand, the geometrical point of view deals with parametric curves and with shape properties of the control polygon which are preserved by the corresponding (rational) curve. We shall explore which characteristics of the system $(u_0, \ldots, u_n)$ lead to shape preserving properties. A usual technique for extracting shape properties from a control polygon will be projection onto an affine subspace of $\mathbb{R}^s$. In this setting, a typical shape preserving property can be described as follows: If the projections of the control points are structured in a particular way, then the projections of the curve enjoy a corresponding property. More precisely, if $h : \mathbb{R}^s \to \mathbb{R}^k$ is any affine mapping, we construct the matrix H of projected points, whose rows are formed by $h(P_0)^T, \ldots, h(P_n)^T$ and a system of functions $(y_0(t), \ldots, y_k(t))^T := h(\gamma(t))$. Since $(f_0, \ldots, f_n)$ form a partition of the unity, we may write

$$(y_0(t), \ldots, y_k(t)) = (f_0(t), \ldots, f_n(t))H.$$

Therefore the relationship between the projected points and the projected curve is determined by the properties of $(f_0(t), \ldots, f_n(t))$. The dimension k of the affine space onto which we project depends on the kind of shape preserving property we are describing. For instance, we need $k = 1$ for monotonicity and $k = 2$ for convexity. These cases will be analyzed in sections 2 and 3 respectively. In Section 4 the case $k > 2$ is analyzed under a functional point of view.

Totally positive systems satisfy previous convexity properties. In fact, in Karlin's book [**16**] it was shown that TP_2 (resp., TP_{k+1}) systems (see Definition 2.2) provide some sufficient conditions for monotonicity (resp., k-convexity) preservation. Let us recall that a system of functions $(u_0, \ldots, u_n)$ is *totally positive* (TP) if all the corresponding collocation matrices of $u_0, \ldots, u_n$ at $t_0, \ldots, t_m$

$$M\begin{pmatrix} u_0, \ldots, u_n \\ t_0, \ldots, t_m \end{pmatrix} := (u_j(t_i))_{i=0,\ldots,m;j=0,\ldots,n}, \tag{1.2}$$

$a \leq t_0 < t_1 < \cdots < t_m \leq b$, have nonnegative minors. More shape preserving properties of totally positive systems can be seen in [**11, 15, 12, 6, 7, 5**]. In Section 5 we characterize totally positive systems of functions in terms of geometric convexity preserving properties of the rational curves. Proposition 5.6 allows us to simplify the approach provided in [**4**].

In contrast with the nice behaviour of the univariate representations, the multivariate case is an open question. Even the Bernstein bases on a triangle do not possess convexity preserving properties in the usual sense. However, some autors have shown that some related properties such as polyhedral convexity or axial convexity are preserved by the Bernstein-Bézier representations. For more details about multivariate considerations, the reader is referred to [**9, 14, 17**].

§2. Monotonicity Preserving Representations

Let $(f_0, \ldots, f_n)$ be a blending system. Then $(f_0, \ldots, f_n)$ is *monotonicity preserving* if the fact that the projections of the control polygon onto a line are ordered implies that the projection of the points of the corresponding curve onto that line are also ordered: $\alpha_0 \leq \cdots \leq \alpha_n$ implies that $\sum_{i=0}^{n} \alpha_i f_i(t)$ is an increasing function. If $\alpha_0 < \cdots < \alpha_n$ implies that $\sum_{i=0}^{n} \alpha_i f_i(t)$ is a strictly increasing function, then $(f_0, \ldots, f_n)$ is said to be a *strictly monotonicity preserving* system.

From Proposition 2.3 of [**4**], we may state the following result:

Proposition 2.1. *Let $(f_0, \ldots, f_n)$ be a blending system of functions defined on $[a, b]$ and let*

$$g_i := \sum_{j=i}^{n} f_j, \quad i = 0, 1, \ldots, n. \tag{2.1}$$

Then:

(i) *$(f_0, \ldots, f_n)$ is monotonicity preserving if and only if $g_1, \ldots, g_n$ are increasing functions.*

(ii) *$(f_0, \ldots, f_n)$ is strictly monotonicity preserving if and only if $g_1, \ldots, g_n$ are increasing and $\sum_{i=1}^{n} g_i$ is a strictly increasing function.*

Let us see a property which is satisfied by monotonicity preserving systems of functions.

The length of a curve $\gamma : [a, b] \to \mathbb{R}^s$ is given by

$$L(\gamma) := \sup_{\substack{t_0 < \cdots < t_N \in [a,b] \\ N \in \mathbb{N}}} \sum_{k=1}^{N} \|\gamma(t_k) - \gamma(t_{k-1})\|,$$

where $\|\cdot\|$ is the euclidean vector norm.

The length of the control polygon $P_0 \cdots P_n$ is

$$L(P_0 \cdots P_n) = \sum_{i=1}^{n} \|P_i - P_{i-1}\|.$$

A blending system is *length diminishing* if $L(\sum_{i=0}^{n} f_i P_i) \leq L(P_0 \cdots P_n)$ for any control points $P_0, \ldots, P_n \in \mathbb{R}^s$. In Corollary 4.3 of [**8**] we showed that: *any blending monotonicity preserving system is length diminishing.*

In the next result, which corresponds to Theorem 2.6 of [**4**], we characterize the systems of functions $(u_0, \dots, u_n)$ such that all systems (1.1) obtained with any positive weights are monotonicity preserving.

We shall need the following definition:

Definition 2.2. *The system $(u_0, \dots, u_n)$ is called TP_r, $1 \le r \le n+1$, if all $k \times k$ minors, $k = 1, \dots, r$ of any collocation matrix (1.2) of $u_0, \dots, u_n$ are nonnegative.*

Let us observe that if $(u_0, \dots, u_n)$ is TP_{n+1} then it is a totally positive system.

Theorem 2.3. *Let $(u_0, \dots, u_n)$ be a system of nonnegative functions such that $\sum_{i=0}^n u_i > 0$. Then $(u_0, \dots, u_n)$ is TP_2 if and only if for all $w_0, \dots, w_n > 0$ the systems (1.1) are monotonicity preserving. Furthermore, if $(u_0, \dots, u_n)$ is TP_2 and for some choice of $w_0, \dots, w_n > 0$ and of $\alpha_0 < \cdots < \alpha_n$ the function $(\sum_{i=0}^n \alpha_i w_i u_i)/(\sum_{i=0}^n w_i u_i)$ is strictly increasing, then all the systems (1.1) are strictly monotonicity preserving for all $w_0, \dots, w_n > 0$.*

Let us observe that, if we choose $w_i = 1$ for $i = 0, \dots, n$, we may derive from the previous result the following consequence:

Corollary 2.4. *A blending totally positive system $(f_0, \dots, f_n)$ is monotonicity preserving.*

§3. Convexity Preserving Representations

A function $u \in C^2[a,b]$ is convex if $u''(t) \ge 0$, for all $t \in [a,b]$. This concept can be generalized to nonsmooth functions: a function $u : [a,b] \longrightarrow \mathbb{R}$ is *convex* if all second order divided differences (at different points t_0, t_1, t_2) are nonnegative, that is, $u[t_0, t_1, t_2] \ge 0$ for all $t_0 < t_1 < t_2$ in $[a,b]$.

Let $u_0, \dots, u_n$ be a linear independent system of functions defined on $[a,b]$. Any function u in the linear span of $u_0, \dots, u_n$ can be written in the form $u = \sum_{i=0}^n c_i u_i$. We are interested in recognizing convex functions u from some property of the coefficients. A discrete version of convexity for $(c_0, \dots, c_n)$ can be described as follows: $c[i, i+1, i+2] = (c_{i+2} - 2c_{i+1} + c_i)/2 \ge 0$, i.e., the second difference $\Delta^2 c_i \ge 0, i = 0, 1, \dots, n-2$. The desired property will be that $u = \sum_{i=0}^n c_i u_i$ is convex provided that $(c_0, \dots, c_n)$ is a convex sequence.

Definition 3.1. *A system of functions $(u_0, \dots, u_n)$ is called convexity preserving if for any convex sequence $c = (c_0, \dots, c_n)$, the function $\sum_{i=0}^n c_i u_i$ is convex.*

In Remark 6.2 of [**2**], we obtained the following characterization of convexity preserving system of functions:

Theorem 3.2. *The system of functions $(u_0, \dots, u_n)$ is convexity preserving if and only if $u_0 + \cdots + u_n$ and $u_1 + 2u_2 + \cdots + nu_n$ are both polynomials of degree less* *than or equal to* 1, *and the functions* $u_2 + 2u_3 + \cdots + (n-1)u_n, \dots, u_{n-1} + 2u_n, u_n$ *are convex.*

Let us give some additional conditions to a blending system $(f_0, \dots, f_n)$ in order to obtain interpretations of convexity preserving system in terms of CAGD (Computer Aided Geometric Design). Any function $f = \sum_{i=0}^n c_i f_i$ can be expressed as a parametric curve: $\gamma(t) := \begin{pmatrix} t \\ f(t) \end{pmatrix}$. If t can be obtained as a linear combination of $f_0, \dots, f_n$,

$$t = \sum_{i=o}^{n} \alpha_i f_i(t), \quad a = \alpha_0 < \alpha_1 < \ldots < \alpha_n = b \ , \tag{3.1}$$

then we may write $\gamma(t) = \sum_{i=0}^n \begin{pmatrix} \alpha_i \\ c_i \end{pmatrix} f_i(t)$, and so, $\begin{pmatrix} \alpha_0 \\ c_0 \end{pmatrix}, \dots, \begin{pmatrix} \alpha_n \\ c_n \end{pmatrix}$ is the control polygon of the graph of f. Let us study now under which conditions on $(\alpha_0, \dots, \alpha_n)$ the fact that $\begin{pmatrix} \alpha_0 \\ c_0 \end{pmatrix}, \dots, \begin{pmatrix} \alpha_n \\ c_n \end{pmatrix}$ is a convex polygon is equivalent to $(c_0, \dots, c_n)$ being a convex sequence. So, for convexity preserving systems of blending functions, the convexity of control polygon would imply the convexity of the curve. The required property is

$$(c_{i+2} - c_{i+1})(\alpha_{i+1} - \alpha_i) \geq (c_{i+1} - c_i)(\alpha_{i+2} - \alpha_{i+1}) \Longleftrightarrow \Delta^2 c_i \geq 0 \ ,$$

$\forall i = 2, \dots, n$ and $\forall c = (c_0, \dots, c_n)$.

Considering the sequences $(0, 1, \dots, n)$ and $(n, n-1, \dots, 0)$, this property is easily seen to be equivalent to $\alpha_{i+1} - \alpha_i = h$, $\forall i = 1, \dots, n$, that is, the α's are equidistant. So, $\alpha_i = \frac{n-i}{n} a + \frac{i}{n} b$. In Corollary 6.5 (ii) of [**2**], convexity preserving blending system of continuous functions for which the constraint

$$\sum_{i=0}^{n} (\frac{n-i}{n} a + \frac{i}{n} b) f_i(t) = t \tag{3.2}$$

holds, were characterized. As we shall see in Theorem 3.5, the hypothesis of continuity can be removed:

Theorem 3.3. *Let* $(f_0, \dots, f_n)$ *be a blending system of functions satisfying (3.2). Then* $(f_0, \dots, f_n)$ *is convexity preserving if and only if* $\sum_{i=j+1}^n (i-j) f_i$ *is convex for all* $j \in \{1, \dots, n-1\}$.

Let us observe that if $t = \sum_{i=0}^n \alpha_i f_i(t)$, with $a = \alpha_0 < \alpha_1 < \ldots < \alpha_n = b$, the fact that the control polygon $\begin{pmatrix} \alpha_0 \\ c_0 \end{pmatrix} \cdots \begin{pmatrix} \alpha_n \\ c_n \end{pmatrix}$ is convex is equivalent to

$$\frac{c_{i+2} - c_{i+1}}{\alpha_{i+2} - \alpha_{i+1}} - \frac{c_{i+1} - c_i}{\alpha_{i+1} - \alpha_i} \geq 0. \tag{3.3}$$

This motivates the following definition:

Definition 3.4. *Let $\alpha = (\alpha_0, \dots \alpha_n)$ such that $a = \alpha_0 < \alpha_1 < \dots < \alpha_n = b$. Let $(f_0, \dots, f_n)$ be a blending system such that (3.1) holds. Then $(f_0, \dots, f_n)$ is convexity preserving with respect to α if for any sequence $(c_0, \dots, c_n)$ satisfying (3.3), the function $\sum_{i=0}^n c_i f_i$ is convex.*

In Theorem 3.1 of [**3**] we characterized convexity preserving systems of functions with respect to α. Let us restate that result removing the hypothesis of linear independence of the system of functions which was not used in the proof.

Theorem 3.5. *Let $(f_0, \dots, f_n)$ be a blending system of functions satisfying (3.1). Then $(f_0, \dots, f_n)$ is convexity preserving if and only if $\sum_{i=j+1}^n (\alpha_i - \alpha_j) f_i$, $j = 1, \dots, n-1$, are convex functions.*

In the previous setting, only bidimensional control polygons with fixed abscissae at $\alpha_0, \dots, \alpha_n$ were allowed. The next definition will deal with general control polygons, including higher dimensional cases. Let $P_0 \cdots P_n$, $P_i \in \mathbb{R}^s$ for $i = 0, 1, \dots, n$, be a control polygon. Let us assume that the projection of the control points $P_0, \dots, P_n$ onto a line is an ordered sequence of different points and that there exists a plane containing that line sucht that the projection of $P_0 \cdots P_n$ onto this plane is a convex polygon. We want that the projection of the curve $\gamma(t) = \sum_{i=0}^n P_i f_i(t)$ onto the plane to be also convex. This can be formulated as follows: for any $\alpha_0 < \alpha_1 < \cdots < \alpha_n$ and $c_0, \dots, c_n \in \mathbb{R}$ such that (3.3) holds, we want that the parametric curve $\begin{pmatrix} \alpha(t) \\ c(t) \end{pmatrix}$ (with $c(t) = \sum_{i=0}^n c_i f_i(t)$ and $\alpha(t) = \sum_{i=0}^n \alpha_i f_i(t)$) is convex. If $\alpha(t)$ is a strictly increasing function, then $\begin{pmatrix} \alpha(t) \\ c(t) \end{pmatrix}$ is a convex curve if and only if $c \circ \alpha^{-1}$ is a convex function. In general, we may not expect that $\alpha(t)$ is increasing unless $(f_0, \dots, f_n)$ is a strictly monotonicity preserving system.

Definition 3.6. *A strictly monotonicity preserving system $(u_0, \dots, u_n)$ is geometrically convexity preserving (resp., geometrically strictly convexity preserving) if for any $\alpha_0 < \cdots < \alpha_n$ and $c_0, \dots, c_n \in \mathbb{R}$ such that (3.3) holds then the function $c \circ \alpha^{-1}$ is convex (resp., strictly convex), where*

$$(\alpha(t), c(t)) := (u_0(t), \dots, u_n(t)) \begin{pmatrix} \alpha_0 & c_0 \\ \vdots & \vdots \\ \alpha_n & c_n \end{pmatrix}.$$

Tchebychev systems, which are closely related with generalized convexity (see Section 4, Chapter 1 of [**16**]), are now defined.

Definitions 3.7. *The system $(u_0, \dots, u_n)$ is Tchebychev (resp., weak Tchebychev) if all its square collocation matrices $M\begin{pmatrix} u_0, \dots, u_n \\ t_0, \dots, t_n \end{pmatrix}$ have positive (resp., nonnegative) determinant. A space with a Tchebychev basis is called a Tchebychev space.*

If $(u_0, \ldots, u_k)$ *is Tchebychev for all* $k = 0, \ldots, n$, *then we say that* $(u_0, \ldots, u_n)$ *is complete Tchebychev*

In Theorem 3.5 of [**4**], geometrically convexity preserving systems were characterized.

Theorem 3.8. *Let* $(f_0, \ldots, f_n)$ *be a blending strictly monotonicity preserving system and* $g_j := \sum_{i=j}^n f_i$, $j = 1, \ldots, n$. *Then* $(f_0, \ldots, f_n)$ *is geometrically convexity preserving if and only if* $(1, g_i, g_j)$ *is a weak Tchebychev system for all* $i < j$.

In Corollary 4.6 of [**4**] we gave a necessary and sufficient condition for the geometric convexity preserving property of all systems (1.1), generating rational curves.

Theorem 3.9. *Let* $(u_0, \ldots, u_n)$ *be a system of nonnegative functions such that* $\sum_{i=0}^n u_i > 0$ *and* $(\sum_{i=1}^n iu_i) / (\sum_{i=0}^n u_i)$ *is a strictly increasing function. Then all systems (1.1) are geometrically convexity preserving for all* $w_0, \ldots, w_n > 0$ *if and only if* $(u_0, \ldots, u_n)$ *is* TP_3.

Finally, applying Corollary 4.8 of [**4**] for $k = 2$, we obtain:

Theorem 3.10. *Let* $(u_0, \ldots, u_n)$ *be a system of nonnegative functions such that* $\sum_{i=0}^n u_i > 0$. *Then all systems (1.1) are geometrically strictly convexity preserving for all* $w_0, \ldots, w_n > 0$ *if and only if* $(u_0, \ldots, u_n)$ *is* TP_3 *and there exists a 3-dimensional Tchebychev subspace of* span$\{u_0, \ldots, u_n\}$.

§ 4. Functional Viewpoint

From now on we shall deal with the problem of higher order convexity preservation. In this section we shall adopt a functional point of view: we want to know if a function is k-convex in terms of their coefficients with respect to a given basis.

Definition 4.1. *A function* $u : [a, b] \to \mathbb{R}$ *is* k*-convex if all* k*-th order divided differences of* u *at distinct ponts* $t_0, \ldots, t_k$ *are nonnegative:* $u[t_0, \ldots, t_k] \geq 0$. *A vector* $c = (c_0, \ldots, c_n) \in \mathbb{R}^{n+1}$ *is* k*-convex if the* k*-th progressive differences*

$$\Delta^k c_i := \sum_{j=0}^{k} (-1)^{k-j} \binom{k}{j} c_{i+j}, \quad i = 0, \ldots, n-k,$$

are nonnegative.

If u is a k-convex function then, for every polynomial p of degree less than or equal to $k-1$, $u - p$ has fewer than k strict changes of sign. In particular, the graph of a 2-convex function crosses a line no more than two times and the graph of a 3-convex function cannot cross a parabola with axis parallel to the Y-axis more than three times.

Definition 4.2. *A system* $(u_0, \ldots, u_n)$ *is* k*-convexity preserving if for any* k*-convex vector* $(c_0, \ldots, c_n)$*, the function* $\sum_{i=0}^{n} c_i u_i$ *is* k*-convex. A matrix* $A \in \mathbb{R}^{(n+1)\times(n+1)}$ *is* k*-convexity preserving if* Ac *is a* k*-convex vector for any* k*-convex* $c \in \mathbb{R}^{n+1}$.

The k-convexity of a system of continuous system of functions can be reduced to show the same property for collocation matrices (1.2) at equidistant points (see Proposition 2.10 of [**2**]).

Let us introduce the following notations which will be useful for analyzing k-convexity preserving matrices.

$$E_j := \begin{pmatrix} I & 0 \\ 0 & E \end{pmatrix} \in \mathbb{R}^{n\times n},$$

where the block I stands for the $(j-1)\times(j-1)$ identity matrix and E for the $(n-j+1)\times(n-j+1)$ lower triangular matrix

$$E := \begin{pmatrix} 1 & 0 & \cdots & 0 \\ \vdots & \ddots & \ddots & \vdots \\ 1 & \cdots & 1 & 0 \\ 1 & \cdots & 1 & 1 \end{pmatrix},$$

with 1's on and below the main diagonal.

In Theorem 3.1 of [**2**], the following test for recognizing k-convexity preserving matrices was obtained.

Theorem 4.3. *Let* A *be an* $n \times n$ *matrix. Then* A *is a* k*-convexity preserving matrix if and only if*

$$(E_1 \cdots E_k)^{-1} A (E_1 \cdots E_k) = \begin{pmatrix} B_{11} & B_{12} \\ 0 & B_{22} \end{pmatrix},$$

where B_{22} *is an* $(n-k)\times(n-k)$ *matrix such that* $B_{22} \geq 0$.

This result can be applied for obtaining a test for the identification of k-convexity preserving systems (see Proposition 6.1 of [**2**]):

Theorem 4.4. *Let* $u_0, \ldots, u_n : [a,b] \to \mathbb{R}$ *and let*

$$(v_0, \ldots, v_n) := (u_0, \ldots, u_n) E_1 \cdots E_k.$$

Then $(u_0, \ldots, u_n)$ *is* k*-convexity preserving if and only if* $v_0, \ldots, v_{k-1}$ *are polynomials of degree less than* $k-1$ *and* $v_k, \ldots, v_n$ *are* k*-convex functions.*

§5. Geometric Viewpoint

In this section we shall deal with a geometric viewpoint: if the control polygon is k-convex in the (geometric) sense that we shall indicate, the curve will be also k-convex.

Let us recall first the case $k = 2$: convex planar curves and polygons.

A polygon $\begin{pmatrix}\alpha_0\\ c_0\end{pmatrix}, \ldots, \begin{pmatrix}\alpha_n\\ c_n\end{pmatrix}$ is *(globally) convex* if

$$(c_k - c_j)(\alpha_j - \alpha_i) \geq (c_j - c_i)(\alpha_k - \alpha_j), \quad \forall\, i < j < k \,,$$

i.e., all 3×3 minors of the matrix

$$\begin{pmatrix} 1 & \alpha_0 & c_0 \\ \vdots & \vdots & \vdots \\ 1 & \alpha_n & c_n \end{pmatrix} \tag{5.1}$$

are nonnegative. A curve is said to be *(globally) convex* if $(1, \alpha(t), c(t))$ is a weak Tchebychev system, that is, $\det M \begin{pmatrix} 1 & \alpha(t) & c(t) \\ t_0 & t_1 & t_2 \end{pmatrix} \geq 0$.

The previous condition means that a curve is convex if and only if all triangles formed with points $\begin{pmatrix}\alpha(t_0)\\ c(t_0)\end{pmatrix}, \begin{pmatrix}\alpha(t_1)\\ c(t_1)\end{pmatrix}, \begin{pmatrix}\alpha(t_2)\\ c(t_2)\end{pmatrix}$, $t_0 < t_1 < t_2$, are positively oriented. Let us observe that a convex curve or polygon may turn an angle greater than π. For example, $\gamma(t) = \begin{pmatrix}\cos t\\ \sin t\end{pmatrix}$, $t \in [0, 3\pi/2]$ is convex.

A *locally convex* polygon, $\begin{pmatrix}\alpha_0\\ c_0\end{pmatrix}, \ldots, \begin{pmatrix}\alpha_n\\ c_n\end{pmatrix}$ is a polygon satisfying:

$$(c_{i+2} - c_{i+1})(\alpha_{i+1} - \alpha_i) \geq (c_{i+1} - c_i)(\alpha_{i+2} - \alpha_{i+1}), \quad i = 0, \ldots, n-2,$$

i.e., all 3×3 minors of the matrix (5.1) using consecutive rows are nonnegative. For smooth curves, local convexity is defined by the nonnegativity of the curvature. For nonsmooth curves, local convexity can be defined as follows: Each t has neighbourhood in which $(1, \alpha(t), c(t))$ is weak Tchebychev. Clearly, global convexity implies local convexity. From Theorem 3.2 of Chapter 2 [**16**] it can be easily deduced that if $\alpha(t)$ is an strictly increasing function, then global convexity is equivalent to local convexity. In this framework, curves cannot rotate an angle greather than π. This setting can be generalized to the case in which a certain linear combination of α and c is strictly increasing. This property means that there exists a line on which the projection of the curve is strictly increasing (compare with Definition 3.6 and its previous remarks). This will be the context of the geometric viewpoint which will be taking into account for the generalizations for $k > 2$.

In order to apply Theorem 3.2 of Chapter 2 of [**16**] for $k > 2$, we need to impose, in addition to the convexity, that certain related systems are Tchebychev. We also need the following matricial definitions:

Definitions 5.1. *We say that an $m \times k$ $(m \geq k)$ matrix is weak left STP (resp., left STP) if all $i \times i$ minors $(1 \leq i < k)$ using initial columns are positive and all $k \times k$ minors are nonnegative (resp., positive).*

Let us observe that, if $m = k$, the concept of left STP matrix coincides with the concept of lowerly STP matrix given in [**10**], and that a matrix is left STP if and only if all its square submatrices of maximal order are lowerly STP.

Definitions 5.2. *A curve $\gamma(t) : [a, b] \to \mathbb{R}^k$ is called geometrically k-convex (resp., strictly geometrically k-convex) if there exists a one-to-one affine mapping $h : \mathbb{R}^k \to \mathbb{R}^k$ such that for all $l = 1, \ldots, k-1$, the systems $(1, y_1(t), \ldots, y_l(t))$ are Tchebychev and the system $(1, y_1(t), \ldots, y_k(t))$ is weak Tchebychev (resp., Tchebychev), where $(y_1(t), \ldots, y_k(t))^T := (h \circ \gamma)(t)$. A polygon $P_0 \cdots P_n$, $P_0, \ldots, P_n \in \mathbb{R}^k$, is called geometrically k-convex (resp., strictly geometrically k-convex) if there exists an affine mapping $h : \mathbb{R}^k \to \mathbb{R}^k$ such that the matrix whose rows are formed by $(1, h(P_0)^T), \ldots, (1, h(P_n)^T)$ is weak left STP (resp., left STP).*

Let us observe that the previous definition of a k-convex polygon is the discrete version of the definition given of a k-convex curve. In fact, if a system of functions $(y_0, \ldots, y_k)$ is weak Tchebychev (resp., Tchebychev) and for all $l = 0, \ldots, k-1$ the systems $(y_0, \ldots, y_l)$ are Tchebychev, then all collocation matrices of $M\begin{pmatrix} y_0, \ldots, y_k \\ t_0, \ldots, t_m \end{pmatrix}$ $(m \geq k)$ are weak left STP (resp., left STP).

Remark 5.3. Let us observe that if a k-dimensional curve or polygon is geometrically k-convex (resp., strictly geometrically k-convex) then there exists a projection onto $\mathbb{R}^r$, $r < k$ such that the projected curve or polygon is strictly geometrically r-convex. For instance, in the case of geometrically k-convex curves, we know that there exists a one-to-one affine mapping T such that $(y_1, \ldots, y_l)$ is Tchebychev for all $l = 1, \ldots, k-1$, where $(y_1(t), \ldots, y_k(t))^T := T \circ \gamma(t)$. Now taking $\pi : \mathbb{R}^k \to \mathbb{R}^r$ given by $\pi(y_1, \ldots, y_k) = (y_1, \ldots, y_r)$, then $\pi \circ T$ is a projection such that $\pi \circ T \circ \gamma(t) := (y_1(t), \ldots, y_r(t))^T$ is a strictly r-convex curve.

Definition 5.4. *A blending system $(f_0, \ldots, f_n)$ is geometrically k-convexity preserving (resp., geometrically strictly k-convexity preserving) if for any k-convex (resp., strictly k-convex) polygon $P_0 \cdots P_n$, $P_0, \ldots, P_n \in \mathbb{R}^k$, the corresponding curve $\gamma(t) = \sum_{i=0}^n P_i f_i(t)$ is k-convex (resp., strictly k-convex).*

Let us assume that a geometrically k-convexity preserving blending system is given. Let us assume further that $P_0 \cdots P_n$ is a control polygon in $\mathbb{R}^s$, $s \geq k$ and that there exists a projection $\pi : \mathbb{R}^s \to \mathbb{R}^k$ such that $\pi(P_0) \cdots \pi(P_n)$ is geometrically k-convex. Then the projection $\pi \circ \gamma$ of the s-dimensional curve $\gamma(t) = \sum_{i=0}^n P_i f_i(t)$ onto $\mathbb{R}^k$ is given by

$$\pi \circ \gamma(t) = \sum_{i=0}^{n} \pi(P_i) f_i(t), \tag{5.2}$$

and so this projection has to be a geometrically k-convex curve. Therefore, a *geometrically k-convexity preserving blending system transforms polygons into curves in such a way that if a given projection of the polygon is geometrically k-convex then the corresponding projection of the curve is also geometrically k-convex.*

Another useful information, which can easily derived from the definition and (5.2) in the case of $\pi : \mathbb{R}^k \to \mathbb{R}^k$ being any bijective affine mapping, can be stated as follows:

Proposition 5.5. *A system $(f_0, \ldots, f_n)$ is geometrically k-convexity preserving (resp., strictly geometrically k-convexity preserving) if and only if for any weak left STP (resp., left STP) $(n+1) \times (k+1)$ matrix A whose first column is $(1, \ldots, 1)^T$, the system $(1, y_1, \ldots, y_k)$ given by*

$$(1, y_1(t), \ldots, y_k(t)) := (f_0(t), \ldots, f_n(t))A \ ,$$

is weak Tchebychev (resp., Tchebychev) and all systems $(y_1, \ldots y_l)$ are Tchebychev for $l = 1, \ldots, k-1$.

From Proposition 5.5 we can derive an interesting consequence: Definition 5.4 of geometrically k-convexity preserving and strictly geometrically k-convexity preserving systems coincide with Definition 4.3 of [**4**]. In fact, Proposition 5.6 shows that Definition 4.3 of [**4**] was redundant.

Let us introduce some notations which will be used in the following theorem. Given $k, n \in \mathbb{N}, k \leq n$, we define

$$Q_{k,n} := \{(\alpha_1, \ldots, \alpha_k) | \alpha_i \in \mathbb{N}, 1 \leq \alpha_1 < \cdots < \alpha_k \leq n\} \ ,$$

and for $\alpha, \beta \in Q_{k,n}$, $A[\alpha|\beta]$ is by definition the $k \times k$ submatrix of A containing rows numbered by α and columns numbered by β.

Proposition 5.6. *Any geometrically k-convexity preserving (resp., strictly geometrically k-convexity preserving) system is also strictly geometrically r-convexity preserving, for any $r < k$.*

Proof: Let us see first that if a matrix $A \in (n+1) \times k$ is a left STP matrix, we may add one column to A and get a matrix $\tilde{A} \in (n+1) \times (k+1)$ which is left STP. Let

$$\rho := 2. \max_{\alpha, \beta \in Q_{k,n+1}} \frac{\det A[\alpha|1, \ldots, k]}{\det A[\beta|1, \ldots, k]}. \tag{5.3}$$

Taking $\alpha = \beta$ in (5.3) we see that $\rho \geq 2$. Now let us define $\tilde{A}$ as the matrix whose first k columns coincide with that those of A and whose last column is $(1, \rho, \ldots, \rho^n)$. If $\gamma \in Q_{k+1,n+1}$, $\gamma_1 < \ldots < \gamma_{k+1}$ we obtain that

$$\begin{aligned} \det \tilde{A}[\gamma|1, \ldots, k+1] = \\ &(\rho^{\gamma_{k+1}} \det A[\gamma \backslash \{\gamma_{k+1}\}|1, \ldots, k] - \rho^{\gamma_k} \det A[\gamma \backslash \{\gamma_k\}|1, \ldots, k]) + \\ &(\rho^{\gamma_{k-1}} A[\gamma \backslash \{\gamma_{k-1}\}|1, \ldots, k] - \rho^{\gamma_{k-2}} \det A[\gamma \backslash \{\gamma_{k-2}\}|1, \ldots, k]) + \cdots \end{aligned}$$

Now each term

$$\rho^{\gamma_{h+1}} \det A[\gamma\backslash\{\gamma_{h+1}\}|1,\ldots,k] - \rho^{\gamma_h} \det A[\gamma\backslash\{\gamma_h\}|1,\ldots,k] \geq$$
$$A[\gamma\backslash\{\gamma_{h+1}\}|1,\ldots,k](\rho^{\gamma_{h+1}} - \rho^{\gamma_h}\frac{\rho}{2})$$

has to be positive because $\gamma_{h+1} \geq \gamma_h + 1$ and $\rho^{\gamma_{h+1}} \geq \rho^{\gamma_h+1} > 0$, and so we have shown that $\tilde{A}$ is left STP.

In the special case that the first column of the left STP matrix A is $(1,\ldots,1)^T$, we shall obtain using Proposition 5.5 that all subsystems $(y_1,\ldots,y_l)$, $l = 1,\ldots,k-1$, of $y_1,\ldots,y_k$ given by

$$(1, y_1(t),\ldots,y_k(t)) := (f_0(t),\ldots,f_n(t))\tilde{A} \ ,$$

are Tchebychev. Since

$$(1, y_1(t),\ldots,y_{k-1}(t)) := (f_0(t),\ldots,f_n(t))A,$$

we obtain from Proposition 5.5 that $(f_0,\ldots,f_n)$ is strictly geometrically $(k-1)$-convexity preserving.

Now, using a recursive procedure, it is easily shown that $(f_0,\ldots,f_n)$ is strictly geometrically r-convexity preserving for all $r < k$. ■

A characterization of geometrically k-convexity preserving and strictly geometrically k-convexity preserving systems of functions was provided in Corollary 4.8 of [**4**]:

Proposition 5.7. *Let $k \leq n$, and $(u_0,\ldots,u_n)$ be a system of nonnegative functions such that $\sum_{i=0}^n u_i > 0$. Then*

(i) The systems (1.1) are geometrically k-convexity preserving for all positive weights $w_0,\ldots,w_n$, if and only if $(u_0,\ldots,u_n)$ is TP_{k+1} and there exists a k-dimensional Tchebychev subspace of span$\{u_0,\ldots,u_n\}$.

(ii) The systems (1.1) are geometrically strictly k-convexity preserving for all positive weights $w_0,\ldots,w_n$, if and only if $(u_0,\ldots,u_n)$ is TP_{k+1} and there exists a $(k+1)$-dimensional Tchebychev subspace of span$\{u_0,\ldots,u_n\}$.

For $k = n$ we obtain the following consequence.

Corollary 5.8. *Let $(u_0,\ldots,u_n)$ be a system of nonnegative functions such that $\sum_{i=0}^n u_i > 0$. Then*

(i) The systems (1.1) are geometrically n-convexity preserving for all positive weights $w_0,\ldots,w_n$, if and only if $(u_0,\ldots,u_n)$ is a totally positive system and there exists an n-dimensional Tchebychev subspace of span$\{u_0,\ldots,u_n\}$.

(ii) The systems (1.1) are geometrically strictly n-convexity preserving for all positive weights $w_0, \ldots, w_n$, if and only if $(u_0, \ldots, u_n)$ is a totally positive and Tchebychev system.

Now, let us mention some remarkable examples of bases generating rational curves which preserve the convexity properties of the control polygon.

Let us begin with the case of rational Bézier curves obtained from the Bernstein basis $(B_0^n, \ldots, B_n^n)$,

$$B_i^n(t) := \binom{n}{i}(1-t)^{n-i}t^i, \quad t \in [0,1].$$

It is well known that $(B_0^n, \ldots, B_n^n)$ is a totally positive and Tchebychev system such that $\sum_{i=0}^n B_i^n(t) = 1$. Therefore, all bases generating rational Bézier curves

$$\left(\frac{w_0 B_0^n}{\sum_{i=0}^n w_i B_i^n}, \ldots, \frac{w_n B_n^n}{\sum_{i=0}^n w_i B_i^n}\right), \qquad w_0, \ldots, w_n > 0,$$

are geometrically strictly k-convexity preserving for all $k \le n$.

On the other hand, we know that a basis of polynomials $(p_0, \ldots, p_n)$ of degree less than or equal to n on $[0,1]$ is totally positive if and only if it can be written as

$$(p_0, \ldots, p_n) = (B_0^n, \ldots, B_n^n)K, \tag{5.4}$$

where K is a nonsingular stochastic totally positive matrix.

Therefore these results provide a procedure for constructing (from the Bernstein basis) all bases $(p_0, \ldots, p_n)$ such that the corresponding systems

$$\left(\frac{w_0 p_0}{\sum_{i=0}^n w_i p_i}, \ldots, \frac{w_n p_n}{\sum_{i=0}^n w_i p_i}\right) \tag{5.5}$$

generate rational polynomial curves which are geometrically strictly k-convexity preserving for all $k \le n$. This construction can be performed in a simple way because any nonsingular stochastic totally positive matrix is a product of bidiagonal nonnegative stochastic matrices (see Theorem 1 of [**13**]).

We consider now the case of rational B-spline curves. Let

$$\tau_0 = \cdots = \tau_{k-1} < \tau_k \le \cdots \le \tau_n < \tau_{n+1} = \cdots = \tau_{n+k},$$

$k \le n$, be an extended sequence of knots such that $\tau_i < \tau_{i+k}$, $i = 0, \ldots, n$. The B-spline basis

$$N_{i,k}(t) := (\tau_{i+k} - \tau_i)[\tau_i, \ldots, \tau_{i+k}](\cdot - t)_+^{k-1}, \quad i = 0, \ldots, n,$$

is used for generating rational B-spline curves on $[\tau_{k-1}, \tau_{n+1}]$ by means of the blending system

$$\left(\frac{w_0 N_{0,k}}{\sum_{i=0}^{n} w_i N_{i,k}}, \ldots, \frac{w_n N_{n,k}}{\sum_{i=0}^{n} w_i N_{i,k}}\right), \quad w_0, \ldots, w_n > 0. \tag{5.6}$$

It is also well known that the B-spline basis $(N_{0,k}, \ldots, N_{n,k})$ is totally positive and that $\sum_{i=0}^{n} N_{i,k}(t) = 1$. Since the space of all polynomials of degree less than or equal to $k-1$ is a Tchebychev subspace of the space generated by $(N_{0,k}, \ldots, N_{n,k})$, we derive from Proposition 5.7 that the bases (5.6) are geometrically k-convexity preserving.

Finally, taking into account that B-spline basis plays a similar role to that of Bernstein basis in (5.4) with respect to the totally positive polynomial spline bases (see Section 4 of [**7**]), we could also derive in a similar way to the previous polynomial case an easy construction of bases generating rational polynomial spline curves which are geometrically k-convexity preserving for all $k \leq n$.

Acknowledgements. Research partially supported by DGICYT PB93-0310 and by the EU project CHRX-CT94-0522.

References

1. Ando, T., Totally positive matrices, Linear Algebra Appl. **90** (1987), 165-219.
2. Carnicer, J.M., M. García-Esnaola and J.M. Peña, Generalized convexity preserving transformations, to appear in Computer-Aided Geom. Design .
3. Carnicer, J.M., M. García-Esnaola and J.M. Peña, Convex curves from convex control polygons, in *Mathematical Methods in Computer Aided Geometric Design III*, M. Dæhlen, T. Lyche and L. L. Schumaker (eds.), Vanderbilt University Press, Nashville, 1995, 63-72.
4. Carnicer, J.M., M. García-Esnaola and J.M. Peña, Convexity of rational curves and total positivity, to appear in Journal of Computational and Applied Mathematics.
5. Carnicer, J.M., T.N.T. Goodman and J.M. Peña, A generalization of the variation diminishing property, Advances in Computational Mathematics **3** (1995), 375-394.
6. Carnicer, J.M. and J.M. Peña, Shape preserving representations and optimality of the Bernstein basis, Advances in Computational Mathematics **1** (1993), 173-196.
7. Carnicer, J.M. and J.M. Peña, Totally positive bases for shape preserving curve design and optimality of B-splines, Computer-Aided Geom. Design **11** (1994), 635-656.
8. Carnicer, J.M. and J. M. Peña, Monotonicity preserving representations, in *Curves and Surfaces in Geometric Design*, P.-J. Laurent, A. Le Méhauté, and L. L. Schumaker (eds.), A K Peters, Wellesley, 1994, 83-90.

9. Dahmen, W., Convexity and Bernstein-Bézier polynomials, in *Curves and Surfaces*, P.-J. Laurent, A. Le Méhauté, and L. L. Schumaker (eds.), Academic Press, New York, 1991, 107-134.
10. Gasca, M. and J.M. Peña, Total positivity, QR-factorization and Neville elimination, SIAM J. Matrix Anal. Appl. **14** (1993), 1132-1140.
11. Goodman, T.N.T., Shape preserving representations, in *Mathematical Methods in Computer Aided Geometric Design*, T. Lyche and L. L. Schumaker (eds.), Academic Press, New York, 1989, 333-357.
12. Goodman, T.N.T., Inflections on curves in two and three dimensions, Computer-Aided Geom. Design **8** (1991), 37-50.
13. Goodman, T.N.T and C.A. Micchelli, Corner cutting algorithms for the Bézier representation of free form curves, Linear Algebra Appl. **99** (1988), 225-252.
14. Goodman, T.N.T. and J. Peters, Bézier nets, convexity and subdivision on higher-dimensional simplices, Computer-Aided Geom. Design **12** (1995), 53-65.
15. Goodman, T.N.T. and H.B. Said, Shape preserving properties of the generalized Ball basis, Computer-Aided Geom. Design **8** (1991), 115-121.
16. Karlin, S., *Total positivity*, Stanford University Press, California, 1968.
17. Sauer, T., Multivariate Bernstein polynomials and convexity, Computer-Aided Geom. Design **8** (1991), 465-478.

J. M. Carnicer
Departamento de Matemática Aplicada.
Universidad de Zaragoza.
Edificio de Matemáticas, 50009 Zaragoza
SPAIN

carnicer@posta.unizar.es

M. García-Esnaola.
Departamento de Matemática Aplicada.
Universidad de Zaragoza.
Centro Politécnico Superior, 50015 Zaragoza
SPAIN

mgesnaola@mcps.unizar.es

J. M. Peña
Departamento de Matemática Aplicada.
Universidad de Zaragoza.
Edificio de Matemáticas, 50009 Zaragoza
SPAIN

jmpena@posta.unizar.es

Stability and Inter-Dependence of Matrix Subdivision Schemes

Albert Cohen, Nira Dyn and David Levin

Abstract. Subdivision schemes with matrix masks are a natural extension of the well studied case of subdivision schemes with scalar masks. Such schemes arise in the analysis of multivariate scalar schemes, in subdivision processes corresponding to shift-invariant spaces generated by more than one function, in geometric modeling where each component of the curve/surface is designed by a different linear combination of the control points, and in the case of schemes which interpolate function and derivatives values simultaneously. The limit of a matrix subdivision scheme can be expressed as a combination of shifts of a refinable matrix function Φ. It is shown that if Φ is stable in the sense of a new stability notion for matrix valued functions, then the scheme is uniformly convergent. Also it is shown that the stability of a maximal submatrix of Φ is related with the linear dependence of its rows, and hence of any vector valued function generated by the subdivision scheme. Finally it is shown that by proper renormalization of the process relative to the vanishing rows of Φ, it is possible to generate vector limit functions with components, which are the first derivative of certain linear combinations of the other components. The same approach allows to analyze the smoothness of Φ.

§1. Introduction

Matrix subdivision schemes play an important role in the analysis of multivariate subdivision schemes [7], in the construction of multiple-knot splines and in Hermite type subdivision [11]. They also have strong connections with multiresolution approximation of multiplicity more than one ([15],[16],[19],[20]) and multi-wavelets ([17],[18],[21]).

The purpose of this work is to understand the characteristics of the different types of matrix subdivision schemes.

Advanced Topics in Multivariate Approximation
F. Fontanella, K. Jetter and P. J. Laurent (eds.) pp. 33–45.

ISBN 981-02-2852-X

A uniform stationary matrix subdivision scheme is defined by a set of real $n \times n$ matrix coefficients $\{A_j : j \in \mathbb{Z}\}$, with a finite number of non-zero A_j's, generating control points $f^k = \{f_j^k : j \in \mathbb{Z}\}$, $k \geq 0$, in $\mathbb{R}^n$ recursively by

$$f_i^{k+1} = \sum_{j \in \mathbb{Z}} A_{i-2j} f_j^k \ , \ i \in \mathbb{Z} \ . \tag{1.1}$$

In (1.1) we find two rules, or masks, for even and for odd i, involving the non-zero matrix coefficients A_j with even and with odd j's respectively.

We say, that $\mathcal{S}$ is a convergent subdivision scheme, or that $\mathcal{S}$ is C^0, if for every set of control points $f^0 = \{f_j^0 \in \mathbb{R}^n : j \in \mathbb{Z}\}$ there is a continuous function $f : \mathbb{R} \to \mathbb{R}^n$ such that

$$\lim_{k \to \infty} \sup_{j \in \mathbb{Z}} \|(\mathcal{S}^k f^0)_j - f(2^{-k} j)\| = 0 \ , \tag{1.2}$$

and $f \not\equiv 0$ for at least one initial data f^0. We denote the above function f by $\mathcal{S}^\infty f^0$, and call it a limit function of $\mathcal{S}$ or a function generated by $\mathcal{S}$. In matrix subdivision schemes theory it is also convenient to consider $n \times n$ matrix valued functions, generated by applying the scheme to sets of matrix control points $F^0 = \{F_j^0 \in \mathbb{R}^n \times \mathbb{R}^n : j \in \mathbb{Z}\}$. The basic limit vector functions of a convergent matrix subdivision scheme are the rows of $\Phi = \mathcal{S}^\infty \Delta_0$ where $\Delta_i = \{\delta_{j,i} I\}_{j \in \mathbb{Z}}$, $I = I_{n \times n}$.

As in scalar subdivision schemes, the matrix function Φ satisfies a two-scale refinement equation

$$\Phi(x) = \sum_{j \in \mathbb{Z}} \Phi(2x - j) A_j \ . \tag{1.3}$$

Hence, each row of Φ, i.e., each basic limit vector function satisfies a refinement equation with the A_j's as coefficients. By the finite support of the 'mask' $\{A_j : j \in \mathbb{Z}\}$ it follows that elements of Φ are functions of compact support. The interest in the basic limit vector functions is due to the fact that the vector limit functions of the subdivision process are in the span of the integer shifts of the columns of Φ. That is, if we start the subdivision process (1.1) with an initial set $f^0 = \{f_i^0 \in \mathbb{R}^n : i \in \mathbb{Z}\}$ we get

$$f = \mathcal{S}^\infty f^0 = \mathcal{S}^\infty \sum_{i \in \mathbb{Z}} \Delta_i f_i^0 = \sum_{i \in \mathbb{Z}} \Phi(\cdot - i) f_i^0 \ . \tag{1.4}$$

Thus the k-th component of the limit vector function f is in the span of the integer shifts of the k-th basic limit vector function,

$$f_k(x) = \sum_{i \in \mathbb{Z}} \sum_{\ell=1}^{n} \Phi_{k,\ell}(x - i)(f_i^0)_\ell \ . \tag{1.5}$$

Therefore, it is important to know if

$$U_k = \{\Phi_{k,\ell}(\cdot - i) : 1 \leq \ell \leq n \ , \ i \in \mathbb{Z}\} \ , \tag{1.6}$$

is a stable basis, say in $L^2(\mathbb{R})$ or $L^\infty(\mathbb{R})$. Another interesting issue is the inter-dependence of the coordinates in the limit vector function f, which plays an important role when the coordinates of the vector subdivision are intertwined as in certain signal processing applications [21]: assuming that the limit function is not identically zero, at least for some initial data f^0, we would like to know when are the functions f_k linearly dependent.

In §2, we show that the number of inter-dependence relations between the coordinates of f is related to a generalized stability notion for the matrix function Φ. We also show that this stability property implies convergence of the subdivision scheme.

We restrict to the study of convergence and stability in the L^∞ sense. More general L^p results of the same type, as well as specific criterions on the mask that ensure stability and regularity of Φ, will appear in [22].

Note that the relations between stability and rank property have been studied in the L^p framework in [15], but in the case of stable refinable vectors that corresponds, as we shall see (Corollary 2.5), to having inter-dependence between every pair of coordinates of f. Also in the case of refinable vectors, conditions on the mask for stability have been obtained in [23].

Inter-dependence can also be expressed by the fact that a coordinate in the limit vector function is uniformly zero. We analyze such a situation in §3 and show that a proper renormalization of the process relative to the vanishing rows of Φ generates vector limit functions with components related to the first derivative of certain linear combinations of the other components.

§2. Relations between Inter-Dependence and Stability

We start with the assumption that the scheme (1.1) is C^0, i.e., convergent. The relation (1.1) represents two rules, one generating the even entries $\{f_{2i}^{k+1}\}_{i\in\mathbb{Z}}$, using the even coefficients $\{A_{2i}\}_{i\in\mathbb{Z}}$, and the other generating the odd entries $\{f_{2i+1}^{k+1}\}_{i\in\mathbb{Z}}$, using the odd coefficients $\{A_{2i+1}\}_{i\in\mathbb{Z}}$. Important to this investigation are the matrices

$$B_0 = \sum_{i\in\mathbb{Z}} A_{2i} \ , \quad B_1 = \sum_{i\in\mathbb{Z}} A_{2i+1} \ , \quad M(0) = \tfrac{1}{2}(B_0 + B_1) \ , \tag{2.1}$$

where $M(\omega) = \frac{1}{2}\sum_{j\in\mathbb{Z}} A_j e^{-ij\omega}$.

The following two Theorems are the matrix analogue of the necessary condition for scalar C^0 schemes, namely, that the sum of the even coefficients and the sum of the odd coefficients are both 1. The proof of the first Theorem is as the proof of the scalar case (Proposition 2.2 in [7]).

Theorem 2.1. *If $\mathcal{S}$ is C^0 then B_0 and B_1 have a common eigenvector with an eigenvalue 1.*

Let W denote the space of all common eigenvectors of B_0 and B_1 corresponding to the eigenvalue 1, and let m be its dimension. Let us also introduce the matrix $V = \left(v^{(1)}...v^{(m)}\right)$, where $\{v^{(1)}, ..., v^{(m)}\}$ is a basis of W.

Theorem 2.2. *If $\mathcal{S}$ is C^0 then W is also the space of eigenvectors corresponding to the eigenvalue 1 of the matrix $M(0)$. Moreover, $M(0)$ has $n-m$ eigenvalues with modulus less than 1 .*

Proof: Clearly, W is included in the space $\tilde{W}$ of eigenvectors with eigenvalue 1 of $M(0)$. Now we observe that

$$J_k = 2^{-k} \sum_{i \in \mathbb{Z}} (\mathcal{S}^k \Delta_0)_i \longrightarrow J = \int_{\mathbb{R}} \Phi(x) dx \ ,$$

as k goes to $+\infty$, where $\Phi = \mathcal{S}^\infty \Delta_0$, and that $2^k J_k$ is the sum of the matrices of the k-iterated scheme. Thus $J_k = M(0)^k$ and $M(0)^k \to J$. Since $M(0)^\infty$ exists, its columns are in $\tilde{W}$ while those of J are in W. Therefore $\tilde{W} = W$ and $M(0)$ has $n-m$ eigenvalues with modulus less than 1 . ∎

In the following we assume that W is the eigenspace of $M(0)$ corresponding to the eigenvalue 1 .

Theorem 2.3. *Let $m < n$, then any $m+1$ elements of $f = \mathcal{S}^\infty f^0$ are linearly dependent.*

Proof: By the continuity of the limit it follows from (1.1) that $B_0 f(x) = B_1 f(x) = f(x)$ for any $x \in \mathbb{R}$. Thus $f(x) \in W$. The matrix V is of rank m, hence any $m+1$ rows of V are linearly dependent. Since $f(x) = Vu(x)$ with $u(x) \in \mathbb{R}^m$, any $m+1$ elements of f are linearly dependent. ∎

Definition 2.1. *We say that $h_1, ..., h_r$ constitute a stable basis if their integer shifts constitute a Riesz basis of $span\{h_j(\cdot - i) : 1 \le j \le r \ , \ i \in \mathbb{Z}\}$.*

Theorem 2.4.
A. *The 1-periodic matrix function $\tilde{\Phi} = \sum_{i \in \mathbb{Z}} \Phi(\cdot - i)$ satisfies $\tilde{\Phi} V = V$ where $\Phi = \mathcal{S}^\infty \Delta_0$.*
B. *Let $dim W = m$, $2 \le m \le n$, then within each basic limit vector function of $\mathcal{S}$ there exist $n-m+2$ elements which do not constitute a stable basis.*

Proof: The property $\tilde{\Phi} V = V$ follows directly from (1.4) when taking the constant initial data $f^0 = v^{(j)}$, $0 \le j \le m$, since by (2.1) $\mathcal{S}^\infty v^{(j)} = v^{(j)}$. For part B let us assume, w.l.o.g., that the upper $m \times m$ part of the $n \times m$ matrix V is $I = I_{m \times m}$, i.e., $V = \left(I \mid U\right)^t$. Viewing the equality $\tilde{\Phi} V = V$ we have

$$\sum_{i \in \mathbb{Z}} \sum_{\ell=1}^{n} \Phi_{k,\ell}(x-i) V_{\ell,j} = V_{k,j} \ , \ 1 \le k \le n \ , \ 1 \le j \le m \ , \qquad (2.2)$$

where for each k only $n-m+1$ elements of the kth row of Φ take part in the sum. Now, if $V_{k,j}=0$ then obviously those $n-m+1$ elements cannot constitute a stable basis. If $V_{k,j}\neq 0$ for $1\leq j\leq m$ we take a linear combination of the equalities (2.2) for $j=1,2$ to get a zero in the r.h.s., using $n-m+2$ elements of the kth row of Φ. ∎

Corollary 2.5. *If one basic limit vector function is stable then* $m=1$ *and any limit function of the subdivision is of the form* $f(x)=h(x)v$ *with* $v\in I\!R^n$.

Since by Corollary 2.5 only in the special case $m=1$ a basic limit vector function can constitute a stable basis, we suggest to consider the stability of Φ as a matrix function.

Definition 2.2. *We say that an* $m\times n$ *matrix function* $G(x)$ *is stable if its integer shifts constitute a stable basis (in* L_∞*) of span*$\{\Phi(\cdot-i)\,:\, i\in Z\!\!Z\}$. *i.e., for any* $\{q_k\}\in \ell_{\infty,I\!R^n}$

$$c_1\sup_{k\in Z\!\!Z}\{\|q_k\|\}\leq \sup_{x\in I\!R}\{\|\sum_{k\in Z\!\!Z}G(x-k)q_k\|\}\leq c_2\sup_{k\in Z\!\!Z}\{\|q_k\|\}\,. \tag{2.3}$$

In (2.3), $\|\cdot\|$ can be any norm in $I\!R^n$ since all of them are equivalent.

It should be noted that a matrix function is stable if a matrix consisting of a subset of its rows is stable. With this definition at hand, the proof of Theorem 2.4 can be further exploited to give:

Theorem 2.6. *Any matrix function consisting of* $k<m$ *rows of* $\Phi=\mathcal{S}^\infty\Delta_0$ *is not stable.*

As a direct consequence of Theorems 2.3 and 2.6 we get,

Corollary 2.7. *Let* $\Phi=\mathcal{S}^\infty\Delta_0$ *be stable, and let* m *be the minimal number of rows of* Φ *which constitute a stable matrix. Then any* $m+1$ *components in any vector valued function generated by* $\mathcal{S}$ *are linearly dependent.*

In the rest of this section we consider refinement equations of the type (1.3), with masks satisfying the requirements of Theorems 2.1 and 2.2, and prove that if a compactly supported continuous solution of such a refinement equation is stable, then the corresponding subdivision scheme is uniformly convergent. This result extends a known result of a similar nature for scalar schemes [2].

Proposition 2.8. *Let* Ψ *be any compactly supported* $C(I\!R)$ *solution of the refinement equation (1.3), then* $\Psi(x)=\widehat{\Psi}(0)VN(x)$ *where* N *is an* $m\times n$ *matrix function.*

Proof: Applying the Fourier transform to (1.3) we obtain

$$\widehat{\Psi}(\omega)=\widehat{\Psi}(\omega/2)M(\omega/2), \tag{2.4}$$

and by iteration,

$$\widehat{\Psi}(\omega) = \widehat{\Psi}(0)[\prod_{k=1}^{\infty} M(2^{-k}\omega)]. \tag{2.5}$$

The convergence of

$$P(\omega) = \prod_{k=1}^{\infty} M(2^{-k}\omega) = \lim_{n\to\infty} [M(2^{-n}\omega)\cdots M(\omega/2)],$$

is ensured by the spectral properties of $M(0)$. Note that $P(0) = M(0)^{\infty}$ is a projection operator onto the subspace W. Moreover, by definition,

$$P(\omega) = \lim_{n\to\infty} [P(2^{-n}\omega)M(2^{-n}\omega)\cdots M(\omega/2)] = P(0)P(\omega),$$

and hence by (2.5)

$$\widehat{\Psi}(\omega) = \widehat{\Psi}(0)P(0)P(\omega). \tag{2.6}$$

We conclude by noting that $P(0)P(\omega)$ can be written in a unique way as $VQ(\omega)$, where $Q(\omega)$ is an $m \times n$ matrix. ■

Proposition 2.9. *Let Ψ be a compactly supported stable $C(\mathbb{R})$ solution of the refinement equation (1.3). Then there exists a non-singular $n \times n$ matrix C, such that $\Phi = C\Psi$ is a stable solution of (1.3) satisfying $\widetilde{\Phi}(x)V = V$ and $\widehat{\Phi}(0)V = V$. We call Φ a normalized solution.*

Proof: The refinement equation (1.3) implies

$$\widetilde{\Psi}(x) = \sum_{i\in\mathbb{Z}} \Psi(x-i) = \sum_{i\in\mathbb{Z}}\sum_{j\in\mathbb{Z}} \Psi(2x-2i-j)A_j$$

$$= \sum_{\ell\in\mathbb{Z}} \Psi(2x-\ell) \sum_{i\in\mathbb{Z}} A_{\ell-2i} .$$

Therefore, $\widetilde{\Psi}(x)V = \widetilde{\Psi}(2x)V$, implying that $\widetilde{\Psi}(x)V = \widetilde{\Psi}(0)V$, by the continuity of $\widetilde{\Psi}$. Now, if Ψ is stable, then for any vector u

$$\sum_{i\in\mathbb{Z}} \Psi(x-i)Vu = \widetilde{\Psi}(0)Vu \neq 0 .$$

Hence, $rank\big(\widetilde{\Psi}(0)V\big) = m$ and there exists a non-singular matrix C such that $C\widetilde{\Psi}(0)V = V$, and hence $C\widetilde{\Psi}(x)V = V$. The stability of $C\Psi$ follows from that of Ψ and the non-singularity of C. Moreover, since $\widetilde{\Psi}(x)V$ is a periodic matrix function which is constant, then $\widetilde{\Psi}(x)V = \widehat{\Psi}(0)V$, which completes the proof of the proposition. ■

Theorem 2.10. *Let Φ be a continuous compactly supported stable solution of the refinement equation (1.3). If the necessary conditions in Theorems 2.1-2.2 are satisfied then the corresponding subdivision scheme (1.1) is convergent.*

Proof: We can assume that Φ is a normalized stable solution of the refinement equation. It is enough to check the convergence with the initial condition $F^0 = \Delta_0$. Using (1.3) we find out that

$$\Phi(x) = \sum_{i\in\mathbb{Z}} \Phi(2^k x - i)(\mathcal{S}^k\Delta_0)_i \ . \tag{2.7}$$

Also, using $\Phi(x) = VN(x)$ and $\widetilde{\Phi}V = V$, we have

$$\sum_{i\in\mathbb{Z}} \Phi(2^k x - i)\Phi(x) = \widetilde{\Phi}(2^k x)VN(x) = VN(x) = \Phi(x) \ . \tag{2.8}$$

Thus it follows that

$$\sum_{i\in\mathbb{Z}} \Phi(2^k x - i)[(\mathcal{S}^k\Delta_0)_i - \Phi(x)] = 0,$$

and thus

$$\sum_{i\in\mathbb{Z}} \Phi(2^k x - i)[(\mathcal{S}^k\Delta_0)_i - \Phi(2^{-k}i)] + \sum_{i\in\mathbb{Z}} \Phi(2^k x - i)[\Phi(2^{-k}i) - \Phi(x)] = 0. \tag{2.9}$$

By the continuity of Φ and its compact support, for any $\varepsilon > 0$ $\exists K$ s.t. for $k > K$

$$\|\sum_{i\in\mathbb{Z}} \Phi(2^k x - i)[\Phi(2^{-k}i) - \Phi(x)]\|_\infty < \varepsilon \ . \tag{2.10}$$

Now, from (2.9) and (2.10)

$$\|\sum_{i\in\mathbb{Z}} \Phi(2^k x - i)[(\mathcal{S}^k\Delta_0)_i - \Phi(2^{-k}i)]\|_\infty \le \varepsilon \text{ for } k > K \ , \tag{2.11}$$

and by the stability of Φ

$$\|\mathcal{S}^k\Delta_0 - \Phi(2^{-k}\cdot)\|_\infty \le \varepsilon \text{ for } k > K \ . \tag{2.12}$$

Hence $\mathcal{S}$ is convergent in the sense of (1.2) and $\mathcal{S}^\infty\Delta_0 = \Phi$. ■

§3. Analysis of Smoothness and Differential Inter-Dependence

It is convenient for the analysis of matrix subdivision schemes to introduce the following transformation:

Let Q be an $n \times (n-m)$ matrix such that $T = \big(V \mid Q\big)$ is a non-singular $n \times n$ matrix. Define the matrix mask

$$\overline{A}_j = T^{-1} A_j T \ , \quad j \in \mathbb{Z} \ , \tag{3.1}$$

and consider the subdivision scheme $\overline{\mathcal{S}}$ and the refinement equation corresponding to this matrix mask. Now the matrices

$$\overline{B}_0 = \sum_{i \in \mathbb{Z}} \overline{A}_{2i} \ , \quad \overline{B}_1 = \sum_{i \in \mathbb{Z}} \overline{A}_{2i+1} \ , \tag{3.2}$$

have the standard unit vectors $e^{(j)}$, $1 \le j \le m$ as eigenvectors with eigenvalue 1, and correspondingly we define $\overline{V} = \big(e^{(1)} ... e^{(m)}\big)$. Thus

$$\overline{B}_i = \big(e^{(1)} ... e^{(m)}, E_i\big) \ , \quad i = 0, 1 \ , \tag{3.3}$$

where E_i is an $(n-m) \times n$ matrix. An appropriate choice of the matrix Q is that consisting of the $n-m$ eigenvectors and possibly generalized eigenvectors of $M(0)$ corresponding to eigenvalues with modulus less than 1 . With this choice of Q , $\overline{M}(0)$ is a diagonal matrix or a Jordan matrix.

Let Ψ be a solution of (1.3), then $\overline{\Psi} = T^{-1} \Psi T$ satisfies

$$\overline{\Psi}(x) = \sum_{j \in \mathbb{Z}} \overline{\Psi}(2x - j) \overline{A}_j \ , \tag{3.4}$$

and if Ψ is stable then $\overline{\Psi}$ is stable. Also, if we perform the subdivision

$$\overline{f}_i^{k+1} = \sum_{j \in \mathbb{Z}} \overline{A}_{i-2j} \overline{f}_j^{k} \ , \ i \in \mathbb{Z} \ . \tag{3.5}$$

starting with the initial vector $\overline{f}^0 = T^{-1} f^0$ then

$$T \overline{\mathcal{S}}^{\infty} \overline{f}^0 = \mathcal{S}^{\infty} f^0 \ . \tag{3.6}$$

Thus, to simplify notations, we can assume that the original mask is such that $V = \big(e^{(1)} ... e^{(m)}\big)$. In this case, the last $n-m$ rows of Φ are identically zero. With this special form of V we can easily derive appropriate difference and divided difference schemes for the scheme (1.1) for analyzing its convergence and the smoothness of its limit vector functions. As in [11] we use a corresponding matrix-valued Laurent polynomial

$$A(z) = \sum_{i \in \mathbb{Z}} A_i z^i \ ,$$

to represent the matrix subdivision scheme (1.1). Now we define the Laurent polynomial

$$D(z) = 2E(z)A(z)E(z^2)^{-1} = \sum_{i \in \mathbb{Z}} D_i z^i \ , \tag{3.7}$$

where $E(z) = \big((z^{-1}-1)e^{(1)} ... (z^{-1}-1)e^{(m)}, e^{(m+1)} ... e^{(n)}\big)$. The corresponding mask $\{D_i\}$ is of finite support by (3.3). It generates a stationary matrix subdivision scheme $\mathcal{S}_D$ transforming the values $\{g_i^k\}$,

$$g_i^k = 2^k \left((f_{i+1}^k)_1 - (f_i^k)_1, ..., (f_{i+1}^k)_m - (f_i^k)_m, (f_i^k)_{m+1}, ..., (f_i^k)_n\right)^t \tag{3.8}$$

into the corresponding values at level $k+1$, namely $\{g_i^{k+1}\}$.

As in the scalar case we have the following relation between $\mathcal{S}$ and $\mathcal{S}_D$.

Theorem 3.1. *If $\frac{1}{2}\mathcal{S}_D$ is contractive then $\mathcal{S}$ is C^0 . If $\mathcal{S}_D$ is C^0 then $\mathcal{S}$ is C^1. Moreover with the initial data G^0 consisting of the first m rows of $\Delta_{-1} - \Delta_0$ and the last $n-m$ rows of Δ_0 , the first m rows of $\mathcal{S}_D^\infty G^0$ are the derivatives of the corresponding elements in the first m rows of Φ . Also, $\Phi_D = \mathcal{S}_D^\infty \Delta_0$ is related to the non-trivial part of Φ by*

$$\Phi_{i,j} = (\Phi_D)_{i,j} * \chi_{[0,1]} \ , \quad 1 \le i,j \le m \ , \tag{3.9a}$$

$$\Phi'_{i,j} = (\Phi_D)_{i,j} \ , \quad 1 \le i \le m \ , \quad m+1 \le j \le n \ , \tag{3.9b}$$

where $\chi_{[0,1]}$ is the characteristic function of the interval $[0,1]$.

In the following we analyze properties of $\mathcal{S}_D$ determined by those of $\mathcal{S}$.

Theorem 3.2. *Let m' denote the dimension of the common eigenspace W_D of $B_0^D = \sum_{i\in\mathbb{Z}} D_{2i}$ and $B_1^D = \sum_{i\in\mathbb{Z}} D_{2i+1}$ corresponding to the eigenvalue 1. If $\mathcal{S}_D$ is C^0 , then $m' \ge m$, and the first m rows of any basis V_D of W_D are linearly independent.*

Proof: Starting from the initial data G^0 as defined in Theorem 3.1, we get by Theorem 3.1 that $\mathcal{S}_D^\infty G^0$ is a matrix function whose first m rows are the derivatives of the first m rows of $\Phi = \mathcal{S}^\infty \Delta_0$. Suppose the first m rows of $\mathcal{S}_D^\infty G^0$ are linearly dependent then we get

$$\sum_{i=1}^{m} \alpha_i \Phi'_{i,j} = 0 \ , \quad 1 \le j \le n \ , \tag{3.10}$$

for some nonzero α_i, $i = 1, \cdots, m$. Thus $\sum_{i=1}^{m} \alpha_i \Phi_{i,j} = c_j$, $1 \le j \le n$, and since these functions are of compact support, it is possible only if all are identically zero, which contradicts the form of V. Thus the first m rows in a limit matrix function generated by $\mathcal{S}_D$ are linearly independent, implying that $m' \ge m$, and that the first m rows in any basis V_D of W_D are linearly independent. ■

Theorem 3.3. *Assume $\mathcal{S}$ is C^1, and that Φ is stable. Then $\mathcal{S}_D$ is C^0 and the first m rows of Φ_D constitute a stable matrix. Moreover $m' = m$.*

Proof: To prove that $\mathcal{S}_D$ is C^0 , it is sufficient by Theorem 2.10 to construct a continuous compactly supported stable solution of the corresponding functional equation. Let us define the first m rows of Φ_D by (3.9), and denote it by $\Phi_D^{[m]}$. It is easy to see that since Φ is of compact support (3.9a) can be rewritten as

$$(\Phi_D)_{i,j} = \sum_{\ell=0}^{\infty} \Phi'_{i,j}(\cdot - \ell) . \tag{3.11}$$

Thus, the elements of $\Phi_D^{[m]}$ are in $C^0(\mathbb{R})$. To show the stability of $\Phi_D^{[m]}$ assume that it is not stable. Then there is $\alpha = \{\alpha_j \in \mathbb{R}^n \ : \ j \in \mathbb{Z}\}$, satisfying $0 < \sup_{j\in\mathbb{Z}} \|\alpha_j\| < \infty$, such that

$$\sum_{j\in\mathbb{Z}} \Phi_D^{[m]}(\cdot - j)\alpha_j = 0 . \tag{3.12}$$

In case all the first m components of all the vectors $\{\alpha_i \ : \ i \in \mathbb{Z}\}$ are identically zero we get by integrating (3.12) from $-\infty$ to x that

$$\sum_{j\in\mathbb{Z}} \Phi^{[m]}(\cdot - j)\alpha_j = c , \tag{3.13}$$

where $\Phi^{[m]}$ stands for the first m rows of Φ . But since Φ is of compact support $c = 0$ in (3.13) in contradiction to the stability of Φ . If some of the first m components of the vectors $\{\alpha_j \ : \ j \in \mathbb{Z}\}$ are non-zero, we integrate (3.12) from $x - 1$ to x and get

$$\sum_{j\in\mathbb{Z}} \Phi^{[m]}(\cdot - j)\gamma_j = 0 , \tag{3.14}$$

with γ_j having the same m components of α_j and the last $n - m$ components of $\alpha_j - \alpha_{j-1}$. It is clear that $0 < \sup_{j\in\mathbb{Z}} \|\gamma_j\|$, and hence (3.14) contradicts the stability of Φ . Thus $\Phi_D^{[m]}$ is stable.

In the following we show that $\Phi_D^{[m]}$ solves the functional equation corresponding to $\mathcal{S}_D$. Since Φ satisfies (1.3) we obtain that

$$\hat{\Phi}(\omega) = \hat{\Phi}(\omega/2)M(\omega/2) . \tag{3.15}$$

In view of the definition of the mask of $\mathcal{S}_D$ by (3.7), we obtain that $\omega\hat{\Phi}^{[m]}(\omega)\mathcal{E}(\omega)^{-1}$ satisfies a functional equation of the type (3.15) with $M_D(\omega) = 2\mathcal{E}(\omega)M(w)\mathcal{E}(2\omega)^{-1}$, where $\mathcal{E}(\omega) = E(e^{-i\omega})$. It follows from (3.9) and the form of $E(z)$ that

$$\omega\hat{\Phi}^{[m]}(\omega)\mathcal{E}(2\omega)^{-1} = \hat{\Phi}_D^{[m]}(\omega) \ , \tag{3.16}$$

and hence $\Phi_D^{[m]}$ solves the functional equation corresponding to $\mathcal{S}_D$. Now, defining $\Phi_D = V_D\Phi_D^{[m]}$, we obtain a stable continuous compactly supported solution of the functional equation corresponding to $\mathcal{S}_D$, with the first m rows of the solution being a stable matrix function. This together with Theorems 2.6 and 3.2 implies that $m' = m$. ■

As a consequence of the above three theorems we get

Corollary 3.4. *Assume $\mathcal{S}$ is C^1, and that Φ is stable. Let*

$$h^k = T_k\mathcal{S}^k f^0 \ , \quad k \in \mathbb{Z}_+ \ ,$$

where T_k is an $n \times n$ diagonal matrix with elements

$$(T_k)_{i,i} = 1 \ , \quad 1 \le i \le m \ , \quad (T_k)_{i,i} = 2^k \ , \quad m+1 \le i \le n \ . \tag{3.17}$$

Then the sequence $\{h^k\}_{k\in\mathbb{Z}_+}$ converges uniformly in the sense of (1.2) to a continuous limit vector function h. The first m components of h are in $C^1(\mathbb{R})$. Moreover there is a matrix $\beta \in \mathbb{R}^{n-m} \times \mathbb{R}^m$, dependent on $\mathcal{S}$ only, such that

$$h_i = \sum_{k=1}^{m} \beta_{i,k}h_k' \ , \quad m+1 \le i \le n \ . \tag{3.18}$$

Proof: By Theorem 3.3, $\mathcal{S}_D$ is C^0 . Let G^0 be as in Theorem 3.1, then $\mathcal{S}_D^\infty G^0$ exists and is continuous. Also we get in view of (3.8), that the elements of the first m rows of $\mathcal{S}_D^\infty G^0$ are the derivatives of the corresponding elements in the limit matrix function H generated by the sequence $H^k = T_k\mathcal{S}^k\Delta_0$, $k \in \mathbb{Z}_+$, while the last $n-m$ rows of $\mathcal{S}_D^\infty G^0$ consist of the corresponding rows of H . This together with Theorem 3.2 and Theorem 3.3 guarantees the existence of a matrix $\beta \in \mathbb{R}^{n-m} \times \mathbb{R}^m$, such that

$$H_{i,j} = \sum_{k=1}^{m} \beta_{i,k}H_{k,j}' \ , \quad m+1 \le i \le n \ , \quad 1 \le j \le n \ , \tag{3.19}$$

implying (3.18). ■

One should note that this analysis can be pushed further: if the renormalized scheme is also C^1 with one of its coordinates converging to zero, a proper renormalization will make this coordinate converge to a combination of the second derivatives of the non-zero components of the original scheme, and the same property will hold for higher derivatives. Also higher smoothness of the non-zero components of the original scheme can be derived, by a repeated application of (3.7), with appropriate matrices $A(z)$ and $E(z)$, corresponding to the structure of the derived scheme in the previous step.

Specific criterions on the masks that ensure the smoothness of the limit function are given in [16] and [23] in the case of refinable vectors and in [22] for refinable matrices.

References

1. Boehm, W., G. Farin and J. Kahmann, A survey of curve and surface methods in CAGD, Computer-Aided Geom. Design **1** (1984), 1–60.
2. Cavaretta, A. S., W. Dahmen and C.A. Micchelli, *Stationary Subdivision*, Memoirs of AMS, No. 453, 1991.
3. Cohen, E., T. Lyche and R. Riesenfeld, Discrete B-splines and subdivision techniques in computer-aided geometric design and computer graphics, Computer Graphics and Image Processing **14** (1980), 87–111.
4. Daubechies, I. and C. Lagarias, Two scale difference equations II. Local regularity, infinite products of matrices and fractals, SIAM J. Math. Anal. **23** (1992), 1031–1079.
5. Deslauriers, G. and S. Dubuc, Interpolation dyadique, in *Fractals, Dimensions non Entières et Applications*, G. Cherbit (ed.), Masson, Paris, 1989, 44–55.
6. Dubuc, S., Interpolation through an iterative scheme, J. Math. Anal. Appl. **114** (1986), 185–204.
7. Dyn, N., Subdivision schemes in computer-aided geometric design, in *Advances in Numerical Analysis II, Wavelets, Subdivision Algorithms, and Radial Basis Functions*, W.A. Light (ed.), Clarendon Press, Oxford, 1992, 36–104.
8. Dyn, N., J.A. Gregory and D. Levin, A four-point interpolatory subdivision scheme for curve design, Computer-Aided Geom. Design **4** (1987), 257–268.
9. Dyn, N., J. A. Gregory and D. Levin, Analysis of linear binary subdivision schemes for curve design, Constr. Approx. **7** (1991), 127–147.
10. Dyn, N. and D. Levin, Interpolating subdivision schemes for the generation of curves and surfaces, in *Multivariate Interpolation and Approximation*, W. Haussmann and K. Jetter (eds.), Birkhäuser-Verlag, Basel, 1990, 91–106.
11. Dyn, N. and D. Levin, Analysis of Hermite-type subdivision schemes, in *Approximation Theory VIII*, C. K. Chui and L. L. Schumaker (eds.), World Scientific, Singapore, 1995.
12. Merrien, J.L., A family of Hermite interpolants by bisection algorithms, Numerical Algorithms **2** (1992), 187–200.

13. Micchelli, C.A. and H. Prautzsch, Uniform refinement of curves, Linear Alg. Appl. **114/115** (1989), 841–870.
14. de Rahm, G., Sur une courbe plane, J. Math. Pures Appl. **35** (1956), 25–42.
15. Dahmen, W. and C.A. Micchelli, Biorthogonal wavelet expansions, to appear in Constr. Approx. (1996).
16. Cohen, A., I. Daubechies and G. Plonka, On the regularity of refinable vectors, submitted to Journal of Fourier Analysis and Applications, 1995.
17. Goodman, T.N.T. and S.L. Lee, Wavelets of multiplicity r, Trans. Amer. Math. Soc. **342** (1994), 307–324.
18. Donovan, G., J.S. Geronimo, D.P. Hardin and P.R. Massopust, Construction of orthogonal wavelets using fractal interpolation functions, preprint, Georgia Inst. Tech., 1994.
19. Hervé, L., Multi–resolution analysis of multiplicity d. Application to dyadic interpolation, Appl. Comput. Harmonic Anal. **1** (1994), 299–315.
20. Plonka, G., Approximation order provided by refinable function vectors, preprint, Universität Rostock, 1995.
21. Strang, G. and V. Strela, Orthogonal multiwavelets with vanishing moments, SPIE Proceedings, Orlando 1994, Optical Engineering.
22. Cohen, A., N. Dyn and D. Levin, Stability and rank properties of matrix subdivision schemes, preprint, Tel-Aviv University, 1996.
23. Shen, Z., Refinable function vectors, preprint, Singapore University, 1995.

Albert Cohen
Université Pierre et Marie Curie
4 Place Jussieu
75005 Paris FRANCE

cohen@ann.jussieu.fr

Nira Dyn and David Levin
School of Mathematical Sciences
Tel-Aviv University
Tel-Aviv 69978 ISRAEL

niradyn@math.tau.ac.il
levin@math.tau.ac.il

Guided Multi-Dimensional Reconstruction from Cross-Sections

Daniel Cohen-Or and David Levin

Abstract. Given parallel $(n-1)$-dimensional cross-sections of an n-dimensional body, one would like to reconstruct the body. The method based upon distance field interpolation gives a robust solution to this problem in its ability to deal with any topology in any dimension. Still this method may give undesired solutions to the problem if the changes from one cross-section to the next are big relative to the size of the details in the cross-sections. In the present work we suggest a way to improve the distance field interpolation method. For this we first assume that the data to the problem are the cross-sections and some specified points on them, prescribing geometric links between the different cross-sections. These links may be manually user-defined, or automatically defined by matching features in the cross-sections. The distance field interpolation method is modified so that the interpolation is guided by those links. We describe some rules for defining a smooth least-distorting mapping which realizes the desired links, and present the corresponding guided interpolation.

§1. Introduction

The problem of reconstructing an n-dimensional body from its cross-sections appears in many applications, from biomedical imaging to 2D and 3D animation. In 3D animation the problem can be viewed as a surface blending problem, namely, the continuous evolution of a surface from a source surface, through intermediate surfaces, into a target surface.

The problem of blending two surfaces, even for polyhedral ones, is not simple. In order to blend between two polyhedral models, a correspondence between their structure (vertices) has to be established [6]. However, a correspondence alone does not guarantee a smooth transition from the source model to the target one, and the *vertices' paths* have to avoid unpleasant situations such as self-intersections. 2D shape blending techniques that deal with the vertices' path problem are not easy to extend to 3D [8,10].

Advanced Topics in Multivariate Approximation
F. Fontanella, K. Jetter and P. J. Laurent (eds.) pp. 47–55.

ISBN 981-02-2852-X

The problem of 2D shape blending can also be considered as a problem of body reconstruction from cross-sections [7,9]. Some of the reconstruction methods use bivariate interpolation by finding a correspondence between the contours of two cross-sections. Other methods use a univariate interpolation between values of corresponding 3D grid points [5]. When the cross-sections are not dense, they may exhibit significant changes in the topology of the boundary contours. A method that handles easily that issue is the *Distance Field Interpolation* (DFI) presented and analyzed in [7]. The DFI is a general set-valued interpolation method for the reconstruction of an n-dimensional model from a sequence of its $(n-1)$-dimensional cross-sections. It uses a univariate interpolation of the *distance field*, and it performs well for most general topology in any dimension. The method is also defined in a discrete space and generates intermediate discrete contours of arbitrary complexity. The DFI method has been adapted successfully in biomedical reconstruction applications [5]. Yet, the DFI method is not performing well if the original body, from which the cross-sections are taken, is too 'twisted'. That is, if there are significant differences between adjacent cross-sections. For example, if two adjacent cross-sections are not aligned due to rotation or translation.

The reconstruction procedure presented in this paper is based on the DFI method. However, it is extended in the sense that the interpolation, and thus the blended surface, follow a warp transformation guided by means of a user-defined control. The control is defined by a point-to-point correspondence between prescribed *anchor points* on the given intermediate cross-sections. In order to achieve least-distorted in-between shapes, we decomposed the warp transformation W into a *rigid* (rotational) transformation R and an *elastic* transformation E, which are separately interpolated. The rigid-elastic decomposition of the warp function, and its particular interpolation are so chosen to minimize the distortion of the intermediate surfaces. The rigid part is used to rotate and translate the source object to a matching general position of the target object, while the finer features of the object are slowly evolved by the elastic part.

In Section 2 we review the DFI method for multi-dimensional reconstruction from cross-sections. In Section 3 we introduce the problem of reconstruction guided by anchor points and consider the application of a warp transformation. The warp transformation, its decomposition and parameterization, and the new warp-guided DFI method are constructed. Implementation details and numerical experiments are described in [3].

§2. Distance Field Interpolation for n-Dimensional Reconstruction

Reconstruction of 3D objects from cross-sections is of major importance in many applications, particularly in biomedical imaging. One class of reconstruction methods builds a polygonal surface by assuming the cross-sections consist of closed polygonal contours [2]. Another class of methods uses a volumetric approach. The entire volume (3D voxel array) is reconstructed from the gray levels of the cross-sections.

The missing data is reconstructed by some interpolation of the gray values of the cross-sections. The surface is then defined and reconstructed by iso-surfacing methods. These methods do suffer from discontinuities in the reconstructed surface, and thus, do not yield smooth blending transformations.

The method of Distance Field Interpolation (DFI) presented in [7] better reconstructs the surface by interpolating distance values rather than the gray values. This method has been adapted successfully in biomedical reconstruction applications [5].

The DFI method can be described in terms of n-dimensional surface reconstruction from $(n-1)$-dimensional cross-sections $(n > 1)$. Consider an object $\Omega \subset \mathbb{R}^n$, discretized or not.

Given a finite set of $(n-1)$-dimensional cross-sections of the object,

$$\Omega_{t_j} = \{x = (x^1, \ldots, x^{n-1}) \mid (x^1, \ldots, x^{n-1}, t_j) \in \Omega\} \ , \quad t_0 < t_1 \ldots < t_M \ , \qquad (1)$$

we would like to reconstruct Ω. Throughout the paper we use lower indices for numbering entities and upper indices to denote the components of a vector. Let us define the *distance field* D_{t_j} at level t_j, $1 \leq j \leq M$: For $x = (x^1 \ldots x^{n-1})$

$$D_{t_j}(x) = \begin{cases} -dist(x, \partial\Omega_{t_j}) \ , & \text{if } x \in \Omega_{t_j} \ , \\ dist(x, \partial\Omega_{t_j}) \ , & \text{otherwise} \ , \end{cases} \qquad (2)$$

where $\partial\Omega_{t_j}$ denotes the boundary of Ω_{t_j}, and *dist* denotes the Euclidean distance in $\mathbb{R}^{n-1}$. Now, using univariate interpolation (with respect to the parameter t) we interpolate, between the cross-sections, the distance values of points having the same first $n-1$ coordinates. The resulting interpolant approximates the $n-1$ Euclidean distance between $y = (x^1, \ldots, x^{n-1}, t)$ and $\partial\Omega_t = \partial(\Omega \mid_{y^n = t})$.

Once this approximated distance field is available, the surface of the object can be determined by the zero points of the distance field, or, in the discretized version, as the boundary between the positive and negative valued lattice points, while the volume itself (its *interior*) is defined as the set of all negative valued points.

Formally, the DFI method defines an approximate domain $\widetilde{\Omega} \approx \Omega$ as follows: For $y = (x^1, \ldots, x^{n-1}, t)$ we first define the interpolant $d_x(t)$ $(x = (x^1 \ldots x^{n-1}))$ by univariate interpolation of the values $\{D_{t_j}(x)\}_{j=0}^{M}$. Now,

$$\widetilde{\Omega} = \{y \mid d_x(t) \leq 0 \} \ . \qquad (3)$$

Evidently, this method can be adapted for blending, where two or more cross-sections are given, to generate the intermediate models. For more details on the choice of the interpolation method, and for approximation rate analysis see [7]. The DFI method suffers from two major deficiencies: The first one is the low approximation order near boundary points of Ω with a tangent orthogonal to $(0, 0, \ldots, 1)$.

To the best of our knowledge, this problem is not resolved yet. The second problem is also related to the approximation power of the method. Here we refer to the case where the ratio between the distances $|t_{j+1} - t_j|$ and the 'width' of the cross-sections Ω_{t_j} is large with respect to the slope of the boundary of Ω in the direction $(0, 0, \ldots, 1)$. The approach taken in this work to overcome this problem is the combination of the DFI with a proper point-to-point warp transformation.

§3. Guided DFI Blending

Let us consider for simplicity the blending of two objects, $S = \Omega_0$ and $T = \Omega_1$. The DFI gives us a way for blending by interpolating (linearly) between the distance functions D_0 and D_1, as described in Section (2). We call this blending process the no-warp DFI blending. No-warp DFI blending is quite restricted, and it may produce non-satisfactory results. To demonstrate this, consider the case when T is just a rotation of S, and S is just a thin rod. The no-warp DFI blending gives at $t = \frac{1}{2}$ either a very small object, or an empty object, which is not the naturally-expected result. On the other hand, the no-warp DFI blending works nicely if S is a thin rod and T is just the rod S missing its middle third.

Consider now the possibility that T is obtained by a transformation (warp) of S, i.e., $T = W_1(S)$. Further, let us assume that T is gradually evolving from S, through a continuum of objects $W_t(S)$, where $\{W_t\}_{t\in[0,1]}$ are smooth transformations, smoothly changing with t, where $W_0 \equiv I$ is the identity transformation. In such a case the blending can be the warp itself, this is the pure-warp blending. Pure-warp blending is also very restricted, for example, it does not allow the possibility of changes in the genus of the objects.

We are going to hybridize the two different methods into a more powerful tool. The first step of the hybrid method is finding a smooth warp $\{W_t\}_{t\in[0,1]}$ such that $W_1(S) \approx T$ in some chosen sense. Then the DFI method is applied, now guided by the warp $\{W_t\}$. Note that the warp is operating in $\mathbb{R}^{n-1}$ and the interpolation of the distance field is performed along the $n-th$ dimension. At this point we go back to the case of $M+1$ given objects, considered as cross-sections of Ω, as in Eq. (1). First we find a smooth warp $\{W_t\}_{t\in[0,1]}$ such that

$$W_{t_j}(\Omega_0) \approx \Omega_{t_j} \quad , \ j = 0, \ldots, M \ . \tag{4}$$

The signed distance function at each level are defined here also by Eq. (2). For $x = (x^1, \ldots, x^{n-1})$ we then interpolate the values $\{D_{t_j}(W_{t_j}(x))\}_{j=0}^{M}$, denoting the interpolant $d_x(t)$. The approximated domain $\widetilde{\Omega}$ defined by the hybrid warped DFI procedure is

$$\widetilde{\Omega} = \{y = (W_t(x), t) \ \mid \ d_x(t) \le 0 \ \} \ . \tag{5}$$

Obtaining the Approximated Warp W_t.

The way we choose to define the approximated warp W_t is derived from several applications. On each level t_j we assume we are given N points $\{p_{j,i}\}_{i=1}^N$ whose mapping is predetermined. Later we refer to these points as *anchor points*. The warp is to be determined so that

$$W_{t_j}(p_{0,i}) = p_{j,i} \ , \quad 1 \le i \le N \ . \tag{6}$$

The warp function maps the source (level zero) anchor points to their corresponding target points at levels $\{t_j\}_{j=0}^M$, while other points in the source domain comply to those constraints. A warp defined by anchor points is attractive because it is applicable in any dimension in a most natural way. From the user's point of view, this method is intuitive and very easy to handle. It also has the advantage of being robust and very stable, in the sense that the results are not very sensitive to the small perturbations in the anchor points.

We now describe the definition of the warp transformation based on anchor points, starting with the case $n = 3$, i.e., the cross-sections are in $\mathbb{R}^2$. As stated earlier in the introduction, in order to achieve control over the shapes of the in-between objects the warp transformation W_t is decomposed into a *rigid* part R and an *elastic* part E. To explain the motivation here, let us view a very simple case: Consider the blending of two objects in $\mathbb{R}^2$, $S = \Omega_0$ and $T = \Omega_1$, where $T = R_\theta S$ and R_θ is a rotation in angle θ. There are two obvious options for defining a warp W_t. One is the *linear interpolation* warp $W_t = (1-t)I + tR_\theta$. The other is the *linear rotation* warp $W_t = R_{t\theta}$. Both warps vary smoothly from the identity I at $t = 0$ to R_θ at $t = 1$. However, the second option is the preferable one since it is an isometry for any t, i.e., non-distorting. Now consider the case where T is obtained from S by a rotation R_θ, a translation c, and an *elastic* transformation E, i.e., $T = E(R_\theta S + c)$. In the following we will choose the rotation R_θ and the translation c so that E is as close as possible to I, in some sense. The warp W_t is defined by

$$W_t(x) = ((1-t)I + tE)(R_{t\theta}x + tc) \ . \tag{7}$$

This approach can be applied to the case of $M+1$ objects in $\mathbb{R}^2$, $\{\Omega_{t_j}\}_{j=0}^M$. First we find a sequence of rotations $\{R^{(j)} = R_{\theta_j}\}_{j=0}^M$, a sequence of elastic transformations $\{E_j\}_{j=0}^M$, and translations $\{c_j\}_{j=0}^M$, with $\theta_0 = 0$, $E_{t_0} = I$ and $c_0=0$. Now we define

$$\theta(t) = \sum_{j=0}^{M} b_j(t)\theta_j \ , \quad E_t = \sum_{j=0}^{M} b_j(t)E_j \ , \quad c(t) = \sum_{j=0}^{M} b_j(t)c_j \ , \tag{8}$$

where $\{b_j(t)\}_{j=0}^M$ are Lagrange-type basis functions for interpolation at the levels $t_0, \ldots, t_M$. Finally, the warp transformation W_t is defined as

$$W_t(x) = E_t(R_{\theta(t)}x + c(t)) \ . \tag{9}$$

In the following we describe a way for obtaining the rotations, the elastic transformations and the translations, and we consider the main case $n = 4$, namely blending of 3D objects.

The Rigid Transformation in $\mathbb{R}^3$.

We start with given sets of anchor points in $\mathbb{R}^3$, $\{p_{j,i}\}_{i=1}^N$, corresponding to the levels $\{t_j\}$, $j = 0, \ldots, M$. The rigid part of the transformation from level t_0 to level t_j is defined by the rotation $R^{(j)}$ and the translation c_j which minimizes

$$Q = \sum_{i=1}^{N} \|R^{(j)} p_{0,i} + c_j - p_{j,i}\|^2 \ . \tag{10}$$

Imagine that the points $\{p_{0,i}\}$ are connected by identical elastic springs to the corresponding points $\{p_{j,i}\}$. The form Q represents the elastic energy of this system, and the rigid transformation which minimizes Q, brings Ω_0 to an equilibrium position in the springs' system.

The above least-square fitting problem of the two sets of anchor points in $\mathbb{R}^3$ is solved using an explicit algorithm, which involves the *singular value decomposition* (SVD) of a 3×3 matrix [1]. As a result, we obtain a 3×3 rotation matrix characterized by three angles, $R^{(j)} = R_{\theta_j, \phi_j, \psi_j}$, and a translation vector c_j. The warp transformation in the 3D case is defined by

$$W_t(x) = E_t(R_{\theta(t),\phi(t),\psi(t)} x + c(t)) \ , \tag{11}$$

with

$$\phi(t) = \sum_{j=0}^{M} b_j(t) \phi_j \ , \quad \psi(t) = \sum_{j=0}^{M} b_j(t) \psi_j \ . \tag{12}$$

The Elastic Warp in $\mathbb{R}^3$.

Once we have computed the rigid transformation, say in $\mathbb{R}^3$, we move to calculating the elastic transformation. In accordance with Eq. (6) and Eq. (9), we look for a transformation E_j, in general a non-linear transformation, $E_j : \mathbb{R}^3 \to \mathbb{R}^3$, such that

$$E_j(R^{(j)} p_{0,i} + c_j) = p_{j,i} \ , \ 1 \le i \le N \ . \tag{13}$$

This is a multivariate scattered data interpolation problem, which we suggest to solve by using *radial basis functions* [4]. In $\mathbb{R}^3$ this means solving three interpolation problems in $\mathbb{R}^3$, for each component of the vector equation defined in Eq. (13).

Radial basis functions have proven to be an effective tool in multivariate interpolation problems of scattered data: Given the scattered d-dimensional data

$$(x_i, F_i), \quad x_i \in \mathbb{R}^d, \quad F_i \in \mathbb{R} \ , \quad i = 1, 2, \ldots, N,$$

we look for an interpolatory function $S(x)$ of the form

$$S(x) = \sum_{i=1}^{N} a_i g\left(\| x - x_i \|\right), \tag{14}$$

where $\| \cdot \|$ denotes the usual Euclidean norm on $\mathbb{R}^d$ and $g : \mathbb{R}^+ \rightarrow \mathbb{R}$. A function of this type is usually referred to as a *pure radial sum*. Using radial functions reflects the fact that the scattered data has no preferred orientation, and their choice is dimension-dependent [4]. Other multivariate approximations, generalizing the univariate splines, use augmented radial sums, where the sum in Eq. (14) is augmented by a low degree polynomial.

Our discussion is mainly concerned with the 3-dimensional case ($d = 3$). We define the elastic transformation $E : \mathbb{R}^3 \rightarrow \mathbb{R}^3$ as:

$$E(q) = L(q) + Z(q) , \tag{15}$$

where L is a linear transformation and Z is of the form

$$Z(q) = (S_1(q), S_2(q), S_3(q)) , \tag{16}$$

and

$$S_k(q) = \sum_{i=1}^{N} a_i^k g(\|q - q_i\|) , \; 1 \leq k \leq 3 , \; q = (q^1, q^2, q^3)^T \in \mathbb{R}^3 .$$

$g : \mathbb{R}^+ \rightarrow \mathbb{R}$ is a univariate function, $\| \cdot \|$ is the Euclidean norm on $\mathbb{R}^3$, and $\{q_i\}$ for $i = 1, 2, \ldots, N$ are the source anchor points *after* the rigid transformation has been applied to them.

The above definition of the transformation E includes a linear part L. This linear part is added because pure radial sums often yield poor approximation of the transformation for points away from the anchor points. Moreover, the linear part is natural here since we would like to reconstruct exactly those transformations E which are linear, and especially the identity transformation.

A proper choice of g is also important. Choosing g as a function with a fast decay results in a finer local influence of the radial part, which can be used in order to obtain different effects for various parts of the object. Other natural choices of the function g are $g(r) = r^2 \log(r)$ in $\mathbb{R}^2$ and $g(r) = r^3$ in $\mathbb{R}^3$ [4]. In the following we describe the computation of the elastic warp in $\mathbb{R}^3$.

After computing the rigid part, namely, the rotation $R^{(j)}$ and the translation c_j, the elastic mapping E_j is defined by the following N interpolation conditions:

$$E_j(R^{(j)} p_{0,i} + c_j) \equiv E_j(q_i) = p_{j,i} \equiv y_i \quad , \quad i = 1, \ldots, N , \tag{17}$$

where $q_i = (q_i^1, q_i^2, q_i^3)^T$, and $y_i = (y_i^1, y_i^2, y_i^3)^T$. This interpolation problem is always solvable if we use an augmented radial approximation (with a properly chosen g) of the form

$$E_j(q) = \sum_{i=1}^{N} a_i g\left(\|q - q_i\|\right) + qA + \alpha_4 \ , \tag{18}$$

where

$$A = (\alpha_1, \alpha_2, \alpha_3)^T \ , \ a_i \in \mathbb{R}^3 \ , \ 1 \le i \le N \ , \ \alpha_\ell = (\alpha_\ell^1, \alpha_\ell^2, \alpha_\ell^2)^T \in \mathbb{R}^3 \ , \ 1 \le \ell \le 4 \ .$$

Thus E_j is determined by $N+4$ coefficients in $\mathbb{R}^3$. The computation of those coefficients involves the solution of 3 square linear systems of size $N+4$ each, where N conditions are derived by the interpolation requirements, as defined in (17), and the additional compatibility conditions are

$$\sum_{i=1}^{N} a_i^k = \sum_{i=1}^{N} a_i^k q_i^1 = \sum_{i=1}^{N} a_i^k q_i^2 = \sum_{i=1}^{N} a_i^k q_i^3 = 0 \ , \ k = 1,2,3 \ .$$

These conditions guarantee that the transformation is affine reducible, i.e., the transformation is affine whenever possible. The system of equations for the vectors of unknowns $u_k = (a_1^k, ..., a_N^k)^T$ and $v_k = (\alpha_1^k, \alpha_2^k, \alpha_3^k, \alpha_4^k)^T$, k=1,2,3, is

$$\begin{cases} Gu_k + Hv_k = b_k \ , \\ H^T u_k = 0 \ , \end{cases} \tag{19}$$

where $b_k = (y_1^k, ..., y_N^k)^T$, $G = \{\ g(\|q_i - q_j\|)\ \}_{i,j=1}^N$ and H is an $N \times 4$ matrix with an i-th row $\{q_i^1, q_i^2, q_i^3, 1\}$, $1 \le i \le N$.

Now that the warp transformation W_t is fully defined, the warped DFI approximation is obtained by (5), that is, the approximated domain $\widetilde{\Omega}_t$ at time t, defined by the hybrid warped DFI procedure, is defined by

$$\widetilde{\Omega}_t = \{y = (W_t(x), t) \ \mid \ d_x(t) \le 0\ \} \ , \tag{20}$$

with W_t given by (11) and d_x as defined in Section 2.

Practical Implementation.

The formal definition (20) seems rather simple, however, it is quite involved to realize. To make the warped DFI approximation an applicable tool in computer graphics, efficient algorithms for computing distance fields and for boundary detection, in 2D and in 3D, should be developed and employed. Implementation details and numerical experiments with the warped DFI method are described in [3]. Examples of 2D and 3D animations generated by the presented method can be found in the URL address: http://www.math.tau.ac.il/~daniel/Blend/blend.html/.

References

1. Arun, K. S., T. S. Huang and S. D. Blostein, Least-squares fitting of two 3-D point sets, IEEE Trans. On Pattern Analysis and Machine Intelligence **9** (1987), 698-700.
2. Barequet, G. and M. Sharir, Piecewise linear interpolation between polygonal slices, in *Proc. 10th Ann. ACM Symp. on Computational Geometry*, Stony Brook, NY, 1994, 93-102.
3. Cohen-Or, D., D.Levin and A. Solomovici, Surface metamorphosis, Technical Report 309/96, Computer Science Department, Tel-Aviv University, January 1996.
4. Dyn, N. Interpolation and approximation by radial and related functions, in *Approximation Theory VI*, C.K. Chui, L.L. Schumaker and J.D. Ward (eds.), Academic Press, 1987, 85-100.
5. Herman, G. T., J. Zheng and C. A. Bucholtz, Shape-based interpolation, IEEE Computer Graphics and Appl. **12** (1992), 69-79.
6. Kent, J. R., W. E. Carlson and R. E. Parent, Shape transformation for polyhedral objects, Computer Graphics **26** (1992), 47-54.
7. Levin, D., Multidimensional reconstruction by set-valued approximation, IMA J.Numerical Analysis **6** (1986), 173-184.
8. Sederberg, T. W., P. Gao, G. Wang and H. Mu, 2-D Shape blending: an antrinsic solution to the vertex path problem, in *Proceedings of SIGGRAPH '93*, **27** (1993), 15-18.
9. Sederberg, T. W. and E. Greenwood, A physically based approach to 2D shape blending, in *Proceedings of SIGGRAPH '92*, **26** (1992), 25-34.
10. Shapira, M. and A. Rappoport, Shape blending using the star-skeleton representation, IEEE Computer Graphics and Appl. **15** (1995), 44-50.

Daniel Cohen-Or and David Levin
School of Mathematical Sciences
Tel-Aviv University
Tel-Aviv 69978
Israel

daniel@math.tau.ac.il
levin@math.tau.ac.il

Multiscale Decompositions on the Interval Based on DCT–I Transforms

Uwe Depczynski and Kurt Jetter

Abstract. We describe a multiscale time–frequency analysis on bounded intervals based on the fast discrete cosine transform DCT–I. General methods of this type and their connection with Sturm–Liouville boundary value problems were introduced and analyzed in [2]. Here we restrict our considerations to the analyzing spaces spanned by systems of type (1.7). We study time–localizing transforms in the subsequent frequency bands, and we compare our analysis with other recommended methods from the literature.

§1. Introduction

The question of how to adapt wavelet decompositions to signals on a bounded interval without disturbing the analysis through boundary effects has been discussed since wavelets and multiresolution analyses have become an efficient tool in signal analysis. This discussion seems to have to come to an end with the publication of [1] showing that an analysis by Daubechies' wavelets can be adapted to intervals by introducing boundary wavelets so as to maintain small supports, as well as approximation power and stability. However, these constructions are restricted to univariate signals (except for tensor product extensions to the multivariate case); and there still seems to be a lack of reliable tests of how good these constructions are for real data and corresponding time-frequency analysis.

The construction in [3,4,6] starts from a different point of view, namely the fact that expansions in terms of Chebyshev polynomials are often a good thing to look at, when bounded intervals are considered and if no apparent periodicities are involved. These papers show that much of the notation and some properties of a multiresolution analysis can be formally extended to this situation, and fast algorithms are developed which are based on various modifications of the discrete

Advanced Topics in Multivariate Approximation
F. Fontanella, K. Jetter and P. J. Laurent (eds.) pp. 57–70.

ISBN 981-02-2852-X

cosine transform (DCT). This construction leads to multivariate versions on the sphere as well ([7]).

The approach in [2] starts from yet another idea, namely the fact that one may adapt the expansion of a signal to the underlying boundary behaviour. For univariate signals this leads in a natural way to expansions according to a complete system $f_0, f_1, \ldots$ of eigenfunctions of a Sturm–Liouville boundary value problem, which is orthogonal with respect to a weighted L_2-norm. Here the boundary behaviour is matched by the eigenfunctions and thus will not disturb the analysis at all. Since for this class of problems the "frequency" of f_j increases with the index j, there is a good reason that we can get a satisfactory frequency analysis by looking at the corresponding parts of the signal in a dyadic collection of spaces

$$V_j := \operatorname{span}\{f_0, \ldots, f_{2^j}\} \quad \text{and} \quad W_j := \operatorname{span}\{f_{2^j+1}, \ldots, f_{2^{j+1}}\}\,. \tag{1.1}$$

We may call these the *low frequency space* V_j and the *high frequency space* W_j, *at level* $j \in \mathbb{N}$.

This latter approach (as the former one) leaves the question of whether there is any chance to get a time analysis simultaneously with the frequency analysis, so as to capture and localize transient features of a signal. The idea is to fix the spaces V_j and W_j, but to give up orthogonality of their bases (while keeping uniform stability) and to look at new bases

$$\mathbf{\Phi_j} := (\phi_{j,0}, \ldots, \phi_{j,2^j})^T \quad \text{and} \quad \mathbf{\Psi_j} := (\psi_{j,0}, \ldots, \psi_{j,2^j-1})^T\,, \tag{1.2}$$

of V_j and W_j, where

$$\begin{pmatrix} \mathbf{\Phi_j} \\ \mathbf{\Psi_j} \end{pmatrix} = \begin{pmatrix} C_j & 0 \\ 0 & D_j \end{pmatrix} \mathbf{f_{j+1}}\,, \quad j \in \mathbb{N}\,, \tag{1.3}$$

with $\mathbf{f_{j+1}} := (f_0, \ldots, f_{2^{j+1}})^T$, and

$$C_j = (c_{k,i}^{(j)}) \in \mathbb{R}^{(2^j+1)\times(2^j+1)}\,, \quad D_j = (d_{k,i}^{(j)}) \in \mathbb{R}^{2^j \times 2^j}\,, \tag{1.4}$$

two non-singular matrices chosen so as to produce "localized" new basis functions. From the orthogonality of the system (1.1) it is then clear that the *dual transformation* to (1.3),

$$\begin{pmatrix} \tilde{\mathbf{\Phi}}_\mathbf{j} \\ \tilde{\mathbf{\Psi}}_\mathbf{j} \end{pmatrix} = \begin{pmatrix} \tilde{C}_j & 0 \\ 0 & \tilde{D}_j \end{pmatrix} \mathbf{f_{j+1}} \quad \text{with} \quad \tilde{C}_j = (C_j^{-1})^T,\ \tilde{D}_j = (D_j^{-1})^T\,, \tag{1.5}$$

will guarantee that the pairs $(\mathbf{\Phi_j}, \tilde{\mathbf{\Phi}}_\mathbf{j})$ and $(\mathbf{\Psi_j}, \tilde{\mathbf{\Psi}}_\mathbf{j})$ yield biorthogonal systems in the level spaces.

It is the scope of this paper to show that, with further assumptions on the system (1.1) the construction of localized and stable bases is indeed possible, and that we find a very satisfactory time-frequency analysis. We will discuss transformation matrices based on modified DCT-matrices (we will thus rely on the sometimes good localization properties of the DCT), and we will compute some very tight Riesz bounds for the new bases. We will also address results on localization in terms of interpolation properties; and we will give some examples showing that our constructions compare favorably with other approaches in the literature.

While the general ideas could be given in terms of an arbitrary Hilbert space $L_{2,\rho}[a,b]$ of measurable functions $f : [a,b] \to \mathbb{R}$ with scalar product

$$< f,g > := < f,g >_\rho := \int_a^b f(x)\,\overline{g(x)}\,\rho(x)\,dx \ , \qquad (1.6)$$

and corresponding ρ-weighted norm $\|f\| := \|f\|_{2,\rho} := \sqrt{\int_a^b |f(x)|^2\,\rho(x)\,dx}$, we find it convenient to look at a more specific case which will allow us later on to use the efficiency of the DCT-I transform. More general results and their relation to Sturm–Liouville boundary value problems can be found in [2]. Thus we will assume that the system (1.1) is given by

$$f_0(x) = \frac{1}{\sqrt{2}}\,r(x) \quad \text{and} \quad f_n(x) = r(x)\,\cos(n\,\vartheta(x)) \ , \ n = 1,2,\dots \ , \qquad (1.7)$$

with a fixed *amplitude function* $r \in C^2[a,b]$, which is strictly positive on the interval $[a,b]$, and a fixed *phase function* $\vartheta : [a,b] \to [0,\pi]$, which is bijective and twice continuously differentiable as well. This phase function must be carefully chosen since it is connected with the weight function ρ which, according to (1.6), might lead to strong damping effects. The Kilgore-Prestin polynomial wavelets [3] e.g., and the construction of Tasche et al. [6, 4] based on the Chebyshev polynomials of the first kind can be interpreted in our framework by putting $[a,b] = [-1,+1]$, $r(x) \equiv \sqrt{2/\pi}$, $\vartheta = \arccos$, and $\rho(x) = 1/\sqrt{1-x^2}$.

§2. Localizing Transform in V_j

We heavily rely on the matrix of the type-I DCT on vectors of length $2^j + 1$,

$$\mathcal{C}_j := \left(\cos(\frac{k\ell\pi}{2^j})\right)_{k,\ell=0}^{2^j} \ \operatorname{diag}(\frac{1}{2}, 1, \dots, 1, \frac{1}{2}) \ ,$$

which is slightly modified in order to define the first block in our transform (1.3).

Definition 1. *For $j \in \mathbb{N}$ we consider the following basis transformation in V_j:*

$$\mathbf{\Phi_j} := C_j\ \mathbf{f_j} \quad \textit{with} \quad C_j = \Gamma_j^{-1}\ \mathcal{C}_j\ F_j^{-1}\ , \tag{2.1}$$

where $F_j = \mathrm{diag}(\frac{1}{\sqrt{2}}, 1, 1, \ldots, 1, \frac{1}{2})$, $\Gamma_j = \mathrm{diag}(\gamma_j, \gamma_{j-1}, \gamma_{j-1}, \ldots, \gamma_{j-1})$ *and* $\gamma_j := \sqrt{(2^{j+1}+1)/2}$.

This transformation for the system (1.7) can be written in the following way by refering to the knots $x_{j,k}$ defined by

$$\vartheta(x_{j,k}) := \frac{k}{2^j}\pi\ , \quad k = 0, \ldots, 2^j\ . \tag{2.2}$$

We have

$$\phi_{j,k} = \nu_{j,k} \sum_{i=0}^{2^j} f_i(x_{j,k})\, f_i\ , \quad k = 0, \ldots, 2^j\ , \tag{2.3}$$

with the normalization factors

$$\nu_{j,k} = \left(\sum_{\ell=0}^{2^j} |f_\ell(x_{j,k})|^2 \right)^{-1/2} = \begin{cases} \dfrac{1}{r(x_{j,k})\sqrt{2^j+1}} & \text{if } k = 0, 2^j, \\ \dfrac{1}{r(x_{j,k})\sqrt{2^{j-1}+1}} & \text{if } 0 < k < 2^j. \end{cases}$$

From this we get the following *interpolating identity* by using the orthogonality of the system (1.7):

$$< \phi_{j,k}, \phi_{j,\ell} >_\rho = \nu_{j,k}\, \phi_{j,\ell}(x_{j,k}) = \nu_{j,\ell}\, \phi_{j,k}(x_{j,\ell})\ . \tag{2.4}$$

The Gramian $G_j(\Phi) = C_j\ C_j^T = (\langle \phi_{j,k}, \phi_{j,\ell} \rangle)$ of the new basis $\phi_{j,0}, \ldots, \phi_{j,2^j}$ can be explicitly given. We have the following result from [2], the proof of which is elementary but technical:

Theorem 1. *The Gramian of the basis $\mathbf{\Phi_j} = \{\phi_{j,0}, \ldots, \phi_{j,2^j}\}$ of the level space V_j has the entries*

$$< \phi_{j,k}, \phi_{j,\ell} >_\rho = \begin{cases} 1 & \textit{if } k = \ell, \\ \dfrac{1}{2^{j+1}+1} & \textit{if } (k,\ell) = (0, 2^j),\ (k,\ell) = (2^j, 0), \\ \dfrac{(-1)^{k+\ell}}{2^j+1} & \textit{if } 0 < k, \ell < 2^j,\ k \neq \ell, \\ \dfrac{(-1)^{k+\ell}}{\sqrt{2^{j+1}+1}\sqrt{2^j+1}} & \textit{otherwise.} \end{cases}$$

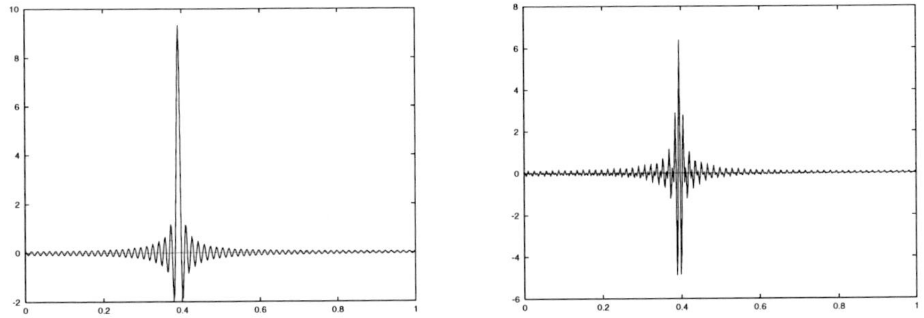

Fig. 1. Functions $\phi_{7,50}$ (left) and $\psi_{7,50}$ (right).

From this we see that the elements of the new basis have norm equal to one, and by equation (2.4) the new basis is localized: We have

$$\phi_{j,k}(x_{j,k}) \approx 2^{j/2} \quad \text{and} \quad \phi_{j,k}(x_{j,\ell}) \approx 2^{-j/2} \ , \ k \neq \ell \ . \tag{2.5}$$

This *localizing property* is depicted in Figure 1.

In order to describe the *stability* of the new basis in terms of the estimates

$$A_j^\phi \sum_{k=0}^{2^j} |c_k|^2 \leq \| \sum_{k=0}^{2^j} c_k \phi_{j,k} \|_\rho^2 \leq B_j^\phi \sum_{k=0}^{2^j} |c_k|^2 \ , \tag{2.6}$$

with the so-called Riesz bounds $0 < A_j^\phi \leq B_j^\phi < \infty$, we use the well-known fact that the optimal Riesz bounds are given by the smallest and the largest eigenvalue of the Gramian, $\lambda_{\min}(G_j(\phi))$ and $\lambda_{\max}(G_j(\phi))$, respectively. We have these uniform bounds from [2], but we give a shorter, compact proof:

Theorem 2. *The Riesz bounds of the basis* $\mathbf{\Phi_j} = \{\phi_{j,0}, \ldots, \phi_{j,2^j}\}$ *of the level space* V_j *can be estimated as follows, uniformly with respect to* $j \in \mathbb{N}$*:*

$$A_j^\phi \geq \frac{33 - \sqrt{129}}{30} = 0.7214\ldots \quad \text{and} \quad B_j^\phi < 2 \ .$$

For $j \geq 4$ *we have the bound*

$$A_j^\phi \geq 1 - \frac{7}{5} \cdot \frac{1}{2^j - 1} - \frac{1}{\sqrt{2^j}} \cdot \frac{1}{\sqrt{2^j - 1}} \quad .$$

Proof: The upper bound follows from a simple application of Gerschgorin's circle theorem:

$$\lambda_{\max}(G_j(\phi)) \leq 1 + 2^j \cdot \frac{1}{2^j + 1} < 2 \quad \text{for all } j \in \mathbb{N}.$$

The lower bound for $j = 1, 2, 3$ is obtained from direct calculations:

$$\lambda_{\min}(G_1(\phi)) = \frac{33 - \sqrt{129}}{30}\,, \ \lambda_{\min}(G_2(\phi)) = \frac{4}{5}\,, \ \lambda_{\min}(G_3(\phi)) = \frac{8}{9}\,.$$

For $j \geq 4$ we write $G_j = X_j + Y_j + Z_j$ in terms of three real-symmetric matrices $X_j = \left((x^{(j)}_{k,\ell})\right)$, $Y_j = \left((y^{(j)}_{k,\ell})\right)$, $Z_j = \left((z^{(j)}_{k,\ell})\right)$ of dimension $2^j + 1$, and we use the estimate

$$\lambda_{\min}(G_j(\phi)) \geq \lambda_{\min}(X_j) + \lambda_{\min}(Y_j) + \lambda_{\min}(Z_j)\,. \tag{2.7}$$

According to Theorem 1 we can take (for $k, \ell = 0, \ldots, 2^j$)

$$x^{(j)}_{k,\ell} = \begin{cases} 1 & \text{for } k = \ell, \\ (-1)^{k+\ell}\alpha & \text{for } k \neq \ell, \end{cases}$$

$$\text{with} \quad \alpha = \frac{1}{2^j + 1}\,,$$

$$y^{(j)}_{k,\ell} = \begin{cases} \beta & \text{for } (k,\ell) = (0, 2^j) \text{ or } (k,\ell) = (2^j, 0), \\ 0 & \text{otherwise}, \end{cases}$$

$$\text{with} \quad \beta = \frac{1}{2^{j+1} + 1} - \frac{1}{2^j + 1} \quad \text{and}$$

$$z^{(j)}_{k,\ell} = \begin{cases} 0 & \text{for } (k \equiv 0(2^j) \wedge \ell \equiv 0(2^j)) \text{ or } 0 < k, \ell < 2^j, \\ (-1)^{k+\ell}\gamma & \text{otherwise}, \end{cases}$$

$$\text{with} \quad \gamma = \frac{1}{\sqrt{2^{j+1} + 1}\sqrt{2^j + 1}} - \frac{1}{2^j + 1}.$$

The spectrum of these matrices can be determined as follows; here, e_k, $k = 0, \ldots, 2^j$, denotes the canonical basis of unit vectors in $\mathbb{R}^{2^j+1}$.

We have $X_j = \alpha\, e \cdot e^T + (1 - \alpha)\, I$ with $e := \sum_{k=0}^{2^j} (-1)^k e_k$. Therefore, e is an eigenvector for the eigenvalue $\lambda = 1 + 2^j\alpha$, and the subspace orthogonal to e is eigenspace for the eigenvalue $\lambda = 1 - \alpha$. Hence

$$\lambda_{\min}(X_j) = \min(1 + 2^j\alpha, 1 - \alpha) = 1 - \alpha\,.$$

For the matrix Y_j we have the spectral decomposition $Y_j = \beta\,(x_1 \cdot x_1^T - x_2 \cdot x_2^T)$ with $x_1 := (e_o + e_{2^j})/\sqrt{2}$ and $x_2 := (e_0 - e_{2^j})/\sqrt{2}$. From this,

$$\lambda_{\min}(Y_j) = -|\beta| = -\beta\,.$$

Finally, Z_j has the spectral decomposition $Z_j = \sqrt{2^{j+1}}\,\gamma\,(y_1 \cdot y_1^T - y_2 \cdot y_2^T)$ with $y_1 := \frac{1}{2} \cdot (1, -\delta, \delta, -\delta, \ldots, -\delta, 1)^T$ and $y_2 := \frac{1}{2} \cdot (1, \delta, -\delta, \delta, \ldots, \delta, 1)^T$, where $\delta = \sqrt{2^{1-j}}$. Therefore,

$$\lambda_{\min}(Z_j) = -\sqrt{2^{j+1}}\,|\gamma| = -\sqrt{2^{j+1}}\,\gamma\,. \quad \blacksquare$$

Remark. A similar analysis can be given for the basis $\tilde{\mathbf{\Phi}}_{\mathbf{j}} = \{\tilde{\phi}_{j,0}, \ldots, \tilde{\phi}_{j,2^j}\}$ of V_j which is determined from the first block in (1.5), and which is biorthogonal to $\mathbf{\Phi}_{\mathbf{j}}$. From

$$\delta_{k,\ell} = < \phi_{j,\ell}, \tilde{\phi}_{j,k} >_\rho = \nu_{j,\ell}\, \tilde{\phi}_{j,k}(x_{j,\ell}) \ , \quad k, \ell = 0, \ldots, 2^j \ , \tag{2.8}$$

we see that these functions are *fundamental interpolatory* with respect to the knots $x_{j,\ell}$. In the special case $r = \sqrt{2/\pi}$ and $\vartheta = \arccos$ these are Tasche's "scaling functions" (up to normalization factors), [6].

§3. Localizing Transform for W_j

Here, we consider the second block in the transform (1.3), with the matrix D_j being a modification of the type-I DCT $\mathcal{C}_{j+1}$, of order $j+1$. It is opportune to introduce the column vector $\mathbf{g_j} = (f_{2^j+1}, \ldots, f_{2^{j+1}})^T$ so that $\mathbf{f_{j+1}} = (\mathbf{f_j}^T, \mathbf{g_j}^T)^T$.

Definition 2. *For $j \in \mathbb{N}$ we consider the following basis transformation in W_j:*

$$\mathbf{\Psi_j} := D_j\ \mathbf{g_j} \quad \textit{with} \quad D_j = \sqrt{\frac{2}{2^j+1}}\ V_j\ \mathcal{C}_{j+1}\ U_j \ , \tag{3.1}$$

where

$$U_j = \begin{pmatrix} 0 & & & & \\ \hline 1 & & & & \\ & 1 & & & \\ & & \ddots & & \\ & & & 1 & \\ & & & & 2 \end{pmatrix} \in \mathbb{R}^{(2^{j+1}+1)\times 2^j}$$

and

$$V_j = \begin{pmatrix} 0 & 1 & 0 & \cdots & \cdots & \cdots & 0 \\ 0 & 0 & 0 & 1 & 0 & \cdots & 0 \\ & & \ddots & \ddots & \ddots & & \\ 0 & \cdots & 0 & 1 & 0 & 0 & 0 \\ 0 & \cdots & \cdots & \cdots & 0 & 1 & 0 \end{pmatrix} \in \mathbb{R}^{2^j\times(2^{j+1}+1)} \ .$$

Thus D_j originates from $\mathcal{C}_{j+1}$ (up to normalization) by deleting all even-numbered rows $0, 2, \ldots, 2^{j+1}$ (pre-multiplication by V_j) and deleting columns $0, 1, \ldots, 2^j$ (post-multiplication by U_j). The transform can be also written in terms of

$$\psi_{j,k} = \mu_{j,k} \sum_{i=2^j+1}^{2^{j+1}} f_i(x_{j+1,2k+1})\, f_i \ , \quad k = 0, \ldots, 2^j - 1 \ , \tag{3.2}$$

with the knots $x_{j+1,2k+1}$ defined as in (2.2) – note that only the odd-numbered knots at level $j+1$ are used –, and the normalization factors given by

$$\mu_{j,k} = \left(\sum_{\ell=2^j+1}^{2^{j+1}} |f_\ell(x_{j+1,2k+1})|^2 \right)^{-1/2} = \frac{\sqrt{2}}{r(x_{j+1,2k+1})\sqrt{2^j+1}}.$$

The interpolating identity takes here the form

$$< \psi_{j,k}, \psi_{j,\ell} >_\rho = \mu_{j,k}\, \psi_{j,\ell}(x_{j+1,2k+1}) = \mu_{j,\ell}\, \psi_{j,k}(x_{j+1,2\ell+1}) \,. \tag{3.3}$$

Theorem 3. [2] *The Gramian* $G_j(\psi) = D_j\ D_j^T = (< \psi_{j,k}, \psi_{j,\ell} >_\rho)$ *of the basis* $\mathbf{\Psi_j} = \{\psi_{j,0}, \ldots, \psi_{j,2^j-1}\}$ *of the level space* W_j *has the entries*

$$< \psi_{j,k}, \psi_{j,\ell} >_\rho = \begin{cases} 1 & \text{for } k = \ell, \\ \dfrac{1}{2^j+1} & \text{for } k \neq \ell. \end{cases}$$

This result leads to the following *localizing property* (cf. Figure 1)

$$\psi_{j,k}(x_{j+1,2k+1}) \approx 2^{j/2} \quad \text{and} \quad \psi_{j,k}(x_{j+1,2\ell+1}) \approx 2^{-j/2}\ , \ k \neq \ell\ ,$$

by using equation (3.3). It is also the basis for the following stability result:

Theorem 4. *The exact Riesz bounds of the basis* $\mathbf{\Psi_j} = \{\psi_{j,0}, \ldots, \psi_{j,2^j-1}\}$ *of the level space* W_j *are given by*

$$A_j^\psi = 1 - \frac{1}{2^j+1} \quad \text{and} \quad B_j^\psi = 2\left(1 - \frac{1}{2^j+1}\right).$$

Proof: As in the proof of Theorem 2 we give a simple decomposition for the Gramian $G_j(\psi)$. Let $e = (1,1,\ldots,1)^T \in \mathbb{R}^{2^j}$. Then $G_j(\psi) = \alpha\ e \cdot e^T + (1-\alpha)\ I$ with $\alpha = \frac{1}{2^j+1}$, showing that e is an eigenvector for the eigenvalue $\lambda = 1 + (2^j - 1)\alpha$, and the subspace orthogonal to e is an eigenspace for the eigenvalue $\lambda = 1 - \alpha$. Hence

$$\lambda_{\min}(G_j(\psi)) = A_j^\psi = 1 - \alpha$$

and

$$\lambda_{\max}(G_j(\psi)) = B_j^\psi = 1 + (2^j - 1)\alpha\ . \quad \blacksquare$$

We see that the quotient $B_j^\psi / A_j^\psi = 2$ is independent of j and that we have the uniform bounds

$$A^\psi := \inf_{j \in \mathbb{N}} A_j^\psi = 2/3 \quad \text{and} \quad B^\psi := \sup_{j \in \mathbb{N}} B_j^\psi = 2\,.$$

Remark. For the dual basis $\tilde{\mathbf{\Psi}}_\mathbf{j}$ as determined in (1.5) we have similar results. We only state the following *property of fundamental interpolation*

$$\delta_{k,\ell} = < \psi_{j,\ell}, \tilde{\psi}_{j,k} >_\rho = \mu_{j,\ell}\, \tilde{\psi}_{j,k}(x_{j+1,2\ell+1})\ ,$$

which for the special case $r = \sqrt{2/\pi}$ and $\vartheta = \arccos$ recovers this property for Tasche's "wavelets".

§4. Multiscale Decomposition

Suppose $f \in V_{j+1}$ is given in terms of the localizing basis $\mathbf{\Phi_{j+1}}$,

$$f = \sum_{k=0}^{2^{j+1}} \alpha_k^{(j+1)} \phi_{j+1,k} . \tag{4.1}$$

In order to perform a multiscale decomposition, we would like to separate, step by step, the low frequency part (represented in terms of the basis $\mathbf{\Phi_j}$) from the high frequency part (represented by the basis $\mathbf{\Psi_j}$), i.e., we would like to compute the decomposition

$$f = \underbrace{\sum_{k=0}^{2^j} \alpha_k^{(j)} \phi_{j,k}}_{\in V_j} + \underbrace{\sum_{k=0}^{2^j-1} \beta_k^{(j)} \psi_{j,k}}_{\in W_j} . \tag{4.2}$$

Denoting

$$\mathbf{a_j} = (\alpha_0^{(j)}, \ldots, \alpha_{2^j}^{(j)})^T \quad \text{and} \quad \mathbf{b_j} = (\beta_0^{(j)}, \ldots, \beta_{2^j-1}^{(j)})^T , \tag{4.3}$$

our aim is to *compute, for the given vector* $\mathbf{a_{j+1}}$, *the two vectors* $\mathbf{a_j}$ *and* $\mathbf{b_j}$.

Such a step can be explained in simple matrix operations as follows: We have, by (4.1) and (1.3):

$$f = \mathbf{a_{j+1}}^T \, \mathbf{\Phi_{j+1}} = \mathbf{a_{j+1}}^T \, C_{j+1} \, \mathbf{f_{j+1}} ,$$

while, by (4.2) and (1.3):

$$f = \begin{pmatrix} \mathbf{a_j} \\ \mathbf{b_j} \end{pmatrix}^T \begin{pmatrix} \mathbf{\Phi_j} \\ \mathbf{\Psi_j} \end{pmatrix} = \begin{pmatrix} \mathbf{a_j} \\ \mathbf{b_j} \end{pmatrix}^T \begin{pmatrix} C_j & 0 \\ 0 & D_j \end{pmatrix} \mathbf{f_{j+1}} .$$

Comparing these we find

$$\mathbf{a_{j+1}} = (C_{j+1}^T)^{-1} \begin{pmatrix} C_j^T & 0 \\ 0 & D_j^T \end{pmatrix} \begin{pmatrix} \mathbf{a_j} \\ \mathbf{b_j} \end{pmatrix} \tag{4.4}$$

and

$$\begin{pmatrix} \mathbf{a_j} \\ \mathbf{b_j} \end{pmatrix} = \begin{pmatrix} (C_j^T)^{-1} & 0 \\ 0 & (D_j^T)^{-1} \end{pmatrix} C_{j+1}^T \, \mathbf{a_{j+1}} . \tag{4.5}$$

This latter equation describes the *decomposition step* and is the solution of our task, while the former one describes the corresponding *reconstruction step* of going from equation (4.2) to equation (4.1) in terms of the corresponding coefficient vectors.

Let us discuss the complexity of these operations. From (2.1) and (3.1) we see that multiplication by C_j or D_j has the same complexity as the type-I DCT of order j and $j+1$, respectively, which is $\mathcal{O}(j2^j)$. Also, the multiplication by the transposed matrices C_j^T or D_j^T has this complexity, since the transpose of the DCT matrix is only a slight modification of the DCT matrix itself. Finally, we have

$$(C_j^T)^{-1} = 2^{1-j}\, \Gamma_j\, C_j^T\, F_j$$

and

$$(D_j^T)^{-1} = 2^{-j}\sqrt{(2^j+1)/2}\; V_j\, M_{j+1}^{-1}\, C_{j+1}\, M_{j+1}\, U_j\ ,$$

with $M_{j+1} = \mathrm{diag}(2,1,1,\ldots,1,2)$. *This shows that both operations (4.4) and (4.5) have the complexity* $\mathcal{O}(j2^j)$.

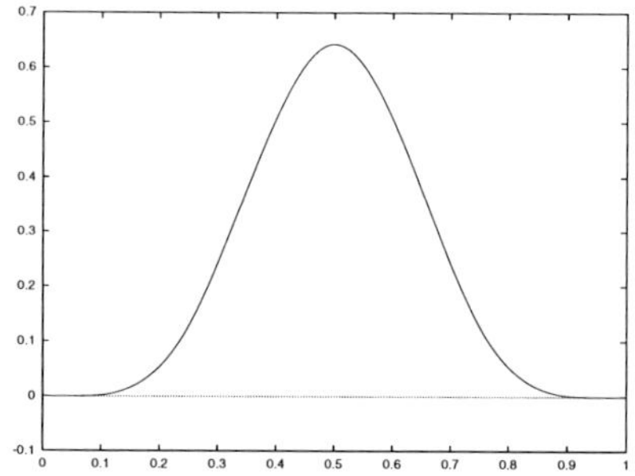

Fig. 2. Cubic spline function N_4, all knots inside $(0,1)$.

§5. Examples

In this section we give a few typical examples of a time–frequency analysis with the following analyzing functions:

1. **Sturm–Liouville Wavelets.** Here we use our analysis based on the system (1.7) with $[a,b] = [0,1]$, $\vartheta(x) = \pi x$, and $r \equiv \sqrt{2}$. For this case the weight function ρ is constant equal one. Corresponding functions $\phi_{7,50}$ and $\psi_{7,50}$ are shown in Figure 1.
2. **Chebyshev Wavelets.** (Kilgore and Prestin [3]) This is a special case of Sturm–Liouville wavelets, where $[a,b] = [-1,+1]$, $\vartheta(x) = \arccos(x)$, and $r \equiv \sqrt{2/\pi}$. Here $\rho(x) = 1/\sqrt{1-x^2}$.
3. **Periodic Spline Wavelets.** (Plonka and Tasche [5]) We use polynomial spline functions of order 6 (*i.e.*, piecewise polynomials of local degree 5).
4. **Cohen–Daubechies–Vial Wavelets.** We take $N = 6$ and use data preconditioning as described in the paper [1]. These wavelets reconstruct polynomials of degree 5.

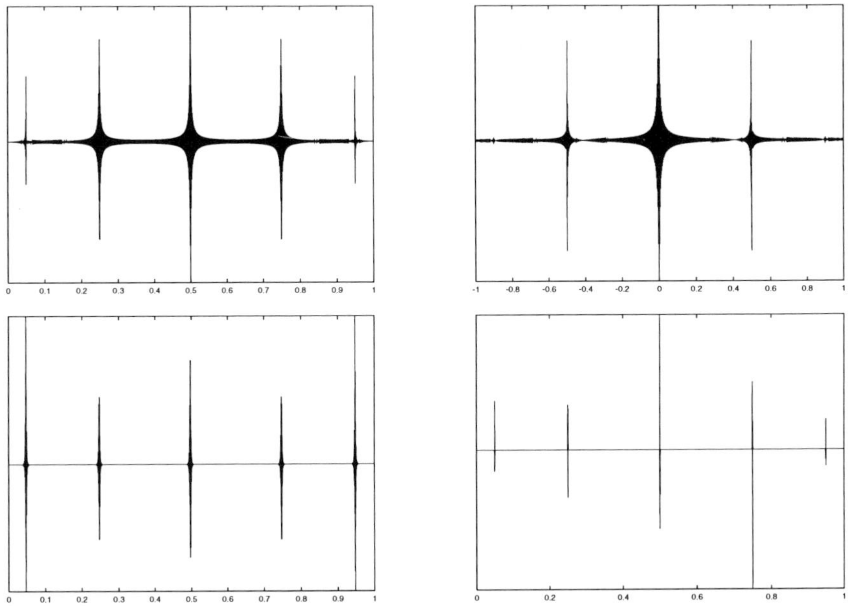

Fig. 3. Coefficients for the highest frequency part, cases 1 – 4.

Our time-frequency analysis tests how good these four analyzing systems separate the frequency bands, and how good they locate those parts of the signal in the time domain, where the signal is less smooth. Due to the lack of space in this paper, our two examples are so constructed in order to test on discontinuities in the third derivative.

We also mention that we restrict our considerations to only two types of Sturm-Liouville Wavelets, although many other types could be used as well. The two types selected show the effects due to the non-regular weight function in case of Chebyshev Wavelets, as will become clear in the examples. Numerical examples using other types of Sturm-Liouville Wavelets can be found in [2].

Example 1 (Spline Functions): Our first example is a cubic B–spline N_4 with knots $0.05, 0.25, 0.5, 0.75, 0.95$ inside the interval $[0, 1]$ (cf. Figure 2). This function is twice continuously differentiable with discontinuities of the third derivative at the given knots. These discontinuities are recognized by all four analyzing function systems, with the Chebyshev wavelets showing some typical damping of the analysis near the boundary (cf. Figure 3).

Example 2 (Chirps): Here we take $f(x) := N_4(x) \cdot \sin(2\pi(1 + 4x)^3)$, $x \in [0, 1]$, with N_4 the cubic B-spline of Example 1 and the second factor a *chirp*. What we expect here is the following: The frequency analysis should yield frequency bands moving from left to the right in the subsequent W-spaces, while the time analysis should still capture the location of the discontinuities in the third derivative originating from the spline. It turns out that all four analyzing systems give a satisfactory frequency analysis, but the discontinuities are only recognized by the Sturm–Liouville wavelets (with the special case of the Chebyshev wavelets showing again damping effects near the boundary).

Figures 4 – 7 show the coefficients in the six spaces W_j of highest frequency with increasing frequency from upper left to lower right.

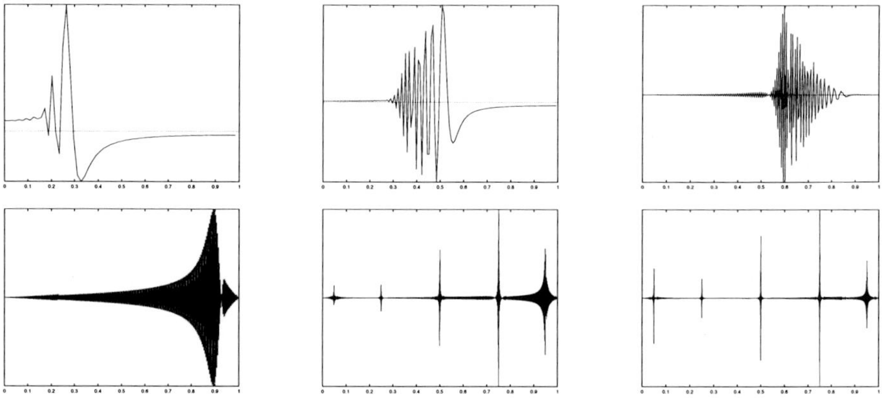

Fig. 4. Analysis of $f(x)$, Sturm–Liouville wavelets.

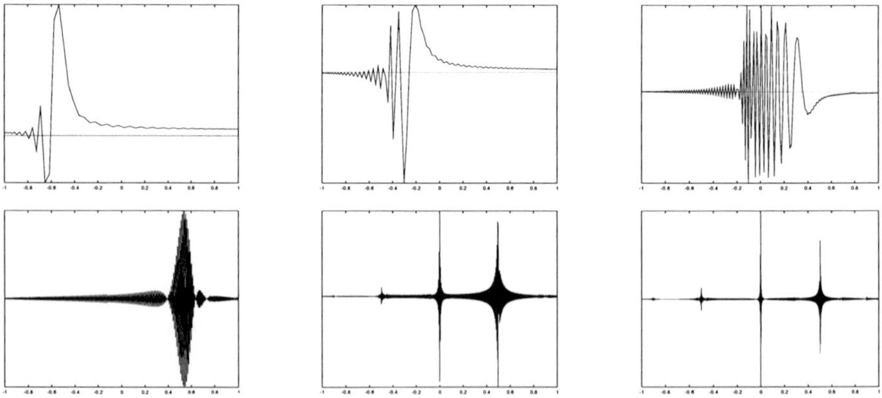

Fig. 5. Analysis of $f(x)$, Chebyshev wavelets.

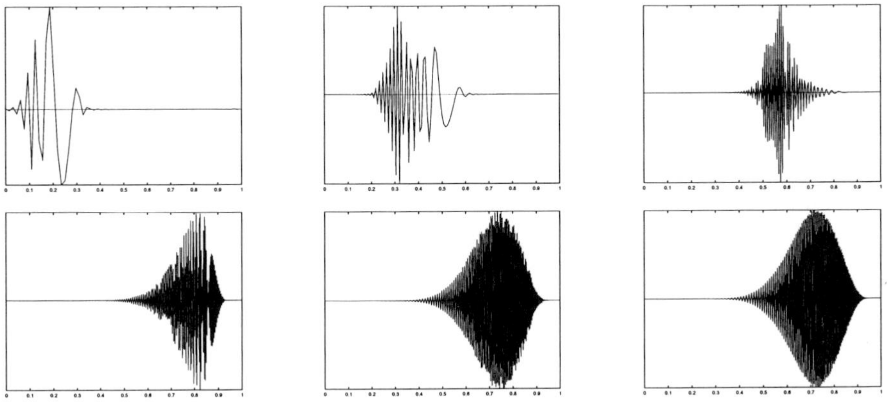

Fig. 6. Analysis of $f(x)$, periodic spline wavelets.

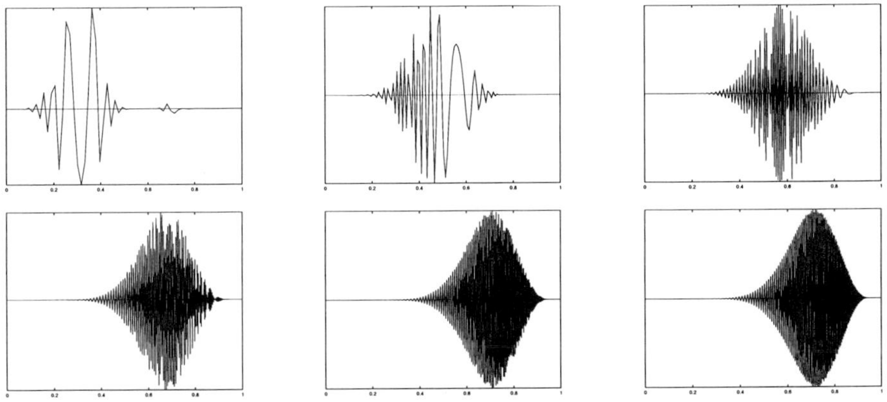

Fig. 7. Analysis of $f(x)$, Cohen–Daubechies–Vial wavelets.

Acknowledgements. Support from the Deutsche Forschungsgemeinschaft for the first author is gratefully acknowledged. We wish to thank Peter Schroeder, Fern-Universität Hagen, for making his software for CDV wavelets available.

References

1. Cohen, A., I. Daubechies and P. Vial, Multiresolution analysis, wavelets and fast algorithms on the interval, Appl. Comput. Harm. Anal. **1** (1993), 54–81.
2. Depczynski, U., Konstruktion von waveletartigen Zerlegungen auf kompakten Intervallen mit Hilfe der Eigenlösungen Sturm–Liouvillescher Randwertpro-

bleme, PhD Thesis, Univ. Duisburg, 1995.
3. Kilgore, T. and J. Prestin, Polynomial wavelets on the interval, preprint 94/3, Universität Rostock, 1994.
4. Plonka, G., K. Selig and M. Tasche, On the construction of wavelets on a bounded interval, Adv. Comput. Math., to appear.
5. Plonka, G. and M. Tasche, On the computation of periodic spline wavelets, Appl. Comput. Harm. Anal. **2** (1995), 1–14.
6. Potts, D., G. Steidl and M. Tasche, Kernels of spherical harmonics and spherical frames, this proceedings.
7. Tasche, M., Polynomial wavelets on [−1, +1], in *Approximation Theory, Wavelets and Applications*, S. P. Singh (ed.), Kluwer Academic Publ., Dordrecht, 1995, 497–512.

Uwe Depczynski and Kurt Jetter
Universität Hohenheim
Institut für Angewandte Mathematik und Statistik
D–70593 Stuttgart (Germany)

depczyns@uni-hohenheim.de
kjetter@uni-hohenheim.de

On the Dimension of the C^1 Spline Space for the Morgan–Scott Triangulation from the Blossoming Approach

Yu Yu Feng, Jernej Kozak and Ming Zhang

Abstract. In this paper, the problem of the dimension of the C^1 spline space for the two-dimensional Morgan-Scott triangulation Δ is discussed. As working tool the blossoming approach is applied, and the known result for dim $S_2^1(\Delta)$ reproved. Let T_0, T_1, T_2 be vertices of the supporting triangle, and T_0', T_1', T_2' vertices of the inner triangle with T_i' lying opposite to T_i. Then dim $S_2^1(\Delta) = 7$ iff all the lines T_iT_i' (or their prolongations) are concurrent. Otherwise dim $S_2^1(\Delta) = 6$. As the main result, the blossoming approach is used to solve the case $n \geq 3$. In the general position of the triangle $T_0'T_1'T_2'$ one obtains dim $S_n^1(\Delta) = \frac{1}{2}n(7n-15)+7$. The dimension is increased by 1 iff all the lines T_iT_i' meet at any vertex T_j'.

§1. Introduction

Let Δ be the Morgan–Scott triangulation (Fig.1) that divides the supporting triangle $T_0T_1T_2$ in seven nondegenerate triangles, and let $S_n^r(\Delta)$ be the space of C^r continuous piecewise polynomial functions of total degree n over Δ. The question of how many degrees of freedom has $S_n^r(\Delta)$ arose quite a lot of attention in the recent years. The answer known only for particular values of r, n is not a simple one since it relates the degree of the spline space, the continuity requirements as well as the geometry of Δ itself. The triangulation Δ itself was introduced to point out that the dimension of multivariate spline spaces of lower degree can be related to the geometric properties of the partition.

Let us recall some results. In [3, 8] it was shown that the dimension of $S_2^1(\Delta)$ may be six or seven, depending on the geometry of the triangulation. The precise

Advanced Topics in Multivariate Approximation
F. Fontanella, K. Jetter and P. J. Laurent (eds.) pp. 71–86.

ISBN 981-02-2852-X

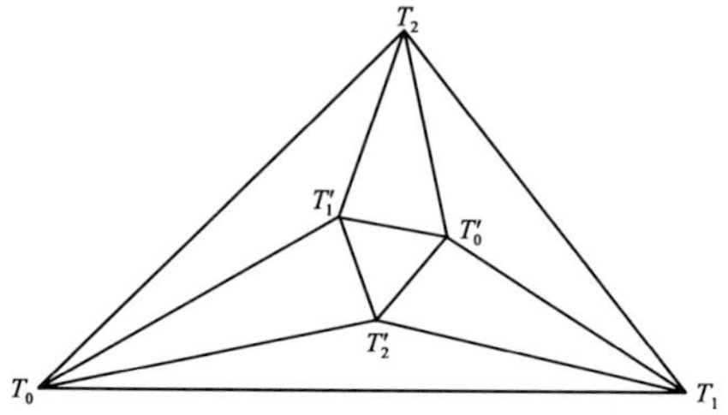

Fig. 1. The Morgan–Scott triangulation.

geometric condition that guarantees existence of a spline $s \in S_2^1(\Delta)$ that is not a polynomial ($\dim S_2^1(\Delta) = 7$) was not uncovered. However, it was pointed out that if Δ is symmetric, the dimension of $S_2^1(\Delta)$ should be seven. Then [1] pointed out that the dimension of $S_2^1(\Delta)$ may be seven even though Δ is not symmetric. Recently [8] gave the final answer for the case $S_2^1(\Delta)$. The dimension of the spline space is seven iff all the lines T_iT_i' (or their prolongations) are concurrent.

In this paper we shall use a different approach to the problem of $\dim S_n^r(\Delta)$. The continuity conditions will be expressed by relations between blossoms of the polynomial pieces. This will produce a new proof of [8]. The main new result is the application of the blossoming approach to $\dim S_n^1(\Delta)$ for the case $n \geq 3$. As it turns out this case differs substantially from the case $n = 2$. One obtains almost always

$$\dim S_n^1(\Delta) = \frac{1}{2}n(7n - 15) + 7.$$

The dimension is increased to $\frac{1}{2}n(7n - 15) + 8$ if and only if the lines T_iT_i' meet at any vertex T_j', a very particular geometric case.

§2. Blossoms of Multivariate Polynomials

In this section, we recall the definition and some basic properties of the blossom. They can be found in [4, 5, 2]. The basic idea of the blossom approach is the fact that there is a one–to–one correspondence between polynomials of total degree at most n and a certain class of polynomials of n variables. This enables one to study relations between polynomials by studying reflected relations between their blossoms which often turns out to be an easier task.

Let us be precise. Given a vector space E, a function $f : E \to \mathbb{R}$ is said to be *affine* if it satisfies $f(\sum_i u_i x_i) = \sum_i u_i f(x_i)$ for all $x_i \in E$ and all $u_i \in \mathbb{R}$ such that $\sum_i u_i = 1$. A function $f : E^n \to \mathbb{R}$ is said to be *n-affine* if it is affine with respect to each argument. It is said to be *symmetric* if it preserves its value under any permutation of its arguments. Let $\pi_n(\mathbb{R}^m)$ denote the space of all m–variable polynomials of *total* degree $\leq n$. For any $p \in \pi_n(\mathbb{R}^m)$ there exists a unique symmetric n-affine polynomial

$$\mathbb{B}_n(p)(x^{(1)}, x^{(2)}, \cdots, x^{(n)}), \qquad x^{(i)} \in \mathbb{R}^m,$$

that pointwise satisfies the *diagonal* property

$$\mathbb{B}_n(p)(x, x, \cdots, x) = p(x), \qquad x \in \mathbb{R}^m.$$

This defines $\mathbb{B}_n(p)$, the *blossom* of p. To give its explicit form let us introduce now some notation. Let $\mathbb{Z}_+$ denote the set of nonnegative integers. Throughout the paper small Greek letters will denote *multiindex vectors* i.e. vectors with nonnegative integer components. For any multiindices

$$\alpha = (\alpha_1, \alpha_2, \cdots, \alpha_m) \in \mathbb{Z}_+^m, \quad \beta = (\beta_1, \beta_2, \cdots, \beta_m) \in \mathbb{Z}_+^m,$$

and a vector $x = (x_1, x_2, \cdots, x_m) \in \mathbb{R}^m$, let

$$|\alpha| := \alpha_1 + \alpha_2 + \cdots + \alpha_m, \quad \alpha! := \alpha_1!\alpha_2! \cdots \alpha_m!\,, \quad x^\alpha := x_1^{\alpha_1} x_2^{\alpha_2} \cdots x_m^{\alpha_m},$$

$$\binom{n}{\alpha} := \begin{cases} \frac{n!}{\alpha!(n-|\alpha|)!}, & |\alpha| \le n, \\ 0, & |\alpha| > n. \end{cases}$$

Let $\alpha \le \beta$ denote the relation $\le$ componentwise, i.e. $\alpha_i \le \beta_i$ for all i, and let $\alpha < \beta$ be $\alpha \le \beta$ with at least one $\alpha_i < \beta_i$. Then

$$\binom{\alpha}{\beta} := \begin{cases} \prod_{j=1}^{m} \binom{\alpha_j}{\beta_j}, & \beta \le \alpha, \\ 0, & \text{otherwise}. \end{cases}$$

The *multivariate elementary symmetric function* σ is given by

$$\sigma_\alpha(x^{(1)}, x^{(2)}, \cdots, x^{(n)}) := \sum_{e^{i_1}+e^{i_2}+\cdots+e^{i_n}=\alpha} (x^{(1)})^{e^{i_1}} (x^{(2)})^{e^{i_2}} \cdots (x^{(n)})^{e^{i_n}}. \tag{1}$$

Here $i_1, i_2, \cdots, i_n \in \{0, 1, \cdots, m\}$, $x^{(i)} \in \mathbb{R}^m$,

$$e^j := (\underbrace{0, 0, \cdots, 0}_{j-1}, 1, 0, 0, \cdots, 0) \in \mathbb{R}^m, \; e^0 := (0, 0, \cdots, 0) \in \mathbb{R}^m.$$

Let $D := \left(\frac{\partial}{\partial x_1}, \frac{\partial}{\partial x_2}, \cdots, \frac{\partial}{\partial x_m}\right)$. For any polynomial $p \in \pi_n(\mathbb{R}^m)$ one has the *Taylor* expansion

$$p(x) = \sum_{|\alpha| \le n} \frac{1}{\alpha!} D^\alpha p(0) x^\alpha. \tag{2}$$

Since the blossoms of x^α are straightforward to compute,

$$\mathbb{B}(x^\alpha)(x^{(1)}, x^{(2)}, \cdots, x^{(n)}) = \binom{n}{\alpha}^{-1} \sigma_\alpha(x^{(1)}, x^{(2)}, \cdots, x^{(n)}), \tag{3}$$

the expansion (2) yields the blossom $\mathbb{B}$ of p as

$$\mathbb{B}(p)(x^{(1)}, x^{(2)}, \cdots, x^{(n)}) = \sum_{|\alpha| \leq n} \frac{1}{\alpha!} D^\alpha p_n(0) \binom{n}{\alpha}^{-1} \sigma_\alpha(x^{(1)}, x^{(2)} \cdots, x^{(n)}). \tag{4}$$

It is easy to prove that the multivariate elementary symmetric functions (1) have the following properties

$$\begin{gathered} \sigma_\alpha(x, x, \cdots, x) = \binom{n}{\alpha} x^\alpha, \\ \sigma_\alpha(ax^{(1)}, ax^{(2)}, \cdots, ax^{(n)}) = a^{|\alpha|} \sigma_\alpha(x^{(1)}, x^{(2)}, \cdots, x^{(n)}), \quad a \in \mathbb{R}, \\ \sigma_\alpha(x^{(1)}, x^{(2)}, \cdots, x^{(r)}, x^{(r+1)}, x^{(r+2)}, \cdots, x^{(n)}) = \\ \sum_{\beta \leq \alpha} \sigma_\beta(x^{(1)}, x^{(2)}, \cdots, x^{(r)}) \sigma_{\alpha-\beta}(x^{(r+1)}, x^{(r+2)}, \cdots, x^{(n)}). \end{gathered} \tag{5}$$

In order to express smoothness conditions as a relation between blossoms we recall the following lemma ([2]). Let $p \in \pi_n(\mathbb{R}^m)$ and $\gamma \in \mathbb{Z}_+^m$. Then

$$D^\gamma p(x) = \frac{n!}{(n-|\gamma|)!} \sum_{\eta \leq \gamma} \binom{\gamma}{\eta} (-1)^{|\gamma - \eta|}$$

$$\mathbb{B}(p) \left(\underbrace{x, x, \cdots, x}_{n-|\eta|}, \underbrace{x+e^1, x+e^1, \cdots, x+e^1}_{\eta_1}, \cdots, \underbrace{x+e^m, x+e^m, \cdots, x+e^m}_{\eta_m} \right).$$

This produces the following corollary ([2]). Let p, $q \in \pi_n(\mathbb{R}^m)$ be restricted to nondegenerate simplices

$$V_p := \langle v^{(0)}, v^{(1)}, \cdots, v^{(k)}, v_p^{(k+1)}, v_p^{(k+2)}, \cdots, v_p^{(n)} \rangle,$$
$$V_q := \langle v^{(0)}, v^{(1)}, \cdots, v^{(k)}, v_q^{(k+1)}, v_q^{(k+2)}, \cdots, v_q^{(n)} \rangle.$$

The piecewise polynomial function, composed of p and q, is in $C^r(V_p \cup V_q)$ if and only if

$$\mathbb{B}(p) \left(\underbrace{x, x, \cdots, x}_{n-r}, w^1, w^2, \cdots, w^r \right) = \mathbb{B}(q) \left(\underbrace{x, x, \cdots, x}_{n-r}, w^1, w^2, \cdots, w^r \right), \tag{6}$$

for any $w^i \in \mathbb{R}^m$ and $x \in V_p \cap V_q$. Of course, due to the nature of polynomials, if p, q agree up to r-th derivative on $V_p \cap V_q$, they agree on all of the k-dimensional hyperplane in which $V_p \cap V_q$ can be embedded.

§3. Smoothness Conditions in Blossoming Form

Let us elaborate (6) further for our case $m = 2$, $k = 1$ where simply continuity across a straight line can be considered. Let ℓ be a straight line that divides $\mathbb{R}^2$ into two halfplanes H_1 and H_2, the supports of polynomials $p, q \in \pi_n(\mathbb{R}^2)$, and

$$p(x) = \sum_{|\alpha|\leq n} a_\alpha \binom{n}{\alpha} x^\alpha, \qquad q(x) = \sum_{|\alpha|\leq n} b_\alpha \binom{n}{\alpha} x^\alpha. \tag{7}$$

Theorem 3.1. *Let f be a piecewise polynomial function composed of p and q. Then $f \in C^r(\mathbb{R}^2)$ if and only if*

$$\sum_{|\beta|=i} (a_{\alpha+\beta} - b_{\alpha+\beta}) \binom{i}{\beta} v^\beta = 0 , \tag{8}$$

for $i = 0, 1, \cdots, n-r$, $|\alpha| \leq r$. Here $v = (v_1, v_2)$ denotes the unit direction vector of ℓ.

Proof: (4) applied to (7), and (6) reveals that $f \in C^r(\mathbb{R}^2)$ if and only if

$$\sum_{|\alpha|\leq n} (a_\alpha - b_\alpha)\sigma_\alpha(x, x, \cdots, x, w^1, w^2, \cdots, w^r) = 0 , \tag{9}$$

for any $x \in \ell$, $w^1, w^2, \cdots, w^r \in \mathbb{R}^2$. By using properties of σ_α listed in (5), (9) can be rewritten as

$$\sum_{|\beta|\leq r} \left(\sum_{\alpha\geq\beta, |\alpha-\beta|\leq n-r} (a_\alpha - b_\alpha)\sigma_{\alpha-\beta}(x, x, \cdots, x) \right) \sigma_\beta(w^1, w^2, \cdots, w^r) = 0 . \tag{10}$$

But elementary symmetric functions are linearly independent, and (10) is equivalent to

$$\sum_{\alpha\geq\beta, |\alpha-\beta|\leq n-r} (a_\alpha - b_\alpha)\sigma_{\alpha-\beta}(x, x, \cdots, x) = 0 , \tag{11}$$

for any $x \in \ell$ and $|\beta| \leq r$. The change of summation indices $\beta := \alpha - \beta$, $\alpha := \beta$ and (5) simplify (11) to

$$\sum_{|\beta|\leq n-r} (a_{\alpha+\beta} - b_{\alpha+\beta}) \binom{n-r}{\beta} x^\beta = 0 , \tag{12}$$

for any $x \in \ell$ and $|\alpha| \leq r$. But ℓ is a line, and (12) is further equivalent to

$$\sum_{|\beta|\leq n-r} (a_{\alpha+\beta} - b_{\alpha+\beta}) \binom{n-r}{\beta} (sv + u)^\beta = 0 , \tag{13}$$

for any $s \in \mathbb{R}$, $|\alpha| \le r$, and some fixed vector u. One can take the coordinate origin on the line ℓ, and (13) reduces to

$$\sum_{|\beta|=i} (a_{\alpha+\beta} - b_{\alpha+\beta}) \binom{n-r}{\beta} v^\beta = 0 , \tag{14}$$

for $i = 0, 1, \cdots, n-r$, $|\alpha| \le r$. Since $\binom{n-r}{\beta} = \binom{n-r}{i}\binom{i}{\beta}$, (14) yields

$$\sum_{|\beta|=i} (a_{\alpha+\beta} - b_{\alpha+\beta}) \binom{i}{\beta} v^\beta = 0, \qquad i = 0, 1, \cdots, n-r, \ |\alpha| \le r.$$

This completes the proof of the theorem. ■

The equations (8) provided by the theorem 3.1 are not independent. Take $|\alpha| \le r-1$, $i \ge 1$, and consider (8) for $\xi := \alpha + (1,0)$, $\eta := \alpha + (0,1)$,

$$\sum_{|\beta|=i-1} x_{\xi+\beta} \binom{i-1}{\beta} v^\beta = 0 , \tag{15}$$

$$\sum_{|\beta|=i-1} x_{\eta+\beta} \binom{i-1}{\beta} v^\beta = 0 . \tag{16}$$

By multiplying equations (15) and (16) by v_1, v_2 respectively we obtain

$$\sum_{k=0}^{i-1} x_{\alpha+(i-k,k)} \binom{i-1}{k} v_1^{i-k} v_2^k = 0 , \tag{17}$$

$$\sum_{k=1}^{i} x_{\alpha+(i-k,k)} \binom{i-1}{k-1} v_1^{i-k} v_2^k = 0 . \tag{18}$$

Sum of (17) and (18) can be written as

$$\sum_{|\beta|=i} x_{\alpha+\beta} \binom{i}{\beta} v^\beta = 0 . \tag{19}$$

This reveals that equations in (8) can be written, for given i, as linear combinations of equations with $i := i-1$. The only exceptions are the boundary values $i = 0$ and $|\alpha| = r$. Thus we have the following conclusion:

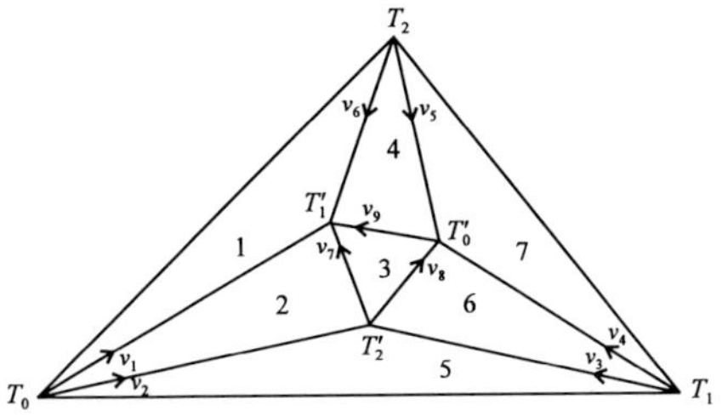

Fig. 2. Morgan–Scott triangulation with denoted triangles.

Theorem 3.2. *Let the assumptions of the theorem 3.1 hold. Then $f \in C^r(\mathbb{R}^2)$ if and only if*

$$\sum_{|\beta|=i} x_{\alpha+\beta} \binom{i}{\beta} v^\beta = 0 \ , \tag{20}$$

for $i = 0, |\alpha| \le r$ and $i = 1, 2, \cdots, n-r$, $|\alpha| = r$. The equations (20) are linearly independent.

Proof: We only have to prove the last claim. Note that conditions in (20) can be dependent only if the index $\alpha + \beta$ is of the same length since the vectors $x_{\alpha+\beta}$, $x_{\alpha'+\beta'}$, $|\alpha+\beta| \ne |\alpha'+\beta'|$, have nothing in common. When $i = 0$, (20) reduces to $x_\alpha = 0, |\alpha| \le r$, each equation taking a different component of x. It is obvious that these $\binom{r+2}{2}$ equations are linearly independent. To complete, we only need to prove that the linear equations

$$\sum_{|\beta|=i} x_{\alpha+\beta} \binom{i}{\beta} v^\beta = 0 \ , \qquad |\alpha| = r \ , \tag{21}$$

are independent for each fixed $i \ge 1$. If the components of x are ordered lexicographically the corresponding matrix reads as

$$\begin{pmatrix} \binom{i}{0} v_1^i & \binom{i}{1} v_1^{i-1} v_2 & \cdots & \binom{i}{i} v_2^i & 0 & \cdots & 0 \\ 0 & \binom{i}{0} v_1^i & \binom{i}{1} v_1^{i-1} v_2 & \cdots & \binom{i}{i} v_2^i & & \\ \vdots & & & & \vdots & & \\ \vdots & & & & \vdots & & \\ 0 & 0 & \cdots & \binom{i}{0} v_1^i & \binom{i}{1} v_1^{i-1} v_2 & \cdots & \binom{i}{i} v_2^i \end{pmatrix} . \tag{22}$$

Since v is a unit length direction vector, v_1 and v_2 can't vanish simultaneously. This implies that the rank of matrix (22) is $r+1$. The theorem is proved. ■

§4. The Dimension of $S_2^1(\Delta)$

Let the Morgan–Scott triangulation be denoted as in Fig. 2. The unit direction vectors of T_0T_1', T_0T_2', T_1T_2', T_1T_0', T_2T_0', T_2T_1' will be denoted as $v_1, v_2, \cdots, v_6$, and of $T_2'T_1'$, $T_2'T_0'$, $T_0'T_1'$ as v_7, v_8, v_9 respectively. We will assume that all the slopes $v_j := (a_j, b_j)$ at inner vertices T_i' are different. The smoothness conditions for two adjacent triangles with a common edge vector $v := (\tilde{v}_1, \tilde{v}_2)$ must be written in general form (13) since the origin can't be placed on all of the common edges in Δ, i.e.

$$\sum_{|\beta|\le 1} (a_{\alpha+\beta} - b_{\alpha+\beta})(sv + u)^\beta = 0 \ , \tag{24}$$

with $u := (u_1, u_2)$ a constant vector. Setting $x_\alpha := a_\alpha - b_\alpha$, (24) could be rewritten as

$$x_\alpha + \sum_{|\beta|=1} x_{\alpha+\beta}(sv^\beta + u^\beta) = 0 \ , \tag{25}$$

which must hold for any $s \in \mathbb{R}$, and $|\alpha| \le 1$. That is

$$x_0 + \sum_{|\beta|=1} x_\beta u^\beta = 0 \ , \tag{26}$$

$$\sum_{|\beta|=1} x_\beta v^\beta = 0 \ , \tag{27}$$

$$x_\alpha + \sum_{|\beta|=1} x_{\alpha+\beta} u^\beta = 0 \ , \quad |\alpha| = 1 \ , \tag{28}$$

$$\sum_{|\beta|=1} x_{\alpha+\beta} v^\beta = 0 \ , \quad |\alpha| = 1 \ . \tag{29}$$

The conditions (26) to (29) are linearly dependent. To see this put $\xi := (1,0)$, $\eta := (0,1)$ as α in (28), (29) to compute

$$\tilde{v}_1(x_\xi + \sum_{|\beta|=1} x_{\xi+\beta}u^\beta) + \tilde{v}_2(x_\eta + \sum_{|\beta|=1} x_{\eta+\beta}u^\beta)$$

$$-u_1 \sum_{|\beta|=1} x_{\xi+\beta}v^\beta - u_2 \sum_{|\beta|=1} x_{\eta+\beta}v^\beta = \sum_{|\beta|=1} x_\beta v^\beta \ . \tag{30}$$

Hence, the conditions (27) are linear combinations of (28) and (29). There are 9 common edges in Δ, and corresponding directions $v_j = (a_j, b_j)$, and constant vectors $u_j := (c_j, d_j)$. Now if

$$a_i^j := (a_{(i,0)}^j, a_{(i-1,1)}^j, \cdots, a_{(0,i)}^j) \ \in \mathbb{R}^{i+1}$$

denotes coefficients that correspond to x^α, $|\alpha| = i$, over triangle j, and the vector A of all 42 coefficients is ordered as

$$A := (a_0^1, a_0^2, \cdots, a_0^7, a_1^1, a_1^2, \cdots, a_1^7, a_2^1, a_2^2, \cdots, a_2^7)^T, \tag{31}$$

the smoothness conditions (26), (28), and (29), written for all common edges, reduce to

$$MA = 0 , \tag{32}$$

with

$$M := \begin{pmatrix} M_{11} & M_{12}(c,d) & \\ & M_{22} & M_{23}(c,d) \\ & & M_{23}(a,b) \end{pmatrix} . \tag{33}$$

Throughout the paper we will assume that the blocks (or elements) in matrices that are left out are zero blocks of the proper dimension. In (33) M_{ij} are matrices of the order $9i \times 7j$ and have the form

$$M_{ij} := \begin{pmatrix} Q_1 & -Q_1 & & & & & \\ & Q_7 & -Q_7 & & & & \\ & & Q_9 & -Q_9 & & & \\ -Q_6 & & & Q_6 & & & \\ & -Q_2 & & & Q_2 & & \\ & & & & -Q_3 & Q_3 & \\ & & -Q_8 & & & Q_8 & \\ & & & & & -Q_4 & Q_4 \\ & & & -Q_5 & & & Q_5 \end{pmatrix} . \tag{34}$$

Vectors c, d depend on the choice of the origin, a, b denote direction data, and Q_k are matrices of the order $i \times j$,

$$Q_k := \begin{cases} 1, & \text{in } M_{11}, \\ \begin{pmatrix} 1 & 0 \\ 0 & 1 \end{pmatrix}, & \text{in } M_{22}, \\ (c_k \;\; d_k), & \text{in } M_{12}(c,d), \\ \begin{pmatrix} c_k & d_k & 0 \\ 0 & c_k & d_k \end{pmatrix}, & \text{in } M_{23}(c,d), \\ \begin{pmatrix} a_k & b_k & 0 \\ 0 & a_k & b_k \end{pmatrix}, & \text{in } M_{23}(a,b). \end{cases} \tag{35}$$

If we take T_1' as coordinate origin then (c_k, d_k) in (35) reads

$$(c_k, d_k) = \begin{cases} (0,0), & k = 1,7,9,6, \\ t_7(a_7, b_7), & k = 2,3, \\ t_9(a_9, b_9), & k = 4,5,8, \end{cases} \tag{36}$$

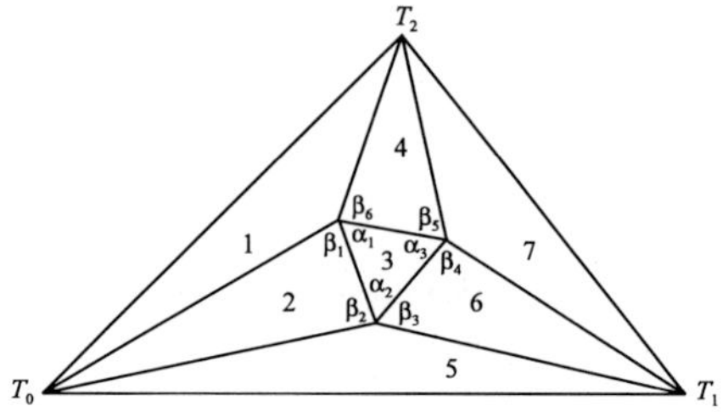

Fig. 3. The partition Δ with angles between vectors v_i.

with t_7, t_9 being the length of $T_1'T_2'$ and $T_1'T_0'$. The block columns of (33) sum up to block 0. Thus rank $M_{ij} \leq \min(9i, 6j) = 6j$. Quite clearly rank $M_{ii} = 6i$, $i = 1, 2$, and it is not difficult to see $\dim S_2^1(\Delta) = 42 - \text{rank } M = 24 - \text{rank } M_{23}(a,b)$. M_{ij} can be simplified (by rank preserving transformations) to $\bar{M}_{ij}$ where

$$\bar{M}_{i,j} := \begin{pmatrix} Q_1 & & & & & \\ & Q_7 & & & & \\ & & Q_9 & & & \\ Q_6 & Q_6 & Q_6 & & & \\ & Q_2 & & Q_2 & Q_2 & \\ & & & Q_3 & & \\ & & & & Q_8 & \\ & & & & & Q_4 \\ & & -Q_5 & & Q_5 & Q_5 \end{pmatrix}. \tag{38}$$

Let $[v_k, v_j] := a_k b_j - a_j b_k$ denote a planar vector product. The following lemma computes rank $M_{23}(a,b)$.

Lemma 4.1. *Let all the directions v_i at inner vertices T_j' be different. Then rank $M_{23}(a,b) \geq 17$, with equality if and only if*

$$[v_3, v_8][v_8, v_2][v_4, v_9][v_9, v_5][v_7, v_1][v_6, v_7] = [v_4, v_8][v_8, v_5][v_3, v_7][v_7, v_2][v_9, v_1][v_6, v_9]. \tag{39}$$

Proof: Since rank $M_{23}(a,b) = \text{rank } \bar{M}_{23}(a,b)$ it is enough to compute $\det \bar{M}_{23}(a,b)$. A straightforward conputation yields

$$\det \bar{M}_{23}(a,b) = [v_1, v_6][v_5, v_4][v_2, v_3]\Big([v_3, v_8][v_8, v_2][v_7, v_1][v_6, v_7][v_4, v_9][v_9, v_5] - [v_4, v_8][v_8, v_5][v_3, v_7][v_7, v_2][v_9, v_1][v_6, v_9]\Big). \tag{40}$$

By assumption, $[v_i, v_j] \neq 0$ if $i \neq j$. Thus $\det \bar{M}_{23}(a,b)$ can vanish only if the (39) holds. ∎

Let us explain the geometric meaning of the condition (39). Since v_i are unit vectors, $[v_k, v_j] = \sin(v_k, v_j)$. Let the angles α_i, β_i be as on Fig. 3. Then the condition (39) can be written as

$$\begin{aligned}\frac{\sin(\alpha_1+\beta_1)}{\sin\beta_1}\cdot\frac{\sin\beta_6}{\sin(\alpha_1+\beta_6)}\cdot\frac{\sin\beta_2}{\sin(\alpha_2+\beta_2)}\\ \cdot\frac{\sin(\alpha_2+\beta_3)}{\sin\beta_3}\cdot\frac{\sin\beta_4}{\sin(\alpha_3+\beta_4)}\cdot\frac{\sin(\alpha_3+\beta_5)}{\sin\beta_5}=1\ .\end{aligned}\tag{41}$$

Let R_i be intersection points of the line T_iT_i' and edge of the inner triangle that does not include T_i'. Elementary geometry gives

$$\frac{\sin(\alpha_1+\beta_1)}{\sin\beta_1}=\frac{T_0T_0'}{T_0R_0}\cdot\frac{R_0T_1'}{T_0'T_1'}\ ,\tag{42}$$

and similarly for the other quotients. Thus the condition can be rewritten as

$$\frac{R_0T_1'}{T_2'R_0}\cdot\frac{R_1T_2'}{T_0'R_1}\cdot\frac{R_2T_0'}{T_1'R_2}=1\ .\tag{43}$$

But AE, BF, CD are concurrent lines if and only if

$$\frac{AD}{DB}\cdot\frac{BE}{EC}\cdot\frac{CF}{FA}=1\ .\tag{44}$$

This concludes the proof of the following theorem:

Theorem 4.1. $\dim S_2^1(\Delta) = 6$ *or* 7, *and it is* 7 *if and only if the segments* $T_j'T_j$, $j = 0,1,2$ *(or their prolongations) are concurrent lines.*

§5. The Dimension of $S_n^1(\Delta)$, $n \ge 3$

Let us consider the case $n = 3$ first. (13) reveals the smoothness conditions for two adjacent triangles as

$$\sum_{|\beta|\le 2} x_{\alpha+\beta}\binom{2}{\beta}(sv+u)^\beta = 0\ ,\tag{45}$$

for any $s \in \mathbb{R}$ and $|\alpha| \le 1$. We first cancel out the dependent relations.

Lemma 5.1. *The smoothness conditions (45) are equivalent to*

$$x_0+\sum_{|\beta|=1} x_\beta\binom{2}{\beta}u^\beta+\sum_{|\beta|=2} x_\beta\binom{2}{\beta}u^\beta = 0\ ,\tag{46}$$

$$x_\alpha + \sum_{|\beta|=1} x_{\alpha+\beta} \binom{2}{\beta} u^\beta + \sum_{|\beta|=2} x_{\alpha+\beta} \binom{2}{\beta} u^\beta = 0, \quad |\alpha| = 1 , \tag{47}$$

$$\sum_{|\beta|=1} x_{\alpha+\beta} v^\beta + \sum_{|\beta|=2} x_{\alpha+\beta} \sum_{0<\tau<\beta} v^\tau u^{\beta-\tau} = 0 , \quad |\alpha| = 1 , \tag{48}$$

$$\sum_{|\beta|=2} x_{\alpha+\beta} \binom{2}{\beta} v^\beta = 0 , \quad |\alpha| = 1 . \tag{49}$$

Proof: It is obvious that (45) is equivalent to (46) – (49) and

$$\sum_{|\beta|=1} x_\beta v^\beta + \sum_{|\beta|=2} x_\beta \sum_{0<\tau<\beta} v^\tau u^{\beta-\tau} = 0 , \tag{50}$$

$$\sum_{|\beta|=2} x_\beta \binom{2}{\beta} v^\beta = 0 . \tag{51}$$

We claim that the equations (50) and (51) are linear combinations of equations (46)–(49). To see this, let $\xi = (1,0)$, $\eta = (0,1)$. A direct calculation shows that

$$\begin{aligned}\sum_{|\beta|=2} x_\beta \binom{2}{\beta} v^\beta = {} & v_1 \sum_{|\beta|=1} x_{\xi+\beta} v^\beta + v_1 \sum_{|\beta|=2} x_{\xi+\beta} \sum_{0<\tau<\beta} v^\tau u^{\beta-\tau} \\ & + v_2 \sum_{|\beta|=1} x_{\eta+\beta} v^\beta + v_2 \sum_{|\beta|=2} x_{\eta+\beta} \sum_{0<\tau<\beta} v^\tau u^{\beta-\tau} \\ & - u_1 \sum_{|\beta|=2} x_{\xi+\beta} \binom{2}{\beta} v^\beta - u_2 \sum_{|\beta|=2} x_{\eta+\beta} \binom{2}{\beta} v^\beta ,\end{aligned}$$

and (51) can be expressed by as a linear combination of equations (48) and (49). A similar argument works for (50). This confirms the lemma. ■

In the case $n = 3$, let us extend (31) with a_3^j, so that the corresponding coefficient vector A belongs to $\mathbb{R}^{70}$. The smoothness conditions (46) – (49) written in the matrix form are

$$M_3 A = 0 , \tag{52}$$

with

$$M_3 := \begin{pmatrix} M_{11} & 2M_{12}(c,d) & M_{13}(c,d) & \\ & M_{22} & 2M_{23}(c,d) & M_{24}(c,d) \\ & & M_{23}(a,b) & M_{24}(a,b,c,d) \\ & & & M_{24}(a,b) \end{pmatrix} . \tag{53}$$

The matrices M_{ij} are again of the form (34), with definition of Q_k in (35) extended by

$$Q_k := \begin{cases} \begin{pmatrix} c_k^2 & 2c_kd_k & d_k^2 \end{pmatrix}, & \text{in} \quad M_{13}(c,d), \\ \begin{pmatrix} c_k^2 & 2c_kd_k & d_k^2 & 0 \\ 0 & c_k^2 & 2c_kd_k & d_k^2 \end{pmatrix}, & \text{in} \quad M_{24}(c,d), \\ \begin{pmatrix} a_kc_k & a_kd_k + c_kb_k & b_kd_k & 0 \\ 0 & a_kc_k & a_kd_k + c_kb_k & b_kd_k \end{pmatrix}, & \text{in} \quad M_{24}(a,b,c,d), \\ \begin{pmatrix} a_k^2 & 2a_kb_k & b_k^2 & 0 \\ 0 & a_k^2 & 2a_kb_k & b_k^2 \end{pmatrix}, & \text{in} \quad M_{24}(a,b). \end{cases} \tag{54}$$

Thus

$$\dim S_3^1(\Delta) = 70 - \text{rank } M_3 \ . \tag{55}$$

The first step to compute rank M_3 is as follows.

Lemma 5.2.

$$\text{rank } M_3 = 18 + \text{rank} \begin{pmatrix} M_{23}(a,b) & M_{24}(a,b,c,d) \\ & M_{24}(c,d) \end{pmatrix} . \tag{56}$$

The proof is straightforward, but somewhat tedious, and will be omitted. It requires a careful cancellation in the first two blocks of M_3. It is easy to verify that

$$\text{rank } M_3 = 54 \ ,$$

when rank $M_{23}(a,b) = 18$. But even if rank $M_{23}(a,b) = 17$, we still have the following fact:

Lemma 5.3. *Suppose that the lines T_iT_i', $i = 0, 1, 2$, do not meet at any vertex T_j'. Then*

$$r := \text{rank} \begin{pmatrix} M_{23}(a,b) & M_{24}(a,b,c,d) \\ & M_{24}(a,b) \end{pmatrix} = 36 \ . \tag{57}$$

Proof: We only have to prove that if the lines T_iT_i' are concurrent but do not meet at any vertex T_j', (57) still holds. In this case rank $\bar{M}_{23}(a,b) = 17$. Let

$$m_{ij} := \begin{pmatrix} m_{ij}^1 \\ m_{ij}^2 \end{pmatrix} := \begin{pmatrix} a_ia_j & a_ib_j + a_jb_i & b_ib_j & 0 \\ 0 & a_ia_j & a_ib_j + a_jb_i & b_ib_j \end{pmatrix}, \tag{58}$$

and let bar denote rank preserving transformations already used to transform M_{ij} from (34) to $\bar{M}_{ij}$ in (38). Then

$$\bar{M}_{24}(a,b,c,d) = \begin{pmatrix} & & 0_{8\times 24} & & & \\ & t_7m_{27} & & t_7m_{27} & t_7m_{27} & \\ & & & t_7m_{37} & & \\ 0_{10\times 4} & & & & t_8m_{89} & \\ & & & & & t_9m_{49} \\ & & -t_9m_{59} & & t_9m_{59} & t_9m_{59} \end{pmatrix}_{18\times 24} . \tag{59}$$

Since $t_7(a_7, b_7) = t_8(a_8, b_8) + t_9(a_9, b_9)$, with t_8 being the length of $T_2'T_0'$, $\bar{M}_{24}(a,b,c,d)$ could be further simplified by columns of $\bar{M}_{23}(a,b)$ to the form

$$\begin{pmatrix} & & 0_{8\times 24} & & \\ & t_7 m_{27} & & t_8 m_{28} & \\ 0_{10\times 4} & & & & \\ & & & & \\ & & -t_9 m_{59} & & \end{pmatrix}_{18\times 24} . \tag{61}$$

By row eliminations applied already in (40) we thus obtain $r = 15$ plus rank of the matrix

$$\begin{pmatrix} [v_7,v_1][v_6,v_7] & [v_9,v_1][v_6,v_9] & 0 & & & & 0_{1\times 24} & & \\ [v_3,v_7][v_7,v_2] & 0 & [v_3,v_8][v_8,v_2] & 0_{2\times 4} & R_2 & & 0_{2\times 4} & R_4 & 0_{2\times 4} \\ 0 & -[v_4,v_9][v_9,v_5] & [v_4,v_8][v_8,v_5] & & & R_3 & & & \\ & 0_{18\times 24} & & & & & \bar{M}_{24}(a,b) & & \end{pmatrix}$$

with

$$R_2 := a_3 t_7 m_{27}^1 + b_3 t_7 m_{27}^2,\ R_3 := -a_4 t_9 m_{59}^1 - b_4 t_9 m_{59}^2,\ R_4 := a_3 t_8 m_{28}^1 + b_3 t_8 m_{28}^2 . \tag{62}$$

Let

$$R_5 := [v_3,v_7][v_7,v_2][v_9,v_1][v_6,v_1], \quad R_6 := [v_4,v_9][v_9,v_5][v_7,v_1][v_6,v_7] . \tag{63}$$

Column transformation carries one step further to

$$r = 16 + \operatorname{rank} \begin{pmatrix} R_5 & [v_3,v_8][v_8,v_2] & 0_{2\times 4} & R_2 & & 0_{2\times 4} & R_4 & 0_{2\times 4} \\ R_6 & [v_4,v_8][v_8,v_5] & & & R_3 & & & \\ & 0_{18\times 24} & & & & \bar{M}_{24}(a,b) & & \end{pmatrix}$$

since $R_5[v_4,v_8][v_8,v_5] = R_6[v_3,v_8][v_8,v_2]$. Thus

$$r = 17 + \operatorname{rank} \begin{pmatrix} R_7 \\ \bar{M}_{24}(a,b) \end{pmatrix} , \tag{64}$$

with

$$R_7 := (0, R_2 R_6, -R_3 R_5, 0, R_4 R_6, 0)_{1\times 24} . \tag{65}$$

Recall the form of $\bar{M}_{24}(a,b)$,

$$\bar{M}_{24}(a,b) = \begin{pmatrix} m_{11} & & & & & \\ & m_{77} & & & & \\ & & m_{99} & & & \\ m_{66} & m_{66} & m_{66} & & & \\ & m_{22} & & m_{22} & m_{22} & \\ & & & m_{33} & & \\ & & & & m_{88} & \\ & & & & & m_{44} \\ & & -m_{55} & & m_{55} & m_{55} \end{pmatrix} .$$

It is now easy to see that rank $\begin{pmatrix} m_{44} \\ m_{55} \end{pmatrix} = 4$ as well as rank $\begin{pmatrix} m_{77} \\ R_2 \end{pmatrix} = 3$. This proves that

$$\text{rank} \begin{pmatrix} \bar{M}_{24}(a,b) \\ R_7 \end{pmatrix} = 19 \ ,$$

and (64) completes the proof. ■

Let us summarise (55), (56), and (57).

Theorem 5.1. *Consider the general case, i.e., the lines T_iT_i', $i = 0,1,2$, do not meet at any vertex T_j'. Then* $\dim S_3^1(\Delta) = 16$.

If T_iT_i', $i = 0,1,2$, meet for example at vertex T_1', the points T_0, T_1', T_0' as well as T_2, T_1', T_2' are lying on a straight line. Thus $[v_1, v_9] = [v_6, v_7] = 0$. (63) implies then $R_5 = R_6 = 0$, hence from (65) $R_7 = 0$. Therefore

$$\text{rank} \begin{pmatrix} \bar{M}_{24}(a,b) \\ R_7 \end{pmatrix} = 18,$$

and $\dim S_3^1(\Delta) = 17$. Thus we have proved the following.

Theorem 5.2. $\dim S_3^1(\Delta) = 17$ *if and only if the segments T_iT_i', $i = 0,1,2$, meet at any vertex T_j', $j = 0,1,2$.*

More generally, the facts 5.1 and 5.2 extend to $n \geq 4$.

Theorem 5.3. *Let $n \geq 3$. Then $\frac{1}{2}n(7n-15)+7 \leq \dim S_n^1(\Delta) \leq \frac{1}{2}n(7n-15)+8$, and $\dim S_n^1(\Delta) = \frac{1}{2}n(7n-15)+8$ if and only if the segments T_iT_i', $i = 0,1,2$, meet at any vertex T_j', $j = 0,1,2$.*

Proof: The case $n = 3$ has already been covered. For general n it is easy to see that

$$\begin{aligned} \text{rank } M_n &\geq \text{rank } M_{11} + \text{rank } M_{22} + \\ &\quad + \text{rank} \begin{pmatrix} M_{23}(a,b) & M_{24}(a,b,c,d) \\ & M_{24}(a,b) \end{pmatrix} + 18(n-3) \\ &= 18(n-2) + \text{rank} \begin{pmatrix} M_{23} & M_{24}(a,b,c,d) \\ & M_{24}(a,b) \end{pmatrix} . \end{aligned}$$

In general case the rank of the last matrix is 36, rank $M_n \geq 18n$, and $\dim S_n^1(\Delta) \leq 7\binom{n+2}{2} - 18n = \frac{1}{2}n(7n-15)+7$. Otherwise rank $M_n \geq 18n-1$, $\dim S_n^1(\Delta) \leq \frac{1}{2}n(7n-15)+8$. Both values are thus upper bounds on the dimension of the spline space. But [7] shows that these values are also lower bounds for both particular cases. This confirms the theorem. ■

Acknowledgements. Y. Y. Feng would like to thank the hosts for the pleasant time spent at the Institute for Mathematics, Physics, and Mechanics at Ljubljana, Slovenija. He would also like to thank the organising committee of the workshop for the invitation and the support.

References

1. Chou, Y. S., L. Y. Su and R. H. Wang, The dimensions of bivariate splines over triangulations, ISNM **75** (1985), Birkhäuser-Verlag, Basel, 71–83.
2. Lai, M. J., A characterisation theorem of multivariate splines in the blossoming form, Computer-Aided Geom. Design **8** (1991), 513–521.
3. Morgan, J. and R. Scott, The dimension of piecewise polynomials, manuscript, 1977.
4. Ramshaw, L., Bézier and B-splines as multiaffine maps, in *Theoretical Foundations of Computer Graphics and CAD*, R. A. Earshaw (ed.), NATO ASI Ser. F **40** (1988), Springer-Verlag, Berlin, 757–776.
5. Ramshaw, L., Blossoms are polar forms, Computer-Aided Geom. Design **6** (1989), 323–358.
6. Schumaker, L. L., On the dimension of the space of piecewise polynomials in two variables, in *Multivariate Approximation Theory*, W. Schempp and K. Zeller (eds.), Birkhäuser-Verlag, Basel, 1979, 251–264.
7. Schumaker, L. L., Bounds on the dimension of spaces of multivariate piecewise polynomials, Rocky Mountain J. Math. **14** (1984), 251–265.
8. Shi, X. Q., The singularity of Morgan-Scott triangulation, Computer-Aided Geom. Design **8** (1991), 201–206.

Yu Yu Feng
Department of Mathematics
University of Science and Technology of China
230026 Hefei, Anhui (People's Republic of China)

Jernej Kozak
Department of Mathematics
University of Ljubljana
Jadranska 19
61000 Ljubljana (Slovenija)

Jernej.Kozak@FmF.Uni-Lj.Si

Ming Zhang
Department of Mathematics
University of Science and Technology of China
230026 Hefei, Anhui (People's Republic of China)

Thinning and Approximation of Large Sets of Scattered Data

Michael S. Floater and Armin Iske

Abstract. Having various concrete industrial applications in mind we focus on surface fitting to large scattered data sets. We describe a general method for modelling data which incorporates both filtering using triangulations, and hierarchical interpolation based on compactly supported radial basis functions. The uniformity of the data points plays a significant role. The utility of the method is confirmed by its adaptability to the specific requirements of several realistic applications.

§1. Introduction

This contribution has been inspired by various concrete industrial applications arising from different branches of science dealing with geophysical and meteorologic applications, surface and terrain modelling, oceanography etc. All such applications require data fitting to large scattered data sets. In particular, for $d \geq 1$ we are given a set $S \subset \mathbb{R}^{d+1}$ containing finitely many samples from the $(d+1)$-dimensional Euclidean space $\mathbb{R}^{d+1}$ and we make no assumptions on their spatial distribution. In order to match the expectations of those wishing to approximate the given data, we are supposed to find a model which best reflects their mental interpretation of it.

For the purpose of this paper we require that the given data set has been sampled from a hypersurface which can be mapped injectively onto a d-dimensional hyperplane H by a projection p. This is quite a realistic assumption for the applications that we have in mind. Without loss of generality and for simplicity of notation we shall consider the special case where

$$H = \text{span}\,\{e_1, \ldots, e_d\} \subset \mathbb{R}^{d+1}$$

Advanced Topics in Multivariate Approximation
F. Fontanella, K. Jetter and P. J. Laurent (eds.) pp. 87–96.

ISBN 981-02-2852-X

and p is given by

$$p(x) = (x^{(1)}, \ldots, x^{(d)})^T \text{ for all } x = (x^{(1)}, \ldots, x^{(d+1)})^T \in \mathbb{R}^{d+1}.$$

Here, $\{e_j\}_{j=1}^{d+1}$ denotes as usual the standard basis of $\mathbb{R}^{d+1}$, i.e. $e_j^T \cdot x = x^{(j)}$ for $x \in \mathbb{R}^{d+1}$ and $1 \leq j \leq d+1$. As a consequence, the given problem can be reduced to the approximation of a real-valued (unknown) function $f : \mathbb{R}^d \to \mathbb{R}$ which is sampled at $N \in \mathbb{N}$ pairwise distinct locations from a set $X = \{x_1, \ldots, x_N\} \subset \mathbb{R}^d$, so that

$$f_k = f(x_k) \in \mathbb{R}, \qquad 1 \leq k \leq N,$$

yields the data to be considered for the approximation and the convex hull $\Omega = \text{conv}(X) \subset \mathbb{R}^d$ fixes the domain of interest.

In this paper we shall, for the special case $d = 2$, study a selection of concrete applications from industry involving different types of scattered data. Section 2 will give a classification of these and provide further background information.

As is well known, scattered data typically exhibit a large difference between the two quantities

$$Q(X) = \max_{x \in \Omega} \min_{1 \leq k \leq N} \|x - x_k\|,$$

and

$$q(X) = \min_{1 \leq j < k \leq N} \|x_j - x_k\|/2.$$

Indeed it can often be observed that, due to the peculiarity of certain sampling techniques, in some regions some points are nearly coincident while in others there are large holes. In this case there is tremendous variation in the density (sparsity) of the data. It should be intuitively very clear that especially for a large number N, the stability of any approximation process will be reduced if the ratio

$$\rho(X) = q(X)/Q(X),$$

which we call the *uniformity* of X is extremely low. Note that $0 < \rho(X) \leq 1$ for if j, k satisfy $\|x_j - x_k\|/2 = q(X)$, then $Q(X) \geq \min_{1 \leq \ell \leq N} \|(x_j + x_k)/2 - x_\ell\| = q(X)$.

In Section 3 we will use triangulation methods in order to suggest preprocessing on the given input data which locates and removes sample points whose contribution we consider to be of least significance. Thus the modelling itself will then only operate on a selected subset $Y \subset X$. For this purpose we shall use a specific interpolation scheme on Y involving compactly supported radial basis functions.

Both the data filtering strategy and the interpolation method are part of our earlier paper [3]. These concepts are adapted here to the exceptional requirements of rather delicate concrete applications.

As already confirmed by observations from related papers (see e. g. [4] and references therein) it is a well-known fact that there is no method which turns

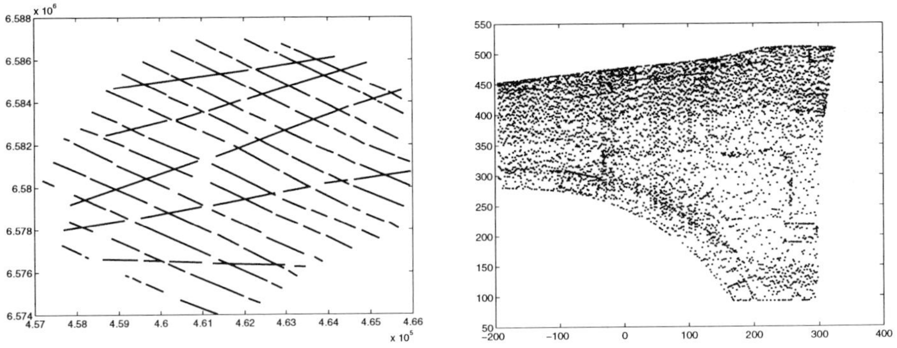

Fig. 1,2. Track data (4706 points) and feature data (5462 points).

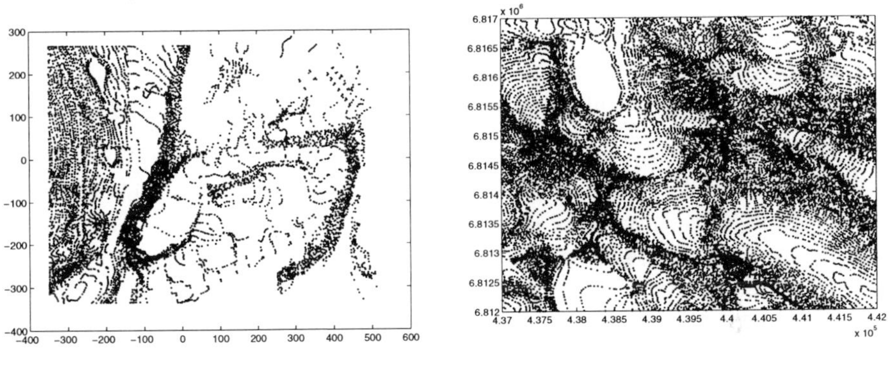

Fig. 3,4. Clustered data (7928 points) and contour data (23092 points).

out to possess best possible performance for all possible applications of this kind. Therefore we are far from tempted to try to improve on the whole range of available methods. Instead we would rather like to illustrate the flexibility and robustness of our method and moreover show the utility of compactly supported radial basis functions in this context. Numerical results are given in Section 4.

§2. Classification of Scattered Data and Applications

In this section we shall consider a specific collection of real world applications which served as motivation for this work. Figures 1-4 show different types of scattered data sets on which we shall report here.

It is rather a matter of taste whether we should consider these data sets as large. The largest sample among these contains 23092 data points and some people might argue that this is rather ridiculous compared to the amount they have in mind. Indeed, in industry there are applications which incorporate point samples many times larger than these [2] and we do not want to set an upper bound here. However, for the presentation of this paper we find it appropriate to work with this selection in order to demonstrate the efficacy of our method. Let us consider each of these as a model problem for possibly larger applications.

- **Track Data**

The sample to be seen in Figure 1 belongs to the broad class of geological data. In particular, here we have seismic lines whose spatial distribution is organized along several specific *tracks* since these data are gathered from ships. This sample set was taken in a North Sea region with the intention of recovering its sea-floor. To be more precise, the main concern of this application is to detect *fault lines* in the surface which represent discontinuities in geological layers caused by severe movements of the earth's crust. Their localization is of vital importance since they provide useful information for geologists about the occurrence of oil reservoirs and their correct volume (see [5]).

- **Feature Data**

Figure 2 shows a highly accurate laser sample of a car wing. The technical process behind this application, known as *reverse engineering*, requires the creation of a digital surface model from the measured data of a physical object, as far as possible automatically. This makes in particular the detection of possible *features* of the design object essential since these reflect its main characteristics. The expected output is a smooth as possible highly accurate approximation to the given data by a surface which, in view of the rendering process, needs to be easily evaluated.

- **Clustered Data**

In Figure 3 we have a sample of an area of terrain around Gjøvik near Lillehammer. The given data set is subject to significant variation of its density (sparsity). While the data is rather sparse in flat regions of the terrain, high density samples (*clusters*) roughly mimic its spatial structure. The sample was produced by using photogrammetry techniques and it was part of the preparation for the 1994 Olympic Winter Games. A model of the area was required for the underground construction of the ice-hockey arena.

- **Contour Data**

The sample to be seen in Figure 4 was taken via satellite. It shows a discrete sample of the *contour* lines of the Hurrungane mountain area in the central part of South Norway.

§3. Data Modelling

In this section we focus on the modelling of the given input data. Bearing in mind the possible types of data of the previous section, it is quite obvious that a successful working method is required to be equipped with enough flexibility for the purpose of its adaptation to the data. In particular, a competitive method needs to possess reasonable performance with respect to (at least) all of the following four properties:

- stability of the calculation process,
- good reproduction quality,
- low computational costs,
- fast evaluation of the solution model.

We cannot optimize all of these four requirements simultaneously. Instead we have to balance them with respect to priorities given by specific expectations of each underlying application.

For this purpose, in this section we describe a method which is designed for scattered data approximation to general large data sets. The method is based on techniques presented in an earlier paper [3], and we adapt it here to the kind of industrial applications we have in mind. In order to keep this presentation self-contained we recall the basic ingredients of the method in [3]. It can roughly be divided into the following two steps.

- data filtering,
- interpolation.

The first one will output a selected subset $Y = \{y_1, \ldots, y_{N_M}\} \subset X$ of the given locations while the latter one will use a hierarchical scheme for the interpolation at locations from Y which we shall briefly describe here:

Having specified a nested sequence of subsets

$$Y_{N_1} \subset Y_{N_2} \subset \cdots \subset Y_{N_M} =: Y \subset X$$

for $k = 1, \ldots, M$, at level k we will use interpolants of the form

$$s_k = \sum_{j=1}^{N_k} c_j^{(k)} \phi_{\alpha_k}(\| \cdot - y_j \|)$$

for the interpolation on Y_{N_k} subject to

$$s_k|_{Y_{N_k}} = f - (s_1 + \ldots + s_{k-1})|_{Y_{N_k}}$$

in order that $s := s_1 + \ldots + s_M$ matches f on Y.

All interpolants are generated by scaled versions of a specific compactly supported radial function $\phi : [0, \infty) \to \mathbb{R}$. To be precise, for a positive real number α_k the k-th interpolant s_k is a linear combination of shifts of the scaled version $\phi_{\alpha_k}(\cdot) := \phi(\cdot / \alpha_k)$.

For the well-posedness of each of the interpolation problems, ϕ is required to be positive definite on $\mathbb{R}^2$:

Definition 3.1. *A continuous radial function $\phi : [0,\infty) \to \mathbb{R}$ is said to be positive definite on $\mathbb{R}^d$, iff the matrix*

$$A = (\phi(\|x_j - x_k\|))_{1\le j,k\le N}$$

is positive definite for all possible $N \in \mathbb{N}$ and all possible choices of sets $X = \{x_1, \ldots, x_N\} \subset \mathbb{R}^d$ containing N pairwise distinct points from the d-dimensional Euclidean space $\mathbb{R}^d$.

Note that the use of compactly supported positive definite radial functions will generate collocation matrices which are not only symmetric and positive definite but also sparse. This allows us to use preferably the conjugate gradient method for the inversion of $A_{\phi,Y_{N_k}}$ at each level k at low computational costs. Moreover due to their local representation, function values of each of the s_k can quickly be accessed. For a selection of suitable compactly supported positive definite radial functions and their construction we recommend the paper [7] by Schaback.

As already observed in [3], when interpolating on the set X or any subset of it, the best possible situation for balancing the stability and approximation behaviour is when the uniformity of the set is maximized. Indeed, if X is the set to be interpolated on, and we combine for some positive real numbers δ and η the two requirements $q(X) \ge \delta$ and $Q(X) \le \eta$, then because $\rho(X) \ge \delta/\eta$, the uniformity of X gives an upper bound for the ratio between δ and η. Since δ is in view of the stability desired to be as high as possible, while η is required to be sufficiently low for the reproduction, $\rho(X)$ is consequently desired to be maximized. If we ignore the effect of points in X near the boundary on the value of $\rho(X)$, we will find that X is optimal, in this sense, when its Delaunay triangulation consists purely of equilateral triangles. We refer to the text book [6] and the article [8] (and references therein) for details on Delaunay triangulations.

Therefore, for simplicity, suppose for the moment that $\Omega = \mathbb{R}^2$ and that $X = \{x_1, x_2, x_3, \ldots\}$ is an infinite sequence of pairwise distinct points in $\mathbb{R}^2$ satisfying

$$Q(X) = \sup_{x\in\mathbf{R}^2} \inf_{k\ge1} \|x - x_k\| < \infty, \qquad q(X) = \inf_{1\le j<k} \|x_j - x_k\|/2 > 0 \,. \tag{1}$$

One quickly deduces from (1) that any ball in $\mathbb{R}^2$ of greater radius than $Q(X)$ contains at least one point of X and that no ball in $\mathbb{R}^2$ contains an infinite number of points of X. Now let $\mathcal{D}(X) = \{T_i\}_i$ be the (infinite) Delaunay triangulation of X, whose edges are denoted by $\{E_j\}_j$. Then, from standard facts concerning Delaunay triangulations [6,8], we observe that

$$Q(X) = \sup_i r(T_i) \,, \qquad q(X) = \inf_j |E_j|/2 \,,$$

where $r(T_i)$ denotes the radius of the circumcircle of triangle T_i and $|E_j|$ is the length of E_j.

Theorem 3.1. *Let $X = \{x_1, x_2, x_3, \ldots\}$ be a sequence of pairwise distinct points in $\mathbb{R}^2$, satisfying (1). Then the uniformity of X, defined by $\rho(X) = q(X)/Q(X)$, is bounded above by*

$$\rho(X) \le \sqrt{3}/2 \ ,$$

where equality holds if and only if every triangle in $\mathcal{D}(X)$ is equilateral.

Proof: Let E_{i_1}, E_{i_2}, E_{i_3}, be the edges belonging to T_i and define

$$\rho(T_i) = \frac{\min\{|E_{i_1}|, |E_{i_2}|, |E_{i_3}|\}}{2r(T_i)} \ .$$

Then from the definition of $\rho(X)$, we find that

$$\rho(X) = \frac{\inf_j |E_j|}{2 \sup_i r(T_i)} = \inf_i \frac{\inf_j |E_j|}{2r(T_i)} \le \inf_i \rho(T_i) \ .$$

Now since $\rho(T_i)$ is invariant under uniform scalings, we may assume that $r(T_i) = 1$. Then we see that $\rho(T_i)$ is maximized when the minimum of the lengths of T_i's three edges is maximized and, resorting to trigonometry, this clearly occurs iff all three lengths are $\sqrt{3}$. Thus $\rho(T_i) \le \sqrt{3}/2$ for all i and so $\rho(X) \le \sqrt{3}/2$ as claimed. Moreover if $\rho(X) = \sqrt{3}/2$, then $\rho(T_i) = \sqrt{3}/2$ for all i and so all triangles are equilateral. ■

For the remainder of this section we shall concentrate on the aforementioned data filtering strategy to be applied on a given finite set X. The method which we refer to as the *thinning algorithm* can be found below. It is based on successive knot removal from X for the purpose of increasing and maintaining the uniformity.

Algorithm 3.1. *(Thinning Algorithm)*

Input: A set X of N scattered centres and a fixed $K < N$.
(1) Let $Y_N = X$ and compute $\mathcal{D}(Y_N)$.
(2) For decreasing $i = N, \ldots, K+1$
  (a) determine the $x \in Y_i$ where the density is highest;
  (b) let $y_i = x$ and $Y_{i-1} = Y_i \setminus \{x\}$;
  (c) compute $\mathcal{D}(Y_{i-1})$.
(3) Let $(y_1, \ldots, y_K)$ be any ordering of Y_K.
Output: An ordering $y_1, \ldots, y_N$ of X.

For step (2a) it remains to define a measure of density for choosing a point x to remove from Y_i. In [3], we chose x to be an endpoint of any shortest edge in $\mathcal{D}(Y_i)$. The choice of which of these endpoints to remove was based on minimizing the longest of their incident edges.

Note that the above algorithm is very stable and easy to implement provided that corresponding code for triangulation methods is available.

The construction of the initial Delaunay triangulation requires asymptotically $\mathcal{O}(N \log N)$ operations and in our current implementation the thinning algorithm requires $\mathcal{O}(N^2)$. However if one used techniques from [1], the maintenance of the location of the closest pair would require only $\mathcal{O}(\log N)$ operations at each step. In that case, the location of the removable point would require only $\mathcal{O}(\log N)$ operations and the thinning algorithm in total only $\mathcal{O}(N \log N)$.

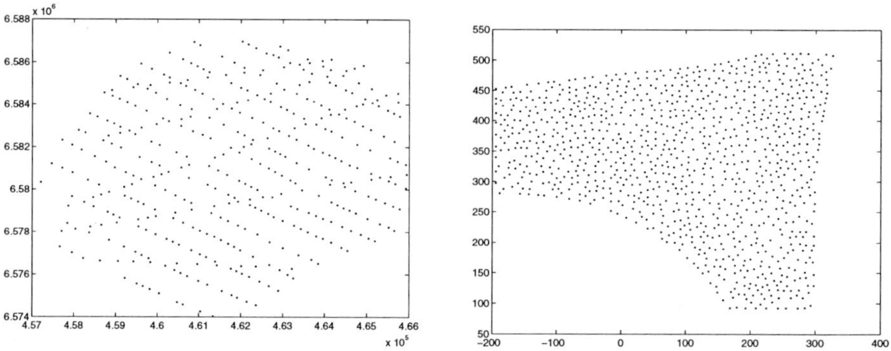

Fig. 5,6. Subset of track data (413 points) and feature data (1000 points).

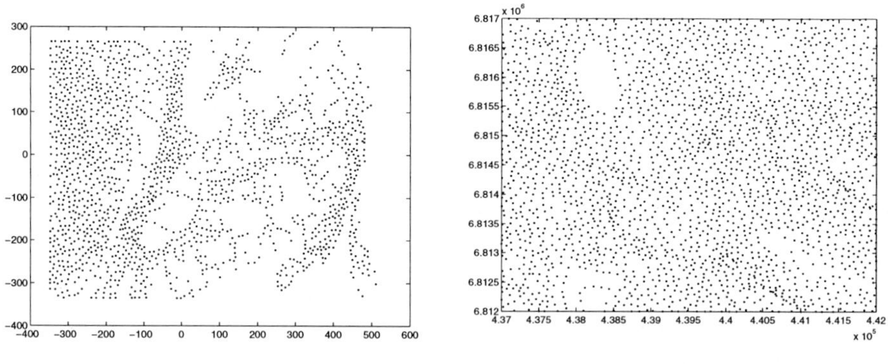

Fig. 7,8. Subset of clustered data (1690 points) and contour data (2352 points).

§4. Numerical Results

In this section we shall provide a small excerpt from our numerical results. In Figures 5-8 we find the subsets Y for each of the data sets from Section 2 after their

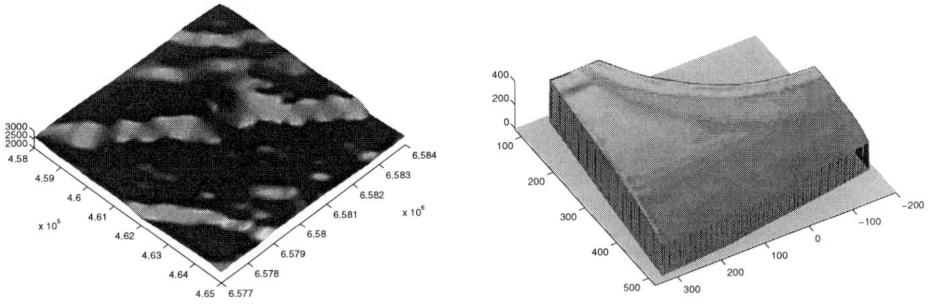

Fig. 9,10. Recovery of the North Sea floor, rendering of the car wing.

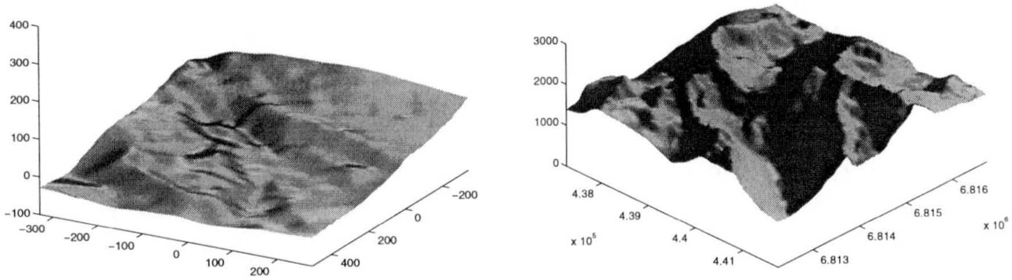

Fig. 11,12. Display of the terrains around Gjøvik and Hurrungane.

filtering. For the interpolation we used $\phi(r) = (1-r)_+^4(4r+1)$ (see [9]) throughout all of our examples. The rendering of the data models can be found in Figures 9-12.

Since the track data itself doesn't contain any finer details of the sea-floor we decided to use one step for the interpolation at Y_{413} where $\alpha_1 = 5000$ is about ten times $Q(Y_{413})$. In contrast, both the clustered and contour data sets incorporate finer details according to the topographical structures of the terrains. For this reason we decided to work with hierarchical interpolation at $M = 3$ steps. The breakpoints N_k were chosen as large as possible such that $\rho(Y_{N_k}) \geq 2^{-(k+1)}$ holds and we set $\alpha_k = 50\, q(Y_{N_k})$. Finally, as to the feature data we note that this application requires first and foremost a highly accurate reproduction of the surface. For these data types we recommend using one step with a large scale for the interpolation on a uniform as possible large subset Y. We chose $N_1 = 1000$ and $\alpha_1 = 200 \approx 50\, q(Y_{1000})$.

Acknowledgements. This work was supported by the European Union projects CHRX-CT94-0522 and ERB CHBG CT 93 0408. We are indebted to several people at SINTEF who supported us during the compilation of this paper. The track data set was provided by Geomatic AS (Oslo) and the contour data by the Norwegian Mapping Authority.

References

1. Bespamyatnikh, S. N., An optimal algorithm for closest pair maintenance, in *Eleventh Annual Symposium on Computational Geometry*, ACM Press, New York, 1995, 152–161.
2. Dæhlen, M. and A. Tveito, Numerical methods and software tools in industrial mathematics, to appear.
3. Floater, M. S. and A. Iske, Multistep scattered data interpolation using compactly supported radial basis functions, to appear in J. Comp. Appl. Math.
4. Franke, R. and G. M. Nielson, Scattered data interpolation and applications: a tutorial and survey, in *Geometric Modeling: Methods and Applications*, H. Hagen and D. Roller (eds.), Springer-Verlag, Berlin, 1991, 131–160.
5. Fremming, N. P., Ø. Hjelle and C. Tarrou, Surface modelling from scattered geological data, in *Numerical Methods and Software Tools in Industrial Mathematics*, M. Dæhlen and A. Tveito (eds.), to appear.
6. Preparata, F. P. and M. I. Shamos, *Computational Geometry*, Springer-Verlag, New York, 1985.
7. Schaback, R., Creating surfaces from scattered data using radial basis functions, in *Mathematical Methods in Computer Aided Geometric Design III*, M. Dæhlen, T. Lyche, and L. L. Schumaker (eds.), Vanderbilt University Press, Nashville, 1995, 477–496.
8. Schumaker, L. L., Triangulation methods, in *Topics in Multivariate Approximation*, C. K. Chui, L. L. Schumaker and F. I. Utreras (eds.), Academic Press, New York, 1987, 219–232.
9. Wendland, H., Piecewise polynomial, positive definite and compactly supported radial functions of minimal degree, in Adv. Comput. Math. **4** (1995), 389-396.

Michael S. Floater
Departamento de Matemática Aplicada
Universidad de Zaragoza
E-50009 Zaragoza
Spain
floater@cc.unizar.es

Armin Iske
SINTEF Anvendt matematikk
Postboks 124, Blindern
N-0314 Oslo
Norway
Armin.Iske@si.sintef.no

Numerical Curves and their Applications to Algebraic Curves

H. Gevorgian, H. Hakopian and A. Sahakian

Abstract. In this paper, Hermite interpolation by bivariate algebraic polynomials and its applications to some problems of the theory of algebraic curves, such as the existence of algebraic curves with given singularities, is considered. The scheme $\mathcal{N} = \{n_1, \ldots, n_s; n\}$ - sequence of multiplicities of nodes associated with the degree of interpolating polynomials is considered. We continue the investigation of canonical decomposition of schemes and define so called maximal schemes. Some numerical results concerning the factorization of schemes are established. This leads to determination of irreducibility or to finding the (exact) number of components of algebraic curves as well as to the characterization of all singular points of a wide family of algebraic curves. Also, the Hilbert function of schemes is discussed. At the end, the problem of regularity of schemes depending on the number of interpolation conditions is considered. The proofs and other details can be found in [6].

§1. Introduction

We define a *scheme* $\mathcal{N} = \{n_1, \ldots, n_s; n\}$ as a collection of nonnegative integers, where $n_1, \ldots, n_s$ are the *members*, n is the *degree* and s is the *length* of $\mathcal{N}$. By $\mathcal{S}$ we denote the set of all schemes. We accept that

$$\{n_1, \ldots, n_s; n\} = \{n_1, \ldots, n_s, 0, \ldots, 0; n\}$$

with arbitrary (finite) number of zeros. So dealing with finite number of schemes from S, we may assume that they have the same length or, when it is necessary, that the length of the given scheme is large enough.

We need some notation. For $\mathcal{N} = \{n_1, \ldots, n_s; n\}$, $\mathcal{M} = \{m_1, \ldots, m_s; m\} \in S$ and $\lambda \in \mathbb{Z}_+$ we denote

$$\mathcal{N} + \mathcal{M} := \{n_1 + m_1, \ldots, n_s + m_s; n + m\}, \quad \lambda\mathcal{N} := \{\lambda n_1, \ldots, \lambda n_s; \lambda n\},$$

Advanced Topics in Multivariate Approximation
F. Fontanella, K. Jetter and P. J. Laurent (eds.) pp. 97–111.

ISBN 981-02-2852-X

$$\mathcal{N}' = \{n_1, \ldots, n_s\}, \quad \text{supp}\,\mathcal{N} := \{\nu : n_\nu \neq 0 \quad 1 \leq \nu \leq s\}.$$

The inequality $\mathcal{N} \leq \mathcal{M}$ means that $n \leq m$ and $n_\nu \leq m_\nu$, $\nu = 1, \ldots, s$.

We call $\mathcal{N} \in \mathcal{S}$ an *interpolation* scheme if the following equality holds:

$$\sum_{\nu=1}^{s} \overline{n_\nu} = \overline{n+1}, \tag{1.1}$$

where $\overline{m} = 0 + \cdots + m$.

By $\mathcal{IS}$ we denote the set of all interpolation schemes.

Interpolation schemes with $s = 1$, i.e., the schemes $\{n+1; n\}$, are called *Taylor* schemes. Let $\pi_n(\mathbb{R}^2)$ be the space of bivariate polynomials of total degree $\leq n$.

Definition 1.1. *For the interpolation scheme* $\mathcal{N} = \{n_1, \ldots, n_s; n\} \in \mathcal{IS}$ and *the node set* $\mathcal{Z} = \{z_\nu = (x_\nu, y_\nu)\}_{\nu=1}^{s} \subset \mathbb{R}^2$ *the (regular) Hermite interpolation problem* $(\mathcal{N}, \mathcal{Z})$ *is to find a unique polynomial* $P \in \pi_n(\mathbb{R}^2)$ *satisfying conditions*

$$\frac{\partial^{i+j} P}{\partial x^i \partial y^j}\Big|_{z_\nu} = \lambda_{i,j,\nu}, \quad i + j < n_\nu, \ \nu = 1, \ldots, s, \tag{1.2}$$

for given collection of values

$$\Lambda = \{\lambda_{i,j,\nu}, \quad i + j < n_\nu, \ \nu = 1, \ldots, s\}.$$

In what follows, we briefly express equalities of the form (1.2) by writing:

$$D^{\mathcal{N}'} P\big|_{\mathcal{Z}} = \Lambda.$$

We assume that there is no interpolation condition at nodes z_ν with $n_\nu = 0$.

Remark 1.2. *The following statements are equivalent for any* $\mathcal{N} \in \mathcal{IS}$*:*

i) $(\mathcal{N}, \mathcal{Z})$ *is singular, i.e. not regular;*

ii) there exists a polynomial P *such that*

$$P \in \pi_n(\mathbb{R}^2), \quad P \neq 0, \quad D^{\mathcal{N}'} P\big|_{\mathcal{Z}} = 0. \tag{1.3}$$

Since the determinant of the system of linear equations (1.2) (with respect to unknown coefficients of P) is a polynomial in variables $x_1, y_1, \ldots, x_s, y_s$, the regularity of the problem $(\mathcal{N}, \mathcal{Z})$ for some $\mathcal{Z}$ implies that it is regular for almost all $\mathcal{Z} \in \mathbb{R}^{2s}$ (with respect to the Lebesgue measure in $\mathbb{R}^{2s}$).

Remark 1.3. *For any $\mathcal{N} = \{n_1, \ldots, n_s; n\} \in \mathcal{S}$, the following statements are equivalent:*

i) $n_\nu \leq n$ for $\nu = 1, \ldots, s$;

ii) there exists a node set $\mathcal{Z} \in \mathbb{R}^{2s}$ and a polynomial P such that (1.3) holds.

This, in particular, implies that the Taylor schemes are the only interpolation schemes which are regular for an arbitrary node set (see [8] and [9] for more general results).

In view of Remark 1.2 the regularity and singularity of $(\mathcal{N}, \mathcal{Z})$ can be defined in the following more general way which enables us to remove the restriction that the scheme $\mathcal{N}$ is an interpolation scheme:

We call the problem $(\mathcal{N}, \mathcal{Z})$ singular, if there exists a polynomial P satisfying (1.3), otherwise it is called regular.

Geometrically, the singularity of $(\mathcal{N}, \mathcal{Z})$ means that there exists an algebraic curve of degree $\leq n$, passing through $\mathcal{Z}$ with multiplicity $\mathcal{N}'$ (*i.e.*, passing through z_ν with multiplicity $n_\nu, \nu = 1, \ldots, s$).

In what follows we will consider the singularity and regularity in this wider sense. Note that in the case of interpolation schemes the given two definitions of regularity and singularity coincide.

Definition 1.4. *We say that the scheme $\mathcal{N}$ is*

i) regular, if $(\mathcal{N}, \mathcal{Z})$ is regular for some node set $\mathcal{Z}$,

ii) singular if $(\mathcal{N}, \mathcal{Z})$ is singular for any node set $\mathcal{Z}$.

The problem of the full description of regular and singular schemes remains still open.

Let us define the following "*less conditions*" class of schemes:

$$LC := \{\mathcal{M} = \{m_1, \ldots, m_s; m\} \in \mathcal{S} : \sum_{\nu=1}^{s} \overline{m}_\nu < \overline{m+1}\}.$$

Definition 1.5. *The scheme $\mathcal{N}$ is called a numerical curve if there is a set $M \subset LC$ such that*

$$\mathcal{N} = \sum_{\mathcal{M} \in M} \mathcal{M}. \tag{1.4}$$

We denote the set of numerical curves by NC.

It is not hard to see that numerical curves are singular schemes.

Conjecture 1.6. ([1], [10]) *Each singular scheme is a numerical curve.*

We have proved in [3] that this conjecture is true under the restriction that there are at most 9 knots whose multiplicities are ≥ 2.

Definition 1.7. *Let* $\mathcal{N} = \{n_1, \ldots, n_s; n\} \in \mathcal{S}$.

i) If $n_1 + n_2 \geq n + 1$, $n_1 \leq n$, $n_2 \leq n$, *then the reduction of* $\mathcal{N}$ *with respect to the first two members is the scheme*

$$\mathcal{N}^\times = \mathcal{N}_{1,2}^\times = \{n - n_2, n - n_1, n_3, \ldots, n_s; 2n - n_1 - n_2\}.$$

ii) If

$$n_i + n_j \leq n, \quad 1 \leq i < j \leq 3, \tag{1.5}$$

then the quadratic transformation of the scheme $\mathcal{N}$ *with respect to the first three members is the scheme (cf. [11], chapter 3, Theorem 7.2)*

$$\mathcal{N}^* = \mathcal{N}_{1,2,3}^* = \{n - n_2 - n_3, n - n_1 - n_3, n - n_1 - n_2, n_4 \ldots, n_s; 2n - n_1 - n_2 - n_3\}.$$

It is not hard to check, that

$$\mathcal{N}^\times = \mathcal{N}_{1,2}^\times = \{n_1 - r, n_2 - r, n_3, \ldots, n_s; n - r\}.$$

$$\mathcal{N}^* = \mathcal{N}_{1,2,3}^* = \{n_1 - t, n_2 - t, n_3 - t, n_4, \ldots, n_s; n - t\},$$

with $r = n_1 + n_2 - n$, $t = n_1 + n_2 + n_3 - n$.

We define reduction or quadratic transformation with respect to the other members in the similar way. We denote

$$\text{supp } Q = \{i, j, k\},$$

if Q is a quadratic transformation with respect to the members in the places i, j, k.

We say that the quadratic transformation Q is *positive* (for $\mathcal{N}$), if $\deg Q(\mathcal{N}) < \deg \mathcal{N}$, *i.e.*, $t > 0$. Note that the positivity of Q (for $\mathcal{N}$) implies that

$$\text{supp } Q \subset \text{supp } \mathcal{N}. \tag{1.6}$$

We have the following

Theorem 1.8. (see [2]) *Let* $\mathcal{N} = \{n_1, \ldots, n_s; n\} \in \mathcal{S}$. *Then*

i) if $n_1 + n_2 = n + r \geq n + 1$, $n_1 \leq n$, $n_2 \leq n$, *then the schemes* $\mathcal{N}$ *and* $\mathcal{N}^\times$ *are both simultaneously singular or not;*

ii) if $n_i + n_j \leq n$, $1 \leq i < j \leq 3$, *then the schemes* $\mathcal{N}$ *and* $\mathcal{N}^*$ *are both simultaneously singular or not.*

This theorem reduces the investigation of an arbitrary scheme to the following two cases:

1) $n_{\nu_0} \geq n + 1$ for some ν_0,

2) $n_i + n_j + n_k \leq n$, $1 \leq i < j < k \leq s$.

In the first case the scheme obviously is regular (Remark 1.3).

Definition 1.9. *A scheme* $\mathcal{N} = \{n_1, \ldots, n_s; n\} \in \mathcal{S}$ *satisfying the second condition above, is called basic.*

The set of basic schemes is denoted by BS.

Conjecture 1.10. [1] *A basic scheme is singular if and only if it belongs to LC.*

We have proved that Conjecture 1.10 is true in the case

$$\sum_{\nu:n_\nu>1} n_\nu \leq 3n \tag{1.7}$$

and the above two conjectures are equivalent (see [3], and [5] for the wider concept of regularity).

Let us define the *intersection product* of schemes as follows:

$$\langle \mathcal{N}, \mathcal{M} \rangle := \sum_{\nu=1}^{s} n_\nu m_\nu - nm.$$

The schemes $\mathcal{N}$ and $\mathcal{M}$ are called *orthogonal* ($\mathcal{N} \perp \mathcal{M}$), if $\langle \mathcal{N}, \mathcal{M} \rangle = 0$. We say that the set of schemes $\mathcal{A}$ is *orthogonal*, if $\langle \mathcal{N}, \mathcal{M} \rangle = 0$, for any distinct $\mathcal{N}, \mathcal{M} \in \mathcal{A}$.

Denote by $\overline{\mathcal{N}}$ the difference of the number of conditions of $\mathcal{N}$ and the dimension of the space of algebraic curves of degree n, *i.e.*, $(\dim \pi_n(\mathrm{I\!R}^2) - 1)$:

$$\overline{\mathcal{N}} := \sum_{\nu=1}^{s} \overline{n}_\nu - \overline{n+1} + 1.$$

Thus, $\overline{\mathcal{N}} \leq 0$ means that $\mathcal{N} \in LC$ and $\overline{\mathcal{N}} = 1$, if $\mathcal{N}$ is an interpolation scheme.

Definition 1.11.

i) The schemes $\mathcal{N}, \mathcal{M}$ are called quadratically equivalent ($\mathcal{N} \sim \mathcal{M}$), if one of them can be obtained from the other by means of quadratic transformations.

ii) We say that the scheme $\mathcal{N}$ reduces to the scheme $\mathcal{M}$ ($\mathcal{N} \to \mathcal{M}$), if $\mathcal{M}$ can be obtained from $\mathcal{N}$ by means of reduction and (or) quadratic transformations.

It follows from Theorem 1.8 that

Remark 1.12. *If the scheme $\mathcal{N}$ reduces to the scheme $\mathcal{M}$, then they are simultaneously regular or not.*

Let us denote by $E_{p,q}$ the scheme $\{e_1, \ldots, e_s; e\}$ with

$$e_p = e_q = e = 1, \quad e_\nu = 0 \; for \;\; \nu = 1, \ldots, s, \;\; \nu \neq p, q.$$

The following definition plays an essential role in our investigation:
The scheme which is quadratically equivalent to some $E_{p,q}$ is called *prime numerical curve*. We denote the class of prime numerical curves by PNC:

$$PNC := \{\mathcal{N} \in \mathcal{S} \; : \; \mathcal{N} \sim E_{p,q} \; for \; some \; p, q\}.$$

Definition 1.13.

i) A scheme which is quadratically equivalent to some basic scheme is called an e-basic scheme.

ii) A scheme which can be reduced to some basic scheme is called an r-basic scheme.

We denote by BS^* and $BS^{\times *}$ the classes of e-basic and r-basic schemes respectively. Of course we have

$$BS \subset BS^* \subset BS^{\times *}.$$

Note that Theorem 1.8 and Remark 1.3. imply

Remark 1.14. *If $\mathcal{N}$ is singular then $\mathcal{N} \in BS^{\times *}$.*

The next theorem establishes the canonical decomposition of r-basic schemes (see [5]):

Theorem 1.15. *Let $\mathcal{N} \in BS^{\times *}$. Then there exists a finite set of prime numerical curves $\Omega_{\mathcal{N}}$, an e-basic scheme $\mathcal{N}^{\downarrow}$ and natural numbers $\mu_A = \mu_{A,\mathcal{N}}$, such that*

$$\mathcal{N} = \sum_{A \in \Omega_{\mathcal{N}}} \mu_A A + \mathcal{N}^{\downarrow}, \quad \{\mathcal{N}^{\downarrow} \cup \Omega_{\mathcal{N}}\} \text{ is orthogonal.} \tag{1.8}$$

Moreover, $\mathcal{N} \to \mathcal{N}^{\downarrow}$ and the decomposition (1.8) is unique.

Note that a prime numerical curve A belongs to $\Omega_{\mathcal{N}}$ in the canonical decomposition (1.8) if and only if $\langle A, \mathcal{N} \rangle > 0$, moreover, we have $\mu_A = \langle A, \mathcal{N} \rangle$.

We have also that

$$\mathcal{N}^{\downarrow} \in LC \text{ and } \langle \mathcal{N}^{\downarrow} \rangle \leq 0, \quad \text{if } \mathcal{N} \in NC. \tag{1.9}$$

§2. Some Numerical Results

We start with a result on factorization of r-basic schemes.

Theorem 2.1. *Suppose the scheme $\mathcal{N} \in BS^*$ with $\overline{\mathcal{N}} = 0$ has a representation*

$$\mathcal{N} = \mathcal{N}_1 + \cdots + \mathcal{N}_q, \quad \mathcal{N}_i \in NC, \quad q \geq 2. \tag{2.1}$$

Then

$$\mathcal{N} \sim \lambda C_0, \tag{2.2}$$

where C_0 is a scheme of degree 3 having nine nonzero members - each equal to 1 and λ is the greatest common divisor of members and degree of $\mathcal{N}$.
Moreover, we have

$$\mathcal{N}_i \sim \lambda_i C_0, \qquad \sum_{i=1}^{q} \lambda_i = \lambda. \tag{2.3}$$

From now on we will use also the canonical decomposition of r-basic scheme $\mathcal{N}$ with slightly changed last summand:

$$\mathcal{N} = \sum_{A \in \Omega_{\mathcal{N}}} \mu_A A + \mu^{\Downarrow} \mathcal{N}^{\Downarrow}, \tag{2.4}$$

where $\mu^{\Downarrow} = \lambda$, $\mathcal{N}^{\Downarrow} = \frac{1}{\lambda}\mathcal{N}^{\downarrow}$, if $\mathcal{N}^{\downarrow} \sim \lambda C_0$ and $\mu^{\Downarrow} = 1$, $\mathcal{N}^{\Downarrow} = \mathcal{N}^{\downarrow}$ otherwise.

The following theorem shows that the canonical decomposition of numerical curve in certain sense is the unique representation as a sum of numerical curves, provided that $\overline{\mathcal{N}^{\downarrow}} = 0$.

Theorem 2.2. *Suppose the scheme $\mathcal{N} \in BS^{\times *}$ with $\overline{\mathcal{N}^{\downarrow}} = 0$ has a representation*

$$\mathcal{N} = \mathcal{N}_1 + \cdots + \mathcal{N}_q, \quad \mathcal{N}_i \in NC, \quad q \geq 2.$$

Then $\mathcal{N}$ is divided by each $\mathcal{N}_i$ i.e.

$$\mathcal{N}_i = \sum_{A \in \Omega_{\mathcal{N}}} \mu_{i,A} A + \mu_i^{\Downarrow} \mathcal{N}^{\Downarrow},$$

with nonnegative coefficients satisfying

$$\sum_{i=1}^{q} \mu_{i,A} = \mu_A, \quad \sum_{i=1}^{q} \mu_i^{\Downarrow} = \mu^{\Downarrow}.$$

The following theorem announced in [7] establishes the interesting fact, that for any $\mathcal{N} \in BS^*$ there is only one (up to the ordering of the members) basic scheme denoted by $\mathcal{N}^b = \{n_1^b, \ldots, n_s^b; n^b\}$, such that $\mathcal{N} \sim \mathcal{N}^b$.

Theorem 2.3. *If the basic schemes $\mathcal{N}$ and $\mathcal{M}$ are quadratically equivalent, then they have the same degrees and the same members, maybe in different order.*

Let us denote by

E - the scheme with zero members and with degree one;

χ_p - the scheme with zero degree and zero members except for p-th member which is one;

$E_p := \chi_p + E$.

In the next theorems we consider the problem of extension of the set $\Omega_{\mathcal{N}}$ in the canonical decomposition and its maximal possible cardinality.

Theorem 2.4. *For any scheme $\mathcal{N} = \{n_1, \dots, n_s; n\}$ with $\mathcal{N} \sim \lambda E_q$ there exist exactly $s - \#\text{supp}\, \mathcal{N}$ prime numerical curves A satisfying*

$$\langle A, \mathcal{N} \rangle = 0, \quad \text{supp}\, A \subset [1, s], \quad \text{supp}\, A \setminus \text{supp}\, \mathcal{N} \neq \emptyset$$

Moreover, these schemes are orthogonal to each other and are of the following form:

$$A = \frac{1}{\lambda}\mathcal{N} + \chi_p, \quad p \in \{1, \dots, s\} \setminus \text{supp}\, \mathcal{N}. \tag{2.5}$$

(The orthogonality of the set (2.5) follows from the equality $\langle \mathcal{N} \rangle = \lambda \langle E_q \rangle = 0$.)

Theorem 2.5. *Let $\{\mathcal{N}\} \cup \mathcal{A}$ be an orthogonal set of schemes with $\mathcal{N} \in BS^*$, $\mathcal{A} \subset PNC$ and*

$$\text{supp}\, \mathcal{A} \subset \text{supp}\, \mathcal{N}, \quad if\ \mathcal{N}^b = \lambda E_q. \tag{2.6}$$

Then

$$\#\mathcal{A} \leq d(\mathcal{N}) := \begin{cases} \#\text{supp}\, \mathcal{N} - 1, & \text{if } \mathcal{N} \sim \{m_1, m_2; m_1 + m_2\}, m_1 m_2 \neq 0; \\ \#\text{supp}\, \mathcal{N} - \#\text{supp}\, \mathcal{N}^b, & \text{otherwise.} \end{cases} \tag{2.7}$$

Moreover, there exists $\mathcal{A}' \subset PNC$ with (2.6) such that $\mathcal{A} \subset \mathcal{A}'$, $\{\mathcal{N}\} \cup \mathcal{A}'$ is orthogonal and $\#\mathcal{A}' = d(\mathcal{N})$.

We denote

$$\text{supp}\, \mathcal{A} := \cup_{A \in \mathcal{A}} \text{supp}\, A,$$

where $\mathcal{A}$ is a set of schemes.

Theorem 2.6. *Let $\mathcal{A}$ be an orthogonal set of prime numerical curves such that* $\text{supp}\, \mathcal{A} \subset U \subset \mathbb{Z}_+$ *and*

$$d_U(\mathcal{A}) := sup\{\#\mathcal{A}' \ :\ \mathcal{A} \subset \mathcal{A}',\ \text{supp}\, \mathcal{A}' \subset U,\ \mathcal{A}' \text{ is orthogonal}\}.$$

Then either

$$d_U(\mathcal{A}) = \#U - 1 \quad or \quad d_U(\mathcal{A}) = \#U.$$

Definition 2.7. *The scheme $\mathcal{N}$ with canonical decomposition (2.4) is called maximal, if $\mathcal{N}^{\downarrow} \not\sim \lambda E_q$ and $\#\Omega_{\mathcal{N}} = d_U(\mathcal{N})$.*

Using Theorem 2.5 and following the proof of the analog of Bezout's theorem for numerical curves ([5], Theorem 5.7) we can improve it as follows:

Theorem 2.8. *Let $\mathcal{N}, \mathcal{M}$ be numerical curves.*

i) If $\langle \mathcal{N}, \mathcal{M} \rangle > 0$, then $\mathcal{N}$ and $\mathcal{M}$ have a common prime numerical curve in their canonical decompositions.

ii) If $\langle \mathcal{N}, \mathcal{M} \rangle \geq 0$ and one of $\mathcal{N}, \mathcal{M}$ is maximal, then $\mathcal{N}$ and $\mathcal{M}$ have a common scheme in their canonical decompositions.

§3. Standard and General Quadratic Transformations of Polynomials

We will call the points of $\mathbb{R}^3$

$$e_1 = (1,0,0), \quad e_2 = (0,1,0), \quad e_3 = (0,0,1)$$

standard fundamental points and the coordinate planes - standard fundamental planes.

Let $P \in \overline{\pi}_n(\mathbb{R}^3)$ be an arbitrary homogeneous polynomial of degree n:

$$P(x) = P(x_1, x_2, x_3) = \sum_{i_1+i_2+i_3=n} a_{i_1 i_2 i_3} x_1^{i_1} x_2^{i_2} x_3^{i_3}. \tag{3.1}$$

It is not hard to check the following

Remark 3.1. *The order of zero of the polynomial P at the fundamental point e_i is the difference of* $\deg P$ *and the highest degree of x_i in (3.1).*

Suppose now that P has zeros of order $\geq n_i$ at the points e_i, $i = 1,2,3$. Then P can be presented in the following form (with fixed i, $(i,j,k,) = (1,2,3)$):

$$P(x) = \sum_{\nu=n_i}^{n} x_i^{n-\nu} A_\nu(x_j, x_k),$$

where A_ν is bivariate homogeneous polynomial of order ν. Therefore we have

$$P(x_2x_3, x_1x_3, x_1x_2) = x_1^{n_1} x_2^{n_2} x_3^{n_3} P^*(x_1, x_2, x_3), \tag{3.2}$$

for some $P^* \in \overline{\pi}_{n-t}(\mathbb{R}^3)$, where

$$t = n_1 + n_2 + n_3 - n, \quad t \in \mathbb{Z}.$$

Now we have the following (cf. [11] and [2]):

Definition 3.2. *The polynomial $Q(P) = P^*$ defined by (3.2) is called the standard (n_1, n_2, n_3)-quadratic transformation of P.*

Setting $x_1 = x_2x_3$, $x_2 = x_1x_3$, $x_3 = x_1x_2$ in (3.2) we get

$$x_1^{n_1^*} x_2^{n_2^*} x_3^{n_3^*} P(x_1, x_2, x_3) = P^*(x_2x_3, x_1x_3, x_1x_2), \tag{3.3}$$

where $n_i^* := n_i - t$. Using Remark 3.1, relations (3.2), (3.3) and the fact, that the degree of x_i, $i = 1,2,3$ in the left hand side of (3.2) is $\leq n$, as well as in the right hand side of (3.3) is $\leq n^* := n - t$, we obtain

Remark 3.3. *The standard quadratic transformation has the following properties:*

1. *The polynomial $Q(P)$ has a zero of order not less than n_i^* at the point e_i, $i = 1, 2, 3$.*
2. *For an arbitrary point*
$$u = (u_1, u_2, u_3) \in \mathbb{R}^3 \quad with \ u_1 u_2 u_3 \neq 0$$
the order of zero of the polynomial P at the point u is the same as the order of zero of the polynomial $Q(P)$ at the point
$$Q(u) := (u_2 u_3, u_1 u_3, u_1 u_2).$$
3. *The order of zero of P at e_i is greater than n_i, if and only if $Q(P)$ has a factor x_i, $(i = 1, 2, 3)$.*
4. *The order of zero of $Q(P)$ at e_i is greater than n_i^*, if and only if P has a factor x_i, $(i = 1, 2, 3)$.*
5. *The polynomial P has a factor different from x_i $(i = 1, 2, 3)$, if and only if $Q(P)$ has such a factor.*
6. *The (n_1^*, n_2^*, n_3^*)-quadratic transformation of $Q(P)$ is P.*

Let now v_1, v_2, v_3 be arbitrary points from $\mathbb{R}^3$ with $vol_3[0, v_1, v_2, v_3] \neq 0$, and let L be a linear transformation of $\mathbb{R}^3$ with $L(e_i) = v_i$ $i = 1, 2, 3$, $(det L \neq 0)$.

Definition 3.4. *The polynomial*
$$Q_{v_1,v_2,v_3}(P) := L^{-1} Q(LP), \quad with \ LP(x) := P(L(x))$$
is called a (general) (n_1, n_2, n_3)-quadratic transformation of the polynomial P with respect to the points v_1, v_2, v_3. In addition, the points v_1, v_2, v_3 and the planes $H_{i,j}$ passing through $0, v_i, v_j$, $1 \leq i < j \leq 3$, are called fundamental for this transformation Q_{v_1,v_2,v_3}.

Let the fundamental planes be given by the equalities:
$$h_{i,j}(x_1, x_2, x_3) = 0, \quad 1 \leq i < j \leq 3.$$

It is not hard to see that
i) a polynomial P has a multiplier $h_{i,j}$, iff LP has a multiplier x_k, $(i, j, k) = (1, 2, 3)$;
ii) a point x belongs to a general fundamental plane, iff $L(x)$ belongs to corresponding standard fundamental plane;
iii) the order of zero of LP at x equals the order of zero of P at $L(x)$.
Using these properties we easily get from Remark 3.3:

Remark 3.5. *The general quadratic transformation has the following properties:*

1. *The polynomial* $Q_{v_1,v_2,v_3}(P)$ *has a zero of order not less than* $n_i^* := n_i - t$ *at the point* v_i, $i = 1,2,3$.
2. *For an arbitrary point* $u = (u_1, u_2, u_3)$ *outside the fundamental planes the order of zero of the polynomial* P *at the point* u *is the same as the order of zero of the polynomial* $Q_{v_1,v_2,v_3}(P)$ *at the point*

$$Q_{v_1,v_2,v_3}(u) := L^{-1}Q(L(u)).$$

3. *The order of zero of* P *at* v_i *is greater than* n_i, *if and only if* $Q_{v_1,v_2,v_3}(P)$ *has a multiplier* $h_{j,k}$, $(i,j,k) = (1,2,3)$.
4. *The order of zero of* $Q_{v_1,v_2,v_3}(P)$ *at* v_i *is greater than* n_i^*, *if and only if* P *has a multiplier* $h_{j,k}$.
5. *The polynomial* P *has a multiplier different from* $h_{i,j}$, *if and only if* $Q_{v_1,v_2,v_3}(P)$ *has such a multiplier.*
6. *The* (n_1^*, n_2^*, n_3^*)*-quadratic transformation of the polynomial* $Q_{v_1,v_2,v_3}(P)$ *with respect to the points* v_1, v_2, v_3 *is* P.

§4. The Hilbert Function of Schemes

For any scheme $\mathcal{N} = \{n_1, \ldots, n_s; n\}$, and $\tau = 0, 1, \ldots$ we denote

$$\mathcal{N}_{+\tau} := \{n_1, \ldots, n_s, n_{s+1}, \ldots, n_{s+\tau}; n\},$$

where $n_{s+1} = \cdots = n_{s+\tau} = 1$. and set

$$R(\mathcal{N}) := min\{\tau : \mathcal{N}_{+\tau} \ is\ regular\}; \quad 0 \le R(\mathcal{N}) \le \overline{n+1}.$$

It is easy to see that $R(\mathcal{N})$ coincides with minimal (with respect to the node set $\mathcal{T}$) dimension of the space of algebraic curves of degree $\le n$ passing through $\mathcal{T}$ with multiplicity $\ge \mathcal{N}'$.

Definition 4.1. *The Hilbert function* $H(\mathcal{N})$ *is defined by the equality:*

$$H(\mathcal{N}) := \overline{n+1} - R(\mathcal{N}), \quad \mathcal{N} \in S.$$

We have $R(\mathcal{N}) = 0$ and $H(\mathcal{N}) = \overline{n+1}$ for a regular scheme $\mathcal{N}$.

Note that $H(\mathcal{N})$ coincides with maximal (with respect to $\mathcal{T}$) number of independent conditions, given by $\mathcal{N}$, i.e. in (1.3).

Let us denote by $mult_x C$ the multiplicity of $x \in \mathbb{R}^3$ in an algebraic curve C and by $mult_{\mathcal{T}} C$ - the sequence $\{mult_x C \ : \ x \in \mathcal{T}\}$, where $\mathcal{T}$ is a node set.

Remark 4.2. *Note that $R(\mathcal{N}) = 1$ means that there is a set $U_{\mathcal{N}} \subset \mathbb{R}^{3s}$ whose complement has measure zero, such that for any node set $\mathcal{T} \in U_{\mathcal{N}}$ there is a unique algebraic curve $\mathcal{N}^{\mathcal{T}}$ of degree n such that*

$$mult_{\mathcal{T}} \mathcal{N}^{\mathcal{T}} \geq \mathcal{N}' \ . \tag{4.1}$$

Indeed, if V is the set of $\mathcal{T}$'s for which there are at least two distinct curves satisfying (4.1), then $(\mathcal{N}_{+1}, \mathcal{T} \cup \{w\})$ is singular for $\mathcal{T} \in V$ and any $w \in \mathbb{R}^3$. Since the scheme $\mathcal{N}_{+1}$ is regular, therefore $vol_{3s} V = 0$.

Theorem 4.3. *For any scheme $\mathcal{N} = \{n_1, \ldots, n_s; n\} \in BS^{\times *}$ we have*

$$\begin{aligned} H(\mathcal{N}) &= \overline{n+1} - R(\mathcal{N})^{\downarrow} = \overline{n+1} - \overline{n^{\downarrow}+1} + \sum_{\nu=1}^{s} \overline{n_\nu^{\downarrow}} \\ &= \sum_{\nu=1}^{s} \overline{n_\nu} - \sum_{A \in \Omega_{\mathcal{N}}} \overline{\mu_A - 1}, \end{aligned} \tag{4.2}$$

the second equality holds provided that Conjecture 1.10 is true.

§5. Some Geometrical Results

In this section we consider applications of numerical curves to algebraic curves. We start with establishing the connection between reducibility of polynomials and schemes.

Theorem 5.1. *Suppose the scheme $\mathcal{N}$ is singular and for any node set $\mathcal{T} \in F \subset \mathbb{R}^{3s}$, where $vol_{3s} F > 0$, there is a reducible polynomial $P = P_{\mathcal{T}}$ satisfying (1.3). Then*

$$\mathcal{N} = \mathcal{N}_1 + \mathcal{N}_2, \tag{5.1}$$

where $\mathcal{N}_i \neq 0$ is singular.

We will say that the quadratic transformation Q is applicable for the node set $\mathcal{T} = \{t_1, \ldots, t_s\} \subset \mathbb{R}^3$ if there is no plane containing O and the nodes

$$\{t_\nu : \nu \in \text{supp } Q\}.$$

In this case we denote

$$Q(\mathcal{T}) := \{Q(t_1), \ldots, Q(t_s)\}.$$

For $A \in PNC$ with $A = Q_1 \cdots Q_l E_{p,q}$ we denote by G_A the collection of node sets $\mathcal{T}$ such that

i) Q_i is applicable for $Q_{i-1} \cdots Q_1 \mathcal{T}$, $i = 1, \ldots, l$;
ii) the plane passing through p-th, q-th nodes of the node set $Q_l \cdots Q_1(\mathcal{T})$ and O does not contain any other its node.

Obviously, the complement of G_A in $\mathbb{R}^{3s}$ has measure zero (in fact, it is a union of surfaces).

Theorem 5.2. *Suppose $A = \{\alpha_1, \ldots, \alpha_s; \alpha\} \in PNC$ and $\mathcal{T} \in G_A$. Then there exists a unique algebraic curve $A^{\mathcal{T}}$ of degree α with*

$$mult_{\mathcal{T}} A^{\mathcal{T}} \geq A'. \tag{5.2}$$

Moreover,

1) $mult_{\mathcal{T}} A^{\mathcal{T}} = A'$.
2) *There is no singular point of $A^{\mathcal{T}}$ outside of $\mathcal{T}$.*
3) *$A^{\mathcal{T}}$ is quadratically equivalent to a line and hence is irreducible.*

In the next theorem we generalize this result for an orthogonal subset of PNC.

Theorem 5.3. *Suppose $\mathcal{A}$ is an orthogonal set of prime numerical curves,*

$$\mathcal{T} \in G_{\mathcal{A}} := \cap_{A \in \mathcal{A}} G_A, \quad \mathcal{M} = \sum_{A \in \mathcal{A}} \lambda_A A,$$

where $\lambda_A \in \mathbb{Z}_+$. Then

i) *There exits a unique algebraic curve $\mathcal{M}^{\mathcal{T}}$ with*

$$mult_{\mathcal{T}} \mathcal{M}^{\mathcal{T}} \geq \mathcal{M}'. \tag{5.3}$$

ii) *For this unique curve we have equality in (5.3), and the curve consists of components $A^{\mathcal{T}}$ with multiplicities λ_A, i.e.,*

$$\mathcal{M}^{\mathcal{T}} = \sum_{A \in \mathcal{A}} \lambda_A A^{\mathcal{T}} . \tag{5.4}$$

iii) *If $x \in \mathcal{M}^{\mathcal{T}} \setminus \mathcal{T}$, then there exists a unique $A \in \mathcal{A}$ such that $x \in A^{\mathcal{T}}$, and moreover,*

$$mult_x \mathcal{M}^{\mathcal{T}} = mult_x A^{\mathcal{T}} = \lambda_A.$$

Remark 5.4. *Note that if $\lambda_A \leq 1$ for all $A \in \mathcal{A}$, then the curve $\mathcal{M}^{\mathcal{T}}$ has no singular point outside of $\mathcal{T}$.*

In the following theorem we discuss the problem of irreducibility of algebraic curves with given singularities.

Theorem 5.5. *Suppose the Conjecture 1.10 is true and $\mathcal{N} \in BS^*$, $\overline{\mathcal{N}} = 0$. Then there is a set $G_{\mathcal{N}} \subset \mathbb{R}^{3s}$ whose complement has measure zero, such that for any node set $\mathcal{T} \in G_{\mathcal{N}}$*

i) *there is a unique algebraic curve $\mathcal{N}^{\mathcal{T}}$ of degree n with*

$$mult_{\mathcal{T}} \mathcal{N}^{\mathcal{T}} \geq \mathcal{N}';$$

ii) *the curve $\mathcal{N}^{\mathcal{T}}$ has a unique irreducible component with multiplicity λ, if $\mathcal{N} \sim \lambda C_0$, $\lambda \geq 2$, and is irreducible otherwise.*

The next theorem concerns the (exact) number of components of algebraic curves with given singularities.

Theorem 5.6. *Suppose that $\mathcal{N}$ is a singular scheme with canonical decomposition*

$$\mathcal{N} = \sum_{A \in \Omega_{\mathcal{N}}} \mu_A A + \mu^{\Downarrow} \mathcal{N}^{\Downarrow}, \quad \mathcal{N}^{\Downarrow} \neq 0 \, ,$$

and let $\mathcal{T} \in G_{\Omega_{\mathcal{N}}}$. Then any curve C of degree $\deg \mathcal{N}$ with

$$mult_{\mathcal{T}} C \geq \mathcal{N}' \tag{5.5}$$

contains all the components $A^{\mathcal{T}}$, $A \in \Omega_{\mathcal{N}}$ with multiplicities μ_A and at least one additional component (not necessarily irreducible) $C^{\Downarrow}$ of degree $\deg \mathcal{N}^{\Downarrow}$ with multiplicity $\mu^{\Downarrow}$ such that

$$mult C^{\Downarrow} \geq (\mathcal{N}^{\Downarrow})'. \tag{5.6}$$

In addition

i) $mult_x C = mult_x C^{\Downarrow}$, if $x \in C^{\Downarrow} \setminus \mathcal{T}$;

ii) $mult_x C = mult_x A^{\mathcal{T}}$, if $x \in C \setminus C^{\Downarrow}$, where $A \in \Omega_{\mathcal{N}}$ with $x \in A^{\mathcal{T}}$.

Moreover, if $\overline{\mathcal{N}^{\downarrow}} = 0$ and Conjecture 1.10 is true, then for any $\mathcal{T} \in G_{\Omega_{\mathcal{N}}} \cap U_{\mathcal{N}^{\Downarrow}}$ we have equality in (5.5) and (5.6), and there is exactly one additional component $C^{\Downarrow}$ (with multiplicity $\mu^{\Downarrow}$) which is irreducible.

At the end we consider the following problem: can the size of $\overline{\mathcal{N}}$ guarantee the regularity of $\mathcal{N}$?

Theorem 5.7. *If Conjecture 1.10 is true, then all the schemes $\mathcal{N} = \{n_1, \ldots, n_s; n\}$ satisfying*

$$\overline{\mathcal{N}} > \overline{n-1}$$

are regular. While for the singular scheme $\mathcal{N} = n\{1,1;1\}$ we have

$$\overline{\mathcal{N}} = \overline{n-1}.$$

Acknowledgements. The research described in this publication was made possible in part by Grant N RYMOOO from the International Science Foundation.

References

1. Gevorgian, H., H. Hakopian and A. Sahakian, On bivariate Hermite interpolation, Mat. Zametki **48** (1990), 137–139 (in Russian).
2. Gevorgian, H., H. Hakopian and A. Sahakian, On the bivariate polynomial interpolation, Mat. Sbornik **183** (1992), 111-126 (in Russian), English transl. in Russian Acad. Sci. Sb. Math. **76** (1993), 211–223.

3. Gevorgian, H., H. Hakopian and A. Sahakian, On the bivariate Hermite interpolation problem, Arbeitspapiere der GMD **692** (1992), Constr. Approx. **11** (1995), 23–36.
4. Gevorgian, H., H. Hakopian and A. Sahakian, Bivariate Hermite interpolation and numerical curves, in *Proc. of the International Conference on Open Problems in Approximation Theory*, Voneshta Voda, Bulgaria, 1993, 85–89.
5. Gevorgian, H., H. Hakopian and A. Sahakian, Bivariate Hermite interpolation and numerical curves, to appear in J. Approx. Th.
6. Gevorgian, H., H. Hakopian and A. Sahakian, Numerical curves and their applications to algebraic curves, submitted to Studia Math.
7. Hakopian, H., Multivariate polynomial interpolation, Habilitation survey, Math. Inst. of Polish Acad. Sci., Warsaw, 1994.
8. Jia, R. Q. and A. Sharma, Solvability of some multivariate interpolation problems, J. Reine Angew. Math. **421** (1991), 73–81.
9. Lorentz, G. G. and R. A. Lorentz, Multivariate Interpolation, in *Lecture Notes in Math.* **1105**, Springer-Verlag, Berlin, 1984, 136–144.
10. Paskov, S.H., Singularity of bivariate interpolation, J. Approx. Th. **75** (1992), 50–67.
11. Walker, R., *Algebraic Curves*, Princeton, New Jersey, 1950.

H. Gevorgian
Mathematics Research Center
Yerevan State University
Alex Manoukian, 1
Yerevan 375049
Armenia
gayane@arminco.com

H. Hakopian
Department of Mathematics
Yerevan State University
Alex Manoukian, 1
Yerevan 375049
Armenia
gayane@arminco.com

A. Sahakian
Mathematical Institute
Armenian National Academy of Sciences
Marshal Baghramian, 24b
Yerevan 375019
Armenia
mathinst@pnas.sci.am

Sectional-Curvature Preserving Skinning Surfaces with a 3D Spine Curve

A. I. Ginnis, P. D. Kaklis and N. C. Gabrielides

Abstract. This work deals with the problem of constructing *sectional-curvature preserving* (*scp*) C^2-continuous surfaces, which interpolate point-sets lying on planes perpendicular to a three-dimensional *spine curve*. The proposed method of solution employs *skinning surfaces*, whose *skeletal lines* and *blending functions* belong to a special family of *polynomial splines of non-uniform degree*, already used for shape-preserving curve interpolation in the plane (see [4], [5]) as well as space (see [3]). Using an appropriate affine transformation, the problem simplifies to a sectional-curvature preservation problem with a straight line as spine curve, a case thoroughly studied in [2].

§1. Introduction

Developing automatic algorithms for *shape-preserving* curve and surface interpolation, is a CAGD domain with intensive basic-research activity and a steadily increasing audience in the CAD/CAM community. While the planar problem is well studied (see [5] and the references cited therein), it seems that a lot of work has still to be done in the areas of shape-preserving interpolatory 3D curves and surfaces; see the literature review in [3] and [2].

Especially in the case of surfaces, the problem becomes really hard when convexity has to be preserved; see pertinent comments in the introduction in [1]. Less difficult is the problem where the designer is interested in preserving the convexity of intersections of the surface with a discrete or continuous family of planes. This is notably true when the interpolation data are point-sets lying on planes, which is likely to occur in practice when the object is characterized by a direction, directly related to its functionality, as it is, e.g., the longitudinal axis of a ship or the fuselage of an air-plane. This problem has been thoroughly treated in [2], where

Advanced Topics in Multivariate Approximation
F. Fontanella, K. Jetter and P. J. Laurent (eds.) pp. 113–123.

ISBN 981-02-2852-X

an automatic algorithm has been developed for constructing C^2-continuous shape-preserving surfaces, which interpolate point-sets lying on planes perpendicular to a straight line.

The purpose of this work is to extend the method in [2], so that it can handle the more general configuration where the interpolation data are lying on planes perpendicular to a general 3D curve.

§2. Formulation of the Problem

Let $\mathbf{c}(u), u \in I_u = [u_1, u_N]$, be an adequately smooth three-dimensional (3D) curve, henceforth referred to as the *spine curve*. Furthermore, let $\mathcal{D}$ be a union of point-sets $\mathcal{D}_i$, $i = 1(1)N$, lying on N planes E_i, which are perpendicular to $\dot{\mathbf{c}}(u_i) := d\mathbf{c}(u_i)/du$ and pass through $\mathbf{c}(u_i)$, with $u_i \in \mathcal{U} = \{u_1, u_2, \ldots, u_N; u_1 < u_2 < \ldots < u_N\}$. Each $\mathcal{D}_i$ consists of M points, that is, $\mathcal{D}_i = \{\mathbf{I}_{i1}, \mathbf{I}_{i2}, \ldots, \mathbf{I}_{iM}\} \in E_i$, with $\mathbf{I}_{i,j+1} \neq \mathbf{I}_{ij}$, $j = 1(1)M-1$ (see Figure 1).

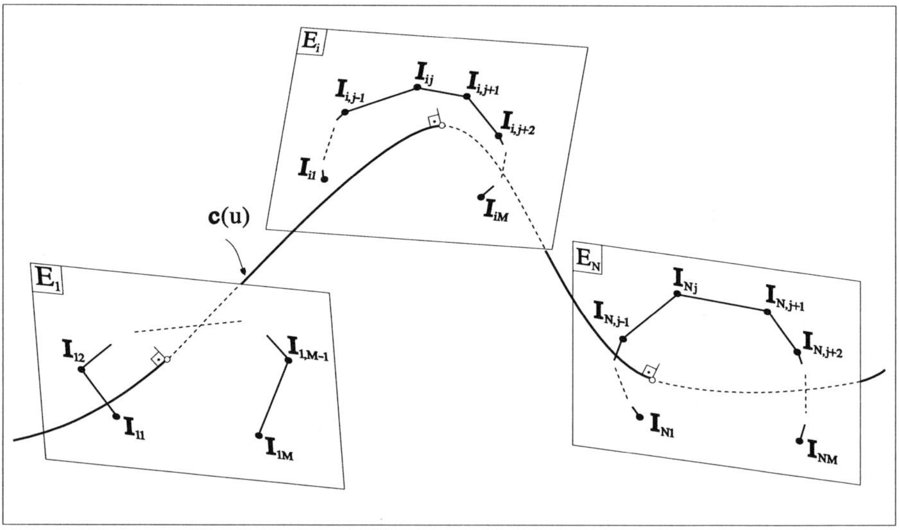

Fig. 1. The interpolation data set $\mathcal{D}$ and the spine curve $\mathbf{c}(u)$.

We are interested in constructing a C^2-continuous surface $\mathbf{P}(u,v)$, $(u,v) \in I\!I = I_u \times I_v$, $I_v = [v_1, v_M]$, which interpolates $\mathcal{D}$ with parametrization $\mathcal{U} \times \mathcal{V}$, $\mathcal{V} = \{v_1, v_2, \ldots, v_M; v_1 < v_2 < \ldots < v_M\}$, and preserves the *shape information* contained in it. Before stating the herein adopted shape-preservation criterion, some further notation is in order. Let $E(u)$ be the plane being perpendicular and passing through the spine curve $\mathbf{c}(u)$ at the parameter value u, and $\kappa_c(u,v)$ be the curvature of the intersection of $\mathbf{P}(u,v)$ with $E(u)$, henceforth referred to as the $\mathbf{c}$-*sectional curvature*. Finally, let

$$\rho_{mr} = \det(\mathbf{s}_{m,r-1}, \mathbf{s}_{mr}, \dot{\mathbf{c}}(u_m)), \qquad \mathbf{s}_{mr} = \frac{\mathbf{I}_{m,r+1} - \mathbf{I}_{mr}}{v_{r+1} - v_r}, \tag{1}$$

be the so-called *convexity indicator* at the point $\mathbf{I}_{mr}$ of $\mathcal{D}_m$, quantifying the local convexity of the planar polygonal line connecting the points of $\mathcal{D}_m$. The sought-for notion of shape-preservation can then be formalized as below :

Definition 1. *Let $\mathbf{P}(u,v)$, $(u,v) \in \mathit{\Pi}$, be a C^2-continuous surface interpolating $\mathcal{D}$ and $\mathbf{c}(u)$, $u \in I_u$, be the associated spine curve. $\mathbf{P}(u,v)$ will be called $\mathbf{c}$-sectional-curvature preserving ($\mathbf{c}$-scp) provided that, if $\rho_{n,r+\ell}$, $l = 0,1$, $n = m, m+1$, share the same sign for some m, r with $1 \le m \le N-1$ and $1 \le r \le M-1$, then its $\mathbf{c}$-sectional curvature $\kappa_c(u,v)$ possesses the same sign in $I_{mr} = [u_m, u_{m+1}] \times [v_r, v_{r+1}]$.*

Note that, assuming that the afore-mentioned convexity indicators are of the same sign, is equivalent to saying that the polygonal arcs $\mathbf{I}_{n,r-1}\mathbf{I}_{nr}\mathbf{I}_{n,r+1}\mathbf{I}_{n,r+2}$, $n = m, m+1$, are both convex and share the same orientation in their respective planes E_n; see Figure 2.

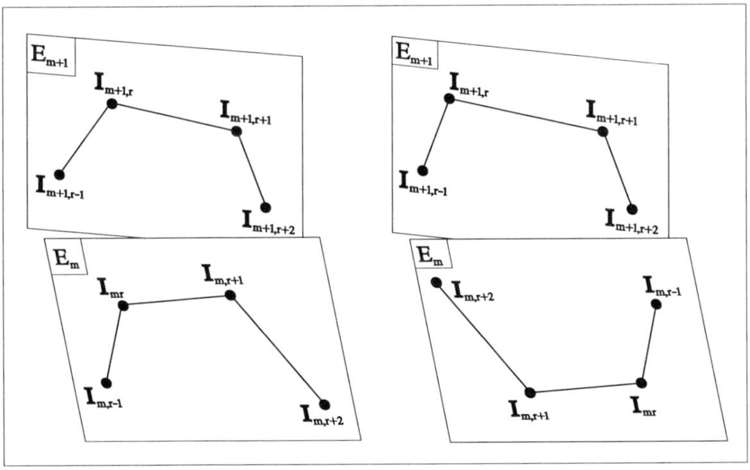

Fig. 2. Two possible configuration for which ρ_{nr}, $\rho_{n,r+1} > 0$, $n = m, m+1$.

By virtue of the above definition, the problem we shall deal with can be accurately formulated as follows :

Problem ($\mathcal{P}$). *Construct a C^2-continuous surface $\mathbf{P}(u,v)$, $(u,v) \in \mathit{\Pi}$, which interpolates $\mathcal{D}$ with parametrization $\mathcal{U} \times \mathcal{V}$ and is $\mathbf{c}$-scp in the sense of Definition 1.*

Note that, for the degenerate case where the spine curve $\mathbf{c}(u)$ is a straight line, problem $(\mathcal{P})$ has been thoroughly studied in [2]. In that work, the solution is obtained in a function space consisting of *skinning surfaces* whose univariate components, i.e, *skeletal lines* and *blending functions*, belong to a special family of polynomial splines of non-uniform degree, already used for shape-preserving curve interpolation in the plane (see [4], [5]) as well as space (see [3]). The study of the asymptotic behaviour of these surfaces and their $\mathbf{c}$-sectional curvature, as the segment degrees increase in some, suitably defined, semi-local manner with respect to a specific parameter subdomain $I\!I_{mr}$, yields sufficient geometric conditions on $\mathcal{D}$ ensuring that, if the segment degrees increase *semi-locally*, the surface will eventually become $\mathbf{c}$-scp in $I\!I_{mr}$; see Theorem 2 in §4.

§3. Transformation of the Problem

In order to exploit the results obtained in [2], this section will investigate the possibility of transforming $(\mathcal{P})$ to a problem $(\mathcal{P}_0)$ in which the spine curve $\mathbf{c}_0$ is a straight line, namely $\mathbf{c}_0(u) = (u,0,0)^T$, $u \in I_u$, and $\mathcal{D}_0$ retains the convexity characteristics of $\mathcal{D}$. Towards this aim, we first seek to construct a one parameter family of homeomorphic mappings $\mathbf{M}(u)$ such that :

$$(\forall\ u \in I_u = [u_1, u_N]) \begin{pmatrix} \mathbf{c}(u) \overset{\mathbf{M}(u)}{\longmapsto} \mathbf{c}_0(u) \\ E(u) \overset{\mathbf{M}(u)}{\longmapsto} E_0(u) \end{pmatrix}, \tag{2}$$

where $E(u)$ and $E_0(u)$ are the planes perpendicular on and passing through $\mathbf{c}(u)$ and $\mathbf{c}_0(u)$, respectively. This can be achieved by defining the affine mappings :

$$\mathbf{M}(u) : E(u) \ni \mathbf{r} \longmapsto \mathbf{r}_0 \in E_0(u), \qquad \mathbf{r}_0 = \mathbf{A}(u)(\mathbf{r} - \mathbf{c}(u)) + \mathbf{c}_0(u), \tag{3}$$

with $\mathbf{A}(u)$ being a family of 3×3 invertible matrices, which map the unit vectors $\mathbf{e}_1 := (1,0,0)^T$, $\mathbf{e}_2 := (0,1,0)^T$ and $\mathbf{e}_3 := (0,0,1)^T$ as below :

$$\mathbf{A}^{-1}(u)\mathbf{e}_1 = \frac{1}{k(u)}\dot{\mathbf{c}}(u),\ k(u) > 0, \quad \mathbf{A}^{-1}(u)\mathbf{e}_i \cdot \dot{\mathbf{c}}(u) = 0,\ i = 2,3. \tag{4}$$

We now turn our attention to the requirement that $\mathbf{M}(u)$ should retain the convexity characteristics of $\mathcal{D}$. In this connection, we state and prove the following

Lemma 1. *Let* $\mathbf{M}(u)$ *be a family of affine mappings which fulfil conditions (3) and (4). Then, for any pair of distinct vectors* $\mathbf{a}$ *and* $\mathbf{b}$ *on* $E(u)$*, the triple product* $\det(\mathbf{a}, \mathbf{b}, \dot{\mathbf{c}}(u))$ *shares the same sign with* $\det(\mathbf{M}(u)\mathbf{a}, \mathbf{M}(u)\mathbf{b}, \mathbf{e}_1)$ *iff* $\det(\mathbf{A}(u)) > 0$.

Proof: As one can easily show by means of (3), the images of the vectors $\mathbf{a}$ and $\mathbf{b}$, under $\mathbf{M}(u)$, are $\mathbf{A}(u)\mathbf{a}$ and $\mathbf{A}(u)\mathbf{b}$, respectively. Then, recalling the first of (4), we have

$$\det\left(\mathbf{M}(u)\mathbf{a}, \mathbf{M}(u)\mathbf{b}, \mathbf{e}_1\right) = \det\left(\mathbf{A}(u)\mathbf{a}, \mathbf{A}(u)\mathbf{b}, \frac{1}{k(u)}\mathbf{A}(u)\dot{\mathbf{c}}(u)\right). \tag{5}$$

Using elementary properties of determinants, the right-hand side in (5) takes the form :

$$\det\left(\mathbf{A}(u)\mathbf{a}, \mathbf{A}(u)\mathbf{b}, \frac{1}{k(u)}\mathbf{A}(u)\dot{\mathbf{c}}(u)\right) = \frac{1}{k(u)}\det\left(\mathbf{A}(u)\right)\det(\mathbf{a}, \mathbf{b}, \dot{\mathbf{c}}(u)) . \quad \blacksquare \qquad (6)$$

Recalling formula (1), we get :

Corollary 1. *If $\mathbf{M}(u)$ fulfil the assumptions of Lemma 1, then the convexity indicators $\rho_{0,mr}$ of the transformed data set*

$$\mathcal{D}_0 = \cup_{i=1}^{N}\mathcal{D}_{0,i}, \quad \mathcal{D}_{0,i} = \mathbf{M}(u_i)\,[\mathcal{D}_i] := \{\mathbf{I}_{0,i1}, \mathbf{I}_{0,i2}, \ldots, \mathbf{I}_{0,iM}\}, \qquad (7)$$

share the same sign with the corresponding convexity indicators ρ_{mr} of the original data $\mathcal{D}$.

We can now proceed to formulate

Problem ($\mathcal{P}_0$). *Let $\mathbf{M}(u)$ be a family of affine mappings fulfilling the assumptions of Lemma 1. Construct a C^2-continuous surface $\mathbf{P}_0(u,v)$, $(u,v) \in I\!I$, which interpolates $\mathcal{D}_0$ (see (7)) with parametrization $\mathcal{U} \times \mathcal{V}$ and is $\mathbf{c}_0$-scp.*

Combining Lemma 1 and Corollary 1 with the definition of curvature for planar curves, we arrive at

Theorem 1. *Let $\mathbf{P}_0(u,v)$, $(u,v) \in I\!I$, be a solution of ($\mathcal{P}_0$). Then $\mathbf{M}^{-1}(u)\mathbf{P}_0(u,v)$ is a solution of ($\mathcal{P}$).*

The above theorem is the basic result of this work, enabling us to exploit the algorithm developed in [1] for solving ($\mathcal{P}_0$); see §4. So, the only issue that remains to be discussed is how one can efficiently define the family of mappings $\mathbf{M}(u)$, namely the family of matrices $\mathbf{A}(u)$ satisfying conditions (4) and possessing positive determinant. Typical choices are :

$$\mathbf{A}^{-1}(u) \text{ is the orthogonal matrix rotating } \mathbf{e}_1 = \frac{\dot{\mathbf{c}}_0(u)}{\|\dot{\mathbf{c}}_0(u)\|} \text{ onto } \frac{\dot{\mathbf{c}}(u)}{\|\dot{\mathbf{c}}(u)\|}, \qquad (8)$$

$$\mathbf{A}^{-1}(u) = \left[\dot{\mathbf{c}}(u),\ \mathbf{v}(u) = \dot{\mathbf{c}}(u) \times \ddot{\mathbf{c}}(u), \dot{\mathbf{c}}(u) \times \mathbf{v}(u)\right] \text{ and} \qquad (9)$$

$$\mathbf{A}^{-1}(u) = \left[\dot{\mathbf{c}}(u),\ \mathbf{v}(u) = \dot{\mathbf{c}}(u) \times \mathbf{v}_0, \dot{\mathbf{c}}(u) \times \mathbf{v}(u)\right]. \qquad (10)$$

Since $\mathbf{P}(u,v) = \mathbf{M}^{-1}(u)\mathbf{P}_0(u,v)$ should be C^2-continuous, choices (8), (10) and (9) induce that the initial spine curve should be, at least, C^3- and C^4- continuous, respectively. Finally, it is worth-noticing that, if the spine curve $\mathbf{c}(u)$ is a polynomial spline and the solution $\mathbf{P}_0(u,v)$ of the transformed problem ($\mathcal{P}_0$) is a polynomial spline surface, then choice (9) or (10) implies that $\mathbf{P}(u,v)$ will be a polynomial spline surface too.

§4. Solution of the Transformed Problem

The transformed problem $(\mathcal{P}_0)$ is treated in [2] by adopting the skinning representation for $\mathbf{P}_0(u,v)$, namely:

$$\mathbf{P}_0(u,v) = \sum_{i=1}^{N} L_i(u)\mathbf{g}_i(v), \qquad (u,v) \in \mathit{\Pi}, \tag{11}$$

where $\mathbf{g}_i(v)$, $v \in I_v$, are the skeletal lines which interpolate the planar data sets $\mathcal{D}_{0,i}$, with parametrization $\mathcal{V}$, i.e.,

$$\mathbf{g}_i(v_j) = \mathbf{I}_{0,ij}, \quad j = 1(1)M, \tag{12}$$

and $L_i(u)$, $u \in I_u$, are the blending functions, satisfying the conditions

$$L_i(u_k) = \delta_{ik}, \quad k = 1(1)N, \tag{13}$$

with δ_{ik} denoting Kronecker's delta. Furthermore

$$L_i(u) \in \Gamma(\Lambda,\mathcal{U}), \quad \mathbf{g_i}(v) \in \Gamma(K_i,\mathcal{V}), \tag{14}$$

where Γ is the family of C^2- continuous splines of non-uniform degree, defined in [3], and $\Lambda = \{\lambda_1, \lambda_2, \ldots, \lambda_{N-1}\}$ and $K_i = \{k_{i1}, k_{i2}, \ldots, k_{i,M-1}\}$ are the associated segment-degree distributions with λ_m, $k_{ij} \geq 3$. Belonging to $\Gamma(\Lambda,\mathcal{U})$, the non-parametric splines $L_i(u)$ admit of the following representation :

$$\begin{aligned} L_i(u) = \delta_{im}(1-t) + \delta_{i,m+1}t + \eta_m^2\left[L''_{im}F_m(1-t) + L''_{i,m+1}F_m(t)\right], \\ t \in [0,1],\ u \in I_{um} = [u_m, u_{m+1}],\ m = 1(1)N-1, \end{aligned} \tag{15a}$$

where $L''_{im} := d^2L_i(u_m)/du^2$ and

$$F_m(t) = \frac{t^{\lambda_m} - t}{\lambda_m(\lambda_m - 1)}, \qquad t = \frac{u - u_m}{\eta_m}, \qquad \eta_m = u_{m+1} - u_m. \tag{15b}$$

For each i, and for type-II' (natural) boundary conditions ($L''_{i1} = L''_{iN} = 0$), the quantities L''_{im} satisfy the linear systems :

$$e_{m-1}L''_{i,m-1} + (d_m + d_{m-1})L''_{im} + e_mL''_{i,m+1} = \frac{\delta_{i,m+1} - \delta_{im}}{\eta_m} - \frac{\delta_{im} - \delta_{i,m-1}}{\eta_{m-1}}, \quad m = 2(1)N-1, \tag{16a}$$

with

$$e_m = \frac{\eta_m}{\lambda_m(\lambda_m - 1)}, \qquad d_m = \frac{\eta_m}{\lambda_m}. \tag{16b}$$

The linear systems (16) are uniquely solvable, for their common left-hand side matrix is strictly diagonally dominant. Analogously, the parametric splines $\mathbf{g}_i(v) \in \Gamma(K_i, \mathcal{V})$ can be represented as

$$\mathbf{g}_i(v) = \mathbf{I}_{ir}(1-s) + \mathbf{I}_{i,r+1}s + \zeta_r^2\left[\ddot{\mathbf{g}}_{ir}\Phi_{ir}(1-s) + \ddot{\mathbf{g}}_{i,r+1}\Phi_{ir}(s)\right], \quad s \in [0,1], v \in I_{vr} = [v_r, v_{r+1}], r = 1(1)M-1, \tag{17a}$$

where

$$\Phi_{ir}(s) = \frac{s^{k_{ir}} - s}{k_{ir}(k_{ir} - 1)}, \qquad s = \frac{v - v_r}{\zeta_r}, \qquad \zeta_r = v_{r+1} - v_r. \tag{17b}$$

The second-order nodal derivatives $\ddot{\mathbf{g}}_{ir} := d^2\mathbf{g}_i(v_r)/dv^2$, $r = 1(1)M$, are evaluated by solving, similar to (16), linear systems, the "boundary equations" of which depend on the specific type of the boundary conditions $\mathcal{B}$ imposed on each one of the skeletal lines, e.g., $\mathcal{B} =$ type $- I$(given tangent vectors), type-II', periodic.

We now proceed to state the basic theoretical outcome of [2], which provides a suitable increase pattern of segment degrees that, under mild constraints on the data, leads to a solution of the transformed problem $(\mathcal{P}_0)$.

Theorem 2. *Let the* V *quantities :*

$$\left\{V^{mr}_{m+\ell,r},\ V^{mr}_{m+\ell,r-1},\ V^{m,r+1}_{m+\ell,r},\ V^{m,r+1}_{m+\ell,r+1}\right\},\ \ell = 1,2,$$

with

$$V^{pj}_{ni} = (\mathbf{s}_{p,j-1} \times \mathbf{s}_{ni} + \mathbf{s}_{ni} \times \mathbf{s}_{pj}) \cdot \mathbf{e}_1,$$

be of the same sign with the convexity indicators :

$$\left\{\rho_{0,m,r+\ell},\ \rho_{0,m+1,r+\ell}\right\},\ \ell = 0,1.$$

Furthermore, let the segment degrees increase semi-locally with respect to the parametric subdomain $\mathbb{I}_{mr}$, *i.e.,*

(i) $\lambda_m \to \infty$ *for some fixed* $m \in \{1, 2, \ldots, N-1\}$,

(ii) $k_{n,r-1}, k_{n,r}, k_{n,r+1} \to \infty$ *for* $n = m,\ m+1$ *and some fixed* $r \in \{2, 3, \ldots, M-2\}$, *with* $k_{mr}/k_{m+1,r} = O_s(1)$ $(\alpha = O_s(\beta) \Leftrightarrow \alpha = O(\beta) \wedge \alpha \neq o(\beta))$, *and*

(iii) $k_{ir} \to \infty$ *for all* $i \neq m, m+1$, *with* $k_{ir} \geq k_{nr}$ *and* $k_{ir}/k_{nr} = O(1)$, $n = m, m+1$.

Then, $\mathbf{P}(u,v)$ *will eventually become* $\mathbf{c}_0$*-scp in* $\mathbb{I}_{mr}$.

§5. A Numerical Example

Exploiting the asymptotic result of Theorem 2, [2] provides the automatic algorithm scpSKIN for constructing C^2-continuous $\mathbf{c}_0$-scp skinning surfaces with $\mathbf{c}_0 = (u,0,0)^T$, $u \in I_u$. So for a given data $\mathcal{D}$ along with a parametrization $\mathcal{U} \times \mathcal{V}$ and a spine curve $\mathbf{c}(u)$, $u \in I_u$, a solution of $(\mathcal{P})$ can be easily obtained in two steps : first, apply scpSKIN on the transformed data set $\mathcal{D}_0 = \cup_{i=1}^{N} \mathbf{M}(u_i)[\mathcal{D}_i]$, with $\mathbf{M}(u)$ being defined by (3) and one of (8) - (10); second apply the inverse of $\mathbf{M}(u)$ on the outcome surface of scpSKIN.

The afore-mentioned scheme is numerically tested for a data set $\mathcal{D}$ consisting of nine (N=9) planar point-sets, which are perpendicular on the helix : $\mathbf{c}(u) = (10\cos(\frac{\pi}{12}u), 10\sin(\frac{\pi}{12}u), 2\pi u)^T$, $u \in [0,24]$. Each point-set $\mathcal{D}_i$ is a convex periodic data with thirteen (M=13) points. The convexity of all $\mathcal{D}_i$ implies that the $\mathbf{c}$-sectional curvature of the surface interpolating $\mathcal{D}$ should be of constant sign, in this case positive. Choosing $\mathbf{M}(u)$ in accordance with (3) and (8) and applying scpSKIN on the transformed data $\mathcal{D}_0$, the algorithm converges after one (1) iteration. The resulting degree distributions of the skeletal lines and the blending functions have as follows : $K_i = \{10,3,10,10,3,10,10,3,10,10,3,10\}$, $i = 1(1)9$, $\Lambda = \{3,3,3,3,3,3,3,3\}$.

Let

$$\mathbf{P}^{tc}(u,v) = \mathbf{M}^{-1}(u) \sum_{i=1}^{9} L_i^c(u)\mathbf{g}_i^c(v), \quad (u,v) \in \Pi,$$

be the so-called "transformed-cubic" skinning interpolant, where $\mathbf{g}_i^c(v)$ are C^2 cubic splines, interpolating $\mathcal{D}_{0,i}$ and $L_i^c(u)$ are C^2 cubic blending functions. As shown in Figs. 3a and 3c, $\mathbf{P}^{tc}(u,v)$ fails to be $\mathbf{c}$-scp. More specifically, Fig. 3c depicts the surface $z = \kappa_c(u,v)$, $(u,v) \in \Pi$, and the plane $z = 0$ (chess pattern), both viewed from a point in the positive half-space $z > 0$. As a result, chess pattern signifies negative sectional curvature, i.e., $\mathbf{c}$-scp failure. Furthermore, the grid lines in Fig. 3c denote the skeletal isoparametrics $u = u_i$, whereas the bullets on them denote the nodes of $\mathcal{V}_i$. Finally the inversely transformed outcome of scpSKIN is free from $\mathbf{c}$-scp failures (see Figs. 3b and 3d).

The graphical output of the analogous experiment, with $\mathbf{M}(u)$ chosen in accordance with (3) and (9), is contained in Figs. 4. The final degree distributions have, in this case, as follows : $K_i = \{4,4,3,3,4,4,4,4,3,3,4,4\}$, $i = 1(1)9$, $\Lambda = \{3,3,3,3,3,3,3,3\}$. In view of the above results, one can legitimately assert that the selection of the matrices $\mathbf{A}(u)$ influences the $\mathbf{c}$-sectional-curvature distribution of both the "transformed-cubic" and the final $\mathbf{c}$-scp interpolant and, thus, their *visual pleasantness* (the surfaces in Figs. 3a, b are *visually better* than those in Figs. 4a, b). Obviously, $\mathbf{A}(u)$ influence the "winding" of the final surface around the spine curve. In this connection, we shall attempt to develop, in the near future, a method for determining $\mathbf{A}(u)$ so that they minimize the "winding" of the final $\mathbf{c}$-scp interpolant.

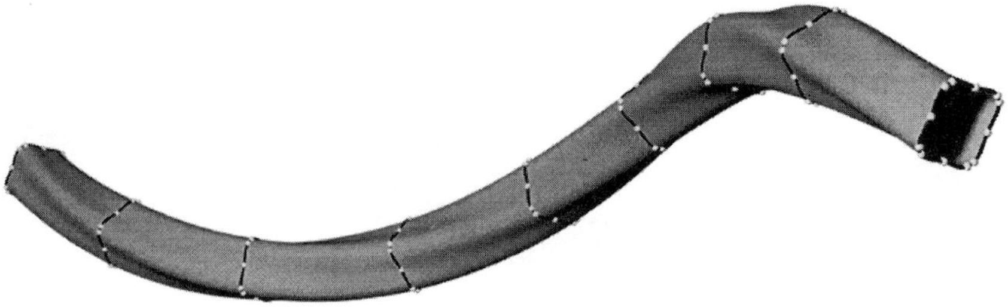

Fig. 3a. The "transformed-cubic" skinning interpolant.

Fig. 3b. The **c**-sectional-curvature preserving skinning interpolant.

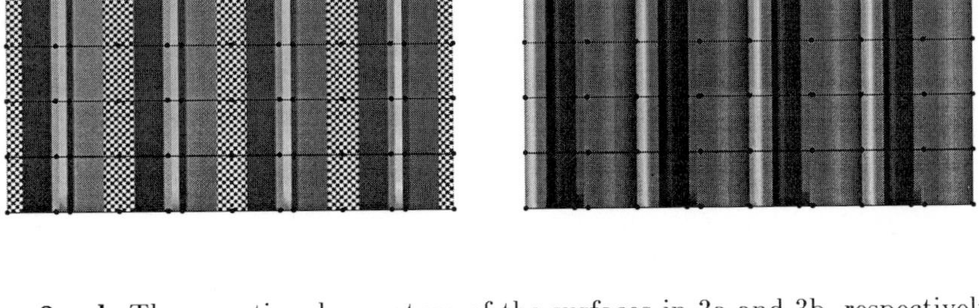

Fig. 3c–d. The **c**-sectional curvature of the surfaces in 3a and 3b, respectively.

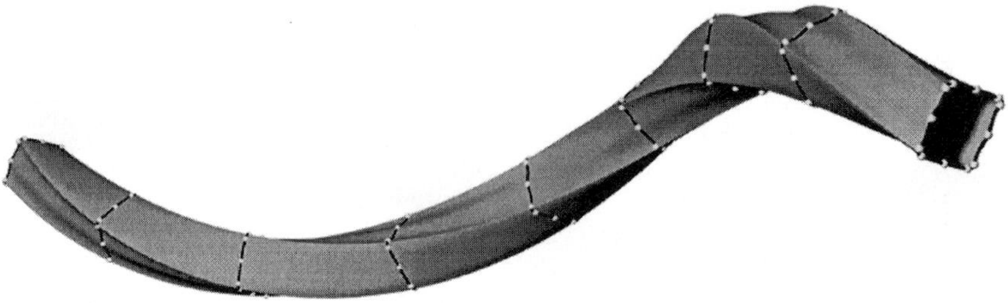

Fig. 4a. The "transformed-cubic" skinning interpolant.

Fig. 4b. The **c**-sectional-curvature preserving skinning interpolant.

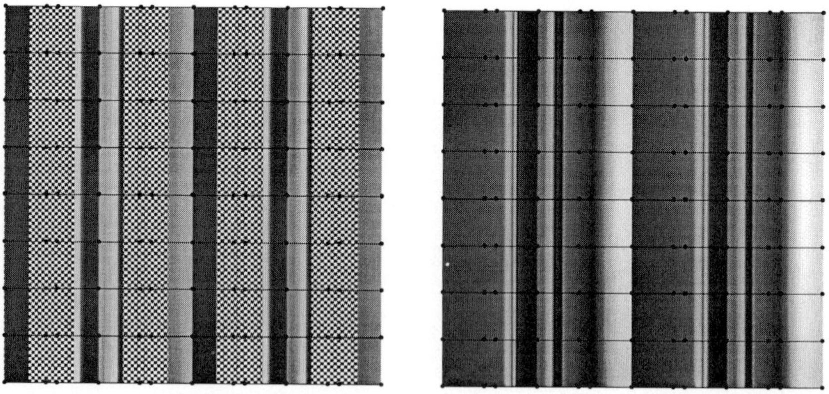

Fig. 4c–d. The **c**-sectional curvature of the surfaces in 4a and 4b, respectively.

References

1. Carnicer, J. M., Multivariate convexity preserving interpolation by smooth functions, Adv. Comput. Math. **3** (1995), 395–404.
2. Kaklis, P. D. and A. I. Ginnis, Sectional-curvature preserving skinning surfaces, 1995, to be published in Computer-Aided Geom. Design .
3. Kaklis, P. D. and M. I. Karavelas, Shape-preserving interpolation in $\mathbb{R}^3$, August 1993, submitted for publication in IMA J. Numer. Anal.
4. Kaklis, P. D. and D. G. Pandelis, Convexity-preserving polynomial splines of non-uniform degree, IMA J. Numer. Anal., **10** (1990), 223–234.
5. Kaklis, P. D. and N. S. Sapidis, Convexity-preserving interpolatory parametric splines of non-uniform polynomial degree, Computer-Aided Geom. Design **12** (1995), 1–26.

A. I. Ginnis, P. D. Kaklis and N. C. Gabrielides
Ship-Design Laboratory
Dept. Naval Architecture & Marine Engineering
National Technical University of Athens
9, Heroon Polytechneiou
157 73 Zografou
Athens (Greece)
alexandros_ginnis@deslab.naval.ntua.gr
kaklis@deslab.naval.ntua.gr
ngabriel@deslab.naval.ntua.gr

Pairs of Refinable Bivariate Splines

Tim Goodman

Abstract. A general formula is obtained for the Fourier transform of a bivariate spline function. It is then shown that if a pair of such functions are refinable, then the Fourier transform must take a particular form involving two terms, each containing a trigonometric polynomial. Examples are given.

§1. Introduction

In constructing multivariate spline functions there is a great wealth of possibilities due to the large variety of choices of partitions and of continuity conditions. Polyhedral splines, see for example [1], provide a wide and useful class of examples. However there is no obvious reason why attention should be restricted to polyhedral splines, and even polyhedral splines have not been much studied apart from box splines and simplex splines.

There is some work on other constructions, for example that of Sablonnière in [7] and in this Proceedings. The intention of this author is to study spline functions which are refinable, i.e. they are finite linear combinations of of translates of dilates of themselves. Such a property is useful in multiresolution techniques such as subdivision algorithms and wavelets. Part of the motivation for this work is in extending the construction in [4] of a piecewise linear wavelet on the 4-direction mesh to higher order smoothness. As pointed out in [2], a straightforward extension by convolution, as is used for box splines, leads to a loss of the Riesz basis property. A further motivation is the paper of Lawton, Lee and Shen [5] in which they characterise single refinable univariate splines and show, in particular, that the integer translates of such a function form a Riesz basis if and only if it is a uniform B-spline. This work has been partially extended to the multivariate case by Sun [8] who shows, in particular, that a single refinable multivariate spline is a linear combination of box splines.

Advanced Topics in Multivariate Approximation
F. Fontanella, K. Jetter and P. J. Laurent (eds.) pp. 125–138.

ISBN 981-02-2852-X

In [5] and [8] it is assumed that there is a single refinable function. However the construction in [4] involves a pair of refinable splines. For this reason the author examines in [3] pairs of univariate refinable splines and characterises such pairs whose integer translates form a Riesz basis, see also [6]. In contrast to the case of a single function [5], this gives rise to new functions with non-standard continuity conditions. In this paper we start an extension of this work to the bivariate case.

In all the above papers on refinable splines, the analysis has been in terms of the Fourier transforms of the spline functions. So in Section 2 we examine the Fourier transforms of bivariate splines. In [3] the analysis depends crucially on the result that the Fourier transform of a pair of refinable splines involves just two terms, each one giving a discontinuity in a different order derivative of the spline functions. An analogous result is given in Section 3 for the bivariate case, but in this case the two terms may both give discontinuities only in the highest order derivatives. This is the case for the continuous linear splines on a 4-direction mesh considered in [4]. This example has now been extended to C^{n-1} splines of degree n on a mesh with $n+3$ directions, but this construction must be postponed to a later paper.

§2. Fourier Transforms of Splines

We shall derive a formula for the Fourier transform $\hat{\phi}$ of a bivariate spline function ϕ, where for u in $\mathbb{R}^2$,

$$\hat{\phi}(u) := \int_{\mathbb{R}^2} \phi(x) e^{-ixu}\, dx.$$

Let w be an arbitrary fixed non-zero vector in $\mathbb{R}^2$. For any point P in $\mathbb{R}^2$ and non-zero vector $v = (v_1, v_2)$, we define $v^\perp := (v_2, -v_1)$ and

$$\phi_w(P, v) := (v^\perp w) \lim_{h\to 0+} \lim_{\epsilon\to 0+} \{\phi(P + hv + \epsilon v^\perp) - \phi(P + hv - \epsilon v^\perp)\}. \tag{2.1}$$

Thus $\phi_w(P, v)$ denotes, in a sense, the jump at P of ϕ in crossing in the direction of w a line in direction v away from P. Note that if w is parallel to v, then $\phi_w(P, v) = 0$.

Now suppose that ϕ is a piecewise polynomial of maximal total degree n on a partition of $\mathbb{R}^2$ with a finite set N of nodes and for each node P in N, a finite number $m(P)$ of edges meeting at P in directions $v_1^P, \dots, v_{m(P)}^P$ away from P.

Theorem 1. *If ϕ is a spline function, as above, with compact support, then $\hat{\phi}$ is continuous and for $u \in \mathbb{R}^2$,*

$$\hat{\phi}(u) = \sum_{P\in N} e^{-iPu} \sum_{k,l\ge 0, k+l\le n} (iwu)^{-l-1} \sum_{j=0}^{m(P)} (iv_j^P u)^{-k-1} D_{v_j^P}^k D_w^l \phi_w(P, v_j^P). \tag{2.2}$$

Here D_v denotes the directional derivative $v_1 D_1 + v_2 D_2$. On recalling (2.1), we note that (2.2) is independent of the magnitudes of w and the vectors v_j^P. We emphasise that w is *any* fixed non-zero vector. By taking w parallel to one of the edges v_j^P, some of the terms in (2.2) will vanish.

In proving Theorem 1 it will be convenient to first prove some preliminary results.

Lemma 1. *Take points A, B in $\mathbb{R}^2$ and let v be a vector in the direction from A to B. Then for any polynomial p of total degree n,*

$$\int_{AB} p(x)e^{-ixu}\,ds(x) = \sum_{k=0}^{n} \frac{|v|}{(ivu)^{k+1}}\{D_v^k p(A)e^{-iAu} - D_v^k p(B)e^{-iBu}\}. \tag{2.3}$$

Proof: Since the right hand side of (2.3) is independent of the magnitude of v, we may assume that v is a unit vector. Then integrating by parts gives

$$\int_{AB} p(x)e^{-ixu}\,ds(x) = \int_{AB} p(x)D_v(\frac{1}{-ivu}e^{-ixu})\,ds(x)$$

$$= \frac{1}{-ivu}[p(x)e^{-ixu}]_A^B + \frac{1}{ivu}\int_{AB} D_v p(x)e^{-ixu}\,ds(x).$$

Applying the same procedure to the last integral and successively repeating this procedure gives (2.3). ■

Lemma 2. *Let Ω be a polygonal region with vertices $P_1, \ldots, P_m$ described anti-clockwise and for $j = 1, \ldots, m$, let v_j denote a vector in the direction from P_j to P_{j+1}, where $P_{m+1} = P_1$. Then for a polynomial p of total degree n,*

$$\int_{\Omega} p(x)e^{-ixu}\,dx =$$

$$\sum_{j=1}^{m} \sum_{k,l\geq 0, k+l\leq n} \frac{v_j^{\perp}w}{(iv_ju)^{k+1}(iwu)^{l+1}}\{D_{v_j}^k D_w^l p(P_{j+1})e^{-iP_{j+1}u} - D_{v_j}^k D_w^l p(P_j)e^{-iP_ju}\}.$$

(2.4).

Proof: We note that for $j = 1, \ldots, m$, the vector $v_j^{\perp}$ is an outward normal from Ω across the edge P_jP_{j+1}. Integrating by parts gives

$$\int_{\Omega} p(x)e^{-ixu}\,dx = \int_{\Omega} p(x)D_w(\frac{1}{-iwu}e^{-ixu})\,dx$$

$$= \frac{-1}{iwu}\sum_{j=1}^{m}\int_{P_jP_{j+1}} p(x)e^{-ixu}\frac{v_j^{\perp}w}{|v_j|}\,ds(x) + \frac{1}{iwu}\int_{\Omega} D_w p(x)e^{-ixu}\,dx$$

$$= \frac{1}{iwu} \sum_{j=1}^{m} \sum_{k=0}^{n} \frac{(v_j^{\perp} w)}{(iv_j u)^{k+1}} \{ D_{v_j}^k p(P_{j+1}) e^{-iP_{j+1}u} - D_{v_j}^k p(P_j) e^{-iP_j u} \}$$

$$+ \frac{1}{iwu} \int_{\Omega} D_w p(x) e^{-ixu} dx,$$

on applying Lemma 1. Applying the same procedure to the last integral and successively repeating this procedure gives (2.4). ■

Proof of Theorem 1: In the situation of Lemma 2, define ψ to be equal to p on Ω and to be zero elsewhere. Then putting $v_0 = v_m$ and recalling (2.1), equation (2.4) can be rewritten as

$$\int_{\mathbf{R}^2} \psi(x) e^{-ixu} dx = \sum_{j=1}^{m} e^{-iP_j u} \sum_{k,l \geq 0, k+l \leq n} (iwu)^{-l-1} \times$$

$$\{ (iv_j u)^{-k-1} D_{v_j}^k D_w^l \psi_w(P_j, v_j) + (-iv_{j-1} u)^{-k-1} D_{-v_{j-1}}^k D_w^l \psi_w(P_j, -v_{j-1}) \}. \quad (2.5)$$

Applying (2.5) on each region on which ϕ is a polynomial and adding gives (2.2). ■

As an example we consider the 3-direction mesh formed by the lines $x = j$, $y = j$, $y - x = j$, $j \in \mathbb{Z}$. In this case $N = \mathbb{Z}^2$ and we take $w = (1,1)$. Then (2.2) becomes

$$\hat{\phi}(u) = \sum_{\nu \in \mathbb{Z}^2} e^{-i\nu u} \sum_{k,l \geq 0, k+l \leq n} (i(u_1 + u_2))^{-l-1} \times$$

$$\{ (iu_1)^{-k-1} [D_1^k (D_1 + D_2)^l \phi_w(\nu, (1,0)) - D_1^k (D_1 + D_2)^l \phi_w(\nu, (-1,0))] +$$
$$(iu_2)^{-k-1} [D_2^k (D_1 + D_2)^l \phi_w(\nu, (0,1)) - D_2^k (D_1 + D_2)^l \phi_w(\nu, (0,-1))] \}. \quad (2.6)$$

Next we give an example to illustrate that a spline function ϕ may have an 'active' knot P and yet the term in the summation in (2.2) corresponding to P vanishes. Let ϕ denote the 'pyramid function' with support on the square with vertices $(1,0), (0,1), (-1,0)$ and $(0,-1)$ which is continuous, satisfies $\phi(0,0) = 1$ and is linear on each of the four triangles formed by the boundary and two diagonals of its support. Then applying (2.2) with $w = (1,0)$ gives

$$\begin{aligned} \hat{\phi}(u) =& (iu_1)^{-2} \{ [(i(u_1 + u_2))^{-1} + (i(u_1 - u_2))^{-1}](e^{iu_1} - e^{-iu_1}) \\ &+ [(i(u_1 + u_2))^{-1} - (i(u_1 - u_2))^{-1} - 2(iu_2)^{-1}](e^{iu_2} - e^{-iu_2}) \} \\ =& \frac{4(u_1 \sin u_2 - u_2 \sin u_1)}{u_1 u_2 (u_1 + u_2)(u_1 - u_2)}. \end{aligned} \quad (2.7)$$

Note that there is no term corresponding to the vertex $(0,0)$. Finally we give two results showing that certain terms in the summation in (2.2) must be non-zero.

Proposition 1. *Suppose that ϕ in Theorem 1 has exact total degree n on some region. Then there is some P in N and some k, l with $k + l = n$ such that the term in the summation in (2.2) involving P, k and l does not vanish identically.*

Proof: We shall suppose that all terms in (2.2) vanish when $k + l = n$ and we shall reach a contradiction. Take any k, l with $k + l = n$. Take any P in N and suppose that $v_0, \ldots, v_m$ are the pairwise linearly independent directions of edges meeting at P. Then the functions $(v_j u)^{-k-1}, j = 0, \ldots, m$ are linearly independent. Thus if there is only one edge parallel to v_j meeting at P, we have $D_{v_j}^k D_w^l \phi_w(P, v_j) = 0$. Noting that $D_{v_j}^k D_w^l \phi$ is piecewise constant, we see that if v_j is not parallel to w, then $D_{v_j}^k D_w^l \phi$ is continuous across this edge. Next suppose that there are two edges parallel to v_j meeting at P. Then $D_{v_j}^k D_w^l \phi_w(P, v_j) - D_{v_j}^k D_w^l \phi_w(P, -v_j) = 0$. Thus if v_j is not parallel to w, the jump of $D_{v_j}^k D_w^l$ across the edge corresponding to v_j equals that across the edge corresponding to $-v_j$. Since this applies to any vertex P, the jump of $D_{v_j}^k D_w^l \phi$ across a line parallel to v_j is constant along the line and, since ϕ has compact support, must be zero.

Therefore we have proved that if v is not parallel to w, then for $k = 0, \ldots, n$, $D_v^k D_w^{n-k} \phi$ is continuous across all lines parallel to v. Thus all derivatives of ϕ of order n are continuous across all lines other than those parallel to w. Since ϕ has compact support and all such derivatives are piecewise constant, it follows that that they must all vanish identically. This contradicts ϕ having exact total degree n on some region. ■

The final result relates the non-zero terms in (2.2) with the support of ϕ.

Proposition 2. *Suppose that ϕ is a spline function as in Theorem 1 on a partition with a set E of edges. Let N' denote all nodes P in N for which the coefficient of e^{-iPu} in the summation in (2.2) does not vanish identically. Let Ω be the smallest closed set which contains all nodes in N' and whose boundary intersects any line containing an edge in E in at most two points. Then ϕ has support in Ω.*

Proof: We first note that Ω lies inside the convex hull of N' and so is bounded. Let L be any line containing an edge of E and which is not parallel to w. If L does not intersect Ω, then L does not contain any node in N' and arguing as in the proof of Proposition 1 shows that ϕ must be continuous across L. Suppose then that L intersects Ω in two (possibly coincident) points. Then L is divided into three connected components: $L \cap \Omega$ and two unbounded components outside Ω. Arguing as in the proof of Proposition 1 shows that ϕ is continuous across each of the unbounded components. Then outside Ω, ϕ is continuous except possibly across a finite number of lines parallel to w. Since ϕ has compact support it follows that ϕ vanishes outside Ω. ■

As already mentioned, the set Ω in Proposition 2 lies in the convex hull of N' and so we can immediately deduce the following.

Corollary 1. *If ϕ and N' are as in Proposition 2, then the support of ϕ lies in the convex hull of N'.*

§3. Refinable Splines

Let $\phi_1, \phi_2 : \mathbb{R}^2 \to \mathbb{R}$ be compactly supported spline functions. We shall suppose that $\phi = (\phi_1, \phi_2)^T$ is *m-refinable*, i.e. for some integer $m \geq 2$, ϕ satisfies a refinement equation

$$\phi(x) = \sum_{j \in \mathbb{Z}^2} A(j)\phi(mx - j), \quad x \in \mathbb{R}^2, \tag{3.1}$$

where $A(j)$ is a 2×2 matrix which vanishes for all but a finite set of j in $\mathbb{Z}^2$. Taking Fourier transforms in (3.1) gives

$$\hat{\phi}(u) = \frac{1}{m^2} C(e^{-\frac{iu}{m}}) \hat{\phi}(\frac{u}{m}), \quad u \in \mathbb{R}^2, \tag{3.2}$$

where

$$C(z) = \sum_{j \in \mathbb{Z}^2} A(j) z^j, \quad |z| = 1. \tag{3.3}$$

We shall suppose henceforth that ϕ is a spline function of maximal total degree n on the partition of $\mathbb{R}^2$ formed by the lines

$$n_j u = r, \quad r \in \mathbb{Z}, \quad j = 0, \ldots, s, \tag{3.4}$$

where $n_j \in \mathbb{Z}^2$, $j = 0, \ldots, s$, are pairwise linearly independent vectors. We note that the partition is invariant under a translation by any element of $\mathbb{Z}^2$ and the nodes have rational coordinates. The set of nodes in $[0, 1) \times [0, 1)$ is finite and we shall denote it by S.

We now apply Theorem 1. Taking $v_j = n_j^{\perp}$, $j = 0, \ldots, s$, and $w = v_0$, (2.2) gives

$$\hat{\phi}(u) = \sum_{P \in S} e^{-iPu} \sum_{j=1}^{s} \sum_{k,l \geq 0, k+l \leq n} \frac{T(u; j, k, l, P)}{(v_j u)^{k+1} (wu)^{l+1}}, \quad u \in \mathbb{R}^2, \tag{3.5}$$

where each term $T(.; j, k, l, P)$ is a trigonometric polynomial, i.e. of form

$$T(u; j, k, l, P) = \sum_{\nu \in \mathbb{Z}^2} a(\nu) e^{-i\nu u}$$

for some coefficients $a(\nu)$ in $\mathbb{R}^2$. We may write (3.5) as

$$\hat{\phi}(u) = \sum_{j=1}^{s} \sum_{k,l \geq 0, k+l \leq n} \frac{T(u; j, k, l)}{(v_j u)^{k+1} (wu)^{l+1}}, \quad u \in \mathbb{R}^2, \tag{3.6}$$

where each term

$$T(u;j,k,l) := \sum_{P\in S} e^{-iPu}T(u;j,k,l,P), \quad u \in \mathbb{R}^2, \tag{3.7}$$

is a trigonometric polynomial in $\frac{u}{M}$ for some integer M.

We shall show that each of the functions $T(.;j,k,l)$ in (3.6) satisfies an equation similar to (3.2). First we need a preliminary result. Both the result and its proof are closely related to Lemma 3 of [8].

Lemma 3. *Suppose that $P_1,\dots,P_r$ are linearly independent homogeneous polynomials on $\mathbb{R}^2$ and $T_1,\dots,T_r$ are trigonometric polynomials on $\mathbb{R}^2$. If $\sum_{j=1}^r T_jP_j$ is identically zero, then $T_1,\dots,T_r$ are identically zero.*

Proof: Define the difference operators

$$\Delta_1 f(x) := f(x+2\pi(1,0)) - f(x), \quad \Delta_2 f(x) := f(x+2\pi(0,1)) - f(x).$$

Then $\Delta_1 T = \Delta_2 T = 0$ for any trigonometric polynomial T. Also if Q is a homogeneous polynomial of degree n, then $\Delta_1 Q$ and $\Delta_2 Q$ are homogeneous polynomials of degree at most $n-1$.

We shall prove the result by induction on r. Clearly it is true for $r=1$. Assume that it is true for $r-1$ and suppose that $\sum_{j=1}^r T_jP_j = 0$. Let P_1 have degree n where, without loss of generality, $P_2,\dots,P_r$ have degree $\le n$. We can choose k, $0\le k\le n$, so that $\Delta_1^k\Delta_2^{n-k}P_1$ is a non-zero constant a_1. Then

$$\begin{aligned}0 &= \Delta_1^k\Delta_2^{n-k}\sum_{j=1}^r T_jP_j\\ &= \sum_{j=1}^r T_j\Delta_1^k\Delta_2^{n-k}P_j\\ &= \sum_{j=1}^r a_jT_j,\end{aligned}$$

for some constants $a_2,\dots,a_r$. Thus

$$\begin{aligned}0 &= \sum_{j=1}^r T_jP_j\\ &= (-\sum_{j=2}^r \frac{a_j}{a_1}T_j)P_1 + \sum_{j=2}^r T_jP_j\\ &= \sum_{j=2}^r T_j(P_j - \frac{a_j}{a_1}P_1).\end{aligned}$$

Since $P_1, \dots, P_r$ are linearly independent, $P_j - \frac{a_j}{a_1} P_1$, $j = 2, \dots, r$ are linearly independent. So our inductive hypothesis gives $T_2 = \dots = T_r = 0$ and hence $T_1 = 0$. ■

Clearly Lemma 3 and its proof extend to $\mathbb{R}^s$, $s \geq 3$.

Proposition 3. *Suppose that (3.1) is satisfied. Then for* $1 \leq j \leq s$ *and* $k, l \geq 0$, $k + l \leq n$, *the function* $T(.; j, k, l)$ *in (3.6) satisfies*

$$T(u; j, k, l) = m^{k+l} C(e^{-\frac{iu}{m}}) T(\frac{u}{m}; j, k, l), \quad u \in \mathbb{R}^2, \tag{3.8}$$

where C *is given by (3.3).*

Proof: Substituting (3.6) into (3.2) gives

$$\sum_{j=1}^{s} \sum_{k,l \geq 0, k+l \leq n} \frac{1}{(v_j u)^{k+1} (wu)^{l+1}} \{T(u; j, k, l) - m^{k+l} C(e^{-\frac{iu}{m}}) T(\frac{u}{m}; j, k, l)\} = 0.$$

Putting the above summation over a common denominator, we see that the result follows from Lemma 3 provided that for $r = 0, \dots, n$, the homogeneous polynomials

$$P_{\alpha,i,r}(u) = \frac{(wu)^\alpha}{(v_i u)^{\alpha+1}} \prod_{j=1}^{s} (v_j u)^{r+1}, \quad u \in \mathbb{R}^2, \quad \alpha = 0, \dots, r, \quad i = 1 \dots, s,$$

are linearly independent. We prove this by induction on r. Suppose that for some r, $0 \leq r \leq n$, the result is true for $r - 1$ (a vacuous assumption for $r = 0$). Suppose that

$$\sum_{\alpha=0}^{r} \sum_{i=1}^{s} \lambda_{\alpha,i} P_{\alpha,i,r} = 0.$$

For $i = 1, \dots, s$, putting $u = v_i^\perp$ gives $\lambda_{r,i} = 0$. If $r = 0$, the result is then established. Suppose that $r \geq 1$. We note that for $0 \leq \alpha \leq r - 1$,

$$P_{\alpha,i,r}(u) = \prod_{j=1}^{s} (v_j u) P_{\alpha,i,r-1}(u), \quad u \in \mathbb{R}^2,$$

and so

$$\sum_{\alpha=0}^{r-1} \sum_{i=1}^{s} \lambda_{\alpha,i} P_{\alpha,i,r-1} = 0.$$

By our inductive hypothesis,

$$\lambda_{\alpha,i} = 0, \quad \alpha = 0, \dots, r - 1, \quad i = 1, \dots, s,$$

and so the result is established. ■

We now prove our main result.

Theorem 2. *Suppose that ϕ is, as above, an m-refinable spline function on the partition formed by (3.4) with exact maximal total degree n. Then for some n', $0 \le n' \le n$, there are functions T_1, T_2 of form $T_j(u) = \sum_{P \in S} e^{-iPu} t_j(u)$, for trigonometric polynomials t_j, $j = 1, 2$, such that*

$$\hat{\phi}(u) = T_1(u)P_1(u)\prod_{j=0}^{s}(v_j u)^{-n-1} + T_2(u)P_2(u)\prod_{j=0}^{s}(v_j u)^{-n'-1}, \tag{3.9}$$

where P_1 and P_2 are homogeneous polynomials of degree $sn + s - 1$ and $sn' + s - 1$, respectively.

Proof: We can suppose that in the summation (3.6) there are two functions $T(.; j, k, l)$ and $T(.; j'k'l')$ which do not vanish identically, such that either $k + l \neq k' + l'$ or $k + l = k' + l'$ and they are linearly independent. For if this does not hold, then (3.6) gives (3.9) with only one of the two terms. Recalling Proposition 1, we can also assume that $k + l = n$. We write $k' + l' = n'$, $T(.; j, k, l) = T_1$, $T(.; j', k', l') = T_2$.

We claim that there is some u for which the 2-vectors $T_1(u)$ and $T_2(u)$ are linearly independent. For suppose that this is not the case. Then there is a scalar function $\lambda(u)$ such that for all u with $T_1(u) \neq 0$,

$$T_2(u) = \lambda(u)T_1(u). \tag{3.10}$$

From (3.8),

$$T_1(u) = m^n C(e^{-\frac{iu}{m}})T_1(\frac{u}{m}), \quad T_2(u) = m^{n'} C(e^{-\frac{iu}{m}})T_2(\frac{u}{m}). \tag{3.11}$$

Applying (3.10) gives

$$\begin{aligned}\lambda(u)T_1(u) &= T_2(u) \\ &= m^{n'} C(e^{-\frac{iu}{m}})T_2(\frac{u}{m}) \\ &= m^{n'} C(e^{-\frac{iu}{m}})\lambda(\frac{u}{m})T_1(\frac{u}{m}) \\ &= m^{n'-n}\lambda(\frac{u}{m})T_1(u).\end{aligned} \tag{3.12}$$

Without loss of generality we may assume that $(T_1)_1 \neq 0$. If $(T_2)_1 = 0$, then by (3.10), $\lambda(u) = 0$ for almost all u and so $T_2 = 0$, which is not the case. So $(T_2)_1 \neq 0$. Expanding $(T_1)_1$ and $(T_2)_1$ in Taylor series about $u = 0$, we write

$$T_2(u)_1 = R_2(u) + O(|u|^{K+1}),$$

$$T_1(u)_1 = R_1(u) + O(|u|^{L+1}),$$

where R_2 and R_1 are homogeneous polynomials of exact degree K and L respectively. For almost all u, $(T_j)_1(u) \neq 0$ and $R_j(u) \neq 0$, $j = 1, 2$. For any such u, (3.12) gives

$$\begin{aligned}
\lambda(u) &= m^{n'-n}\lambda(\frac{u}{m}) \\
&= m^{r(n'-n)}\lambda(\frac{u}{m^r}), \quad r = 1, 2, \ldots \\
&= m^{r(n'-n)}\frac{R_2(\frac{u}{m^r}) + O(m^{-r(K+1)})}{R_1(\frac{u}{m^r}) + O(m^{-r(L+1)})} \quad \text{as} \quad r \to \infty \\
&= m^{r(n'-n)}\frac{m^{-rK}R_2(u) + O(m^{-r(K+1)})}{m^{-rL}R_1(u) + O(m^{-r(L+1)})} \\
&= m^{r(n'-n+L-K)}\frac{R_2(u) + o(1)}{R_1(u) + o(1)}.
\end{aligned}$$

Since $0 < \lambda(u) < \infty$,

$$n' - n + L - K = 0 \tag{3.13}$$

and by (3.10),

$$T_2(u)R_1(u) = T_1(u)R_2(u). \tag{3.14}$$

By continuity (3.14) holds for all u. Applying Lemma 3 to the first components of (3.14) shows that R_1 and R_2 are linearly dependent. In particular $K = L$ and so $n' = n$ by (3.13). It follows from (3.14) that $T_1(u)$ and $T_2(u)$ are linearly dependent, which contradicts our original supposition.

Now let T denote the 2×2 matrix $[T_1, T_2]$. We have shown that $\det T$ does not vanish identically. By (3.8), for $u \in \mathbb{R}^2$,

$$\begin{aligned}
T(u) &= C(e^{-\frac{iu}{m}})T(\frac{u}{m})\begin{bmatrix} m^n & 0 \\ 0 & m^{n'} \end{bmatrix} \\
&= C(e^{-\frac{iu}{m}})C(e^{-\frac{iu}{m^2}})\ldots C(e^{-\frac{iu}{m^r}})T(\frac{u}{m^r})\begin{bmatrix} m^{rn} & 0 \\ 0 & m^{rn'} \end{bmatrix},
\end{aligned}$$

for $r = 1, 2, \ldots$. Also from (3.2), for $r = 1, 2, \ldots$, $u \in \mathbb{R}^2$,

$$\hat{\phi}(u) = \frac{1}{m^{2r}}C(e^{-\frac{iu}{m}})C(e^{-\frac{iu}{m^2}})\ldots C(e^{-\frac{iu}{m^r}})\hat{\phi}(\frac{u}{m^r}).$$

So for u with $\det T(u) \neq 0$, we have for $r = 1, 2, \ldots$, $\det T(\frac{u}{m^r}) \neq 0$ and

$$\hat{\phi}(u) = \frac{1}{m^{2r}}T(u)\begin{bmatrix} m^{-rn} & 0 \\ 0 & m^{-rn'} \end{bmatrix}T(\frac{u}{m^r})^{-1}\hat{\phi}(\frac{u}{m^r}).$$

Hence

$$\lim_{r\to\infty}\begin{bmatrix} m^{-r(n+2)} & 0 \\ 0 & m^{-r(n'+2)}\end{bmatrix} T(\frac{u}{m^r})^{-1}\hat{\phi}(\frac{u}{m^r}) = T(u)^{-1}\hat{\phi}(u). \tag{3.15}$$

Since ϕ has compact support, $\hat{\phi}$ is analytic and we can write, as $u \to 0$,

$$T(u)^{-1}\hat{\phi}(u) = \{Q(u) + O(|u|^{N+1})\}^{-1}\begin{bmatrix} Q_1(u) + O(|u|^{N_1+1}) \\ Q_2(u) + O(|u|^{N_2+1})\end{bmatrix}, \tag{3.16}$$

where Q, Q_1, Q_2 are homogeneous polynomials of exact degree N, N_1, N_2 respectively. (It may be that the first or second components of $T^{-1}\hat{\phi}$, but not both, vanish identically, but this is ignored in our present analysis and will be considered later.) For almost all u we have $Q_1(u)$, $Q_2(u)$, $Q(u)$ and both components of $T(u)^{-1}\hat{\phi}(u)$ are non-zero. For any such u, (3.15) and (3.16) give

$$\begin{aligned}(T(u)^{-1}\hat{\phi}(u))_1 &= \lim_{r\to\infty} m^{-r(n+2)}\frac{Q_1(\frac{u}{m^r}) + O(m^{-r(N_1+1)})}{Q(\frac{u}{m^r}) + O(m^{-r(N+1)})} \\ &= \lim_{r\to\infty} m^{-r(n+2+N_1-N)}\frac{Q_1(u) + o(1)}{Q(u) + o(1)}\end{aligned}$$

with a corresponding formula for $(T(u)^{-1}\hat{\phi}(u))_2$. Thus

$$n + 2 + N_1 - N = 0, \quad n' + 2 + N_2 - N = 0 \tag{3.17}$$

and

$$T(u)^{-1}\hat{\phi}(u) = \frac{1}{Q(u)}\begin{bmatrix} Q_1(u) \\ Q_2(u)\end{bmatrix},$$

which gives

$$\hat{\phi}(u) = \frac{1}{Q(u)}T(u)\begin{bmatrix} Q_1(u) \\ Q_2(u)\end{bmatrix}. \tag{3.18}$$

By continuity, (3.18) holds for all u with $Q(u) \neq 0$. If the first (or second) component of $T(u)^{-1}\hat{\phi}$ vanishes identically, then we take P_1 (or P_2) to be zero.

We now compare (3.18) with (3.6). From (3.17) we know deg $Q-$ deg $Q_1 = n+2$, deg $Q-$ deg $Q_2 = n'+2$. Replacing u by ru and letting $r \to \infty$ gives the following.

If $n \neq n'$, then for $u \in \mathbb{R}^2$,

$$\frac{T_1(u)Q_1(u)}{Q(u)} = \sum_{j=1}^{s}\sum_{k=0}^{n}\frac{T(u;j,k,n-k)}{(v_j u)^{k+1}(wu)^{n-k+1}}, \tag{3.19}$$

$$\frac{T_2(u)Q_2(u)}{Q(u)} = \sum_{j=1}^{s}\sum_{k=0}^{n'} \frac{T(u;j,k,n'-k)}{(v_j u)^{k+1}(wu)^{n'-k+1}}. \tag{3.20}$$

If $n = n'$, then for $u \in \mathbb{R}^2$,

$$\frac{T_1(u)Q_1(u) + T_2(u)Q_2(u)}{Q(u)} = \sum_{j=1}^{s}\sum_{k=0}^{n} \frac{T(u;j,k,n-k)}{(v_j u)^{k+1}(wu)^{n-k+1}}. \tag{3.21}$$

Now suppose that $n \neq n'$. Multiplying through by the denominator in (3.19) and applying Lemma 3 as in the proof of Proposition 3, shows that $Q_1(u)(wu)^{n+1} \prod_{j=1}^{s}(v_j u)^{n+1}$ is a linear combination of

$$Q(u)(wu)^k(v_i u)^{-k-1}\prod_{j=1}^{s}(v_j u)^{n+1}, \quad k = 0,\ldots,n, \quad i = 1,\ldots,s. \tag{3.22}$$

Recalling that $w = v_0$, we have

$$\frac{Q_1(u)}{Q(u)} = P_1(u)\prod_{j=0}^{s}(v_j u)^{-n-1},$$

where P_1 is a homogeneous polynomial of degree $sn + s - 1$. Deriving a similar result for (3.20) and substituting into (3.18) gives (3.9).

Finally suppose that $n = n'$. Then applying our above procedure to (3.21) shows that there are constants a, b, not both zero, so that $(aQ_1(u)+bQ_2(u))(wu)^{n+1} \prod_{j=1}^{s}(v_j u)^{n+1}$ is a linear combination of the terms in (3.22). Without loss of generality we may assume $a \neq 0$. Then

$$\begin{aligned}
&\frac{Q_2(u)(aT_2(u) - bT_1(u))}{Q(u)} \\
&\quad = \frac{a(T_1(u)Q_1(u) + T_2(u)Q_2(u)) - T_1(u)(aQ_1(u) + bQ_2(u))}{Q(u)} \\
&\quad = \sum_{j=1}^{s}\sum_{k=0}^{n} \frac{aT(u;j,k,n-k) - T_1(u)c_{j,k}}{(v_j u)^{k+1}(wu)^{n-k+1}},
\end{aligned}$$

for some constants $c_{j,k}$, by (3.21). Since T_2 and T_1 are linearly independent, $aT_2 - bT_1 \neq 0$, and so our previous argument shows that $Q_2(u)(wu)^{n+1}\prod_{j=1}^{s}(v_j u)^{n+1}$ is a linear combination of the terms in (3.22) and hence $Q_1(u)(wu)^{n+1}\prod_{j=1}^{s}(v_j u)^{n+1}$ is also. Thus we have

$$\frac{Q_j(u)}{Q(u)} = P_j(u)\prod_{j=0}^{s}(v_j u)^{-n-1}, \quad j = 1,2,$$

where P_1 and P_2 are homogeneous polynomials of degree $sn+s-1$. So again (3.18) gives (3.9). ■

To finish this paper we give two examples.

Example 1.

Let ϕ_1 be the characteristic function of the triangle with edges $x_2 = 0$, $x_1 = 1$ and $x_1 = x_2$, and let ϕ_2 be the characteristic function of the triangle with edges $x_1 = 0$, $x_2 = 1$, $x_1 = x_2$. Here the refinement equation (3.1) with $m = 2$ is

$$\phi(x) = \begin{bmatrix} 1 & 0 \\ 0 & 1 \end{bmatrix} (\phi(2x)+\phi(2x-(1,1)))+\begin{bmatrix} 1 & 1 \\ 0 & 0 \end{bmatrix} \phi(2x-(1,0))+\begin{bmatrix} 0 & 0 \\ 1 & 1 \end{bmatrix} \phi(2x-(0,1)),$$

so that from (3.3),

$$C(z) = \begin{bmatrix} 1+z_1+z_1z_2 & z_1 \\ z_2 & 1+z_2+z_1z_2 \end{bmatrix}.$$

From (2.6) we have

$$\hat{\phi}_1(u) = \frac{(u_1+u_2)(e^{-iu_1}-1) - u_1(e^{-i(u_1+u_2)}-1)}{u_1u_2(u_1+u_2)},$$

$$\hat{\phi}_2(u) = \hat{\phi}_1(u_2, u_1).$$

Example 2.

Here ϕ_1 is the pyramid function described after the proof of Theorem 1. We let ϕ_2 be the 'pyramid function' with support on $[0,1]^2$, i.e.

$$\phi_2(x) = \phi_1(x_1+x_2-1, x_1-x_2).$$

Then in (3.3) we have

$$C(z) = \begin{bmatrix} 1+\frac{1}{2}(z_1+z_1^{-1}+z_2+z_2^{-1}) & \frac{1}{2}(1+z_1^{-1})(1+z_2^{-1}) \\ z_1z_2 & \frac{1}{2}(1+z_1)(1+z_2) \end{bmatrix}.$$

Here $\hat{\phi}_1$ is given by (2.7) and

$$\hat{\phi}_2 = \frac{1}{2}e^{-\frac{1}{2}i(u_1+u_2)}\hat{\phi}_1\left(\frac{1}{2}(u_1+u_2), \frac{1}{2}(u_1-u_2)\right).$$

These two examples will be extended to splines with higher degree on partitions with more directions in a later paper.

References

1. Goodman, T. N. T., Polyhedral splines, in *Computation of Curves and Surfaces*, W. Dahmen, M. Gasca and C. A. Micchelli (eds.), Kluwer, NATO ASI Series, Dordrecht, 1990, 347-382.
2. Goodman, T. N. T., Construction of wavelets with multiplicity, Rendiconti di Mathematica (Serie VII) **14** (1994), 665-691.
3. Goodman, T. N. T., Characterising pairs of refinable splines, preprint.
4. Goodman, T.N.T., S. L. Lee and W. S. Tang, Wavelet bases for a set of commuting unitary operators, Adv. Comput. Math. **1** (1993), 109-125.
5. Lawton, W., S.L. Lee and Z. Shen, Complete characterization of refinable splines, Adv. Comput. Math. **3** (1995), 137-145.
6. Plonka, G., Two-scale symbol and autocorrelation symbol for B-splines with multiple knots, Adv. Comput. Math. **3** (1995), 1-22.
7. Sablonniere, P., Some families of splines with hexagonal support on a three direction mesh, in *Mathematical Methods in Computer Aided Geometric Design III*, M. Daehlen, T. Lyche and L. L. Schumaker (eds.), Academic Press, New York, 1995, 467-475.
8. Sun, Q., Refinable functions with compact support, preprint.

Tim N.T. Goodman
Department of Mathematics and Computer Science
University of Dundee
Dundee DD1 4HN
Scotland
tgoodman@mcs.dundee.ac.uk

The Subdivision and Transition Operators Associated with a Refinement Equation

Rong-Qing Jia

Abstract. The subdivision and transition operators are introduced in connection with a refinement equation. Eigenvalues and invariant subspaces of these operators are investigated. The spectral properties of the subdivision and transition operators are then used to study approximation and smoothness properties of the solution of the refinement equation.

§1. Introduction

In this paper we are concerned with refinement equations of the form

$$\phi = \sum_{\alpha \in \mathbb{Z}^s} a(\alpha)\phi(2 \cdot - \alpha), \tag{1.1}$$

where ϕ is the unknown function defined on the s-dimensional Euclidean space $\mathbb{R}^s$, and a is a finitely supported sequence on $\mathbb{Z}^s$, called the *refinement mask*.

If

$$\sum_{\alpha \in \mathbb{Z}^s} a(\alpha) = 2^s, \tag{1.2}$$

then it is known that there exists a nontrivial compactly supported distribution ϕ satisfying the refinement equation (1.1) and the solutions are unique up to a constant factor (see [1]). Throughout this paper, we assume that the condition (1.2) is satisfied.

The Fourier transform of an integrable function f on $\mathbb{R}^s$ is defined by

$$\hat{f}(\xi) = \int_{\mathbb{R}^s} f(x)e^{-ix\cdot\xi}\,dx, \qquad \xi \in \mathbb{R}^s,$$

Advanced Topics in Multivariate Approximation
F. Fontanella, K. Jetter and P. J. Laurent (eds.) pp. 139–154.

ISBN 981-02-2852-X

where $x \cdot \xi$ denotes the inner product of two vectors x and ξ in $\mathbb{R}^s$. This definition can be naturally extended to include compactly supported distributions. Thus, if (1.2) is satisfied, then there exists a unique compactly supported distribution ϕ satisfying the refinement equation (1.1) subject to the condition $\hat{\phi}(0) = 1$. This distribution is said to be the *normalized solution* to the refinement equation with mask a.

A *sequence* on $\mathbb{Z}^s$ is a function from $\mathbb{Z}^s$ to $\mathbb{C}$. The *support* of a sequence b on $\mathbb{Z}^s$ is defined by

$$\operatorname{supp} b := \{\alpha \in \mathbb{Z}^s : b(\alpha) \neq 0\}.$$

If $\operatorname{supp} b$ is a finite set, then we say that b is *finitely supported*. Let $\ell(\mathbb{Z}^s)$ denote the linear space of all sequences on $\mathbb{Z}^s$, and let $\ell_0(\mathbb{Z}^s)$ denote the linear space of all finitely supported sequences on $\mathbb{Z}^s$.

In order to study the refinement equation, we introduce two linear operators associated with the mask a. One is the subdivision operator S_a, and the other is the transition operator T_a. The *subdivision operator* S_a is the linear operator on $\ell(\mathbb{Z}^s)$ defined by

$$S_a u(\alpha) := \sum_{\beta \in \mathbb{Z}^s} a(\alpha - 2\beta) u(\beta), \qquad \alpha \in \mathbb{Z}^s,$$

where $u \in \ell(\mathbb{Z}^s)$. The *transition operator* T_a is the linear operator on $\ell_0(\mathbb{Z}^s)$ defined by

$$T_a v(\alpha) := \sum_{\beta \in \mathbb{Z}^s} a(2\alpha - \beta) v(\beta), \qquad \alpha \in \mathbb{Z}^s, \tag{1.3}$$

where $v \in \ell_0(\mathbb{Z}^s)$. The subdivision and transition operators play an important role in our study of approximation and smoothness properties of the solutions of the refinement equation.

Before proceeding further, we introduce some notation. A *multi-index* is an s-tuple $\mu = (\mu_1, \ldots, \mu_s)$ with its components being nonnegative integers. The *length* of μ is $|\mu| := \mu_1 + \cdots + \mu_s$, and the factorial of μ is $\mu! := \mu_1! \cdots \mu_s!$. For two multi-indices $\mu = (\mu_1, \ldots, \mu_s)$ and $\nu = (\nu_1, \ldots, \nu_s)$, we write $\nu \leq \mu$ if $\nu_j \leq \mu_j$ for $j = 1, \ldots, s$. If $\nu \leq \mu$, then we define

$$\binom{\mu}{\nu} := \frac{\mu!}{\nu!(\mu - \nu)!}.$$

For $j = 1, \ldots, s$, D_j denotes the partial derivative with respect to the jth coordinate. For $\mu = (\mu_1, \ldots, \mu_s)$, D^μ is the differential operator $D_1^{\mu_1} \cdots D_s^{\mu_s}$. Moreover, p_μ denotes the monomial given by

$$p_\mu(x) := x_1^{\mu_1} \cdots x_s^{\mu_s}, \qquad x = (x_1, \ldots, x_s) \in \mathbb{R}^s.$$

The *total degree* of p_μ is $|\mu|$. For a nonnegative integer k, we denote by Π_k the linear span of $\{p_\mu : |\mu| \le k\}$. Then $\Pi := \cup_{k=0}^\infty \Pi_k$ is the linear space of all polynomials of s variables.

For a compactly supported distribution ϕ on $\mathbb{R}^s$ and a sequence $b \in \ell(\mathbb{Z}^s)$, the *semi-convolution* of ϕ with b is defined by

$$\phi *' b := \sum_{\alpha \in \mathbb{Z}^s} \phi(\cdot - \alpha) b(\alpha).$$

Let $\mathbb{S}(\phi)$ denote the linear space $\{\phi *' b : b \in \ell(\mathbb{Z}^s)\}$. We call $\mathbb{S}(\phi)$ the *shift-invariant space* generated by ϕ. By $K(\phi)$ we denote the linear space given by

$$K(\phi) := \{b \in \ell(\mathbb{Z}^s) : \phi *' b = 0\}.$$

Also, we define

$$N(\phi) := \{\xi \in \mathbb{R}^s : \hat{\phi}(\xi + 2\pi\beta) = 0 \ \forall \beta \in \mathbb{Z}^s\}.$$

If ϕ is a compactly supported function in $L_p(\mathbb{R}^s)$ $(1 \le p \le \infty)$, then the shifts of ϕ are stable if and only if $N(\phi)$ is the empty set (see [7]).

Now let ϕ be a compactly supported function in $L_p(\mathbb{R}^s)$ $(1 \le p \le \infty)$. The norm in $L_p(\mathbb{R}^s)$ is denoted by $\|\cdot\|_p$. Let $S := \mathbb{S}(\phi) \cap L_p(\mathbb{R}^s)$, and for $h > 0$ let $S^h := \{g(\cdot/h) : g \in S\}$. For a positive integer k, we say that S (or $\mathbb{S}(\phi)$) provides *approximation order* k if for each sufficiently smooth function f in $L_p(\mathbb{R}^s)$, there exists a constant $C > 0$ such that

$$\inf_{g \in S^h} \|f - g\|_p \le C\, h^k \qquad \forall h > 0.$$

Under the assumption $\hat{\phi}(0) \ne 0$, it was proved by the author in [5] that the following three statements are equivalent:

(a) $\mathbb{S}(\phi)$ provides approximation order k.

(b) ϕ satisfies the moment conditions of order k, that is, $D^\mu \hat{\phi}(2\pi\beta) = 0$ for all μ with $|\mu| \le k - 1$ and all $\beta \in \mathbb{Z}^s \setminus \{0\}$.

(c) $\mathbb{S}(\phi)$ contains Π_{k-1}.

If $\mathbb{S}(\phi)$ contains Π_{k-1}, then we say that ϕ has *accuracy* k (see [4]). Note that this concept is valid even if ϕ is a compactly supported distribution.

Here is a brief outline of this paper. In Section 2 we investigate the spectral properties of the subdivision and transition operators. In Section 3 we give a characterization for the accuracy of a refinable function in terms of the refinement mask. In Section 4 we study invariant subspaces of the subdivision and transition operators and indicate the role played by these invariant subspaces in the study of smoothness of the solution of the refinement equation.

§2. The Subdivision and Transition Operators

In this section we study the spectral properties of the subdivision and transition operators.

Let X be a linear space over $\mathbb{C}$. The linear space of all linear functionals on X is called the *algebraic dual* of X, and is denoted by X'. If $f \in X'$ and $x \in X$, we write $\langle f, x \rangle$ for $f(x)$, the image of x under the mapping f. It is easily seen that $\ell(\mathbb{Z}^s)$ is the algebraic dual of $\ell_0(\mathbb{Z}^s)$ with $\langle u, v \rangle$ given by

$$\langle u, v \rangle := \sum_{\alpha \in \mathbb{Z}^s} u(\alpha) v(\alpha), \qquad u \in \ell(\mathbb{Z}^s),\ v \in \ell_0(\mathbb{Z}^s).$$

Let X and Y be two linear spaces over $\mathbb{C}$, and let T be a linear mapping from X to Y. The *kernel* of T, denoted by $\ker(T)$, is the subspace of X consisting of all the elements $x \in X$ such that $Tx = 0$. The *algebraic adjoint* T' is the linear mapping from Y' to X' such that for each $g \in Y'$, $T'g$ is the element of X' determined by

$$\langle T'g, x \rangle = \langle g, Tx \rangle \qquad \forall\, x \in X.$$

Evidently, the subdivision operator S_a is the algebraic adjoint of the transition operator T_{a^*}, where a^* denotes the sequence given by $a^*(\alpha) = a(-\alpha)$, $\alpha \in \mathbb{Z}^s$. Indeed, for $u \in \ell(\mathbb{Z}^s)$ and $v \in \ell_0(\mathbb{Z}^s)$, both $\langle S_a u, v \rangle$ and $\langle u, T_{a^*} v \rangle$ are equal to

$$\sum_{\alpha \in \mathbb{Z}^s} \sum_{\beta \in \mathbb{Z}^s} u(\alpha) a(-2\alpha + \beta) v(\beta).$$

Let X be a linear space over $\mathbb{C}$, and let X' be the algebraic dual of X. For a subspace E of X, the *annihilator* $E^\perp$ is the linear subspace of X' given by

$$E^\perp := \{ f \in X' : \langle f, x \rangle = 0\ \forall\, x \in E \}.$$

For a subspace F of X', the *annihilator* ${}^\perp F$ is given by

$${}^\perp F := \{ x \in X : \langle f, x \rangle = 0\ \forall\, f \in F \}.$$

Consider the quotient space $X'/E^\perp$. For an element f in X', we denote by $\overline{f}$ the element $f + E^\perp$ in $X'/E^\perp$. It is easily seen that the linear mapping from $X'/E^\perp$ to E' given by

$$\overline{f} \mapsto f|_E$$

is an isomorphism. Thus, we may identify $X'/E^\perp$ with E' by defining

$$\langle \overline{f}, x \rangle := \langle f, x \rangle \qquad \text{for } \overline{f} \in X'/E^\perp \text{ and } x \in E.$$

Let T be a linear mapping on X, and let E be an invariant subspace of T. Then $E^\perp$ is an invariant subspace of T'. Indeed, for $f \in E^\perp$ and $x \in E$, we have

$\langle T'f, x\rangle = \langle f, Tx\rangle = 0$; hence $T'f \in E^\perp$. Let $\overline{T'}$ be the quotient mapping induced by T' on $X'/E^\perp$:

$$\overline{T'}(\overline{f}) := \overline{T'f}.$$

For $\overline{f} \in X'/E^\perp$ and $x \in E$, we have

$$\langle \overline{T'}(\overline{f}), x\rangle = \langle \overline{T'f}, x\rangle = \langle T'f, x\rangle = \langle f, Tx\rangle = \langle \overline{f}, Tx\rangle.$$

This shows that $\overline{T'}$ is the algebraic adjoint of $T|_E$.

In order to study the spectral properties of the transition operator T_a, we shall use the concept of admissible sets. For a bounded subset K of $\mathbb{R}^s$, we denote by $\ell(K)$ the linear subspace of $\ell_0(\mathbb{Z}^s)$ consisting of all sequences supported on $K \cap \mathbb{Z}^s$. Let T be a linear operator on $\ell_0(\mathbb{Z}^s)$. A bounded subset K of $\mathbb{R}^s$ is said to be *admissible* for T, if $\ell(K)$ is invariant under T. See [3] for a study of admissible sets for the transition operator associated with a multivariate refinement equation.

The norm of an element $x = (x_1, \ldots, x_s)$ in $\mathbb{R}^s$ is defined by

$$|x| := |x_1| + \cdots + |x_s|.$$

For $r > 0$, we denote by B_r the closed ball $\{x \in \mathbb{R}^s : |x| \le r\}$. Let T_a be the transition operator associated with an element $a \in \ell_0(\mathbb{Z}^s)$. Set

$$N_a := \max\{|\alpha| : a(\alpha) \ne 0\}. \tag{2.1}$$

We claim that T_a maps $\ell(B_N)$ to $\ell(B_{(N+N_a)/2})$ for every integer $N \ge N_a$. Indeed, if $|\alpha| > (N + N_a)/2$ and $|\beta| \le N$, then $a(2\alpha - \beta) = 0$, because

$$|2\alpha - \beta| \ge 2|\alpha| - |\beta| > (N + N_a) - N = N_a.$$

But $v \in \ell(B_N)$ implies $v(\beta) = 0$ for $|\beta| > N$; hence by (1.3) we obtain $T_a v(\alpha) = 0$ for $|\alpha| > (N + N_a)/2$. This verifies our claim. In particular, B_N is an admissible set for T_a.

Denote by I_0 and I the identity mapping on $\ell_0(\mathbb{Z}^s)$ and $\ell(\mathbb{Z}^s)$, respectively. We claim that $\ker(T_a - \sigma I_0) \subseteq \ell(B_{N_a})$ for every $\sigma \in \mathbb{C} \setminus \{0\}$. To see this, let v be a nonzero element in $\ell_0(\mathbb{Z}^s)$ such that $T_a v = \sigma v$, and let $N := \max\{|\alpha| : v(\alpha) \ne 0\}$. If $N > N_a$, then $v = T_a v/\sigma$ lies in $\ell(B_{(N+N_a)/2})$, which contradicts the choice of N. This shows that $\ker(T_a - \sigma I_0)$ is contained in the finite dimensional linear space $\ell(B_{N_a})$. In particular, T_a has only a finite number of nonzero eigenvalues.

Theorem 2.1. *For $\sigma \in \mathbb{C} \setminus \{0\}$, the linear spaces $\ker(S_a - \sigma I)$ and $\ker(T_a - \sigma I_0)$ have the same dimension. In particular, σ is an eigenvalue of S_a if and only if it is an eigenvalue of T_a.*

Proof: Let N_a be the integer as defined in (2.1). Then $E := \ell(B_{N_a})$ is an invariant subspace of T_{a^*}, and $\ell(\mathbb{Z}^s)/E^\perp$ is the algebraic dual of E. Moreover, the quotient mapping $\overline{S_a - \sigma I}$ on $\ell(\mathbb{Z}^s)/E^\perp$ is the algebraic adjoint of $(T_{a^*} - \sigma I_0)|_E$. Since E is finite dimensional, we see that $\ker(\overline{S_a - \sigma I})$ and $\ker((T_{a^*} - \sigma I_0)|_E)$ have the same dimension. But $\ker((T_{a^*} - \sigma I_0)|_E) = \ker(T_{a^*} - \sigma I_0)$. Also, $\ker(T_{a^*} - \sigma I_0)$ and $\ker(T_a - \sigma I_0)$ have the same dimension. Therefore, $\ker(\overline{S_a - \sigma I})$ and $\ker(T_a - \sigma I_0)$ have the same dimension.

It remains to prove that $\ker(\overline{S_a - \sigma I})$ and $\ker(S_a - \sigma I)$ have the same dimension. For this purpose, it suffices to show that the linear mapping from $\ker(S_a - \sigma I)$ to $\ker(\overline{S_a - \sigma I})$ given by $u \mapsto \overline{u}$ is an isomorphism. Since $\ker(\overline{S_a - \sigma I})$ is finite dimensional, we only have to prove that this mapping is one-to-one. Thus, let u be an element in $\ker(S_a - \sigma I)$ such that $\overline{u} = 0$. Then we have $S_a u = \sigma u$ and $u \in E^\perp$. We wish to prove that $u = 0$. If $u \neq 0$, set $N := \min\{|\alpha| : u(\alpha) \neq 0\}$. Since $u \in E^\perp = \ell(B_{N_a})^\perp$ implies $u(\alpha) = 0$ for all $|\alpha| \le N_a$, we must have $N > N_a$. It follows that $u \in \ell(B_{(N+N_a)/2})^\perp$. Note that T_{a^*} maps $\ell(B_N)$ to $\ell(B_{(N+N_a)/2})$. From the relation $\langle S_a u, v\rangle = \langle u, T_{a^*} v\rangle$ we deduce that S_a maps $\ell(B_{(N+N_a)/2})^\perp$ to $\ell(B_N)^\perp$. Hence $u = S_a u/\sigma$ lies in $\ell(B_N)^\perp$, which contradicts the choice of N. This completes the proof of the theorem. ■

Suppose ϕ satisfies the refinement equation (1.1) with mask $a \in \ell_0(\mathbb{Z}^s)$. For $b \in \ell(\mathbb{Z}^s)$, we have

$$\sum_{\alpha \in \mathbb{Z}^s} b(\alpha)\phi(\cdot - \alpha) = \sum_{\alpha \in \mathbb{Z}^s} (S_a b)(\alpha)\phi(2 \cdot - \alpha). \tag{2.2}$$

Indeed, since ϕ satisfies the refinement equation (1.1), we have

$$\sum_{\alpha \in \mathbb{Z}^s} b(\alpha)\phi(\cdot - \alpha) = \sum_{\alpha \in \mathbb{Z}^s} b(\alpha) \sum_{\beta \in \mathbb{Z}^s} a(\beta)\phi(2 \cdot - 2\alpha - \beta) = \sum_{\gamma \in \mathbb{Z}^s} c(\gamma)\phi(2 \cdot - \gamma),$$

where

$$c(\gamma) = \sum_{\alpha \in \mathbb{Z}^s} a(\gamma - 2\alpha) b(\alpha), \qquad \gamma \in \mathbb{Z}^s.$$

Hence $c = S_a b$. This verifies (2.2).

Theorem 2.2. *Let ϕ be the normalized solution of the refinement equation (1.1) with mask a. If ϕ has accuracy k, then the subdivision operator S_a and the transition operator T_a have eigenvalues $1, 1/2, \ldots, (1/2)^{k-1}$. Moreover, for each $j \in \{0, 1, \ldots, k-1\}$, the multiplicity of the eigenvalue $(1/2)^j$ is at least*

$$\binom{j+s-1}{s-1}.$$

Proof: Since ϕ has accuracy k, $\mathbb{S}(\phi)$ contains the monomial $p_\mu : x \mapsto x^\mu$ ($x \in \mathbb{R}^s$) for all μ with $|\mu| \le k-1$. Let μ be a multi-index with $|\mu| = j \le k-1$. Then there exists a sequence b_μ on $\mathbb{Z}^s$ such that $p_\mu = \phi *' b_\mu$. By (2.2) we have

$$p_\mu = \sum_{\alpha \in \mathbb{Z}^s} b_\mu(\alpha)\phi(\cdot - \alpha) = \sum_{\alpha \in \mathbb{Z}^s} (S_a b_\mu)(\alpha)\phi(2 \cdot - \alpha).$$

On the other hand,

$$p_\mu = p_\mu(2\,\cdot)/2^j = \sum_{\alpha\in\mathbb{Z}^s} (1/2)^j\, b_\mu(\alpha)\phi(2\cdot-\alpha).$$

Comparing the above two equations gives

$$\sum_{\alpha\in\mathbb{Z}^s} (S_a b_\mu - b_\mu/2^j)(\alpha)\phi(2\cdot-\alpha) = 0.$$

Recall that $K(\phi)$ is the linear space $\{b \in \ell(\mathbb{Z}^s) : \phi*'b = 0\}$. Thus, the previous argument shows that

$$(S_a - (1/2)^j I)b_\mu \in K(\phi). \tag{2.3}$$

By (2.2) we see that $K(\phi)$ is invariant under S_a.

Let N_a be the integer as given in (2.1). Let $N := N_a + 1$ and $E := \ell(B_N)$. Then E is an invariant subspace of T_{a*}, and hence $E^\perp$ is an invariant subspace of S_a. Consider the quotient space $\ell(\mathbb{Z}^s)/E^\perp$. As the algebraic dual of the finite dimensional space E, $\ell(\mathbb{Z}^s)/E^\perp$ is finite dimensional. Let $\overline{S_a}$ be the quotient mapping on $\ell(\mathbb{Z}^s)/E^\perp$ induced by S_a. Let $F := K(\phi)+E^\perp$ and $\overline{F} := \{f+E^\perp : f \in F\}$. Since F is invariant under S_a, $\overline{F}$ is an invariant subspace of $\ell(\mathbb{Z}^s)/E^\perp$ under $\overline{S_a}$. It follows from (2.3) that

$$(\overline{S_a - (1/2)^j I})\,\overline{b_\mu} \in \overline{F}. \tag{2.4}$$

Let Q be the linear span of $\{\overline{b_\mu} : |\mu| = j\}$. We claim that $Q \cap \overline{F} = \{0\}$ and $\{\overline{b_\mu} : |\mu| = j\}$ forms a basis for Q.

Suppose there are complex numbers c_μ ($|\mu| = j$) such that $\sum_{|\mu|=j} c_\mu \overline{b_\mu} \in \overline{F}$. Then

$$\sum_{|\mu|=j} c_\mu b_\mu \in F = E^\perp + K(\phi).$$

Hence there exist $v \in E^\perp$ and $w \in K(\phi)$ such that $\sum_{|\mu|=j} c_\mu b_\mu = v + w$. It follows that

$$\phi*'v = \phi*'(v+w) = \phi*'\Big(\sum_{|\mu|=j} c_\mu b_\mu\Big) = \sum_{|\mu|=j} c_\mu(\phi*'b_\mu) = \sum_{|\mu|=j} c_\mu p_\mu.$$

We have $p := \sum_{|\mu|=j} c_\mu p_\mu \in \Pi_j$. Since $v \in E^\perp = \ell(B_N)^\perp$, we have $v(\alpha) = 0$ for $|\alpha| \le N = N_a + 1$. Moreover, since a is supported on B_{N_a}, the normalized solution ϕ of (1.1) is also supported on B_{N_a} (see [1]). Hence the polynomial $p = \phi*'v$ vanishes on the set $\{x \in \mathbb{R}^s : |x| < 1\}$, and therefore p is identically zero. Since $p = \sum_{|\mu|=j} c_\mu p_\mu$, we obtain $c_\mu = 0$ for all μ with $|\mu| = j$. This justifies our claim.

We observe that $Q+\overline{F}$ is a finite dimensional invariant subspace of $\overline{S_a}$. Choose a basis W for $\overline{F}$. Then $\{\overline{b_\mu} : |\mu| = j\} \cup W$ is a basis for $Q + \overline{F}$. By (2.4) we see that the matrix of $\overline{S_a}$ with respect to the basis $\{\overline{b_\mu} : |\mu| = j\} \cup W$ has the form

$$\begin{pmatrix} A & C \\ 0 & B \end{pmatrix},$$

where A is the diagonal matrix $\operatorname{diag}\{1/2^j, \ldots, 1/2^j\}$ with its number of rows being the same as the cardinality of the basis $\{\overline{b_\mu} : |\mu| = j\}$, which is $\binom{j+s-1}{s-1}$. Thus, $1/2^j$ is an eigenvalue of $\overline{S_a}$ of multiplicity at least $\binom{j+s-1}{s-1}$. From the proof of Theorem 2.1, we know that $\ker\left(\overline{S_a - (1/2)^j I}\right)$ and $\ker\left(S_a - (1/2)^j I\right)$ have the same dimension. Therefore the multiplicity of the eigenvalue $(1/2)^j$ for S_a is at least $\binom{j+s-1}{s-1}$. By Theorem 2.1, the transition operator T_a has the same property. ■

For the case $s = 1$, Heil, Strang, and Strela in [4] raised the question about whether the existence of the eigenvalues $1, 1/2, \ldots, 1/2^{k-1}$ for S_a is sufficient to ensure that ϕ has accuracy k. In [8], Jia, Riemenschneider, and Zhou gave a negative answer to their question by exhibiting a counterexample as follows. Let a be the sequence on $\mathbb{Z}$ given by $a(0) = 1/2$, $a(1) = 3/4$, $a(2) = 1/2$, $a(3) = 1/4$ and $a(\alpha) = 0$ for $\alpha \in \mathbb{Z} \setminus \{0,1,2,3\}$. Then the subdivision operator S_a has eigenvalues $1, 1/2, 1/4$. The normalized solution ϕ of the refinement equation with mask a lies in $L_2(\mathbb{R})$ with linearly independent shifts. But ϕ does not have accuracy 2.

The following is an example in the two dimensional case ($s = 2$).

Example 2.3. *Let a be the sequence on $\mathbb{Z}^2$ such that $a(\alpha) = 0$ for $\alpha \in \mathbb{Z}^2 \setminus [0,3]^2$ and*

$$\big(a(\alpha_1,\alpha_2)\big)_{0\le\alpha_1\le3,0\le\alpha_2\le3} = \frac{1}{8}\begin{bmatrix} 1+t & 2 & 1-t & 0 \\ 2 & 5+t & 4 & 1-t \\ 1-t & 4 & 5+t & 2 \\ 0 & 1-t & 2 & 1+t \end{bmatrix},$$

where t is a real number. Then the transition operator T_a has eigenvalues

$$1, \frac{1}{2}, \frac{1}{2}, \frac{1+t}{4}, \frac{1}{4}, \frac{1}{4}, \frac{1}{4}, \frac{1}{4}, \frac{1+t}{8}, \frac{1+t}{8}, \frac{1-t}{8}, \frac{1-t}{8}, \frac{1-t}{8}, \frac{1-t}{8}. \tag{2.5}$$

All the other eigenvalues of T_a are 0.

Let us establish this result. For $\beta \in \mathbb{Z}^2$, we denote by δ_β the sequence given by $\delta_\beta(\beta) = 1$ and $\delta_\beta(\alpha) = 0$ for $\alpha \in \mathbb{Z}^2 \setminus \{\beta\}$. Let W be the linear space spanned by $\{\delta_{(1,1)}, \delta_{(2,1)}, \delta_{(1,2)}, \delta_{(2,2)}\}$, V the linear space spanned by $\{\delta_{(1,0)}, \delta_{(0,1)}, \delta_{(3,2)}, \delta_{(2,3)}\}$, and U the linear space spanned by $\{\delta_{(0,0)}, \delta_{(2,0)}, \delta_{(0,2)}, \delta_{(3,1)}, \delta_{(1,3)}, \delta_{(3,3)}\}$. Write T for T_a. We observe that W is invariant under T, and

$$T\begin{bmatrix} \delta_{(1,1)} \\ \delta_{(2,1)} \\ \delta_{(1,2)} \\ \delta_{(2,2)} \end{bmatrix} = \frac{1}{8}\begin{bmatrix} 5+t & 1-t & 1-t & 1+t \\ 2 & 4 & 0 & 2 \\ 2 & 0 & 4 & 2 \\ 1+t & 1-t & 1-t & 5+t \end{bmatrix}\begin{bmatrix} \delta_{(1,1)} \\ \delta_{(2,1)} \\ \delta_{(1,2)} \\ \delta_{(2,2)} \end{bmatrix}.$$

By computation we find that the eigenvalues of $T|_W$ are $1, 1/2, 1/2, (1+t)/4$. Next, $V + W$ is invariant under T, and

$$T\delta_\beta \in (1/4)\,\delta_\beta + W \qquad \text{for } \beta \in \{(1,0),(0,1),(3,2),(2,3)\}.$$

Finally, $U + V + W$ is invariant under T,

$$T\delta_\beta \in ((1+t)/8)\,\delta_\beta + V + W \qquad \text{for } \beta \in \{(0,0),(3,3)\},$$

and

$$T\delta_\beta \in ((1-t)/8)\,\delta_\beta + V + W \qquad \text{for } \beta \in \{(2,0),(0,2),(3,1),(1,3)\}.$$

This shows that $T_a|_{U+V+W}$ has the eigenvalues as given in (2.5). All the other eigenvalues of T_a are 0 (see [3, Theorem 4.2]).

Let ϕ_t be the normalized solution of the refinement equation with mask a corresponding to the parameter t. In the next section we will show that ϕ_t does not have accuracy 3 for $t \in \mathbb{R} \setminus \{-1, 0\}$, even though T_a has eigenvalue $(1/2)^j$ of multiplicity at least $j+1$ for $j = 0, 1, 2$.

§3. Approximation Order

If ϕ is a compactly supported function in $L_p(\mathbb{R}^s)$ $(1 \le p \le \infty)$ with $\hat{\phi}(0) \neq 0$, then $\mathbb{S}(\phi)$ provides L_p-approximation order k if and only if ϕ has accuracy k (see [5, Theorem 6]). In this section we give a characterization of the accuracy of a refinable function in terms of the refinement mask.

Let ϕ be the normalized solution of the refinement equation (1.1) with the mask $a \in \ell_0(\mathbb{Z}^s)$. Taking Fourier transform of both sides of (1.1), we obtain

$$\hat{\phi}(\xi) = H(\xi/2)\hat{\phi}(\xi/2), \qquad \xi \in \mathbb{R}^s, \tag{3.1}$$

where

$$H(\xi) := \sum_{\alpha \in \mathbb{Z}^s} a(\alpha) e^{-i\alpha\cdot\xi}/2^s, \qquad \xi \in \mathbb{R}^s. \tag{3.2}$$

Note that H is 2π-periodic. Also, $H(0) = 1$ by (1.2). We denote by Γ the set of all s-tuples $(\gamma_1, \ldots, \gamma_s)$ for which $\gamma_j = 0$ or 1, $j = 1, \ldots, s$. Recall that $N(\phi)$ is the set of those $\xi \in \mathbb{R}^s$ for which $\hat{\phi}(\xi + 2\pi\beta) = 0$ for all $\beta \in \mathbb{Z}^s$.

Theorem 3.1. *Let k be a positive integer. If*

$$D^\mu H(\pi\varepsilon) = 0 \qquad \forall \varepsilon \in \Gamma \setminus \{0\} \text{ and } |\mu| \le k-1, \tag{3.3}$$

then ϕ has accuracy k. Conversely, if ϕ has accuracy k, and if $N(\phi) \cap (\pi\Gamma) = \emptyset$, then (3.3) holds true.

Proof: Suppose (3.3) holds true. In order to prove that ϕ has accuracy k, it suffices to show that $D^\mu \hat{\phi}(2\pi\beta) = 0$ for all $|\mu| \le k-1$ and $\beta \in \mathbb{Z}^s \setminus \{0\}$, since $\hat{\phi}(0) \neq 0$.

For this purpose, we apply the differential operator D^μ to both sides of (3.1) and obtain

$$D^\mu \hat{\phi}(2\pi\beta) = (1/2)^{|\mu|} \sum_{\nu\le\mu} \binom{\mu}{\nu} D^\nu H(\pi\beta) D^{\mu-\nu}\hat{\phi}(\pi\beta), \qquad \beta \in \mathbb{Z}^s, \tag{3.4}$$

where the Leibniz formula for differentiation has been employed. We shall prove by induction on n that $D^\mu\hat{\phi}(2\pi\beta) = 0$ for all $|\mu| \le k-1$ and $\beta \in (2^n\mathbb{Z}^s) \setminus (2^{n+1}\mathbb{Z}^s)$. For $\beta \in \mathbb{Z}^s \setminus (2\mathbb{Z}^s)$, by (3.3) we have $D^\nu H(\pi\beta) = 0$ for all ν with $|\nu| \le k-1$. This together with (3.4) yields $D^\mu\hat{\phi}(2\pi\beta) = 0$ for $|\mu| \le k-1$. Let $n > 0$ and suppose $D^\mu\hat{\phi}(2\pi\beta) = 0$ for all $|\mu| \le k-1$ and $\beta \in (2^{n-1}\mathbb{Z}^s) \setminus (2^n\mathbb{Z}^s)$. Then for $\beta \in (2^n\mathbb{Z}^s) \setminus (2^{n+1}\mathbb{Z}^s)$, we have $D^\nu\hat{\phi}(\pi\beta) = 0$ for $|\nu| \le k-1$. By (3.4), it follows that $D^\mu\hat{\phi}(2\pi\beta) = 0$ for $|\mu| \le k-1$. This completes the induction procedure. Since $\mathbb{Z}^s \setminus \{0\}$ is the union of $(2^n\mathbb{Z}^s) \setminus (2^{n+1}\mathbb{Z}^s)$, $n = 0, 1, 2, \ldots$, we conclude that $D^\mu\hat{\phi}(2\pi\beta) = 0$ for $|\mu| \le k-1$ and $\beta \in \mathbb{Z}^s \setminus \{0\}$. Therefore ϕ has accuracy k.

Conversely, suppose that ϕ has accuracy k and $N(\phi) \cap (\pi\Gamma) = \emptyset$. Let $\varepsilon \in \Gamma \setminus \{0\}$. Since $N(\phi) \cap (\pi\Gamma) = \emptyset$, we have $\pi\varepsilon \notin N(\phi)$. Hence there exists some $\beta \in \mathbb{Z}^s$ such that $\hat{\phi}(\pi\varepsilon + 2\pi\beta) \ne 0$. By (3.1) we see that $H(\xi) = \hat{\phi}(2\xi)\bigl[1/\hat{\phi}(\xi)\bigr]$ is valid for ξ in some neighborhood of $\pi\varepsilon + 2\pi\beta$. By using the Leibniz formula for differentiation, we obtain

$$D^\mu H(\pi\varepsilon + 2\pi\beta) = \sum_{\nu\le\mu} \binom{\mu}{\nu} 2^{|\nu|} D^\nu\hat{\phi}(2\pi\varepsilon + 4\pi\beta)\, D^{\mu-\nu}\bigl[1/\hat{\phi}\bigr](\pi\varepsilon + 2\pi\beta).$$

But $D^\nu\hat{\phi}(2\pi\varepsilon + 4\pi\beta) = 0$ for $|\nu| \le k-1$, because ϕ has accuracy k. Therefore $D^\mu H(\pi\varepsilon + 2\pi\beta) = 0$ for $|\mu| \le k-1$. Since H is 2π-periodic, this shows that $D^\mu H(\pi\varepsilon) = 0$ for all μ with $|\mu| \le k-1$ and all $\varepsilon \in \Gamma \setminus \{0\}$, as desired. ■

The following lemma gives a condition equivalent to (3.3).

Lemma 3.2. *Let $a \in \ell_0(\mathbb{Z}^s)$ and let H be given as in (3.2). Let p be a polynomial of s variables. Then the following two conditions are equivalent:*
(a) $p(iD)H(\pi\varepsilon) = 0$ for all $\varepsilon \in \Gamma \setminus \{0\}$.
(b) $\sum_{\beta\in\mathbb{Z}^s} a(2\beta)p(2\beta) = \sum_{\beta\in\mathbb{Z}^s} a(2\beta+\gamma)p(2\beta+\gamma)$ for all $\gamma \in \Gamma$.

Proof: For $\varepsilon \in \Gamma$, we deduce from (3.2) that

$$\begin{aligned} 2^s p(iD)H(\pi\varepsilon) &= \sum_{\alpha\in\mathbb{Z}^s} a(\alpha)p(\alpha)\, e^{-i\pi\alpha\cdot\varepsilon} \\ &= \sum_{\gamma\in\Gamma}\sum_{\beta\in\mathbb{Z}^s} a(2\beta+\gamma)\, p(2\beta+\gamma)\, e^{-i\pi\gamma\cdot\varepsilon}. \end{aligned}$$

Let

$$b(\gamma) := \sum_{\beta\in\mathbb{Z}^s} a(2\beta+\gamma)\, p(2\beta+\gamma), \qquad \gamma \in \Gamma.$$

Then it follows that

$$2^s p(iD)H(\pi\varepsilon) = \sum_{\gamma\in\Gamma} b(\gamma)e^{-i\pi\gamma\cdot\varepsilon}. \tag{3.5}$$

Condition (b) says that $b(\gamma) = b(0)$ for all $\gamma \in \Gamma$. Hence by (3.5) we have

$$2^s p(iD)H(\pi\varepsilon) = b(0)\sum_{\gamma\in\Gamma} e^{-i\pi\gamma\cdot\varepsilon} = 0 \qquad \forall \varepsilon \in \Gamma \setminus \{0\}.$$

Therefore (b) implies (a).

Conversely, if condition (a) is satisfied, then by (3.5) we have

$$\sum_{\gamma\in\Gamma} b(\gamma)e^{-i\pi\gamma\cdot\varepsilon} = 0 \qquad \forall \varepsilon \in \Gamma \setminus \{0\}.$$

Let η be an element in Γ. Then it follows that

$$\sum_{\varepsilon\in\Gamma} e^{i\pi\eta\cdot\varepsilon} \sum_{\gamma\in\Gamma} b(\gamma)e^{-i\pi\gamma\cdot\varepsilon} = \sum_{\gamma\in\Gamma} b(\gamma).$$

On the other hand,

$$\sum_{\varepsilon\in\Gamma} e^{i\pi\eta\cdot\varepsilon} \sum_{\gamma\in\Gamma} b(\gamma)e^{-i\pi\gamma\cdot\varepsilon} = \sum_{\gamma\in\Gamma} b(\gamma) \sum_{\varepsilon\in\Gamma} e^{i\pi(\eta-\gamma)\cdot\varepsilon} = 2^s b(\eta).$$

This shows that $2^s b(\eta) = \sum_{\gamma\in\Gamma} b(\gamma)$. Therefore $b(\eta) = b(0)$ for all $\eta \in \Gamma$. In other words, (a) implies (b). ■

Let $a \in \ell_0(\mathbb{Z}^s)$, and let k be a positive integer. we say that a satisfies the *sum rules* of order k if for all $p \in \Pi_{k-1}$

$$\sum_{\beta\in\mathbb{Z}^s} a(2\beta)p(2\beta) = \sum_{\beta\in\mathbb{Z}^s} a(2\beta+\gamma)p(2\beta+\gamma) \qquad \forall \gamma \in \Gamma.$$

Let ϕ be the normalized solution of (1.1) with the refinement mask a satisfying (1.2). Theorem 3.1 and Lemma 3.2 tell us that ϕ has accuracy k if a satisfies the sum rules of order k. The converse statement is true under the additional condition $N(\phi) \cap (\pi\Gamma) = \emptyset$.

The following example illustrates the general theory.

Example 3.3. *Let a be the sequence on $\mathbb{Z}^2$ given in Example 2.3, and let $\phi = \phi_t$ be the normalized solution of the refinement equation with mask a corresponding to the parameter t. Then for $t \in \mathbb{R} \setminus \{-1, 0\}$, ϕ_t has exact accuracy 2, and for $t = 0$ or -1, ϕ_t has exact accuracy 3.*

Let us verify the condition $N(\phi_t) \cap (\pi\Gamma) = \emptyset$ for $t \neq -1$. Note that in the present case $\Gamma = \{(0,0),(1,0),(0,1),(1,1)\}$. Since $\hat{\phi}_t(0) = 1$, we have $0 \notin N(\phi_t)$.

Let H be the function given in (3.2). Then $H(\xi_1, 0) = (1 + e^{-i\xi_1})^3/8$ for $\xi_1 \in \mathbb{R}$. It follows that $\hat{\phi}_t(\pi, 0) = \prod_{k=1}^{\infty} H(\pi/2^k, 0) \neq 0$. Similarly, $\hat{\phi}_t(0, \pi) \neq 0$. This shows that $(\pi, 0) \notin N(\phi_t)$ and $(0, \pi) \notin N(\phi_t)$. Moreover,

$$H(\xi_1, -\xi_1) = (1 + e^{-i\xi_1})(1 + e^{i\xi_1})\big[(1+t) + (1-t)\cos \xi_1\big]/8, \qquad \xi_1 \in \mathbb{R}.$$

If $(\pi, \pi) \in N(\phi_t)$, we would have $\hat{\phi}_t(\pi, -\pi) = 0$ and $\hat{\phi}_t(3\pi, -3\pi) = 0$. Since $\hat{\phi}(\xi) = \prod_{k=1}^{\infty} H(\xi/2^k)$, there exist positive integers m and n such that

$$(1+t) + (1-t)\cos(\pi/2^m) = 0 \quad \text{and} \quad (1+t) + (1-t)\cos(3\pi/2^n) = 0. \tag{3.6}$$

It follows that $t \neq 1$ and $\cos(\pi/2^m) = \cos(3\pi/2^n)$. Hence $\pi/2^m + 3\pi/2^n \in 2\pi\mathbb{Z}$ or $\pi/2^m - 3\pi/2^n \in 2\pi\mathbb{Z}$, which implies $m = n = 1$. This together with (3.6) yields $t = -1$. In other words, for $t \neq -1$, $N(\phi_t) \cap (\pi\Gamma) = \emptyset$.

Recall that p_μ is the monomial $x \mapsto x^\mu$ ($x \in \mathbb{R}^2$). It can be easily verified that

$$\sum_{\beta \in \mathbb{Z}^2} a(2\beta)\, p(2\beta) = \sum_{\beta \in \mathbb{Z}^2} a(2\beta + \gamma)\, p(2\beta + \gamma) \qquad \forall \gamma \in \Gamma \tag{3.7}$$

is valid for $p = p_\mu$ with $\mu = (0,0), (0,1), (1,0), (2,0)$ or $(0,2)$. But for $t \neq 0$ and $\mu = (1,1)$, $p = p_\mu$ does not satisfy (3.7). Therefore, by Theorem 3.1 and Lemma 3.2, ϕ_t has exact accuracy 2 for $t \in \mathbb{R} \setminus \{-1, 0\}$.

For $t = 0$, (3.7) holds true for all $p = p_\mu$ with $|\mu| \leq 2$, but it is not true for p_μ with $\mu = (3,0)$. Therefore, for $t = 0$, ϕ_0 has exact accuracy 3.

It remains to deal with the case $t = -1$. In this case, $(\pi, \pi) \in N(\phi)$, where $\phi := \phi_{-1}$. We observe that the corresponding transition operator T_a does not have eigenvalue $1/8$. Hence ϕ has accuracy at most 3, by Theorem 2.2. In order to prove that ϕ has accuracy 3, it suffices to show that for $|\mu| \leq 2$

$$D^\mu \hat{\phi}(2\pi\beta) = 0 \qquad \forall \beta \in \mathbb{Z}^2 \setminus \{0\}. \tag{3.8}$$

Since (3.7) is valid for $p = p_\mu$ with $|\mu| \leq 2$ except $\mu = (1,1)$, (3.8) holds true for $|\mu| \leq 2$ except $\mu = (1,1)$, by Lemma 3.2. Thus, it remains to prove that (3.8) is true for $\mu = (1,1)$. Suppose $\beta = (\beta_1, \beta_2) \in \mathbb{Z}^2 \setminus (2\mathbb{Z}^2)$. Then $D^\nu H(\pi\beta) = 0$ for $|\nu| \leq 1$. Applying (3.4) to $\mu = (1,1)$, we obtain

$$D_1 D_2 \hat{\phi}(2\pi\beta) = (1/4) D_1 D_2 H(\pi\beta)\, \hat{\phi}(\pi\beta). \tag{3.9}$$

If both β_1 and β_2 are odd, then $\hat{\phi}(\pi\beta) = 0$, since $(\pi, \pi) \in N(\phi)$. If β_1 is even and β_2 is odd, or β_1 is odd and β_2 is even, then $D_1 D_2 H(\pi\beta) = 0$. Thus, $D_1 D_2 \hat{\phi}(2\pi\beta) = 0$ for $\beta \in \mathbb{Z}^2 \setminus (2\mathbb{Z}^2)$, by (3.9). Using (3.4) repeatedly, we obtain $D_1 D_2 \hat{\phi}(2\pi\beta) = 0$ for $\beta \in (2^n\mathbb{Z}^2) \setminus (2^{n+1}\mathbb{Z}^2)$, $n = 0, 1, 2, \ldots$. Therefore, $D_1 D_2 \hat{\phi}(2\pi\beta) = 0$ for all $\beta \in \mathbb{Z}^2 \setminus \{0\}$, as desired.

§4. Invariant Subspaces

In this section we demonstrate that the sum rules for the refinement mask a are equivalent to the existence of certain invariant subspaces for the subdivision and transition operators. We also indicate the role played by these invariant subspaces in the study of smoothness properties of the solution of the refinement equation.

A sequence a on $\mathbb{Z}^s$ is called a *polynomial sequence* if there exists a polynomial p such that $a(\alpha) = p(\alpha)$ for all $\alpha \in \mathbb{Z}^s$. The degree of a is the same as the degree of p. For a nonnegative integer k, let P_k be the linear space of all polynomial sequences of degree at most k. Then P_k is a linear subspace of $\ell(\mathbb{Z}^s)$. Let

$$V_k := {}^{\perp}P_k = \Big\{ v \in \ell_0(\mathbb{Z}^s) : \sum_{\alpha\in\mathbb{Z}^s} p(\alpha)v(\alpha) = 0 \; \forall\, p \in \Pi_k \Big\}.$$

It is easily seen that $P_k = V_k^{\perp}$.

Theorem 4.1. *Let a be an element in $\ell_0(\mathbb{Z}^s)$. Then the following statements are equivalent:*

(a) The sequence a satisfies the sum rules of order $k+1$.
(b) V_k is invariant under the transition operator T_a.
(c) P_k is invariant under the subdivision operator S_a.
(d) $D^\mu H(\pi\varepsilon) = 0$ for all $|\mu| \le k$ and all $\varepsilon \in \Gamma \setminus \{0\}$.

Proof: (a) $\Rightarrow$ (b): Let $p \in \Pi_k$ and $v \in V_k$. We have

$$\sum_{\alpha\in\mathbb{Z}^s} p(\alpha)T_a v(\alpha) = \sum_{\beta\in\mathbb{Z}^s} \Big[\sum_{\alpha\in\mathbb{Z}^s} p(\alpha)a(2\alpha-\beta) \Big] v(\beta).$$

Let $q = p(\cdot/2)$. Then $p = q(2\,\cdot)$. By Taylor's formula, we have

$$q(2\alpha) = q(2\alpha - \beta + \beta) = \sum_{|\mu|\le k} q_\mu(2\alpha-\beta)\beta^\mu,$$

where $q_\mu := D^\mu q/\mu!$. Hence

$$\sum_{\alpha\in\mathbb{Z}^s} p(\alpha)a(2\alpha-\beta) = \sum_{\alpha\in\mathbb{Z}^s} q(2\alpha)a(2\alpha-\beta) = \sum_{|\mu|\le k} c_\mu \beta^\mu,$$

where

$$c_\mu := \sum_{\alpha\in\mathbb{Z}^s} q_\mu(2\alpha-\beta)a(2\alpha-\beta)$$

is independent of β, by condition (a). Thus, we obtain

$$\sum_{\alpha\in\mathbb{Z}^s} p(\alpha)T_a v(\alpha) = \sum_{|\mu|\le k} c_\mu \sum_{\beta\in\mathbb{Z}^s} \beta^\mu v(\beta) = 0,$$

because $v \in V_k$. This shows that $T_a v \in V_k$ for $v \in V_k$. In other words, V_k is invariant under T_a.

(b) $\Rightarrow$ (c): Condition (b) implies that V_k is also invariant under T_{a^*}, where a^* is the sequence given by $a^*(\alpha) = a(-\alpha)$ for $\alpha \in \mathbb{Z}^s$. But S_a is the algebraic adjoint of T_{a^*}. Hence $P_k = V_k^{\perp}$ is invariant under S_a.

(c) $\Rightarrow$ (a): For $p \in \Pi_k$, let $q(\gamma) := \sum_{\beta \in \mathbb{Z}^s} a(2\beta+\gamma)\, p(2\beta+\gamma)$ for $\gamma \in \mathbb{Z}^s$. We claim that q is a polynomial sequence. Indeed, by using Taylor's formula, we have

$$p(2\beta+\gamma) = \sum_{|\mu|\le k} t_\mu(2\beta)\gamma^\mu,$$

where $t_\mu := D^\mu p/\mu!$. Set $q_\mu := t_\mu(-2\,\cdot)$. Then for $\gamma \in \mathbb{Z}^s$,

$$\begin{aligned} q(\gamma) &= \sum_{\beta\in\mathbb{Z}^s} a(2\beta+\gamma)\, p(2\beta+\gamma) \\ &= \sum_{\beta\in\mathbb{Z}^s} \sum_{|\mu|\le k} a(\gamma+2\beta)\, q_\mu(-\beta)\gamma^\mu = \sum_{|\mu|\le k} (S_a q_\mu)(\gamma)\,\gamma^\mu. \end{aligned}$$

By condition (c), $S_a q_\mu$ is a polynomial, hence so is q. We observe that q is 2-periodic, i.e., for any $\eta \in \mathbb{Z}^s$, $q(\gamma+2\eta) = q(\gamma)$. Therefore q must be a constant sequence. This verifies condition (a).

Finally, the equivalence between (a) and (d) was proved in Lemma 3.2. ■

We remark that the equivalence between (c) and (d) was proved by Cavaretta, Dahmen, and Micchelli in [1, p. 98]. In the univariate case ($s = 1$), Daubechies and Lagarias in [2] observed that (a) implies (b), and used the linear spaces V_k in their study of regularity of refinable functions. In the multivariate case, Riemenschneider and Shen in [9] pointed out that (d) implies (b), and employed the linear spaces V_k in their investigation of regularity of multivariate refinable functions.

The following result tells us that a refinable function in $W_1^k(\mathbb{R}^s)$ has accuracy $k+1$. Consequently, if the normalized solution ϕ of the refinement equation (1.1) lies in $W_1^k(\mathbb{R}^s)$, and if $N(\phi)\cap(\pi\Gamma) = \emptyset$, then the refinement mask a satisfies all the conditions in Theorem 4.1.

Theorem 4.2. *Let ϕ be a compactly supported function in $W_1^k(\mathbb{R}^s)$ with $\hat\phi(0) \ne 0$. If ϕ satisfies the refinement equation (1.1), then $\Pi_k \subset \mathbb{S}(\phi)$ and $\mathbb{S}(\phi)$ provides approximation order $k+1$.*

Proof: It suffices to show that $D^\mu\hat\phi(2\pi\beta) = 0$ for all $|\mu| \le k$ and $\beta \in \mathbb{Z}^s \setminus \{0\}$. For $n = 1, 2, \ldots$, it follows from (3.1) that

$$\hat\phi(\xi) = \left[\prod_{j=1}^n H(\xi/2^j)\right]\hat\phi(\xi/2^n), \qquad \xi \in \mathbb{R}^s.$$

Hence

$$\hat{\phi}(2\pi\beta) = \hat{\phi}(2^{n+1}\pi\beta) \qquad \forall \beta \in \mathbb{Z}^s.$$

By the Riemann-Lebesgue lemma, we have

$$\hat{\phi}(2\pi\beta) = \lim_{n\to\infty} \hat{\phi}(2^{n+1}\pi\beta) = 0 \qquad \forall \beta \in \mathbb{Z}^s \setminus \{0\}.$$

Now let $0 < r \le k$. Assuming that $D^\nu \hat{\phi}(2\pi\beta) = 0$ for all ν with $|\nu| < r$ and all $\beta \in \mathbb{Z}^s \setminus \{0\}$, we wish to prove that $D^\mu \hat{\phi}(2\pi\beta) = 0$ for all μ with $|\mu| = r$ and all $\beta \in \mathbb{Z}^s \setminus \{0\}$. Let μ be a multi-index with $|\mu| = r$. Applying the differential operator D^μ to both sides of (3.1), we obtain

$$D^\mu \hat{\phi}(\xi) = \frac{1}{2^r} \sum_{\nu \le \mu} \binom{\mu}{\nu} D^{\mu-\nu} H(\xi/2) D^\nu \hat{\phi}(\xi/2), \qquad \xi \in \mathbb{R}^s.$$

If $\nu \le \mu$ and $\nu \ne \mu$, then $D^\nu \hat{\phi}(2\pi\beta) = 0$ for all $\beta \in \mathbb{Z}^s \setminus \{0\}$ by the induction hypothesis. Hence for $\beta \in \mathbb{Z}^s \setminus \{0\}$ we have

$$D^\mu \hat{\phi}(4\pi\beta) = (1/2)^r D^\mu \hat{\phi}(2\pi\beta).$$

Repeating this process n times, we obtain

$$D^\mu \hat{\phi}(2^{n+1}\pi\beta) = (1/2)^{rn} D^\mu \hat{\phi}(2\pi\beta), \qquad \beta \in \mathbb{Z}^s \setminus \{0\}.$$

It follows that

$$D^\mu \hat{\phi}(2\pi\beta) = 2^{nr} D^\mu \hat{\phi}(2^{n+1}\pi\beta), \qquad \beta \in \mathbb{Z}^s \setminus \{0\}.$$

In order to complete the induction procedure, it suffices to show that

$$\lim_{n\to\infty} 2^{nr} D^\mu \hat{\phi}(2^{n+1}\pi\beta) = 0 \qquad \text{for } \beta \in \mathbb{Z}^s \setminus \{0\}. \tag{4.1}$$

For this purpose, let ϕ_μ be the function defined by

$$\phi_\mu(x) = (-ix)^\mu \phi(x), \qquad x \in \mathbb{R}^s.$$

Then $D^\mu \hat{\phi} = \hat{\phi}_\mu$ and

$$((-iD)^\nu \phi_\mu)\hat{\ }(\xi) = \xi^\nu D^\mu \hat{\phi}(\xi), \qquad \xi \in \mathbb{R}^s.$$

Since $\phi \in W_1^k(\mathbb{R}^s)$ and ϕ is compactly supported, we have $(-iD)^\nu \phi_\mu \in L_1(\mathbb{R}^s)$ for $|\mu|, |\nu| \le k$. Therefore, by the Riemann-Lebesgue lemma, we have

$$\lim_{n\to\infty} (2^{n+1}\pi\beta)^\nu D^\mu \hat{\phi}(2^{n+1}\pi\beta) = 0 \qquad \text{for } \beta \in \mathbb{Z}^s \setminus \{0\}.$$

Fix an element $\beta = (\beta_1, \ldots, \beta_s) \in \mathbb{Z}^s \setminus \{0\}$. Then $\beta_j \ne 0$ for some j. Let ν be the s-tuple such that its jth coordinate equals r and all the other coordinates are 0. Then we have $\beta^\nu = \beta_j^r$, and hence

$$\lim_{n\to\infty} (2^{n+1}\pi\beta_j)^r D^\mu \hat{\phi}(2^{n+1}\pi\beta) = 0.$$

This verifies (4.1). The proof of the theorem is complete. ■

In [1, Theorem 8.3], Cavaretta, Dahmen, and Micchelli proved that a compactly supported refinable function ϕ in $C^k(\mathbb{R}^s)$ has accuracy $k+1$ under the condition $\hat{\phi}(0) \neq 0$. In [10], Ron pointed out that the same result holds true for a compactly supported refinable function in $W_2^k(\mathbb{R}^s)$. However, the essential part of Theorem 4.2 had already been established in [1, Theorem 8.4], whose proof was suggested by me to the authors of [1].

We will give a comprehensive study of smoothness of multivariate refinable functions elsewhere. For the univariate case the reader is referred to [6].

Acknowledgements. Research was partially supported by NSERC Canada under Grant OGP 121336

References

1. Cavaretta, A. S., W. Dahmen and C. A. Micchelli, Stationary Subdivision, Memoirs of Amer. Math. Soc., Volume 93, 1991.
2. Daubechies, I. and J. C. Lagarias, Two-scale difference equations: II. Local regularity, infinite products of matrices and fractals, SIAM J. Math. Anal. **23** (1992), 1031–1079.
3. Han, B. and R. Q. Jia, Multivariate refinement equations and subdivision schemes, manuscript.
4. Heil, C., G. Strang, and V. Strela, Approximation by translates of refinable functions, Numer. Math., to appear.
5. Jia, R. Q., The Toeplitz theorem and its applications to approximation theory and linear pde's, Trans. Amer. Math. Soc. **347** (1995), 2585–2594.
6. Jia, R. Q., Subdivision schemes in L_p spaces, Adv. Comput. Math. **3** (1995), 309–341.
7. Jia, R. Q. and C. A. Micchelli, On linear independence of integer translates of a finite number of functions, Proc. Edinburgh Math. Soc. **36** (1992), 69-85.
8. Jia, R. Q., S. D. Riemenschneider and D. X. Zhou, Approximation by multiple refinable functions, manuscript.
9. Riemenschneider, S. D. and Z. W. Shen, Multidimensional interpolatory subdivision schemes, manuscript.
10. Ron, A., Smooth refinable functions provide good approximation orders, CMS Technical Summary Report #95-09, University of Wisconsin-Madison.

Rong-Qing Jia
Department of Mathematical Sciences
University of Alberta
Edmonton, T6G 2G1
Canada
jia@xihu.math.ualberta.ca

Approximation of Smooth Functions by sk–Splines

A.K. Kushpel, J. Levesley and W.A. Light

Abstract. The space of sk–splines is the linear span of shifts of a single kernel K. Such spline functions have been considered by many different investigators (Tikhomirov, Pinkus, Micchelli, Dyn, Chui, Light, Cheney) under very restrictive conditions on the kernel K. We impose more general conditions and this enables us to give a unified treatment for a wide range of spline functions, having finite, fractional, infinite, analytic, or entire smoothness. For instance, we will consider problems of existence and uniqueness of sk–spline interpolants, give an error analysis for functional sets of different smoothness, present a variational approach, and give some extremal properties of sk–splines. We conclude by outlining a generalisation of sk–splines to compact abelian groups.

§1. Introduction

In the classical sense a spline is a piecewise polynomial. Schoenberg [44] is the father of modern spline theory but such functions go as far back as the time of Leibnitz (1697). Spline functions have interesting extremal properties and provide a reasonable and useful apparatus for approximation. In the framework of this paper it is impossible to mention all directions of this important and rapidly developing theory since there are a hundreds of remarkable results which are worthy of mention.

The extremal role of splines was clarified by Favard [13] in connection with the problem of the best approximation by trigonometric polynomials. In the work of Kolmogorov [20] spline functions appeared in some sharp inequalities for norms of derivatives. Extremal properties of spline functions in the theory of n-widths were discovered first by Tikhomirov [66]. Nikolskii [42] and Schoenberg [47,49] applied spline interpolation to the problem of the best quadrature formulas.

We would like to remark on the celebrated result of Holladay [17] in which cubic splines appear as the solution of some variational problem. This idea was studied in the well known works of Ahlberg, Nielson and Walsh [1,2]. Attéia [5], and Anselone and Laurent [4] introduced abstract spline functions in Hilbert space in the same

Advanced Topics in Multivariate Approximation
F. Fontanella, K. Jetter and P. J. Laurent (eds.) pp. 155–180.

ISBN 981-02-2852-X

way. Radial basis functions, which have recently received a lot of attention (see e.g. [35,37,39,70]), may also be realised as the solution of some variational problem in a Hilbert space, and can be interpreted as Green's functions for suitable linear operators.

The last fifty years have seen an explosion of interest in splines. Polynomial splines and a great number of generalisations have been discussed by many investigators. We just mention here the contributions of Ahlberg, Nielson and Walsh [1,2], Dyn [11], Birkhoff [6], Jerome and Schumaker [18], Kushpel [24–26], Levesley and Kushpel [34], Light and Cheney [35], Lorentz [36], Micchelli [38], Micchelli and Pinkus [40], Schoenberg [46–51], Schumaker [57], Stechkin and Subbotin [58], Subbotin [60–62], and Varga [69].

In this paper we study one such generalisation; the sk–splines. The space of sk–splines is the linear span of shifts of a single kernel K. Such spline functions have been considered by different researchers (Cheney, Chui, Dyn, Light, Micchelli, Pinkus, Tikhomirov and many others) under very restrictive conditions on the kernel K, like sign regularity (CVD, TP, ETP, STP, etc.) or conditionally positive definiteness of the kernel K. We impose more general conditions on the kernel K and this enables us to give a unified treatment for a wide range of spline functions, having finite, fractional, infinite, analytic, or entire smoothness. In this more general setting we are not able to apply, for example, Rolle's theorem (or a sign regularity property), Taylor expansion arguments, Gershgorin's theorem, or the apparatus of divided differences, which are usual for the analysis of polynomial splines.

The sk–splines were introduced, and their basic theory developed, by Kushpel [25–32]. They were then used as tools for estimating n–widths and for the approximation of smooth functions. There are a number of reasons for using sk–splines:

1. In applications we have many equations which have solutions with infinite smoothness. Consequently, the use of polynomial splines will be not effective due to saturation phenomena. However, appropriate sk–splines are able to approximate such functions well.
2. Use of sk–splines in n–width theory enables us to calculate exact values of n–widths in known and new situations in which the application of standard methods meet with difficulties of a fundamental nature.
3. The sk–splines look promising for approximation since sk–splines have good order of convergence on sets of functions with smoothness from the fractional, up to the analytic and entire. In many situations sk–splines give a new examples of optimal subspaces in sense of Kolmogorov's n–widths.

Many of the usual sets of functions in approximation theory can be represented in integral form. As a motivating example we consider sets of periodic functions which can be written as a convolution

$$f(x) \ = \ \alpha + (K * \phi)(x) = \alpha + (2\pi)^{-1} \int_{-\pi}^{\pi} K(x-\tau)\phi(\tau)d\tau,$$

where $\alpha \in \mathrm{I\!R}$. In particular, the Sobolev classes $W_p^r = \mathrm{I\!R} \oplus D_r * U_p$ where $U_p = \{\phi : \|\phi\|_p \leq 1\}$, and $D_r(x) = \sum_{k=1}^{\infty} k^{-r} \cos(kx - r\pi/2)$ is the degree r Bernoulli monospline. This convolution form suggests the use of sums $\alpha + \sum_{k=1}^{n} c_k K(x - x_k)$ for approximation of functional sets $\mathrm{I\!R} \oplus K * U_p$. For the Sobolev classes such sums are the polynomial splines of degree $r - 1$ and defect 1. If the kernel K corresponds to the functional set $W_p^{L_m}$, where L_m is a polynomial differential operator, we obtain the well known L-splines of Micchelli [38,51,54].

Definition 1. *Let $\{0 = x_0 < \ldots < x_{n-1} < 2\pi\}$ be arbitrary points in $[0, 2\pi)$ and $K \in L_1[0, 2\pi)$. An sk–spline is a function of the form*

$$\mathrm{sk}(x) = c_0 + \sum_{k=1}^{n} c_k K(x - x_k),$$

where the real coefficients c_k, $0 \leq k \leq n$, satisfy

$$\sum_{k=1}^{n} c_k = 0.$$

The paper is organized as follows: Section 2 deals with theorems of existence and uniqueness of interpolating sk–splines. In Section 3 we showing some applications in n–width theory. In Section 4 we demonstrate some approximation properties of multi–dimensional sk–splines on a wide range of sets of smooth functions. Section 5 contains some theorems about sk–splines on compact abelian groups and uniform spaces and in Section 6 we present a variational viewpoint on sk–splines.

§2. Existence and Uniqueness of Interpolating sk–Splines

If we would like to use sk–splines for the reconstruction of smooth functions using some discrete information then the problem of spline interpolation naturally arises. In what follows we will assume that the kernel K is continuous.

Let $\Delta_n = \{0 = x_0 < \ldots < x_n < 2\pi\}$ be an arbitrary partition of $[0, 2\pi)$ and define the sk–spline by

$$\mathrm{sk}(x) = c_0 + \sum_{k=1}^{n} c_k K(x - x_k), \quad \sum_{k=1}^{n} c_k = 0,$$

where $c_k \in \mathrm{I\!R}$, $0 \leq k \leq n$. Further, suppose $0 \leq y_0 < y_1 < \ldots < y_n < 2\pi$.

Given $f \in C_{2\pi}$, the problem of sk–spline interpolation is the problem of finding c_k^*, $0 \leq k \leq n$, such that

$$\mathrm{sk}(y_j) = c_0^* + \sum_{k=1}^{n} c_k^* K(y_j - x_k) \;=\; f(x_j), \quad 1 \leq k \leq n.$$

In the case $K(x) = D_r(x)$, $r \in \mathbb{N}$, the existence of splines interpolants has been considered by many people, using such techniques as Rolle's theorem (Korneichuk[22]), or the cyclic variate diminishing (CVD) property of the kernel K (see e.g. [43]). Existence of polynomial splines of low degree was proven using Gershgorin's eigenvalue theorem. If we would like to consider the existence of sk–splines in a more general situations, Gershgorin's theorem may not apply, and the kernel K may not have any sign regular property.

We begin with the case when data points and knots of the sk–spline are equidistant because in this case we can get a more complete result.

Proposition 1. (Kushpel [25,26,30]) *Let $x_k = 2\pi k/n$ and $y_k = x_k + y$ for some fixed $y \in \mathbb{R}$. Further, let*

$$\lambda_j(y) = \frac{2}{n}\sum_{k=1}^{n} e^{ijx_k} K(y - x_k).$$

If $|\lambda_j(y)| > 0$, $1 \le j \le n-1$, then the interpolating sk–spline with knots x_k, $1 \le k \le n$ and interpolation points y_k, $1 \le k \le n$, exists and is unique.

In the next proposition various conditions are given which ensure the satisfaction of the hypotheses of the previous proposition, and hence guarantee the existence of sk–spline interpolants. These conditions are in terms of the Fourier coefficients of the kernel K and are therefore useful for practical application.

Proposition 2. *Let*

$$K(x) = \sum_{k=1}^{\infty} \alpha_k \cos(kx - \beta\pi/2),$$

where $\beta \in \mathbb{R}$, $\alpha_k > 0$, and $\sum_{k=1}^{\infty} \alpha_k < \infty$. Then $|\lambda_j(y)| > 0$, for all $1 \le j \le n-1$ if one of the following holds:

1. $n = 2q-1$, $|\beta| \ne 2p-1$, $q, p \in \mathbb{N}$, $y = 0$;
2. $n = 2q-1$, $|\beta| = 2p-1$, $p, q \in \mathbb{N}$, $y = 0$,

$$\sum_{m=0}^{\infty} \alpha_{mn+j} - \sum_{m=1}^{\infty} \alpha_{mn-j} \ne 0;$$

3. $n = 2q$, $|\beta| \ne 2p-1$, $p, q \in \mathbb{N}$, $y = 0$;
4. $n = 2q$, $|\beta| = 2p-1$, $p, q \in \mathbb{N}$, $y = \pi/n$.
   If α_k has the form $\alpha_k = k^{-1}\xi_k$, where $\xi_k > 0$, $k \in \mathbb{N}$, then $|\lambda_j(y)| > 0$, $1 \le j \le n-1$ if one of the following holds:
5. $n = 2q$, $|\beta| \ne 2p$, $p, q \in \mathbb{N}$, $\Delta\alpha_k \ge 0$, $\Delta^{(2)}\alpha_k \ge 0$, $y = \pi/n$;
6. $n = 2q-1$, $\beta \in \mathbb{R}$, $q \in \mathbb{N}$, $\Delta\alpha_k \ge 0$, $y = \pi/n$;
7. $n \in \mathbb{N}$, $\Delta^{(s)}\alpha_k \ge 0$, $1 \le s \le 3$, $\beta \in [4p, 4p+1] \cup [2+4p, 3+4p]$, $p \in \mathbb{Z}$, $\forall y \in (0, \pi/n)$;

8. $n \in \mathbb{N}$, $\Delta^{(s)}\alpha_k \geq 0$, $1 \leq s \leq 3$, $\forall \beta \in [1+4p, 2+4p] \cup [3+4p, 4+4p]$, $p \in \mathbb{Z}$, $\forall y \in (\pi/n, 2\pi/n)$;
9. $n = 2q$, $q \in \mathbb{N}$, $\Delta^{(s)}\alpha_k \geq 0$, $1 \leq s \leq 3$,

$$y \neq \begin{cases} 0, & \beta = 2p-1, \quad p \in \mathbb{Z}, \\ \frac{\pi}{n}, & \beta = 2p, \quad p \in \mathbb{Z}; \end{cases}$$

10. $n = 2q-1$, $q \in \mathbb{N}$, $\forall y \in \mathbb{R}$.

An overview of results when kernel K satisfies some sign regular properties can be found in [43]. The problem of existence of L-spline interpolants was considered by Micchelli [38], Schoenberg [51], Shevaldin [56] and others.

Remarks 1. *In the case* $\alpha_k = k^{-r}$, $r > 1$, $\beta = r\pi/2$, $r \in \mathbb{N}$, $n = 2p-1$, $p \in \mathbb{R}$ *the problem of existence of spline interpolants was considered by:*

1. *Ahlberg, Nilson and Walsh [1] if* $r = 2p$, $p \in \mathbb{N}$, $y = 0$;
2. *Subbotin [54] if* $r = 2p-1$, $p \in \mathbb{N}$, $y = \pi/n$;
3. *Subbotin and Zhensykbaev [62,71] if* $y \neq \pi(1-(-1)^{r+1})/2n$;
4. *Galkin [15] proved that the interpolating spline exists and is unique for* $r = 2, 3, ..., y = 0$ *if and only if* $n = 2p-1$, $p \in \mathbb{N}$. *Then Zhensykbaev [71] extended this result for the case* $y = \pi(1-(-1)^{r+1})/2n$.
5. *Shevaldin [55] proved Proposition 2 for all fractional* $r > 1$, *and in [54] considered the case* $\alpha_k = \rho^k k^{-1}$, $0 < \rho < 1$.
6. *In the general situation the problem of existence of interpolating sk–splines was considered by Kushpel [25,26,30], Light and Cheney [35], and Serduk and Stepanets [53].*

If the data points are in general position then the we have following statement:

Proposition 3. (Kushpel [25]) *Let* $0 = x_0 < ... < x_{n-1} < 2\pi$ *be an arbitrary partition of* $[0, 2\pi)$ *and*

$$K(x) = \sum_{k=1}^{\infty} \alpha_k \cos kx, \quad \alpha_k > 0, \quad \sum_{k=1}^{\infty} \alpha_k^{1/2} < \infty.$$

Then each $f \in C_{2\pi}$ *has a unique sk–spline interpolant, where* $\{x_s\}_{s=0}^{n-1}$ *are the knots and points of interpolation.*

Many well–known basis systems, including those of Haar and Faber-Schauder, are constructed with splines. It was shown by Figiel and Ciesielski [7,8] (see also [68, p.190]) that on any smooth manifold M^n, the Sobolev spaces $W_p^r(M^n)$ have piecewise polynomial bases. In the space $S^m(\Delta_n)$, of polynomial splines of degree m and defect 1, there are several types of bases. Let $x_s \in \Delta_n$, $1 \leq s \leq n$, and

$$(x)_+^m = \begin{cases} x^m, & \text{if } x \geq 0, \\ 0, & \text{if } x < 0. \end{cases}$$

Then,

$$S^m(\Delta_n) = \mathrm{span}\{1, x, ..., x^m, (x - x_1)_+^m, ..., (x - x_n)_+^m\}.$$

Another basis, more effective for calculations, introduced by Schoenberg in 1946, is the B-spline basis, $\{B_i^m(x)\}_{i=1}^n$. Schoenberg applied the apparatus of divided differences for constructing of B-splines, which are the nonnegative splines of degree m with minimal support, and form a partition of unity. A multi-dimensional analogy of the B-splines, the so-called box splines, were introduced by DeBoor and De Vore [10].

In the more general sk-spline setting we are not able to construct such compactly supported bases. However, as shown in the next proposition, we are able to get useful representations for cardinal sk-splines.

Proposition 4. *Define* $\overline{\mathrm{sk}}_0$ *by*

$$\overline{\mathrm{sk}}_0(y, x) = \begin{cases} 1, & \text{if } x = y, \\ 0, & \text{if } x = y_m,\ 1 \le m \le n-1. \end{cases}$$

Then, under the same conditions as Proposition 1, for every $y \in \mathbb{R}$ *we have:*

$$\overline{\mathrm{sk}}_0(y, x) = \frac{1}{n} + \frac{1}{n} \sum_{j=1}^{n-1} \frac{\rho_j(x)\rho_j(y) + \sigma_j(x)\sigma_j(y)}{|\lambda_j(y)|^2}$$

where $\rho_j = \Re\lambda_j$, *and* $\sigma_j = \Im\lambda_j$.

By making shifts and changes of parameter y we can get different examples of bases in the space of sk-splines. If the kernel K can be written in the form

$$K(x) = \sum_{k=1}^{\infty} \alpha_k \cos(kx - \beta\pi/2)$$

where $\alpha_k = k^{-2s}$, $\beta = 2s$, $s \in \mathbb{N}$ and $y = 0$, then the above proposition follows from the results of Golomb [16]. Zhensykbaev [71,72] extended the results of Golomb to the case $\alpha_k = k^{-2s-1}$, $\beta = 2s - 1$, $s \in \mathbb{N}$, $y = \pi/n$. Proposition 4 was proved by Kushpel [25,26,30] for sk-splines.

In Figure 1 we have some pictures of cardinal sk-splines with 10 equally spaced interpolation points, where the kernel

$$K(x) = \sum_{k=1}^{\infty} \rho^{-k} \cos kx,$$

$0 < \rho < 1$ (rho) as given, is (up to a constant multiple and summand) the Poisson kernel.

In the next section we will show some applications of Propositions 1 and 5 to the problem of sharp estimates of n-widths from below and estimates of convergence of sk-splines for a wide range of smooth functions.

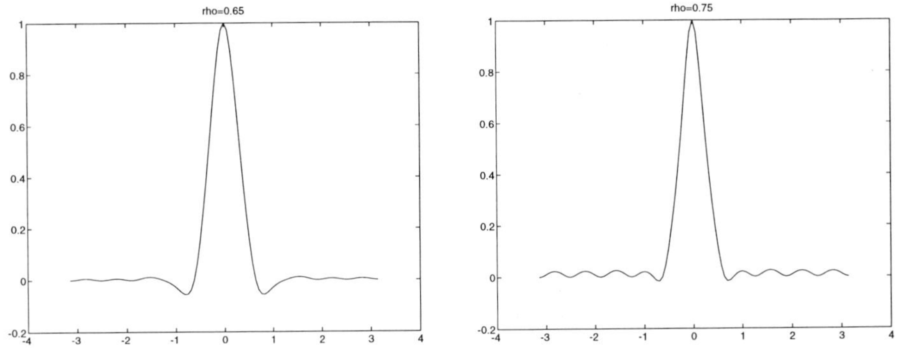

Fig. 1. Cardinal sk–splines using the Poisson kernel.

§3. sk–Splines and Sharp Estimates of n-Widths

Exact results are always much appreciated in approximation theory. When difficulties of a fundamental nature arise, successful endeavors to overcome them usually lead to new ideas and methods which can then be applied in other difficult situations. In this section we present an application of sk–splines to the problem of calculation of n–widths. Let X be a Banach space and A a (convex, compact, centrally symmetric) subset of X. The linear n–width of A in X is defined by

$$\delta_n(A,X) := \inf_{P_n} \sup_{f\in A} \|f - P_n f\|,$$

where P_n varies over all linear operators of rank at most n that map X into itself. The Kolmogorov n–width of A in X is defined by

$$d_n(A,X) := \inf_{X_n} \sup_{f\in A} \inf_{g\in X_n} \|f - g\|,$$

where X_n runs over all subspaces of X of dimension n or less. The Gel'fand n–width of A in X is defined by

$$d^n(A,X) := \inf_{L^n} \sup_{x\in L^n\cap A} \|x\|,$$

where L^n runs over all subspaces of codimension at most n. The Bernstein n–width of A in X is defined by

$$b_n(A,X) := \sup_{X_{n+1}} \sup\{\varepsilon > 0 : (\varepsilon B \cap X_{n+1}) \subset A\},$$

where X_{n+1} is any $(n+1)$-dimensional subspace of X, and $B = \{x \in X : \|x\| \le 1\}$ is the unit ball of X. More information on n–widths is contained in the book by Pinkus [43].

The problem of exact calculation of n–widths usually splits into two parts. To compute an upper bound we fix some n–dimensional subspace $X_n \in X$, and calculate the value

$$E(X_n, A, X) = \sup_{f\in A} \inf_{g\in X_n} \|f-g\|_X \geq d_n(A, X).$$

Usually, a respectable upper bound can be found by approximating by elements of $\mathcal{T}_n$ the space of trignometric polynomials of degree $\leq n$. In the case $X = C_{2\pi}$, where $C_{2\pi}$ is the space of 2π-periodic continuous functions, and $A = K * U_\infty$, $U_\infty = \{f : \|f\|_\infty \leq 1\}$, with K a kernel of special type, sharp upper bounds for $d_{2n-1}(K * U_\infty, C_{2\pi})$ were obtained as far back as 1936–1938 in the well known works of Favard [13], Akhiezer and Krein [3], Szokefalvy–Nady [64,65], Stechkin [59], and Dzadyk [12]. They proved that

$$E(\mathcal{T}_{n-1}, K * U_\infty, L_\infty) := \sup_{f\in A} \inf_{g\in \mathcal{T}_{n-1}} \|f-g\|_\infty = \|f_n\|_\infty,$$

where $f_n(x) = K * \text{sign} \sin nx$,

$$K(x) = \sum_{k=1}^{\infty} \alpha_k \cos\left(kx - \beta\frac{\pi}{2}\right).$$

The difficulty in finding lower bounds for n–widths is that all n–dimensional subspaces have to be considered. The first sharp lower bounds for n-widths in L_∞ were obtained by Tikhomirov [66]. He proved that

$$d_{2n-1}(W^r_\infty, L_\infty) = d_{2n}(W^r_\infty, L_\infty) = n^{-r}\mathcal{K}_r,$$

where

$$\mathcal{K}_r = \sum_{m=0}^{\infty} (-1)^m (2m+1)^{-(r+1)}.$$

Adapting the technique of Tikhomirov, Micchelli, Pinkus, Dyn, Chui, and Smith proved analogous results for functional sets which are generated by sign-regular kernels.

For the formulation of the main result of this section we need to introduce a set $C_{y,n}$ of kernel functions. Before giving a precise mathematical definition of this set we motivate the definition with some geometric ideas.

Let $x_k = k\pi/n$ and $y_k = y + x_k$, $y \in \mathbb{R}$, $1 \leq k \leq 2n$ be knots (numbers by which kernel is shifted) and interpolation points respectively. Let $K_1 = K * D_1$ and let $\overline{\text{sk}}_1(x,y)$ be the cardinal sk_1–spline interpolant, with kernel K_1, which satisfies

$$\overline{\text{sk}}_1(x,y) = \begin{cases} 1, & x = y, \\ 0, & x = y_k, \quad 1 \leq k \leq 2n. \end{cases}$$

Then,

$$\overline{\mathrm{sk}}_1(x,y) = c_0 + \sum_{k=1}^{2n} c_k K_1(x-x_k)$$
$$= c_0 + K * \sum_{k=1}^{2n} c_k D_1(x-x_k),$$

where $\sum_{k=1}^{2n} c_k = 0$, and we recall that D_1 is the degree 1 Bernoulli monospline. Because $\sum_{k=1}^{2n} c_k = 0$, then $\eta_{2n}(x) = \sum_{k=1}^{2n} c_k D_1(x-x_k)$ is a piecewise constant. If the sequence $\{\eta_{2n}((x_{k-1}+x_k)/2)\}$ alternates in sign then $K \in C_{y,n}$.

The following analytic definition is more complete and gives us a recipe for ensuring the property C_y for any kernel which satisfies the very general conditions of Proposition 2. This definition plays a central role in the methods of Kushpel [26,30,31] for obtaining sharp lower bounds for n–widths.

Definition 2. (Kushpel [26,30]) *Let*

$$K_1 = K * D_1, \quad \text{where } D_1(x) = \sum_{k=1}^{\infty} k^{-1} \sin kx, \quad x \in \mathbb{R}.$$

Define $x_k = k\pi/n$, $k = 1, 2, \ldots, 2n$. *Set*

$$\rho_j(y) = n^{-1} \sum_{\nu=1}^{2n} \cos(jx_\nu) K_1(y-x_\nu), \quad \sigma_j(y) = n^{-1} \sum_{\nu=1}^{2n} \sin(jx_\nu) K_1(y-x_\nu),$$

and

$$c_j = \left(\sum_{m=0}^{\infty} \frac{(-1)^m}{2m+j} - \sum_{m=1}^{\infty} \frac{(-1)^m}{2m-j} \right) \Big/ (\rho_j^2(y) + \sigma_j^2(y)).$$

Further, let

$$P_n(y,t) = \frac{1}{2n} \left(2 \sum_{j=1}^{n-1} c_j(\rho_j(y) \sin jt - \sigma_j(y) \cos jt) + c_n \rho_n(y) \sin nt \right).$$

We will say that $K \in C_{y,n}$ *if* $\rho_j^2(y) + \sigma_j^2(y) > 0, 1 \le j \le n-1$ *and*

$$\operatorname{sign} P_n(y, t_k) = (-1)^k e_k \epsilon, \quad 1 \le k \le 2n,$$

where $\epsilon \in \{1, -1\}$ *is a fixed number,* $e_k \in \{0,1\}$, *and* $t_k = (x_k + x_{k-1})/2$, *with* $x_k = k\pi/n$. *If* $K \in C_{y,n}$ *for all* $n \in \mathbb{N}$ *and some sequence* $\mathbf{y} = \{y(n)\}$, *then we shall say that* $K \in C_{\mathbf{y}}$.

Theorem 1. *If* $K \in L_1(\mathbb{T}) \bigcap C_y$ *for* $y = \xi$ *where* $|f_n(\xi)| = \|f_n\|_\infty$ *then*

$$d_{2n}(K * U_\infty, L_\infty) \ge \|f_n\|_\infty .$$

Remarks 2. *We present here several examples of kernels K which fall into the framework of Theorem 1. The first example is given by the Bernoulli monosplines of degree r, D_r, $r \in \mathbb{N}$, which correspond to the Sobolev classes W_∞^r (Tikhomirov [66,67]). The kernel $K_{\mathcal{L}_N}$ from the integral representation of the functional sets $W_\infty^{\mathcal{L}_N}$, investigated by Chui and Smith [9], Shevaldin [54], Nguen Thi Hoa [41], also has property C_y. All the above mentioned kernels are in some sense sign regular. It is possible to show that an arbitrary sign regular kernel K satisfies the condition $C_{\mathbf{y}}$. The corresponding n–widths were considered by Micchelli and Pinkus [40], Dyn [11], Forst [14], and others.*

The first exact result for n–widths of convolution sets with kernels which are not sign regular were obtained by Kushpel [26,30]. It was proved that the kernel

$$P_{\rho,\beta}(x) = \sum_{k=1}^{\infty} \rho^k \cos\left(kx - \beta\frac{\pi}{2}\right)$$

is not sign regular for $\beta = 1$ or 0, but that for $0 \le \rho \le 1/7$ we have

$$d_{2n-1}(P_{\rho,\beta} * U_\infty, L_\infty) = d_{2n}(P_{\rho,\beta} * U_\infty, L_\infty) = d_{2n-1}(P_{\rho,\beta} * U_1, L_1) = \frac{4}{\pi}\arctan \rho^n .$$

Then, using method of Kushpel, Shevaldin [55] improved the range for ρ and considered the case of an arbitrary $\beta \in \mathbb{R}$. He proved that $P_{\rho,\beta} \in C_{\mathbf{y}} \cap L_1(\mathbb{T})$ for

$$0 < \rho \le \begin{cases} 0.2 & \text{if } \beta \in \mathbb{Z}, \\ 0.196881 & \text{otherwise.} \end{cases}$$

For sets of functions with fractional derivatives the situation is still not completely resolved. Let $\alpha > 0$, $\beta \in \mathbb{R}$ and

$$D_{\alpha,\beta}(x) = \sum_{k=1}^{\infty} k^{-\alpha} \cos\left(kx - \beta\frac{\pi}{2}\right)$$

be the Weyl–Nady kernel. If $|\beta - \alpha| \le \epsilon(n,\alpha)$, $\beta \in \mathbb{R}$, then it was shown by Kushpel [31] that for all $n \in \mathbb{N}$, $D_{\alpha,\beta} \in C_{\mathbf{y}} \cap L_1(\mathbb{T})$, and

$$\begin{aligned} d_{2n-1}(D_{\alpha,\beta} * U_\infty, L_\infty) &= d_{2n}(D_{\alpha,\beta} * U_\infty, L_\infty) \\ &= d_{2n-1}(D_{\alpha,\beta} * U_1, L_1) \\ &= \|D_{\alpha,\beta} * \operatorname{sign} \sin n(\cdot)\|_\infty . \end{aligned}$$

Applying sk–spline techniques, Shevaldin [56] proved that for

$$C(\beta) = \begin{cases} 0.249971, & \beta \in \mathbb{Z} , \\ 0.247812, & \text{otherwise,} \end{cases}$$

$n \le C(\beta)(r+1)$, and $\alpha \ge 1$, $D_{\alpha,\beta} \in C_{\mathbf{y}} \cap L_1(\mathbb{T})$, and

$$d_{2n}(D_{\alpha,\beta} * U_\infty, L_\infty) = \|f_n\|_\infty = \|D_{\alpha,\beta} * \text{sign } \sin n(\cdot)\|_\infty .$$

Using standard methods, and Theorem 1, we can show that, for all previous choices of the kernel K, we actually have

$$\begin{aligned} b_{2n-1}(K*U_\infty, L_\infty) &= b_{2n-1}(K*U_1, L_1) = d_{2n}(K*U_\infty, L_\infty) = d_{2n-1}(K*U_\infty, L_\infty) \\ &= d^{2n}(K*U_1, L_1) = d^{2n-1}(K*U_1, L_1) = d_{2n-1}(K*U_1, L_1) = \delta_{2n}(K*U_\infty, L_\infty) \\ &= \delta_{2n-1}(K*U_\infty, L_\infty) = \delta_{2n}(K*U_1, L_1) = \delta_{2n-1}(K*U_1, L_1) = \|f_n\|_\infty . \end{aligned}$$

Serduk [52] applied the methods of Kushpel and proved that, if ϕ is a non–increasing function, then for

$$K(t) = \sum_{k=1}^{\infty} \phi(k) e^{-\alpha k^r} \cos\left(kt - \beta\frac{\pi}{2}\right), \quad 0 < r < 1, \quad \alpha > 0,$$

we have

$$d_{2n}(K*U_\infty, L_\infty) = \|f_n\|_\infty = \|K * \text{sign}\,\{\sin n(\cdot)\}\|_\infty ,$$

if $n < \alpha r C(\beta)$, with

$$C(\beta) = \begin{cases} 0.230150, & \beta \in \mathbb{Z}\ , \\ 0.2300576, & \text{otherwise.} \end{cases}$$

§4. Interpolation on the Torus

In the remainder of this paper we shall be borrowing the one dimensional results of Kushpel [26] in order to prove convergence results for generalised sk–spline interpolants on $\mathbb{T}^d = [0, 2\pi)^d$, the d–dimensional torus. We shall be interpolating functions in convolution classes

$$\mathbb{R} \oplus K * U_p = \{f = \alpha + K*\phi : \ \alpha \in \mathbb{R}, \phi \in U_p\},$$

where

$$U_p = \{\phi \in L_p : \ \|\phi\|_p \le 1\},$$

and

$$K(\mathbf{x}) = \prod_{m=1}^{d} K_m(x_m), \ \ \mathbf{x} = (x_1, x_2, \ldots, x_d),$$

is a product of one dimensional kernels. Thus, by choosing each K_m to have a certain amount of smoothness, we can approximate functions which have different amounts of smoothness in different directions. It is this product construction that will allow us to decompose the total error into a sum of one dimensional errors. In the remainder of this paper C will denote a generic constant whose value is not necessarily the same at each occurrence.

Before we consider the construction of the generalised sk–spline interpolant in more detail we give some examples in order to clarify the sorts of functions we are interpolating.

Example 1. *Suppose that*

$$K_m(x_m) = \sum_{\ell=1}^{\infty} a_{\ell,m} \cos\left(\ell x_m - \frac{\beta_m \pi}{2}\right).$$

Then, for particular choices of the $a_{\ell,m}$ and the constant β_m we obtain the following concrete examples:

1. *$a_{\ell,m} = \ell^{-r_m}$, $\beta_m = r_m$ gives us a Bernoulli monospline in each direction, and in this case $k * U_p$ is the Sobolev class of functions whose $\mathbf{r} = (r_1, r_2, \ldots, r_d)$th derivative is in L_p.*
1. *$a_{\ell,m} = \exp(-\alpha_m \ell^{r^m})$, with $\alpha_m > 0$ and $0 < r_m < 1$, and $\beta_m = 0$. The convolution class in this case contains infinitely differentiable functions.*
1. *$a_{\ell,m} = 2(\rho_m)^\ell$, with $0 < \rho_m < 1$, and $\beta_m = 0$, is (up to an added constant) the Poisson kernel and the convolution class consists of functions which are analytic in each of the coordinate directions.*
1. *$a_{\ell,m} = \exp(-\alpha_m \ell^{r_m})$, with $\alpha_m > 0$ and $r_m > 1$, and the resulting convolution is an entire function of x_m.*

In what follows we will be investigating the rate of convergence of interpolants (which we will describe below) on the grid $\Delta_\mathbf{n}$ with points

$$\mathbf{x_j} = (x_{\mathbf{j},1}, x_{\mathbf{j},2}, \ldots, x_{\mathbf{j},d}), \;\; \mathbf{j} \le 2\mathbf{n},$$

where $x_{\mathbf{j},m} = \dfrac{j_m \pi}{n_m}$, and $\mathbf{j}$ and $\mathbf{n}$ are multiindices.

In [25], for the case $d = 1$, interpolants of the form

$$\mathrm{sk}(x) = c_0 + \sum_{i=1}^{2n} c_j K(x - x_j)$$

are examined, where $\sum_{i=1}^{2n} c_i = 0$ and $K(x) = \sum_{\ell \ge 1} a_\ell \cos \ell x$, with $a_\ell > a_{\ell+1} > 0$ for all $\ell \in \mathbb{N}$. These form a subset of the sk–splines. Writing $\cos \ell x = 1/2(e^{i\ell x} + e^{-i\ell x})$ and noting that the dual group of $\mathbb{T}$ is $\{e^{i\ell\cdot}, \ell \in \mathbb{Z}\}$, we see by Theorem 7 that K is $\{\hat{e}\}$–SCPD (see Definition 4), and hence, by Theorem 8, that, for each m, there is a unique sk–spline interpolant to arbitrary data on $\{x_{\mathbf{j},m} : j_m \le 2n_m\}$.

Let

$$\mathrm{sk}_m(x_m) = \begin{cases} 1, & x_m = 0, \\ 0, & x_m = \frac{j\pi}{n_m}, \; j = 1, 2, \ldots, 2n_m - 1, \end{cases}$$

be the cardinal sk–spline associated with the mth coordinate. We construct a cardinal generalised sk–spline on the grid $\Delta_\mathbf{n}$ with products of such one dimensional cardinal splines:

$$\overline{\mathrm{sk}}(\mathbf{x}) = \prod_{m=1}^{d} \mathrm{sk}_m(x_m).$$

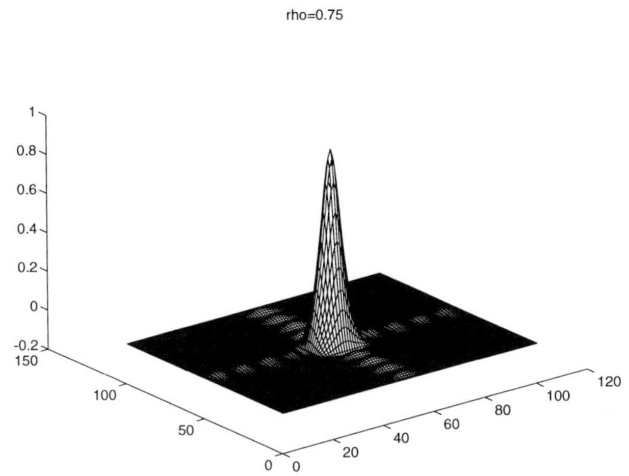

Fig. 2. Two dimensional cardinal sk–spline.

In the next figure we see the cardinal sk–spline, on a 10 by 10 uniform grid of interpolation points, using the Poisson kernel with $\rho = 0.75$ in both directions.

So, using the shifted cardinal sk–spline

$$\overline{\mathrm{sk}}(\mathbf{x} - \mathbf{x_j}) = \prod_{m=1}^{d} \mathrm{sk}_m(x_m - x_{\mathbf{j},m}),$$

we can write the generalised sk–spline interpolant to $f \in \mathbb{R} \oplus K * U_p$ in the form

$$\overline{\mathrm{sk}}(f, \Delta_{\mathbf{n}})(\mathbf{x}) = \sum_{\mathbf{j} \leq 2\mathbf{n}} f(\mathbf{x_j}) \mathrm{sk}(\mathbf{x} - \mathbf{x_j}).$$

The following theorem is proved in Levesley and Kushpel [34].

Theorem 3. *Let $K \in \mathbf{C}(\mathbb{T}^d)$ have the absolutely convergent Fourier series*

$$k(\mathbf{x}) = \prod_{m=1}^{d} K_m(x_m) = \prod_{m=1}^{d} \sum_{\ell=1}^{\infty} a_{\ell,m} \cos(\ell x_m),$$

where $a_{\ell,m} > a_{\ell+1,m} > 0$, for all $\ell \in \mathbb{N}$, $1 \leq m \leq d$, and

$$\sum_{s=1}^{\infty} a_{2n_m s - \ell, m} \leq C a_{2n_m - \ell, m},$$

where C is independent of m, ℓ and $\mathbf{n}$. Then, for $1 \le p \le 2 \le q \le \infty$, with $p^{-1} - q^{-1} \ge 2^{-1}$,

$$\sup_{f \in K * U_p} \|f - \mathrm{sk}(f, \Delta_{\mathbf{n}})\|_q \le \max_{1 \le m \le d} \left(\sum_{\ell \ge n_m} (a_{\ell,m})^{qp(q-p)^{-1}} \right)^{p^{-1}-q^{-1}} .$$

Remarks 3.

1. *If we define $a_{\mathbf{l}}$, $\mathbf{l} \in \mathbb{Z}^d$, by*

$$a_{\mathbf{l}} = \begin{cases} 0, & \mathbf{l}_m = 0, \text{ for any } 1 \le m \le d, \\ 2^{-d} \prod_{m=1}^{d} a_{|\mathbf{l}_m|,m}, & \text{otherwise}, \end{cases}$$

then we can write $K(\mathbf{x}) = \sum_{\mathbf{l} \in \mathbb{Z}^d} a_{\mathbf{l}} e^{i\mathbf{l}\mathbf{x}}$.

2. *If $d = 1$ and $a_\ell = e^{-\alpha \ell^r}$, $\alpha > 0$ and $r \ge 1$, then from Theorem 3 it follows that*

$$\begin{aligned} \epsilon_n(K * U_1, L_\infty) &= \sup\{\|f - \mathrm{sk}(f, \Delta_n)\|_\infty \;:\; f \in K * U_1\} \\ &\le C e^{-\alpha n^r}, \end{aligned}$$

as $n \to \infty$. Hence, because L_p is continuously embedded in L_q for $1 \le q \le p \le \infty$, we have $\epsilon_n(K * U_p, L_q) \le C e^{-\alpha n^r}$, for any $1 \le p, q \le \infty$. It is clear that in this case the dimension of the sk–spline subspace is $2n$. Let d_n and b_n be the n–widths in the sense of Kolmogorov and Bernstein respectively, and λ_n and π_n be, respectively, the linear and projective n–widths (for the definitions of these see e.g. [43],[68]). Since the operator sk : $f \to \mathrm{sk}(f, \Delta_n)$, is a linear projector (due to the uniqueness of the solution of the interpolation problem) we have $b_{2n} \le d_{2n} \le \lambda_{2n} \le \pi_{2n} \le C\epsilon_{2n}$. Using the results of Kushpel [32], we have $b_{2n}(K * U_p, L_q) \ge C e^{-\alpha n^r}$, as $n \to \infty$, for all $1 \le p, q \le \infty$. So we see that sk–splines provide a new example of extremal subspaces for $K * U_p$ in L_q, for all $1 \le p, q \le \infty$.

The next result is connected with the convergence of sk–spline interpolants with number-theoretical knots on unisotropic Sobolev classes. In this case, a special choice of knots give us almost optimal orders of convergence. Let $\mathbf{g} = (g_1, \ldots, g_d) \in \mathbb{Z}^d$, $1 \le g_s \le P - 1$, $1 \le s \le d$, where P is a prime number. Let $\mathbf{x}_k = \mathbf{g}k/P$, $1 \le k \le P$ be the set of knots which are also data points $\Delta_P^{\mathbf{g}}$. The sequence $\{\mathbf{x}_k\}$ is a sequence of minimal discrepency, in the sense described in [24]. The fundamental concepts in this number theoretic method were advanced by Korobov [23], Halton, Hlawka, and others, between 1957 and 1962. More information on the use of number theoretic methods in numerical analysis can be found in Keng and Yuan [19].

We employ the following construction of interpolants:

$$\mathrm{sk}(\mathbf{x}) = c_0 + \sum_{k=1}^{n} c_k K(\mathbf{x} - \mathbf{x}_k) \,,$$

where the Fourier series of the kernel K has the form

$$K(\mathbf{x}) \sim \prod_{s=1}^{d} \sum_{k=1}^{\infty} a_{k,s} \cos kx_s, \quad a_{k,s} = k^{-r_s},$$

corresponds to the Sobolev class

$$W_p^r = K * U_p \oplus \mathbb{R} = \left\{ f \;\middle|\; \left\| \frac{\partial^{|\mathbf{r}|} f}{\partial x_1^{r_1} \ldots \partial x_d^{r_d}} \right\|_p \le 1 \right\},$$

where $\mathbf{r} = (r_1, r_2, \ldots, r_d)$ is a multi–index.

Theorem 4. (Kushpel, Gomes, Ragozin, and Levesley [33]) *Let* $1 < r = r_1 = \ldots = r_\nu < r_{\nu+1} \le \ldots \le r_d$. *Then for* $1/p - 1/q \ge 1/2$, $1 \le p \le 2 \le q \le \infty$ *and some* $\mathbf{g} = (g_1, \ldots, g_d)$, $1 \le g_s \le P - 1$ *we have, as* $P \to \infty$,

$$\sup_{f \in W_p^r} \|f - sk(f, \Delta_P^{\mathbf{g}})\|_q \ll P^{-r+1/p-1/q} (\log P)^{r\nu - (1/p-1/q)},$$

where $a_n \ll b_n$ *if there exists* C, *independent of* n, *such that* $a_n \le C b_n$, *for all* $n \in \mathbb{N}$.

Proof: Letting $\overline{\mathrm{sk}}(x)$ be the cardinal sk–spline which is 1 at 0 and 0 at x_k, $k = 1, 2, \ldots, P-1$, we consider, for any kernel K, the error

$$\begin{aligned}
\Delta(K * U_p, \mathbf{x}) &= \sup_{f \in K * U_p} \left| f(\mathbf{x}) - \sum_{k=1}^{P} f(\mathbf{x}_k) \overline{\mathrm{sk}}(\mathbf{x} - \mathbf{x}_k) \right| \\
&= \sup_{\phi \in K * U_p} \left| K * \phi(\mathbf{x}) - \sum_{k=1}^{P} K * \phi(\mathbf{x}_k) \overline{\mathrm{sk}}(\mathbf{x} - \mathbf{x}_k) \right| \\
&\le \|K(\mathbf{x} - \cdot) - \sum_{k=1}^{P} K(\mathbf{x}_k - \cdot) \overline{\mathrm{sk}}(\mathbf{x} - \mathbf{x}_k)\|_{p'},
\end{aligned}$$

using Hölder's inequality after expressing the convolution in its integral form. Now, employing the Hausdorff–Young inequality (see e.g. [73]), we have

$$\Delta(K * U_p, \mathbf{x}) \le \left(\sum_{\mathbf{l} \in \mathbb{Z}^d} a_{\mathbf{l}}^p | e^{i\mathbf{l}\mathbf{x}} - \sum_{k=1}^{P} e^{i\mathbf{l}\mathbf{x}_k} \overline{\mathrm{sk}}(\mathbf{x} - \mathbf{x}_k)|^p \right)^{\frac{1}{p}}.$$

Thus, the problem becomes one of bounding the error in interpolating the complex exponentials. We do this in two parts, splitting $\mathbb{Z}^d$ into two parts, $\Gamma(Q)$ and its complement, where $Q \in \mathbb{N}$. The choice of $\Gamma(Q)$ for a particular choice of Q is effected in the following way. Define χ by $2^{\chi} \le Q \le 2^{\chi+1}$. For fixed $\mathbf{y} \in \mathbb{R}^d$ and $\mathbf{s} \in \mathbb{N}^d$, let $A_{\mathbf{s}} = \{\mathbf{l} \in \mathbb{Z}^d : 2^{s_k} \le |l_k| < 2^{s_{k+1}},\ k = 1, 2, \ldots, d\}$. Then,

$$\Gamma(Q) = \bigcup_{\mathbf{ys} \le \chi} A_{\mathbf{s}}.$$

Clearly $\mathrm{Card}(A_{\mathbf{s}}) = 2^{\mathbf{se}+d}$, where $\mathbf{e} = (1, 1, \ldots, 1)$. In order to get error estimates we use the easily verifiable fact that

$$|e^{i\mathbf{lx}} - \sum_{k=1}^{P} e^{i\mathbf{lx}_k} \overline{\mathrm{sk}}(\mathbf{x} - \mathbf{x}_k)| \le 2 \min \left\{ \frac{\sigma_{\mathbf{l}}}{a_{\mathbf{l}}}, 1 \right\},$$

where

$$\sigma_{\mathbf{l}} = \sum_{\mathbf{z} \in \mathbb{Z}^d, \mathbf{zg} \neq 0} a_{\mathbf{l}+\mathbf{z}}.$$

Hence,

$$\begin{aligned} \Delta(K * U_p, \mathbf{x}) &\le \left(\left(\sum_{\mathbf{l} \in \mathbb{Z}^d / \Gamma(Q)} + \sum_{\mathbf{l} \in \Gamma(Q)} \right) \left(a_{\mathbf{l}}^p |e^{i\mathbf{lx}} - \sum_{k=1}^{P} e^{i\mathbf{lx}_k} \overline{\mathrm{sk}}(\mathbf{x} - \mathbf{x}_k)|^p \right) \right)^{\frac{1}{p}} \\ &\le 2 \left(\sum_{\mathbf{l} \in \mathbb{Z}^d / \Gamma(Q)} a_{\mathbf{l}}^p \right)^{\frac{1}{p}} + 2 \left(\sum_{\mathbf{l} \in \Gamma(Q)} \sigma_{\mathbf{l}}^p \right)^{\frac{1}{p}}. \end{aligned} \tag{1}$$

We bound the first sum above

$$\begin{aligned} \left(\sum_{\mathbf{l} \in \mathbb{Z}^d / \Gamma(Q)} a_{\mathbf{l}}^p \right)^{\frac{1}{p}} &= \left(\sum_{\mathbf{ys} \ge \chi} \sum_{\mathbf{l} \in A_{\mathbf{s}}} a_{\mathbf{l}}^p \right)^{\frac{1}{p}} \\ &\le \left(\sum_{\mathbf{ys} \ge \chi} 2^{\mathbf{se}+d} \max_{\mathbf{l} \in A_{\mathbf{s}}} a_{\mathbf{l}}^p \right)^{\frac{1}{p}} \\ &\le C \left(\sum_{\mathbf{ys} \ge \chi} 2^{\mathbf{se} - p\mathbf{sr}} \right)^{\frac{1}{p}}, \end{aligned}$$

where, in the last inequality above we have used the fact that $K = D_{\mathbf{r}}$, so that $a_{\mathbf{l}} = \mathbf{l}^{-\mathbf{r}}$. In the case of the anisotropic Sobolev classes generated by the kernel $D_{\mathbf{r}}$, we let $\mathbf{y}' = (p\mathbf{r} - \mathbf{e})/(pr - 1)$ and choose $\mathbf{y}$ to be any multiindex satisfying

$\mathbf{y}_k = \mathbf{y}'_k = 1$, $1 \le k \le \nu$, and $\mathbf{y}_k < \mathbf{y}'_k$, $k = \nu+1, \ldots, d$. This leads to a particular shape for the domain $\Gamma(Q)$. For different function classes we would make a different choice of $\mathbf{y}$, leading to a different shape of domain $\Gamma(Q)$. It can be shown that for such a $\mathbf{y}$,

$$\sum_{\mathbf{ys}\ge\chi} 2^{\mathbf{se}-p\mathbf{sr}} \le 2^{-rp+1}\chi^{\frac{\nu-1}{p}}$$
$$\le Q^{-rp+1}(\log Q)^{\nu-1}.$$

Using the last two inequalities we see that

$$\left(\sum_{\mathbf{l}\in\mathbb{Z}^d/\Gamma(Q)} a_{\mathbf{l}}^p\right)^{\frac{1}{p}} \le Q^{-r+\frac{1}{p}}(\log Q)^{\frac{\nu-1}{p}} \tag{2}$$

To bound the second sum in (1) we adapt the approach detailed in Kuipers and Niederreiter [24, Theorem 5.7 and Example 5.4] to show that, as long as $\Gamma(Q) \subseteq [-P+1, P-1]^d$, then there exists $\mathbf{g} \in G$, as alluded to in the statement of this theorem, such that, for all $\mathbf{l} \in \Gamma(Q)$,

$$\sigma_{\mathbf{l}}^p \le \sum_{g\mathbf{s}=0,\mathbf{s}\ne 0,\|\mathbf{s}\|_\infty<P} a_{\mathbf{s}+\mathbf{l}}^p$$
$$\le CP^{-rp}(\log P)^{\nu}.$$

Therefore, because $\mathrm{Card}(\Gamma(Q)) \le Q(\log Q)^{\nu-1}$,

$$\left(\sum_{\mathbf{l}\in\Gamma(Q)} \sigma_{\mathbf{l}}^p\right)^{\frac{1}{p}} \le CP^{-r}(\log P)^{\frac{\nu}{p}}Q^{\frac{1}{p}}(\log Q)^{\frac{\nu-1}{p}}.$$

The result for $q=\infty$ follows from substituting the last equation and (2) into (1) and choosing $Q = P(\log P)^{-\nu}$. Using duality we get the same result measuring in the p' norm with functions from $K * U_1$, and we get the stated results using interpolation arguments. ■

If $a_n \ll b_n$ and $b_n \ll a_n$ then we write $a_n \asymp b_n$.

Remarks 4. *It is known [65] that the best approximation by trigonometric polynomials with spectrum in the hyperbolic cross Q_P^r is realised by Fourier sums and*

$$E(W_p^r, Q_P^r, L_q) \asymp P^{-r+1/p-1/q}(lnP)^{-(-r+1/p-1/q)(\nu-1)},$$
$$r > 1/p - 1/q, \quad 1 < p \le q < \infty.$$

For linear n–widths $\lambda_P(W_p^r, L_q)$ we have (see [65]) that

$$\lambda_P(W_p^r, L_q)$$
$$\asymp \begin{cases} (P^{-1}ln^{\nu-1}P)^{r-1/p+1/2}, & r > 1/p, 1 < p \le 2 < q < \infty,\ 1/p \ge 1-1/q, \\ (P^{-1}ln^{\nu-1}P)^{r-1/2+1/q}, & r > 1-1/q, 1 < p \le 2 < q < \infty,\ 1/p < 1-1/q. \end{cases}$$

§5. sk–Splines on Compact Abelian Groups and Homogeneous Spaces

In this section we shall state some results concerning generalised sk–splines on compact abelian groups. We shall not prove these results here, but refer the interested reader to Levesley and Kushpel [33].

Suppose a set G has both a group structure and a topological structure, and the group multiplication with a fixed element of the group is continuous with respect to the topology. Then G is a topological group. Furthermore, if the group is compact in this topology and the multiplication is commutative then we have a compact abelian group. There is a unique translation invariant measure on G for which the measure of G is 1. This is the normalised Haar measure μ.

A *character* of the group G is a continuous homomorphism from G onto the unit circle, and with pointwise multiplication, the set of characters is a group $\widehat{G}$, which we will call the *dual group* of G. We call the identity element in this group $\hat{e}$. In what follows a crucial result is that if G is compact, $\widehat{G}$ is discrete; see [45, page 9].

To each $f \in L_1(G)$, we can associate a Fourier series

$$f \sim \sum_{\chi \in \widehat{G}} \alpha_\chi \chi,$$

where

$$\alpha_\chi = \int_G f(x)\overline{\chi(x)}d\mu(x).$$

We now define the generalised sk–spline:

Definition 3. *Let $K \in C(G)$ have the Fourier series*

$$K = \sum_{\chi \in \widehat{G}/\hat{e}} \alpha_\chi \chi.$$

Then, a generalised sk–spline is one of the form

$$sk(x) = c_0 + \sum_{i=1}^{n} c_i K(xx_i^{-1}),$$

where $x_1, x_2, \ldots, x_n \in G$ and $\sum_{i=1}^n c_i = 0$.

Our first theorem concerns the density of generalised sk–splines and presupposes the fundamentality of the characters.

Theorem 5. *Let $\widehat{G}_0 \subseteq \widehat{G}$ be closed under conjugation, and let $K = \sum_{\chi \in \widehat{G} \setminus \widehat{G}_0} \alpha_\chi \chi$, where $\alpha_\chi \neq 0$ for any $\chi \in \widehat{G} \setminus \widehat{G}_0$. Then, every function in $\mathbf{C}(G)$ can be uniformly approximated by a function of the form*

$$f + \sum_{i=1}^{n} c_i K(\cdot\, x_i^{-1}),$$

for some $n \in \mathbb{N}$, and $x_1, x_2, \ldots, x_n \in G$, where f is in the linear span of $\widehat{G}_0$, and

$$\sum_{i=1}^{n} c_i \chi_0(x_i) = 0, \quad \text{for all } \chi_0 \in \widehat{G}_0.$$

In fact, there is a more general result concerning the density of translates of a spherical kernel on an arbitrary compact homogeneous space.

Theorem 6. (Ragozin and Levesley [44]) *Suppose M is a compact manifold in $\mathbb{R}^d$ which admits a transitive group G of orthogonal transformations. Let $K(\mathbf{x}, \mathbf{y})$ be a continuous real symmetric kernel on $M \times M$ which satisfies $K(g \cdot \mathbf{x}, g \cdot \mathbf{y}) = K(\mathbf{x}, \mathbf{y})$ for all $g \in G$. Let H_n be the polynomials of degree n which are orthogonal to those of degree less than n in $L^2(M)$, where integration is with respect to μ, the normalized surface measure on M. Fix a point $\mathbf{p} \in M$. and let $K_{\mathbf{p}}$ be the isometries in G which fix $\mathbf{p}$. Let ${}^{K_{\mathbf{p}}}H_n$ be the zonal polynomials in H_n which satisfy $p(g \cdot \mathbf{x}) = p(\mathbf{x})$ for all $g \in K_{\mathbf{p}}$. Then $\mathcal{K}$ is dense in $\mathbf{C}(M)$ if and only if $T_k(p)(\cdot) = \int_M K(\cdot, \mathbf{y}) p(\mathbf{y}) d\mu(\mathbf{y}) \neq 0$ for any n and any $p \in {}^{K_{\mathbf{p}}}H_n$. In particular for any basis $\{p_{nj}\}$ of ${}^{K_{\mathbf{p}}}H_n$ consisting of eigenvectors with $T_k(p_{nj}) = a_{nj} p_{nj}$, density holds if and only if $a_{nj} \neq 0$ for all n, j.*

Remarks 5. *The above result generalises Theorem 5 and the results of Sun [63], concerning the density of translates of radial functions on the sphere.*

In order to discuss interpolation with generalised sk–splines we first of all need to introduce the notion of conditional positive definiteness:

Definition 4. *Given a subset $\widehat{G}_0 \subseteq \widehat{G}$, a continuous function K is conditionally positive definite with respect to $\widehat{G}_0$, which we abbreviate to $\widehat{G}_0$–CPD if, for any $n \in \mathbb{N}$, and distinct $x_1, x_2, \ldots, x_n \in G$,*

$$\sum_{i,j=1}^{n} c_i \bar{c}_j K(x_i x_j^{-1}) \geq 0,$$

for any $c \in \mathbb{C}^n$ satisfying

$$\sum_{i=1}^{n} c_i \chi(x_i) = 0, \quad \text{for every } \chi \in \widehat{G}_0.$$

If there is a strict inequality in the quadratic form above then we say that g is strictly conditionally positive definite with respect to $\widehat{G}_0$, or g is $\widehat{G}_0$–SCPD.

In the next theorem we characterise a set of $\widehat{G}_0$–SCPD functions and in the final theorem see how, for a special choice of $\widehat{G}_0$, we obtain a generalised sk–spline with which we can interpolate. Note the similarity of this result to Proposition 3.

Theorem 7 Let $\widehat{G}_0 \subseteq \widehat{G}$**.** *Then, if*

$$K = \sum_{\chi \in \widehat{G} \setminus \widehat{G}_0} \alpha_\chi \chi,$$

with $\alpha_\chi > 0$ *and*

$$\sum_{\chi \in \widehat{G} \setminus \widehat{G}_0} \alpha_\chi^{\frac{1}{2}} < \infty,$$

then K *is* $\widehat{G}_0$*–SCPD.*

Theorem 8. *Let K be $\{\hat{e}\}$–SCPD. Then, for arbitrary data $v_1, v_2, \ldots, v_n \in \mathbb{C}$, $n \in \mathbb{N}$, there exist constants $c_0, c_1, \ldots, c_n$, satisfying*

$$\sum_{i=1}^{n} c_i = 0,$$

such that

$$c_0 + \sum_{j=1}^{n} c_i K(x_i x_j^{-1}) = v_i, \; i = 1, 2, \ldots, n.$$

§6. The Variational Approach

In this section we shall generalise the result of Holladay [17] which tells us that the periodic cubic spline is the minimiser, over all periodic interpolants, of the semi-norm

$$\left(\int_0^{2\pi} (g^{(2)}(x))^2 dx \right)^{\frac{1}{2}}.$$

Theorem 9. *Let $(\mathbf{C}(G))'$ be the dual space of $\mathbf{C}(G)$, $B : (\mathbf{C}(G))' \to \mathbf{C}(G)$ be continuous and B^* be its adjoint. Suppose also that $(B^*)^{-1}$ exists and has kernel $\langle \widehat{G}_0 \rangle$ for some $\widehat{G}_0 \subseteq \widehat{G}$. Furthermore, suppose that, given $\{x_i\}_{1 \le i \le n} \subseteq G$ and* *complex data* $\{v_i\}_{1\le i\le n}$, $V = \{f \in B^*(L_2(G)) : f = \sum_{\chi\in\widehat{G}} \alpha_\chi(f)\chi$ and $f(x_i) = v_i,\ 1 \le i \le n\}$. *Then, if there is an element of* V *of the form*

$$g(x) = g_0(x) + B^* B(\sum_{i=1}^{n} c_i \delta(x\, x_i^{-1})),\quad g_0 \in \langle \widehat{G}_0 \rangle,$$

then g *minimises, over all* $f \in V$, *the semi-norm*

$$\left(\int_G |(B^*)^{-1} g(x)|^2 d\mu(x) \right)^{\frac{1}{2}}.$$

Example 2. *In this example we see that the result of Holladay is a special case of the above, when* G *is the unit circle. If we let the operator* B *be convolution with*

$$D_2(x) = \sum_{\ell=1}^{\infty} \ell^{-2} \cos \ell x,$$

then B *is self-adjoint and continuous from* $(\mathbf{C}(G))' \to \mathbf{C}(G)$. *Convolution with* D_2 *is equivalent to two integrations, so that* $(B^*)^{-1}$ *is equivalent to two differentiations, which has kernel, in the periodic functions, consisting of the constants. Setting* $D_4 = D_2 * D_2$ *it is easy to show that all periodic cubic splines can be written in the form* $c_0 + D_4 * (\sum_{i=1}^{n} c_i \delta(\cdot - x_i))$ *with* $0 \le x_1 < x_2 < \dots x_n < 2\pi$. *The above theorem says that the cubic spline interpolant to given data at* $\{x_i\}_{1\le i\le n}$ *is the minimiser over all interpolants with square integrable second derivative, of the semi-norm*

$$\left(\int_0^{2\pi} (f^{(2)}(x))^2 dx \right)^{\frac{1}{2}}.$$

Acknowledgements. This research has been partially supported by The Royal Society, The International Science Foundation, the Ukrainian Government, and The Association of Ukranians in Great Britain.

References

1. Ahlberg, J. H., E. N. Nilson and F. L. Walsh, Best approximation and convergence properties of higher order spline approximations, J. Math. Mech. **14** (1965), 231–244.
2. Ahlberg, J. H., E. N. Nilson and F. L. Walsh, *The Theory of Spline Functions and their Applications*, Academic Press, New York, 1967.

3. Akhiezer, N. I. and M. G. Krein, On the best approximation of differentiable periodic functions by trignometric sums, Dokl. Acad. Nauk. SSSR **15** (1937), 107–112 (in Russian).
4. Ansolone, P. M. and P. J. Laurent, A general method for constructing interpolating or smoothing spline functions, Numer. Math. **12** (1968), 66–82.
5. Attéia, M., Généralisation de la définition et des propriétés des "spline fonctions", Comptes Rendus Acad. Sci. Paris **260** (1965), 3550–3553.
6. Birkhoff, G., Local spline approximation by moments, J. Math. Mech. **13** (1967), 977–990.
7. Ciesielski, Z. and T. Figiel, Spline basis in classical function spaces on compact C^∞ manifolds I, Stud. Math. **76** (1983), 1–58.
8. Ciesielski, Z. and T. Figiel, Spline basis in classical function spaces on compact C^∞ manifolds II, Stud. Math. **76** (1983), 95–136.
9. Chui, C. K., and P. W. Smith, Some nonlinear approximation problems related to n–widths, J. Approx. Th. **13** (1975) 421–430.
10. De Boor C. and R. DeVore, Approximation by smooth multivariate splines, Trans. Amer. Math. Soc. **276** (1983), 775–788.
11. Dyn, N., Perfect splines of minimum norm for monotone norms and norms induced by inner-products, with applications tensor product approximation and n–widths of integral operators, J. Approx. Th. **38** (1983), 105–138.
12. Dzyadak, V. K., Best approximation on classes of periodic functions that are defined by integrals of a linear combination of absolutely monotonic kernels, Mat. Zametki **16** (1974), 691–701 (in Russian).
13. Favard, J., Sur l'approximation des fonctions par les polynomes trigonométriques, Comptes Rendus Acad. Sci. Paris **203** (1936), 1122–1124.
14. Forst, W., Über die Breite von Klassen holomorpher periodischer Funktionen, J. Approx. Th. **19** (1977), 235–239.
15. Galkin P. V., On a solvability problem of periodic spline interpolation, Mat. Zametki **8** (1970), 563–573 (in Russian).
16. Golomb, M., Approximation by periodic spline interpolants on uniform meshes, J. Approx. Th. **1** (1968), 26–65.
17. Holladay, J. C., A smoothest curve approximation, Math. Comp. **2** (1957), 233–243.
18. Jerome, J. M. and L. L. Schumaker, On Lg-splines, J. Approx. Th. **2** (1969), 29–49.
19. Keng, H. L. and W. Yuan, *Applications of Number Theory to Numerical Analysis*, Springer–Verlag, Beijing, 1981.
20. Kolmogorov, A. N., On inequalities between upper bounds of derivatives on an infinite interval, Uch. Zap. Moscow State Univ **3** (1939), 3–16 (in Russian).
21. Korneichuk, N. P., *Splines in Approximation Theory*, Nauka, Moskow, 1984 (in Russian).

22. Korneichuk, N. P., *Exact Constants in Approximation Theory*, Cambridge University Press, Cambridge, 1991.
23. Korobov, N. M., *Trigonometric Sums and their Applications*, Nauka, Moscow, 1989 (in Russian).
24. Kuipers, L. and H. Niederreiter, *Uniform Distribution of Sequences*, John Wiley – Interscience, 1974.
25. Kushpel, A. K., Extremal properties of splines and n–widths in space $C_{2\pi}$, preprint 84.25, Inst. Math. Acad. Sci. of Ukraine SSR, 1984 (in Russian).
26. Kushpel, A. K., sk–splines and exact estimates of n–widths of functional classes in space $C_{2\pi}$, preprint 85.51, Inst. Math. Acad. Sci. of Ukraine SSR, 1985 (in Russian).
27. Kushpel, A. K., Rate of convergence of sk–spline interpolants on classes of convolutions, in *Investigations in Approximation Theory*, Inst. Math. Akad. Nauk Ukrain. SSR, Kiev, 1987 (in Russian).
28. Kushpel, A. K., On a family of extremal subspaces, Ukrain. Math. Zh. **39** (1987) (English translation).
29. Kushpel, A. K., Sharp estimates of the widths of convolution classes in the space C, Trudy Math. Inst. of Acad. Nauk. USSR, Moscow **180** (1987) (English translation).
30. Kushpel, A. K., Sharp estimates of widths of convolution classes, Izvestia Acad. Nauk SSSR **52** (1988), 1315–1332.
31. Kushpel, A. K., Estimates of widths of convolution classes in spaces C and L, Ukrain. Math. Zh. **4** (1989) (English translation).
32. Kushpel, A. K. Widths of classes of analytical functions in the space L_q, Ukrain. Math. Zh. **4** (1989) (English translation).
33. Kushpel, A. K., S. M. Gomes and J. Levesley, Approximation on the torus using number theoretic knots, in preparation.
34. Levesley, J. and A. K. Kushpel, Generalised sk–spline interpolation on compact abelian groups, submitted for publication in Proc. London Math. Soc.
35. Light, W. A. and E. W. Cheney, Interpolation by periodic radial basis functions, J. Math. Anal. Appl. **168** (1992), 111–130.
36. Lorentz, G. G., *Approximation of Functions*, Holt, Rinehart and Winston, New York, 1966.
37. Madych, W. R., and S. A. Nelson, Multivariate interpolation and conditionally positive definite functions, Approx. Theory and its Appl. **4** (1988), 77–89.
38. Micchelli, C. A., Cardinal L-splines, in *Studies in Splines and Approximation Theory*, Academic Press, 1976.
39. Micchelli, C. A. Interpolation of scattered data: distance matrices and conditionally positive definite functions, Constr. Approx. **2** (1986), 11–22.
40. Micchelli, C. A. and A. Pinkus, On n–widths and optimal recovery in M^r, in *Approx. Theory II*, G. G. Lorentz, C. K. Chui and L. L. Schumaker (eds.), Academic Press, New York (1976), 475–478.

41. Nguen, Thi Hoa, Rolle's theorem for differential operators and some extremal problems in approximation theory, Dokl. Akad. Nauk SSSR **295** (1987), 1313–1318.
42. Nikolskii, S. M., *Quadrature Formulas*, Nauka, Moskow, 1974 (in Russian).
43. Pinkus, A., *n–Widths in Approximation Theory*, Springer-Verlag, Berlin, 1985.
44. Ragozin, D. and J. Levesley, The fundamentality of translates of spherical functions on compact homogeneous spaces, submitted to J. Approx. Th.
45. Rudin, W., *Fourier Analysis on Groups*, Interscience, 1962.
46. Schoenberg, I. J., Contributions to the problem of approximation of equidistant data by analytic functions, Quarterly Appl. Math. **4** (1946), 45–99, 112-141.
47. Schoenberg, I. J., Spline functions, convex curves, and mechanical quadratures, Bull. Amer. Math. Soc. **64** (1958), 232–237.
48. Schoenberg, I. J., Spline interpolation and the higher derivatives, Proc. Nat. Acad. U.S.A. **51** (1964), 24–28.
49. Schoenberg, I. J., Spline interpolation and the best quadrature formulae, Bull. Am. Math. Soc. **70** (1964), 143–148.
50. Schoenberg, I. J., On trigonometric spline interpolation, J. Math. Mech. **13** (1964), 795–825.
51. Schoenberg, I. J., On Micchelli's theory of cardinal L-splines, in *Studies in Splines and Approximation Theory*, Academic Press, 1976.
52. Serduk, A. B., Sharp estimates of n–widths of convolution sets, private communication (in Russian).
53. Serduk, A. B. and A. I. Stepanets, On the existence of interpolating sk–splines, Ukrain. Math. Zh., to appear (in Russian).
54. Shevaldin, V. T., Periodical L-spline interpolants with equdistant knots, in *Approximation of Functions by Polynomials and Splines*, Sverdlovsk, UNC Acad. NAUK SSSR, 1985, 140–147.
55. Shevaldin, V. T., n–widths of convolution classes with the Poisson kernel, Mat. Zametki **51** (1992), 126–136 (in Russian).
56. Shevaldin, V. T., Lower estimates of n–widths with bounded fractional derivative, Mat. Zametki **53** (1993), 145–151 (in Russian).
57. Schumaker, L. L., *Spline Functions: Basic Theory*, Springer-Verlag, Berlin, 1970.
58. Stechkin, S. B. and Yu. N. Subbotin, *Splines in Numerical Mathematics*, Nauka, Moscow, 1976 (in Russian).
59. Stechkin, S. B., On the best approximation of certain classes of periodic functions by trigonometric polinomials, Izv. Acad. Nauk SSSR, Ser. Math. **20** (1956), 643–648 (in Russian).
60. Subbotin, Yu. N., On a connection among the finite differences and respective derivatives, Trudy Math. Inst. of Acad. Nauk. USSR **78** (1965), 24–42 (in Russian).

61. Subbotin, Yu. N., Extremal functional interpolation and approximation by splines, Mat. Zametki **16** (1974), 843–854 (in Russian).
62. Subbotin, Yu. N., Interpolating splines, in *Approximation Theory, Proceedings of the Conference in Poznan, August 22–26, 1972*, Warszawa, 1975, 221–234.
63. Sun, X., The fundamentality of the translates of a continuous function on spheres, Numerical Algorithms **8** (1994), 131–134.
64. Szokefalvy–Nady, B., Über gewisse Extremalfragen bei transformierten trigonometrischen Entwicklungen 1, Periodischer Fall, Berichte Math.–Phys. Acad. d. Wiss. Leipzig **90** (1938), 103–134.
65. Szokefalvy–Nady, B., Sur une classe générale de procédés de sommation pour les séries de Fourier, Hung. Acta Math. **1** (1948), 14–52.
66. Tikhomirov, V. M., n–widths of sets in functional spaces and approximation theory, Uspekhi Math. Nauk. **15** (1960), 81–120 (in Russian).
67. Tikhomirov, V. M., Some problems of approximation theory, Doklady Acad. Nauk. SSSR **160**, 774-777 (in Russian).
68. Tikhomirov, V. M., Approximation theory, in *Modern Problems in Mathematics* **14**, VINITI, Moscow, 1987 (in Russian).
69. Varga, R., *Functional Analysis and Approximation Theory in Numerical Analysis*, SIAM, Philadelphia, 1971 (1991 printing).
70. Wu, Z. and R. Schaback, Local error estimates for radial basis function interpolation of scattered data, IMA J. Num. Anal. **13** (1993), 13–27.
71. Zhensykbaev, A. A., Some problems of approximation by splines in functional spaces, dissertation, Dniepropetrovsk, 1973 (in Russian).
72. Zhensykbaev, A. A., Approximation of differentiable periodic functions by splines on a uniform subdivision, Math. Zametki **13** (1973), 807–816 (in Russian).
73. Zygmund, A., *Trignometric Series*, Chelsea, New York, 1952.

A.K. Kushpel
Leicester University
Department of Mathematics and Computer Science
Leicester University
Leicester LE1 7RH
UK
and
Institute of Mathamatics
National Academy of Sciences of the Ukraine
Kiev
Ukraine
ak99@mcs.le.ac.uk

J. Levesley
Leicester University
Department of Mathematics and Computer Science
Leicester University
Leicester LE1 7RH
UK
jl1@mcs.le.ac.uk

W.A. Light
Leicester University
Department of Mathematics and Computer Science
Leicester University
Leicester LE1 7RH
UK
pwl@mcs.le.ac.uk

Shape Preserving Spline Approximation via Local Algorithms

B. I. Kvasov

Abstract. The main goal of this paper is to construct an algorithm for shape preserving spline approximation of complex multivalued surfaces that appear in some applications. We consider firstly two different local 1-D algorithms of shape preserving interpolation and approximation. Then we use the direct product of 1-D splines and special monotonizing parametrization to obtain the surface satisfying given tolerances that inherits some shape preserving properties of 1-D splines.

§1. Introduction

In many practical problems we deal with approximation of discrete data when geometric properties of the data such as positivity, monotonicity, convexity, presence of linear sections, the angles and the bends should be retained. Standard approaches such as spline interpolation, NURBS and others usually fail in these requirements. To obtain the necessary solution many authors [1,2,5,7] introduce some parameters in the structure of the spline. Then they choose these factors in such a way as to satisfy the geometric constraints. The key idea here is to develop algorithms for automatic selection of these parameters. We consider two such local 1-D algorithms based on generalized cubic splines. Then the direct product of 1-D splines and special monotonizing parametrization are used to obtain the shape preserving approximation of complex multivalued surfaces. We formulate the shape preserving properties of 2-D splines and give some numerical examples.

Advanced Topics in Multivariate Approximation
F. Fontanella, K. Jetter and P. J. Laurent (eds.) pp. 181–196.

ISBN 981-02-2852-X

§2. The Class of Shape Preserving Interpolants

Let the sequence of points $V = \{P_i \,|\, i = 0, 1, \ldots, N\}$, $P_i = (x_i, f_i)$, on the plane $\mathbb{R}^2$ be fixed, where $\Delta : a = x_0 < x_1 < \cdots < x_N = b$ forms a partition of the interval $[a, b]$. We shall use the following notation for the first two differences $\Delta_i f = (f_{i+1} - f_i)/h_i$, $h_i = x_{i+1} - x_i$, $i = 0, 1, \ldots, N-1$, and $\delta_i f = \Delta_i f - \Delta_{i-1} f$, $i = 1, 2, \ldots, N-1$. As usual, we shall say that the initial data increase monotonically (decrease monotonically) on the subinterval $[x_n, x_k]$, $n > k$, if $\Delta_i f > 0$ ($\Delta_i f < 0$) for $i = n, \ldots, k-1$. We shall call the data convex down (up) on $[x_n, x_k]$, $k > n+1$ if $\delta_i f > 0$ ($\delta_i f < 0$) for $i = n, \ldots, k-2$.

We consider the problem of searching for a sufficiently smooth function $S(x)$ such that $S(x_i) = f_i$, $i = 0, 1, \ldots, N$, and $S(x)$ preserves the form of the initial data; this will be called a *shape preserving interpolation problem.* It means that $S(x)$ should monotonically increase or decrease if the data have the same behaviour. Analogously, $S(x)$ should also be convex (concave) in intervals where the data show this behaviour.

Evidently the solution of the shape preserving interpolation problem is not unique. We formalize the class of functions in which we search for the solution.

Definition 2.1. *The class $I(V)$ of functions is called shape preserving, if the following conditions hold for any $S(x) \in I(V)$:*

(1) $S(x) \in C^2[a, b]$;

(2) $S(x_i) = f_i$ *for* $i = 0, 1, \ldots, N$;

(3) $S'(x)\Delta_i f \geq 0$, *and* $S'(x) = 0$ *if* $\Delta_i f = 0$, *for all* $x \in [x_i, x_{i+1}]$ *and* $i = 0, 1, \ldots, N-1$;

(4) $S''(x_i)\delta_i f \geq 0$ *for* $i = 1, 2, \ldots, N-1$;
$S''(x)\delta_j f \geq 0$ *for* $x \in [x_i, x_{i+1}]$ *and* $j = i, i+1$, *if* $\delta_i f \delta_{i+1} f \geq 0$;
$S(x)$ *has only one inflection point* $\overline{x}$ *in the interval* (x_i, x_{i+1}) *if* $\delta_i f \delta_{i+1} f < 0$ *and also* $S''(x)\delta_i f \geq 0$ *for* $x \in [x_i, \overline{x}]$;
the number of inflection points in the interval (x_{i-1}, x_{i+1}) *does not exceed the number of sign changes in the sequence* $\delta_{i-1} f, \delta_i f, \delta_{i+1} f$.

Remark. When counting the number of sign changes in the sequence $\delta_{i-1} f, \delta_i f, \delta_{i+1} f$, the zeros are omitted.

The following propositions, which characterize the properties of shape preserving interpolants, are proved by using simple geometric considerations.

Lemma 2.1. *If* $\Delta_{i-1} f \Delta_i f \leq 0$, *then for the function* $S(x)$ *to be shape preserving it is necessary that* $S'(x_i) = 0$.

Lemma 2.2. *If* $\delta_i f = 0$ *and* $\delta_{i-1} f \delta_{i+1} f \geq 0$, *then the unique shape preserving function on the interval* $[x_{i-1}, x_{i+1}]$ *is the straight line passing through the points* P_{i-1}, P_i, P_{i+1}.

Corollary 2.1. *If $\delta_i f = \delta_{i+1} f = 0$, then the unique shape preserving function in the interval $[x_{i-1}, x_{i+1}]$ is the straight line passing through the points P_j, $j = i-1, i, i+1, i+2$.*

Lemma 2.3. *If $\delta_i f = 0$ and $\delta_{i-1} f \delta_{i+1} f < 0$, then for $S(x) \in I(V)$ it is necessary that one of the following conditions be met:*
(1) $S'(x_i)\delta_{i-1} f > \Delta_i f \delta_{i-1} f$ and $S''(x_i) = 0$;
(2) $S'(x) = \Delta_i f$, and $S''(x) = 0$ for all $x \in [x_{i-1}, x_{i+1}]$.

Lemma 2.4. *Let $\delta_i f \neq 0$ and $S''(x_i)S''(x) \geq 0$ for all $x \in [z_1, z_2]$, $z_1, z_2 \in [x_i, x_{i+1}]$. Then for $S(x) \in I(V)$ it is necessary that one of the following conditions be met, where $\Delta_z S = (S(z_2) - S(z_1))/(z_2 - z_1)$:*
(1) $S'(z_1) < \Delta_z S < S'(z_2)$ for $\delta_i f > 0$,
(2) $S'(z_1) > \Delta_z S > S'(z_2)$ for $\delta_i f < 0$,
(3) $S'(x) = \Delta_z S$, $S''(x) = 0$ for all $x \in [z_1, z_2]$.

Lemma 2.4 immediately implies

Corollary 2.2. *If $\delta_i f \delta_{i+1} f > 0$, and $S'(x_j) \neq \Delta_i f$ for $j = i, i+1$, then for $S(x)$ to be shape preserving, it is necessary that the following condition holds:*

$$S'(x_i)\delta_i f < \Delta_i f \delta_i f < S'(x_{i+1})\delta_i f \ .$$

Corollary 2.3. *If $\delta_{i-1} f \delta_i f > 0$ and $\delta_i f \delta_{i+1} f > 0$, then for $S(x)$ to be shape preserving it is necessary that the following inequalities hold:*

$$\min(\Delta_{i-1} f, \Delta_i f) \leq S'(x_i) \leq \max(\Delta_{i-1} f, \Delta_i f) \ .$$

Lemma 2.5. *If $S'(x_i) = 0$, then for $S(x)$ to be shape preserving it is necessary that $S''(x_i)\Delta_i f \geq 0$ and $S''(x_i)\Delta_{i-1} f \leq 0$.*

Theorem 2.1. *For the existence of a shape preserving function it is necessary and sufficient that none of the following conditions hold:*
(1) $\Delta_{i-1} f \Delta_i f \leq 0$, $\Delta_{i-1} f \neq 0$, $\delta_{i-2} f \delta_i f \geq 0$, $\delta_{i-1} f = 0$, $i = 3, \ldots, N-1$,
(2) $\Delta_{i-1} f \Delta_i f \leq 0$, $\Delta_i f \neq 0$, $\delta_i f \delta_{i+2} f \geq 0$, $\delta_{i+1} f = 0$, $i = 1, \ldots, N-3$,
(3) $\delta_i f \neq 0$, $\delta_{i-1} f = \delta_{i+1} f = 0$, $\delta_i f \delta_k f \geq 0$, $k = i-2, i+2$, $i = 3, \ldots, N-3$.

Necessity of this assertion is proved directly by using Lemmas 2.1–2.5. The proof of the sufficiency consists in locally constructing the shape preserving function $S(x)$ which interpolates arbitrary data and for which the conditions (1)–(3) of the Theorem 2.1 are not satisfied.

We define now the admissible values $S_i^{(r)} = S^{(r)}(x_i)$, $r = 1, 2$, at the knots of the mesh Δ. These values should satisfy the following constraints:
For $1 \leq i \leq N-1$: If $\delta_i f \neq 0$, then

$$\min(\Delta_{i-1} f, \Delta_i f) < S_i' < \max(\Delta_{i-1} f, \Delta_i f) \quad \text{and} \quad \delta_i f S_i'' \geq 0 \ . \tag{2.1}$$

For $2 \le i \le N-2$: If $\delta_i f = 0$ and $\delta_{i-1} f \delta_{i+1} f < 0$, then

$$(S_i' - \Delta_i f)\delta_{i-1} f > 0,\ S_i'\Delta_i f \ge 0,\ \text{and} S_i'' = 0\ . \tag{2.2}$$

$$\begin{aligned}&(S_1' - \Delta_1 f)\delta_2 f < 0,\ S_1'\Delta_1 f \ge 0,\ S_1'' = 0 \quad \text{if} \quad \delta_1 f = 0,\\ &(S_{N-1}' - \Delta_{N-1} f)\delta_{N-2} f > 0,\ S_{N-1}'\Delta_{N-1} f \ge 0,\ S_{N-1}'' = 0\\ &\qquad\qquad\qquad\qquad\qquad\qquad \text{if } \delta_{N-1} f = 0;\end{aligned} \tag{2.3}$$

$$\begin{aligned}&(\Delta_0 f - S_0')\delta_1 f > 0,\ S_0'\Delta_0 f \ge 0\,(\Delta_0 f \ne 0),\ S_0''\delta_1 f \ge 0 \ \text{if} \ \delta_1 f \ne 0,\\ &(\Delta_0 f - S_0')\delta_2 f < 0,\ S_0'\Delta_0 f \ge 0\,(\Delta_0 f \ne 0),\ S_0''\delta_2 f \le 0\\ &\qquad\qquad\qquad\qquad\qquad \text{if} \quad \delta_1 f = 0,\ \delta_2 f \ne 0;\end{aligned} \tag{2.4}$$

$$\begin{aligned}&(S_N' - \Delta_{N-1} f)\delta_{N-1} f > 0,\ S_N'\Delta_{N-1} f \ge 0\,(\Delta_{N-1} f \ne 0),\ S_N''\delta_{N-1} f \ge 0\\ &\qquad\qquad\qquad\qquad\qquad \text{if} \quad \delta_{N-1} f \ne 0,\\ &(S_N' - \Delta_{N-1} f)\delta_{N-2} f < 0,\ S_N'\Delta_{N-1} f \ge 0\,(\Delta_{N-1} f \ne 0),\ S_N''\delta_{N-2} f \le 0\\ &\qquad\qquad\qquad\qquad \text{if} \quad \delta_{N-1} f = 0, \delta_{N-2} f \ne 0.\end{aligned} \tag{2.5}$$

For the constructing of the shape preserving function $S(x)$ it is sufficient to eliminate those intervals from the consideration where $S(x)$ is linear, and to define $S(x)$ in an arbitrary subinterval $[x_i, x_{i+1}]$ for the following possible configurations of the data:

(A) $\delta_i f \delta_{i+1} f > 0,\ \ 0 \le i \le N-1$;

(B) $\delta_i f = 0,\ \ \delta_{i-1} f \delta_{i+1} f < 0,\ 1 \le i \le N-1$;

(C) $\delta_i f \delta_{i+1} f < 0,\ \ 1 \le i \le N-2$.

(If $i = 0, N$, then we formally set $\delta_i f = S_i''$).

Case (C) can be reduced to case (B), if the mesh Δ is extended by adding an additional point of inflection on the straight line connecting P_i and P_{i+1}. In cases (A) and (B) the construction of the shape preserving function can be reduced (according to [3]) to the solution of the Hermite interpolation problem in $[x_i, x_{i+1}]$, with the given values $S_j^{(r)}$, $r = 0, 1, 2$ and $j = i, i+1$ satisfying the monotonicity and the convexity requirements in this interval and the additional restrictions

$$\min(S_i', S_{i+1}') < \Delta_i f < \max(S_i', S_{i+1}'); \tag{2.6}$$

$$\Delta_i f S_j' \ge 0, \quad j = i, i+1; \tag{2.7}$$

$$S_j''/(S_{i+1}' - S_i') \ge 0, \quad j = i, i+1. \tag{2.8}$$

According to Definition 2.1 the following relations should be satisfied too:

$$S''(x)S''(x_j) \ge 0, \quad j = i, i+1; \quad x \in [x_i, x_{i+1}]. \tag{2.9}$$

§3. The Solution of the Hermite Interpolation Problem with Constraints

The question of constructing the shape preserving function $S(x)$ locally can be solved by using generalized cubic splines [8,9].

Definition 3.1. *Our generalized cubic spline on the mesh* Δ *will be a function* $S(x) \in C^2[a,b]$, *which in each subinterval* $[x_j, x_{j+1}]$ *can be written as*

$$S(x) = [S_j - \varphi_j(0)h_j^2 S_j''](1-t) + [S_{j+1} - \psi_j(1)h_j^2 S_{j+1}'']t \\ +\varphi_j(t)h_j^2 S_j'' + \psi_j(t)h_j^2 S_{j+1}'', \tag{3.1}$$

where $t = (x - x_j)/h_j$ *and where the functions* $\varphi_j(t)$, $\psi_j(t)$ *satisfy the conditions*

$$\varphi_j^{(r)}(1) = \psi_j^{(r)}(0) = 0, \quad r = 0,1,2; \quad \varphi_j''(0) = \psi_j''(1) = 1.$$

Here, we assume that $\varphi_j''(t)$, $\psi_j''(t)$ *are continuous monotonic functions of the variable* $t \in [0,1]$, *and that*

$$\varphi_j(t) = \varphi(p_j, t), \quad \psi_j(t) = \varphi(q_j, 1-t), \quad p_j, q_j \geq 0. \tag{3.2}$$

In order to solve the Hermite interpolation problem with constraints on the interval $[x_i, x_{i+1}]$ let us define a function

$$S(x) = \begin{cases} S(x, x_i, x_{i1}) & \text{if} \quad x \in [x_i, x_{i1}]; \\ S(x, x_{i1}, x_{i+1}) & \text{if} \quad x \in [x_{i1}, x_{i+1}], \end{cases}$$

which has the form (3.1) on the intervals $[x_i, x_{i1}]$, $[x_{i1}, x_{i+1}]$, and which satisfies the interpolation and smoothness conditions:

$$S^{(r)}(x_j) = f_j^{(r)}, \quad S^{(r)}(x_{i1} - 0) = S^{(r)}(x_{i1} + 0), \quad r = 0,1,2; \quad j = i, i+1.$$

We assume that the inequalities (2.6)–(2.8) are fulfilled, such that according to (2.7) we have $S_i' S_{i+1}' \geq 0$. Let us introduce the notations

$$h_{i1} = x_{i1} - x_i, \quad \mu_{i1} = 1 - \lambda_{i1} = h_{i1}/h_i, \quad \tau_i = \frac{S_{i+1}' - \Delta_i f}{S_{i+1}' - S_i'},$$

$$\alpha_i = \frac{S_{i1} - f_i}{h_{i1}}, \quad \beta_i = \frac{f_{i+1} - S_{i1}}{h_i - h_{i1}}, \quad \sigma_j = \frac{h_i S_j''}{S_{i+1}' - S_i'}, \quad j = i, i+1.$$

Then by the inequalities (2.6) we have

$$\Delta_i f = \tau_i S_i' + (1 - \tau_i) S_{i+1}', \quad \text{with} \quad 0 < \tau_i < 1. \tag{3.3}$$

From (3.1) we obtain

$$\begin{aligned}
\alpha_i &= \frac{1}{\psi_i'(1)}\Big\{h_{i1}T_i^{-1}S_i'' - [\psi_i(1) - \psi_i'(1)]S_i' + \psi_i(1)S_{i1}'\Big\},\\
\beta_i &= \frac{1}{\varphi_{i1}'(0)}\Big\{(h_i - h_{i1})T_{i1}^{-1}S_{i+1}'' - \varphi_{i1}(0)S_{i1}' + [\varphi_{i1}(0) + \varphi_{i1}'(0)]S_{i+1}'\Big\},\\
T_j^{-1} &= [\varphi_j(0) + \varphi_j'(0)][\psi_j(1) - \psi_j'(1)] - \varphi_j(0)\psi_j(1), \quad j = i, i1.
\end{aligned} \tag{3.4}$$

Since the second derivative of the spline is continuous at the knot x_{i1}, we have the equation

$$\begin{aligned}
S_{i1}' = \left[\frac{\lambda_{i1}}{\psi_i'(1)} - \frac{\mu_{i1}}{\varphi_{i1}'(0)}\right]^{-1}&\left[\frac{\lambda_{i1}}{\psi_i'(1)}S_i' - \frac{\mu_{i1}}{\varphi_{i1}'(0)}S_{i+1}'\right.\\
&\left.- \frac{\varphi_i'(0)}{\psi_i'(1)}\lambda_{i1}h_{i1}S_i'' + \frac{\psi_{i1}'(1)}{\varphi_{i1}'(0)}\mu_{i1}(h_i - h_{i1})S_{i+1}''\right].
\end{aligned} \tag{3.5}$$

Now taking into account the identity $\mu_{i1}\alpha_i + \lambda_{i1}\beta_i = \Delta_i f$ and substituting here the expressions for α_i, β_i, S_{i1}' from (3.4) and (3.5), we arrive at the following equation to be satisfied for μ_{i1}:

$$\Phi_i(\mu_{i1}) := A_i\mu_{i1}^3 + B_i\mu_{i1}^2 + C_i\mu_{i1} + D_i = 0. \tag{3.6}$$

If we require $p_{i1} = q_i$,then according to (3.2) the coefficient A_i will vanish, and (3.6) reduces to the quadratic equation

$$\Phi_i(\mu_{i1}) = \frac{1}{\psi_i'(1)}\left[\hat{B}_i\mu_{i1}^2 + \hat{C}_i\mu_{i1} + \hat{D}_i\right] = 0, \tag{3.7}$$

where

$$\begin{aligned}
\hat{B}_i =& [\psi_i(1)\varphi_i'(0) + T_i^{-1}]\sigma_i + [\psi_i(1)\psi_{i1}'(1) - T_{i1}^{-1}]\sigma_{i+1},\\
\hat{C}_i =& -\psi_i(1)\varphi_i'(0)\sigma_i + [-\psi_i(1)\psi_{i1}(1) + 2T_{i1}^{-1}]\sigma_{i+1} + 2\psi_i(1) - \psi_i'(1),\\
\hat{D}_i =& -T_{i1}^{-1}\sigma_{i+1} - \psi_i(1) + \tau_i\psi_i'(1).
\end{aligned}$$

Since

$$\begin{aligned}
\Phi_i(0) =& \tau_i - \frac{1}{\psi_i'(1)}[T_{i1}^{-1}\sigma_{i+1} + \psi_i(1)],\\
\Phi_i(1) =& -(1 - \tau_i) + \frac{1}{\psi_i'(1)}[T_i^{-1}\sigma_i + \psi_i(1)],
\end{aligned}$$

we can find values $\overline{p}_i$, $\overline{q}_i$, $\overline{q}_{i1}$, such that for all $p_i \geq \overline{p}_i$, $q_i \geq \overline{q}_i$, and $q_{i1} \geq \overline{q}_{i1}$, according to (3.3), we have $\Phi_i(0) > 0$ and $\Phi_i(1) < 0$. Therefore, the equation (3.7) has a unique root, say, $\mu_{i1} \in (0, 1)$.

As $p_{i1} = q_i$ we can rewrite the equation (3.5) in the form

$$S'_{i1} = S'_i + \mu_{i1}(S'_{i+1} - S'_i) - \lambda_{i1}\mu_{i1}h_i[\varphi'_i(0)S''_i + \psi'_{i1}(1)S''_{i+1}]. \tag{3.8}$$

When considering $f(x)$ as a sufficiently smooth function, we can assume that $S_j^{(r)} - f_j^{(r)} = O(h_i^{k+1-r})$, $r = 1, 2$; $j = i, i+1$; $k = 2$ or $k = 3$. Then using (3.8) we obtain

$$\begin{aligned}S'_{i1} - f'(x_{i1}) =& S'_i - f'_i - [\varphi'_i(0) + \psi'_{i1}(1)]\lambda_{i1}\mu_{i1}h_i f''_i \\ &+ h_{i1}(h_i - h_{i1})[1/2 - \psi'_{i1}(1)]f'''_i + O(h_i^k).\end{aligned}$$

This implies that the approximation order for the derivative of the spline at the point x_{i1} will increase if according to (3.2) we put $q_{i1} = p_i$.

We consider now the question of the shape preserving properties for the generalized spline $S(x)$ in the interval $[x_i, x_{i+1}]$. The following criterion is valid.

Theorem 3.1. *Subject to the restrictions*

$$\frac{\psi_i(1)}{\psi'_i(1)} - \varphi'_i(0)\sigma_i < 1 - \tau_i, \quad \text{and} \quad \frac{\psi_i(1)}{\psi'_i(1)} - \varphi'_i(0)\sigma_{i+1} < \tau_i, \tag{3.9}$$

the unique shape preserving generalized cubic spline $S(x)$ exists solving the Hermite interpolation problem with restrictions (2.6)–(2.9).

Proof: The conditions (2.6) – (2.8) are satisfied by the construction. The requirement (2.9) means that on $[x_i, x_{i+1}]$, there are no points of inflection for the shape preserving function $S_f(x)$. Let us show that this condition will be fulfilled for the spline $S(x)$, if the following inequalities are valid:

$$\begin{aligned}\min(\alpha_i, \beta_i) &< S'_{i1} < \max(\alpha_i, \beta_i), \\ \min(S'_i, \Delta_i f) &< \alpha_i < \max(S'_i, \Delta_i f), \\ \min(S'_{i+1}, \Delta_i f) &< \beta_i < \max(S'_{i+1}, \Delta_i f).\end{aligned}$$

It is convenient to rewrite these inequalities in the form

$$\begin{aligned}\alpha_i(S'_{i+1} - S'_i)^{-1} &< S'_{i1}(S'_{i+1} - S'_i)^{-1} < \beta_i(S'_{i+1} - S'_i)^{-1}, \\ S'_i(S'_{i+1} - S'_i)^{-1} &< \alpha_i(S'_{i+1} - S'_i)^{-1} < \Delta_i f(S'_{i+1} - S'_i)^{-1}, \\ \Delta_i f(S'_{i+1} - S'_i)^{-1} &< \beta_i(S'_{i+1} - S'_i)^{-1} < S'_{i+1}(S'_{i+1} - S'_i)^{-1}.\end{aligned} \tag{3.10}$$

From (3.4) and (3.8) we find

$$\alpha_i = S'_i + \mu_{i1}(S'_{i+1} - S'_i)\frac{\psi_i(1)}{\psi'_i(1)}\left[1 + \frac{T_i^{-1}}{\psi_i(1)}\sigma_i + \lambda_{i1}\varphi'_i(0)(\sigma_{i+1} - \sigma_i)\right], \tag{3.11}$$

$$\beta_i = S'_{i+1} - \lambda_{i1}(S'_{i+1} - S'_i)\frac{\psi_i(1)}{\psi'_i(1)}\left[1 + \frac{T_i^{-1}}{\psi_i(1)}\sigma_{i+1} - \mu_{i1}\varphi'_i(0)(\sigma_{i+1} - \sigma_i)\right].$$

This enables us to write the conditions (3.10) in the form

$$\frac{\psi_i(1)}{\psi_i'(1)}\left[1+\frac{T_i^{-1}}{\psi_i(1)}\sigma_i+\lambda_{i1}\varphi_i'(0)(\sigma_{i+1}-\sigma_i)\right]<1+\lambda_{i1}\varphi_i'(0)(\sigma_{i+1}-\sigma_i),$$
$$\frac{\psi_i(1)}{\psi_i'(1)}\left[1+\frac{T_i^{-1}}{\psi_i(1)}\sigma_{i+1}-\mu_{i1}\varphi_i'(0)(\sigma_{i+1}-\sigma_i)\right]<1-\mu_{i1}\varphi_i'(0)(\sigma_{i+1}-\sigma_i),$$
$$0<\mu_{i1}\frac{\psi_i(1)}{\psi_i'(1)}\left[1+\frac{T_i^{-1}}{\psi_i(1)}\sigma_i+\lambda_{i1}\varphi_i'(0)(\sigma_{i+1}-\sigma_i)\right]<1-\tau_i,$$
$$0<\lambda_{i1}\frac{\psi_i(1)}{\psi_i'(1)}\left[1+\frac{T_i^{-1}}{\psi_i(1)}\sigma_{i+1}-\mu_{i1}\varphi_i'(0)(\sigma_{i+1}-\sigma_i)\right]<\tau_i.$$

In order to guarantee these inequalities and the conditions $\Phi_i(0)>0$ and $\Phi_i(1)<0$, it is sufficient to choose the parameters p_i, q_i in such a way that the restrictions (3.9) are satisfied. According to (3.1),

$$S_i'=\alpha_i+[\varphi_i(0)+\varphi_i'(0)]h_{i1}S_i''-\psi_i(1)h_{i1}S_{i1}''.$$

Substituting here the expression for α_i from (3.11) we have

$$S_{i1}''=\frac{S_{i+1}'-S_i'}{h_i\psi_i'(1)}\left[1+\varphi_i'(0)(\mu_{i1}\sigma_i+\lambda_{i1}\sigma_{i+1})\right]. \tag{3.12}$$

If the inequalities (3.9) are fulfilled, the expression in square parentheses is positive, and $S_{i1}''(S_{i+1}'-S_i')\geq 0$. Since $S_j''(S_{i+1}'-S_i')\geq 0$, $j=i,i+1$, we conclude that $S_{i1}''S_j''\geq 0$ for $j=i,i+1$.

From (3.1) on the interval $[x_i,x_{i1}]$ we have

$$S''(x)=S_i''\varphi_i''(t)+S_{i1}''\psi_i''(t).$$

Since $\varphi_i''(t),\psi_i''(t)\geq 0$ for $t\in[0,1]$, we get

$$S''(x)S_j''\geq 0,\quad j=i,i1\quad\text{for}\quad x\in[x_i,x_{i1}].$$

An analogous argument can be given for the subinterval $[x_{i1},x_{i+1}]$. As a result the function $S''(x)$ is convex in the interval $[x_i,x_{i+1}]$, and $S'(x)$ is monotone there. Due to the assumption $S_i'S_{i+1}'\geq 0$ the function $S(x)$ has the monotonicity property. The theorem is proved. ■

The given construction completes the proof of the sufficiency conditions of Theorem 2.1 from the previous section.

§4. The Problem of Shape Preserving Approximation

For the grid $\Delta : a = x_0 < x_1 < \cdots < x_N = b$ we suppose that a set of intervals $F = \{F_i \mid i = 0, \ldots, N\}$, $F_i \equiv [f_i - \varepsilon_i, f_i + \varepsilon_i]$, $i = 0, \ldots, N$, is given with prescribed small $\varepsilon_i > 0$. We consider the problem of searching for a sufficiently smooth function $S(x) \in C^2[a,b]$ such that $S(x_i) \in F_i$ for $i = 0, \ldots, N$, and such that $S(x)$ preserves the shape of the initial data; this will be called a *shape preserving approximation problem.*

To formalize this problem we introduce the interval differences [6]

$$\Delta_i F = h_i^{-1}(F_{i+1} - F_i) = [\Delta_i f - e_i, \Delta_i f + e_i], \quad e_i = h_i^{-1}(\varepsilon_i + \varepsilon_{i+1}), \quad i = 0, \ldots, N-1,$$

$$\delta_i F = \Delta_i F - \Delta_{i-1} F = [\delta_i f - E_i, \delta_i f + E_i], \quad E_i = e_{i-1} + e_i, \quad i = 1, \ldots, N-1,$$

$$[a_1, a_2] - [b_1, b_2] = [a_1 - b_2, a_2 - b_1] > 0 \quad \text{if and only if} \quad a_1 > b_2.$$

The initial data are said to increase (decrease) monotonically on a subinterval $[x_R, x_K]$, $K > R$, if $\Delta_i F > 0$ ($\Delta_i F < 0$), $i = R, \ldots, K-1$. The data are called convex downward (upward) on $[x_R, x_K]$, $K > R+1$, if $\delta_i F > 0$ ($\delta_i F < 0$) for $i = R+1, \ldots, K-1$.

We assume that for any i, the intervals $\Delta_i F$, $\delta_i F$ do not contain zeros, i.e. $(\Delta_i f)^2 > e_i^2$, $i = 0, \ldots, N-1$, and $(\delta_i f)^2 > E_i^2$, $i = 1, \ldots, N-1$.

If the values of a function $S(x)$ are such that $S(x_i) \in F_i$ for $i = 0, \ldots, N$, then we have $\Delta_i S \in \Delta_i F$, $i = 0, \ldots, N-1$, and $\delta_i S \in \delta_i F$, $i = 1, \ldots, N-1$. Taking into account the inequalities for the initial data, we obtain $\Delta_i S \, \Delta_i f > 0$, $i = 0, \ldots, N-1$, and $\delta_i S \, \delta_i f > 0$, $i = 1, \ldots, N-1$.

Definition 4.1. *The class $I(\Delta, F)$ of functions is called shape preserving approximants, if for any $S(x) \in I(\Delta, F)$ the following conditions hold:*

1. *$S(x) \in C^2[a,b]$;*
2. *$S(x_i) \in F_i$ for $i = 0, \ldots, N$;*
3. *$S(x)$ is monotonic in $[x_i, x_{i+1}]$, for $i = 1, \ldots, N-2$, if $\Delta_{i-1} f \Delta_i f > 0$ and $\Delta_i f \Delta_{i+1} f > 0$;*
   $S(x)$ is monotonic in $[x_0, x_1]$ if $\Delta_0 f \Delta_1 f > 0$, and is monotonic in $[x_{N-1}, x_N]$ if $\Delta_{N-2} f \Delta_{N-1} f > 0$;
   $S'(x)$ has one sign change in $[x_{i-1}, x_{i+1}]$, for $i = 1, \ldots, N-1$, if $\Delta_{i-1} f \Delta_i f < 0$; the number of sign changes of the function $S'(x)$ in $[a,b]$ coincides with the number of sign changes in the sequence $\Delta_0 f, \Delta_1 f, \ldots, \Delta_{N-1} f$;
4. *$S''(x_i)\delta_i f \geq 0$ for $i = 1, \ldots, N-1$;*
   the number of sign changes of the function $S''(x)$ in $[a,b]$ coincides with the number of sign changes in the sequence $\delta_1 f, \delta_2 f, \ldots, \delta_{N-1} f$.

The shape preserving approximation problem is, by definition, the problem of searching for a function $S(x) \in I(\Delta, F)$. We seek a solution of the shape preserving approximation problem in the form of a generalized cubic spline as given in Definition 3.1.

§5. Algorithm of Shape Preserving Local Approximation

The set of splines according to Definition 3.1 will be denoted by S_3^G. The dimension of this space is $4N - 3(N-1) = N+3$, and for the construction of a basis of B-splines for S_3^G (i.e., a set of nonnegative functions with local supports of minimal length) we have to extend the mesh Δ by adding points $x_{-3} < x_{-2} < x_{-1} < a$ and $b < x_{N+1} < x_{N+2} < x_{N+3}$. Simple calculations permit us to obtain an explicit form for these B-splines:

$$B_i(x) = \begin{cases} B_i''(x_{i-1})\Psi_{i-2}(x), \quad x \in [x_{i-2}, x_{i-1}], \\ B_i''(x_{i-1})[v_{i-1} + v_{i-1}'(x - x_{i-1})] \\ \quad + B_i''(x_{i-1})\Phi_{i-1}(x) + B_i''(x_i)\Psi_{i-1}(x), \quad x \in [x_{i-1}, x_i], \\ -B_i''(x_{i+1})[v_{i+1} + v_{i+1}'(x - x_{i+1})] \\ \quad + B_i''(x_i)\Phi_i(x) + B_i''(x_{i+1})\Psi_i(x), \quad x \in [x_i, x_{i+1}], \\ B_i''(x_{i+1})\Phi_{i+1}(x), \quad x \in [x_{i+1}, x_{i+2}], \\ 0, \quad x \notin [x_{i-2}, x_{i+2}], \end{cases} \tag{5.1}$$

where

$$\Phi_j(x) = \varphi_j\left(\frac{x - x_j}{h_j}\right)h_j^2, \quad \Psi_j(x) = \psi_j\left(\frac{x - x_j}{h_j}\right)h_j^2,$$
$$v_j^{(r)} = \psi_{j-1}^{(r)}(1)h_{j-1}^{2-r} - \varphi_j^{(r)}(0)h_j^{2-r}, \quad r = 0, 1.$$
$$B_i''(x_j) = \frac{y_{i+1} - y_{i-1}}{v_j'\omega_{i-1}'(y_j)}, \quad y_j = x_j - v_j/v_j', \quad j = i-1, i, i+1,$$
$$\omega_{i-1}(x) = (x - y_{i-1})(x - y_i)(x - y_{i+1}).$$

We further consider the case when 'averaged nodes' of B-splines $y_i = x_i - v_i/v_i'$, $i = 0, \dots, N$ coincide with the nodes of main mesh Δ, i.e. $v_i = \psi_{i-1}(1)h_{i-1}^2 - \varphi_i(0)h_i^2 = 0$, $i = 0, \dots, N$ and $x_{-i} = x_0 - ih_0$, $x_{N+i} = x_N + ih_{N-1}$, $i = 1, 2, 3$.

Basis splines $B_i(x)$, $i = -1, \dots, N+1$ have the following properties: $B_i(x) > 0$ if $x \in (x_{i-2}, x_{i+2})$ and $B_i(x) \equiv 0$ otherwise,

$$\sum_{j=-1}^{N+1} B_j(x) \equiv 1 \quad \text{for} \quad x \in [a, b].$$

Any spline $S(x) \in S_3^G$ can be uniquely represented in the form

$$S(x) = \sum_{j=-1}^{N+1} b_j B_j(x) \quad \text{for} \quad x \in [a,b] \tag{5.2}$$

with some constant coefficients b_j.

From (5.1) the expression (5.2) for the spline $S(x)$ in the subinterval $[x_i, x_{i+1}]$ is transformed to the form

$$S(x) = b_i + \Delta_i b(x - x_i) + \varphi_i(t) h_i^2 \delta_i b / v_i' + \psi_i(t) h_i^2 \delta_{i+1} b / v_{i+1}', \tag{5.3}$$

where $\delta_j b = \Delta_j b - \Delta_{j-1} b$, $j = i, i+1$, and $\Delta_j b = (b_{j+1} - b_j)/h_j$. Whence the formulae

$$S(x_i) = b_i + \delta_i b \left(\frac{\psi_{i-1}'(1)}{\psi_{i-1}(1)} \frac{1}{h_{i-1}} - \frac{\varphi_i'(0)}{\varphi_i(0)} \frac{1}{h_i} \right)^{-1}, \tag{5.4}$$

$$S'(x_i) = \frac{1}{v_i'} [\psi_{i-1}'(1) h_{i-1} \Delta_i b - \varphi_i'(0) h_i \Delta_{i-1} b], \tag{5.5}$$

$$S''(x_i) = \delta_i b / v_i' \tag{5.6}$$

follow, and hence

$$\begin{aligned} b_{i-1} &= S(x_i) - h_{i-1} S'(x_i) + h_{i-1}^2 [-\psi_{i-1}(1) + \psi_{i-1}'(1)] S''(x_i), \\ b_i &= S(x_i) - h_i^2 \varphi_i(0) S''(x_i), \\ b_{i+1} &= S(x_i) + h_i S'(x_i) - h_i^2 [\varphi_i(0) + \varphi_i'(0)] S''(x_i), \\ & \qquad\qquad i = 0, \ldots, N. \end{aligned} \tag{5.7}$$

Algorithm 5.1. We compute the coefficients in (5.2) from formulae (5.7), with $S''(x_i)$ being approximated using the second divided difference

$$b_i = f_i - 2h_i^2 (h_{i-1} + h_i)^{-1} \varphi_i(0) \delta_i f, \quad i = 1, \ldots, N-1. \tag{5.8}$$

To determine the coefficients b_i, for $i = -1, 0, N, N+1$, we use the boundary conditions: $S^{(k)}(x_i) = f_i^{(k)}$, $i = 0, N$, $k = 0, 1$. Using the formulae (5.4)–(5.7), we get

$$\begin{aligned} b_{-1} &= b_1 - 2h_0 f_0', \\ b_0 &= f_0 - (f_0 + h_0 f_0' - b_1)[1 + \varphi_0'(0)/\varphi_0(0)]^{-1}, \\ b_N &= f_N - (f_N - h_{N-1} f_N' - b_{N-1}) \\ & \qquad \times [1 - \psi_{N-1}'(1)/\psi_{N-1}(1)]^{-1}, \\ b_{N+1} &= b_{N-1} + 2h_{N-1} f_N'. \end{aligned} \tag{5.9}$$

We find the parameters p_i, q_i, $i = 0, \ldots, N-1$, from the conditions of shape preservation as formulated in Definition 4.1 in two steps. Using the constraints $|b_i - f_i| \le \varepsilon_i$, $i = 1, \ldots, N-1$, which in view of (6.1) are equivalent to

$$2h_i^2(h_{i-1} + h_i)^{-1}\varphi_i(0)|\delta_i f| \le \varepsilon_i, \quad i = 1, \ldots, N-1, \tag{5.10}$$

we first find p_i, and then obtain q_{i-1} from the condition $v_i = 0$ or $\varphi(q_{i-1}, 0)h_{i-1}^2 = \varphi(p_i, 0)h_i^2$.

According to (5.9) the quantities p_0, q_{N-1} are selected so as to satisfy the inequalities

$$\begin{aligned} |b_0 - f_0| &= |f_0 + h_0 f_0' - b_1||1 + \varphi_0'(0)/\varphi_0(0)|^{-1} \le \varepsilon_0, \\ |b_N - f_N| &= |f_N - h_{N-1} f_N' - b_{N-1}| \\ &\quad \times |1 - \psi_{N-1}'(1)/\psi_{N-1}(1)|^{-1} \le \varepsilon_N. \end{aligned} \tag{5.11}$$

Finally we find p_i, q_i from the constraints $|S(x_i) - f_i| \le \varepsilon_i$, $i = 0, \ldots, N$. From (5.4) and (5.8),

$$\begin{aligned} S(x_i) = f_i + H_i^{-1}\Big\{ &-\frac{2h_{i-1}^2\varphi_{i-1}(0)}{h_{i-2} + h_{i-1}}\delta_{i-1} f - \frac{2h_{i+1}^2\varphi_{i+1}(0)}{h_i(h_i + h_{i+1})}\delta_{i+1} f \\ &+ [1 + 2h_{i-1}^{-1}h_i\varphi_i(0) - 2(h_{i-1} + h_i)^{-1}(\psi_{i-1}'(1)h_{i-1} - \varphi_i'(0)h_i)]\delta_i f\Big\}, \end{aligned} \tag{5.12}$$

where

$$H_i = \frac{\psi_{i-1}'(1)}{\psi_{i-1}(1)}\frac{1}{h_{i-1}} - \frac{\varphi_i'(0)}{\varphi_i(0)}\frac{1}{h_i}.$$

Therefore according to the estimate (5.10) by (5.12) we obtain

$$|S(x_i) - f_i| \le H_i^{-1}\theta_i \le \varepsilon_i, \tag{5.13}$$

where $\theta_i = \varepsilon_{i-1}h_{i-1}^{-1} + \frac{4}{3}|\delta_i f| + \varepsilon_{i+1}h_i^{-1}$.

If $h_{i-1} \le h_i$, then from (5.13) we have

$$|S(x_i) - f_i| \le \theta_i\left[\left(\frac{1}{h_{i-1}} + \frac{1}{h_i}\right)\frac{\psi_{i-1}'(1)}{\psi_{i-1}(1)}\right]^{-1} \le \varepsilon_i.$$

Taking into account the constraint $\psi_{i-1}'(1)/\psi_{i-1}(1) \ge 3$ for generalized cubic splines we can put

$$\frac{\psi_{i-1}'(1)}{\psi_{i-1}(1)} - 3 = \max\left(\frac{h_{i-1}h_i}{h_{i-1} + h_i}\frac{\theta_i}{\varepsilon_i} - 3, 0\right), \quad i = 2, \ldots, N-2, \tag{5.14}$$

in order to define q_{i-1}. The value p_i is calculated from the condition $v_i = 0$. The cases where $h_i < h_{i-1}$ and $p_i < q_{i-1}$ are considered in a similar way.

For $i = 1, N-1$, according to (5.9) and (5.11) we come again to the formula (5.14) that permits us to choose the parameters q_0, p_{N-1}. We fulfill the conditions of Definition 4.1 by the final choice of the parameters p_i, q_i, $i = 0, \ldots, N-1$. This gives us the following result.

Theorem 5.1. *Subject to the inequalities*

$$\delta_1 f(\Delta_0 f - f_0') > |\delta_1 f|\varepsilon_1 h_0^{-1}, \quad f_0'\Delta_0 f \geq 0,$$
$$\delta_{N-1} f(f_N' - \Delta_{N-1} f) > |\delta_{N-1} f|\varepsilon_{N-1} h_{N-1}^{-1}, \quad f_N'\Delta_{N-1} f \geq 0,$$

being satisfied, the generalized cubic spline $S(x)$ constructed by Algorithm 5.1 of three-point local approximation is a shape preserving approximant.

Remark 5.1. For $f(x) \equiv 1$ and $f(x) \equiv x$ it can be immediately checked that $b_i = 1$ and $b_i = x_i$ for $i = -1, \ldots, N+1$. Therefore, the shape preserving spline $S(x)$ reproduces straight lines.

§6. Shape Preserving Approximation of Surfaces

Let the domain $G : [c,d] \times [0,1] \subset WU$ be partitioned into N rectangular subdomains by the straight lines $w = w_i$, $i = 0, \ldots, N$, corresponding to the grid $\Delta_w : c = w_0 < w_1 < \cdots < w_N = d$. Assume that a grid $\Delta_u^i : 0 = u_0^i < u_1^i < \cdots < u_{M_i}^i = 1$, $i = 0, 1, \ldots, N$, is given on every straight line $w = w_i$. The number of nodes and their positions within each grid are independent of each other. The values f_{ij} of some function $f(w,u)$ are given with tolerances ε_{ij} at the nodes u_j^i, $j = 0, \ldots, M_i$, $i = 0, \ldots, N$.

A surface of the class $C^{2,2}(G)$, passing through the points $P_{ij} = (w_i, u_j^i, \tilde{f}_{ij})$, where $\tilde{f}_{ij} \in [f_{ij} - \varepsilon_{ij}, f_{ij} + \varepsilon_{ij}]$, $j = 0, \ldots, M_i$, $i = 0, \ldots, N$, can be constructed by generalizing the algorithm of local approximation by splines from Section 5. In addition to being efficient at constructing the surface, this algorithm also preserves the shape of the input data.

This surface is sought in the form of a function:

$$S(w,u) = \sum_{i=-1}^{N+1} b_i(u) B_i(w). \tag{6.1}$$

Here the generalized basis splines $B_i(w)$ are the same as in (5.2). The functions $b_i(u)$, $i = -1, \ldots, N+1$, generalize the local approximation formulae from Sections 5 (Algorithms 5.1), and they are linear combinations of the one-dimensional interpolation shape preserving splines $S_i(u)$, $i = 0, \ldots, N$, as described in Sections 2 and 3. These splines define curves along sections $w = w_i$, $i = 0, \ldots, N$, and they pass through the points (u_j^i, f_{ij}), $j = 0, \ldots, M_i$.

Formally, necessary formulae (Algorithm 6.1) can be obtained by replacing the values $f_j^{(k)}$ in Algorithm 5.1 by the functions $S_j^{(k)}(u)$, $k = 0, 1, 2$, respectively. Similar changes are made to the boundary conditions. For the scheme given below we use the boundary conditions: $S(w_i, u) = S_i(u)$, $\frac{\partial}{\partial w} S(w_i, u) = g_i(u)$, $i = 0, N$, with $g_i(u) = \frac{\partial}{\partial w} f(w_i, u)$.

Algorithm 6.1. We compute the coefficients in (6.1) by the formulae:

$$\begin{aligned}
b_{-1}(u) &= b_1(u) - 2h_0 g_0(u), \\
b_0(u) &= S_0(u) - [S_0(u) + h_0 g_0(u) - b_1(u)] \\
&\qquad \times [1 + \varphi_0'(0)/\varphi_0(0)]^{-1}, \\
b_i(u) &= S_i(u) - 2h_i^2(h_{i-1} + h_i)^{-1}\varphi_i(0)\delta_i S(u), \\
&\qquad i = 1, \dots, N-1, \\
b_N(u) &= S_N(u) - [S_N(u) - h_{N-1} g_N(u) - b_{N-1}(u)] \\
&\qquad \times [1 - \psi_{N-1}'(1)/\psi_{N-1}(1)]^{-1}, \\
b_{N+1}(u) &= b_{N-1}(u) + 2h_{N-1} g_N(u),
\end{aligned} \tag{6.2}$$

where

$$\delta_i S(u) = \Delta_i S(u) - \Delta_{i-1} S(u), \quad \Delta_j S(u) = [S_{j+1}(u) - S_j(u)]/h_j, \ j = i-1, i.$$

The approximating spline $S(w,u)$ possesses the following properties of preserving the shape of the initial data.

Property 6.1. *Let the functions $S_j(u)$, $j = i-1, \dots, i+2$, $1 \le i \le N-1$, be monotonic and/or convex on the interval $[\tilde{u}_m, \tilde{u}_{m+1}]$ and satisfy the conditions*

$$S_j^{(k)}(u)\delta_j f^{(k)}(u) < 0, \quad j \ne 0, N, \quad S_j^{(k)}(u) g_j^{(k)}(u) < 0, \quad j = 0, N,$$

where $k = 1$ and/or $k = 2$, respectively. Then for any fixed $\tilde{w} \in [w_i, w_{i+1}]$, $2 \le i \le N-3$, the generalized spline $S(w,u)$ constructed by Algorithm 6.1 will be monotonic and/or convex in $[\tilde{u}_m, \tilde{u}_{m+1}]$.

Property 6.2. *Let the choice of parameters p_i, q_i, $i = -2, \dots, N+2$, of a generalized spline $S(w,u)$ ensure that, for given functions $E_j(u)$ the following estimate holds true for any $\tilde{S}_j(u)$ such that $\Delta_i \tilde{S}(u)$ and $\delta_i \tilde{S}(u)$ do not change sign on $[0,1]$:*

$$|\tilde{S}_j(u) - S_j(u)| \le E_j(u), \quad j = 0, \dots, N.$$

Then for any fixed u, the spline $S_u(w) = S(w,u)$ is a shape preserving approximant.

To prove this assertion, it is sufficient to take advantage of the relations

$$\frac{\partial^k}{\partial u^k} S(w,u) = \sum_{i=-1}^{N+1} b_i^{(k)}(u) B_i(w), \quad k = 1, 2,$$

use the expressions (6.2) for the coefficients $b_i(u)$, and take into account the finiteness of the B-splines: $B_i(w) > 0$ at $w \in (w_{i-2}, w_{i+2})$ and $B_i(w) \equiv 0$ at $w \notin (w_{i-2}, w_{i+2})$.

A nonunique shape preserving surface given point by point as a family of curvilinear nonintersecting sections can be constructed by introducing the standard parametrization

$$x = S^x(w,u), \quad y = S^y(w,u), \quad z = S^z(w,u). \tag{6.3}$$

We choose the nodes of nonuniform meshes in the directions w and u according to the results [4]. In our case the original points $T_{ij} = (x_{ij}, y_{ij}, z_{ij})$, $j = 0,\ldots,M_i$, $i = 0,\ldots,N$, are considered to belong to the parallelepiped $\prod_{ij} = \{\tilde{\chi}_{ij} | |\tilde{\chi}_{ij} - \chi_{ij}| \le \varepsilon^{\chi}_{ij}\}$, where we put $\chi_{ij} = \chi(w_i, u_j)$ for every coordinate function in (6.3) and ε^{χ}_{ij} is the admissible deviation for the appropriate variable. The resulting surface is obtained as a triple of shape preserving splines constructed by the above algorithm.

§7. Numerical Examples

In the numerical tests the initial data were given point by point as a collection of nonintersecting, in general, curvilinear sections of a 3-D body. At the beginning the system of curves along the initial cross-sections was constructed via the 1-D algorithm of shape preserving interpolation from sections 2 and 3. Along the orthogonal direction the set of generalized local approximation splines was generated. On the resulting surface the system of curvilinear coordinate lines forming the regular mesh was constructed. Along these lines shape properties of the initial data such as convexity, monotonicity, presence of linear sections and other were retained. The defining functions $\varphi_i(t)$, $\psi_i(t)$ for the generalized cubic spline were taken in the form

$$\varphi_i(t) = \varphi(p_i, t) = P_i(1-t)^3/[1 + p_i t(1-t)],$$
$$\psi_i(t) = \varphi(q_i, 1-t), \quad P_i^{-1} = 2(1+p_i)(3+p_i),$$

that correspond to rational cubic splines with quadratic denominator. The initial data and the resulting shape preserving surface are given in figures 1 and 2, correspondingly.

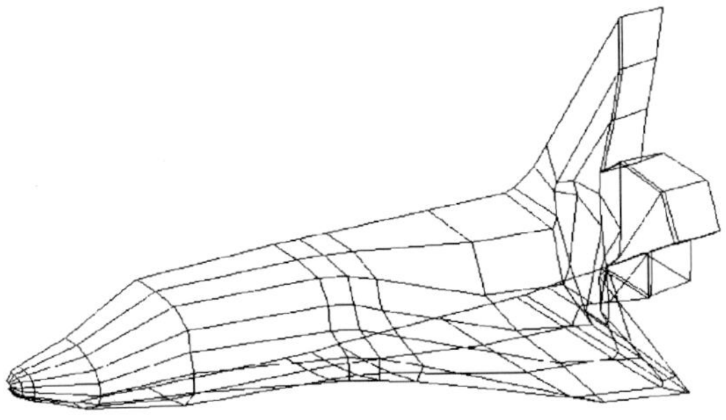

Fig. 1. The initial aircraft data.

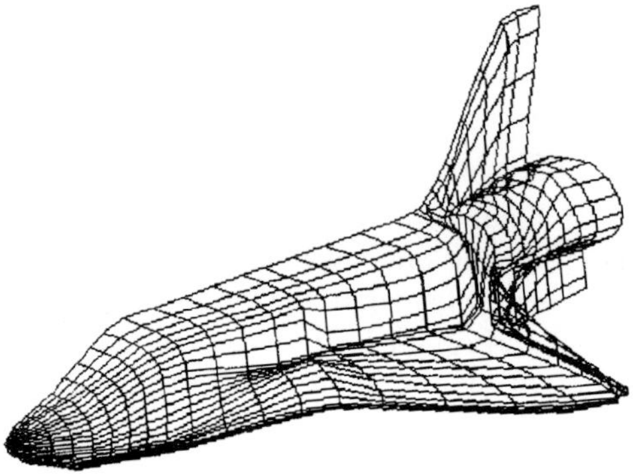

Fig. 2. The resulting shape preserving aircraft surface.

Acknowledgements. Supported in part by the Russian Fundamental Research Fund (code 93-01-00495)

References

1. Beatson, R. K. and H. Wolkowitz, Post-processing cubics for monotonicity, SIAM J. Numer. Anal. **26** (1989), 480–502.
2. Boor, C., *A Practical Guide to Splines*, Springer-Verlag, New York, 1978.
3. Kvasov, B. I., Isogeometric interpolation by generalized splines, Russ. J. Numer. Anal. Math. Modelling **10** (1995).
4. Kvasov, B. I. and S. A. Yatsenko, Monotonizing parametrization for interpolational splines, submitted to Int. J. Numerical Algorithms.
5. McCartin, B. J., Computation of exponential splines, SIAM J. Sci. Stat. Comput. **11** (1990), 242–262.
6. Moore, R. E., *Interval Analysis*, Prentice-Hall, Englewood Cliffs, 1966.
7. Schaback, R., Adaptive rational splines, Constr. Approx. **6** (1990), 167–179.
8. Späth, H., *Eindimensionale Spline-Interpolations-Algorithmen*, R. Oldenbourg-Verlag, München, 1990.
9. Zav'yalov, Ju. S., B. I. Kvasov and V. L. Miroshnichenko, *Methods of Spline Functions*, Nauka, Moscow, 1980 (in Russian).

Boris I. Kvasov
Institute of Computational Technologies
Russian Academy of Sciences
Lavrentyev Avenue 6
630090 Novosibirsk (Russia)

kvasov@ictnw.scnet.nsk.su

Introducing Dilation and Translation for Incomplete Sigma-Pi Neural Networks of Hopfield-Type

Burkhard Lenze

Abstract. In this contribution we continue our investigations on a special family of sigma-pi neural networks. Here, we focus our attention to the case of Hopfield-type networks, although the general idea of introducing dilation and translation applies to any sigma-pi neural net. We show that so-called generalized sigma-pi units including dilation and translation can be used to make a network perform in a similar way as usual Hopfield-type networks do. However, because of the additional degrees of freedom induced by the dilation and translation parameters the generated networks will prove to be more powerful and flexible than the classical ones without causing a significant increase of complexity. This approach especially makes sense in cases where the number of multiplicative synaptic correlations of neural activities is limited for some theoretical or practical reason and a so-called incomplete situation appears. At the end of the paper we will take a look at a special application of our strategy and discuss the solution of parity-type problems as an example.

§1. Introduction

The problem we are going to address in this paper is how to design a special two-layer neural network in order to solve an arbitrary autoassociative recognition problem. The most common and well-known two-layer neural networks which can be used in this context are the so-called Hopfield networks (cf. [8,9,10]). We will discuss these networks in some detail in their general (incomplete) sigma-pi version including dilation and translation (see also [1,2,3,4,5,6,11,15,16] for the classical case) and will prove that they terminate in recall mode after a finite number of steps. With this result in hand we will explicitly apply our strategy to a special situation and come to the so-called hyperbolic-type Hopfield neural networks. Hyperbolic-type networks

Advanced Topics in Multivariate Approximation
F. Fontanella, K. Jetter and P. J. Laurent (eds.) pp. 197–210.

ISBN 981-02-2852-X

with one hidden layer have already been considered in some detail, earlier, and proved to be useful in connection with interpolation and approximation networks, image reconstruction and processing (cf. [12,13,14]). Here, we will take a look at hyperbolic-type networks without hidden layer and apply them in the Hopfield sense in order to find solutions for so-called parity-type problems.

§2. Generalized Sigma-Pi Hopfield Neural Networks

A standard Hopfield neural network consists of a finite number of completely connected neural processing units (often called neurons or simply units), whose dynamic states are essentially ruled by the following signum-type transfer function $T_S : \mathbb{R} \to \{-1, 1, n.d.\}$,

$$T_S(\xi) := \begin{cases} -1, & \xi < 0 \,, \\ n.d., & \xi = 0 \,, \\ 1, & \xi > 0 \,. \end{cases} \tag{2.1}$$

Here, the abbreviation $n.d.$ means that for $\xi = 0$ the function T_S is not defined. In case that the Hopfield neural network has precisely n processing units or neurons the network should be able to associate a finite number of let's say t bipolar coded vectors with itself,

$$(\vec{x}^{(1)}, \vec{x}^{(1)}), (\vec{x}^{(2)}, \vec{x}^{(2)}), \ldots, (\vec{x}^{(t)}, \vec{x}^{(t)}) \in \{-1, 1\}^n \times \{-1, 1\}^n \,. \tag{2.2}$$

More precisely, in case that one of the stored vectors $\vec{x}^{(s)}$ or a slightly disturbed version of it is entered into the network it should be able to generate the correct output $\vec{x}^{(s)}$. In the following, we give a more precise definition of what we mean by a Hopfield neural network. We start with a definition concerning only the topology of the network.

Definition 2.1. *A Hopfield neural network is a neural network consisting of a finite number of units and links with the following properties:*
Each unit is an input and, simultaneously, an output unit and, moreover, each unit is connected with each other unit in both directions by so-called bidirectional links.

For a Hopfield neural network there are obviously no differences between input and output units. In detail, a single Hopfield unit or Hopfield neuron is defined as follows, where we immediately give the most general sigma-pi formulation including dilation and translation and where # denotes the number of elements of the respective set under consideration.

Definition 2.2. *Let $n \in \mathbb{N}$ and Γ be a non-empty collection of subsets of $\{1, \ldots, n\}$. A generalized sigma-pi Hopfield neuron is a function κ,*

$$\kappa : \mathbb{R}^n \to \{-1\}^n \cup \{1\}^n \cup \{n.d.\}^n, \tag{2.3}$$

generated by the signum transfer function $T_S : \mathbb{R} \to \{-1, 1, n.d.\}$, a set of $\#\Gamma$ sigma-pi weights $w_R \in \mathbb{R}$, $R \in \Gamma$, and n dilation and translation weights $\alpha_k, d_k \in \mathbb{R}$,

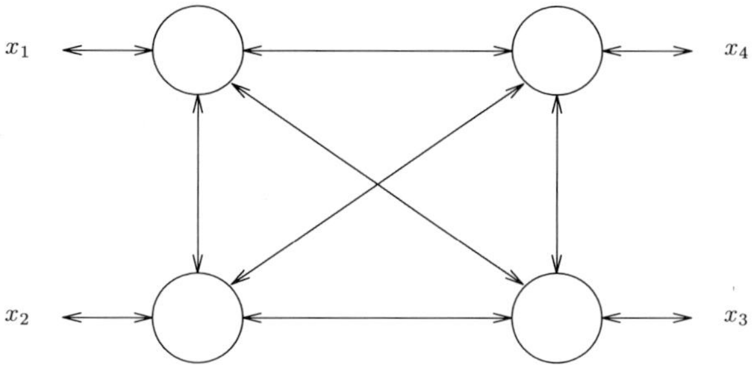

Fig. 1. Structure of a Hopfield neural network with four neurons.

$\alpha_k > 0$, $k \in \{1, \dots, n\}$. *In detail, a given vector* $\vec{x} = (x_1, x_2, \dots, x_n) \in \mathbb{R}^n$ *is mapped onto the vector* $\vec{y} = (y, y, \dots, y) \in \{-1\}^n \cup \{1\}^n \cup \{n.d.\}^n$ *consisting of* n *identical components* y,

$$y := T_S(\sum_{R\in\Gamma} w_R \prod_{k\in R} (\alpha_k x_k - d_k)), \tag{2.4}$$

precisely,

$$\kappa(\vec{x}) := (T_S(\sum_{R\in\Gamma} w_R \prod_{k\in R} (\alpha_k x_k - d_k)), \dots, T_S(\sum_{R\in\Gamma} w_R \prod_{k\in R} (\alpha_k x_k - d_k))) = \vec{y}. \tag{2.5}$$

The sigma-pi neuron is called complete of order r, $r \in \{1, 2, \dots, n\}$, *in case that* Γ *coincides with the set of all subsets of* $\{1, \dots, n\}$ *with at most* r *elements; if no such* r *exists, we say that it is incomplete.*

Remark 2.1. *In some sense it seems to be strange to let a Hopfield neuron map onto a vector* $\vec{y}$ *consisting of* n *identical components as defined in (2.5). The reason for this kind of definition only becomes clear in case that we take a look at a collection of Hopfield neurons arranged together in order to generate a Hopfield network (see Figure 1 for the special case of four Hopfield neurons connected to a Hopfield network and Definition 2.3, below, for further details). As already stated in Definition 2.1 a single Hopfield neuron as part of a Hopfield neural network is completely connected to all other neurons and has one bidirectional link to the outer world. Moreover, in case of concrete signal processing it acts as follows: At a fixed time (deterministic time-discrete update by an external scheduling process) the neuron takes a look at at all signals at its* n *input/output links. This signal vector* $\vec{x}$ *is used in order to calculate the real value* y *via (2.4). Then, this real* *value y is fed back into the network and to the outer world using the n identically initialized bidirectional links; each link is simultaneously triggered with the same signal y. This is the formal reason for defining the action of a Hopfield neuron as done in (2.5) with the vector $\vec{y}$ having n identical components y.*

Remark 2.2. *In case that $\alpha_k = 1$ and $d_k = 0$, $1 \leq k \leq n$, we obtain the usual sigma-pi Hopfield neurons.*

Example. *For $n = 3$, a generalized Hopfield neuron induced by*

$$y := T_S(w_{\{1\}}(\alpha_1 x_1 - d_1) + w_{\{2\}}(\alpha_2 x_2 - d_2) + w_{\{3\}}(\alpha_3 x_3 - d_3) + w_{\{\}}) \tag{2.6}$$

is complete of order 1 and – in view of proper degrees of freedom – coincides with a standard Hopfield neuron of type

$$y := T_S(w_{\{1\}}x_1 + w_{\{2\}}x_2 + w_{\{3\}}x_3 + w_{\{\}}). \tag{2.7}$$

On the other hand, a generalized Hopfield neuron induced by

$$y := T_S(w_{\{1,2,3\}}(\alpha_1 x_1 - d_1)(\alpha_2 x_2 - d_2)(\alpha_3 x_3 - d_3)) \tag{2.8}$$

is incomplete and the additional dilation and translation parameters increase its degree of flexibility while keeping the number of sigma-pi weights resp. the number of multiplicative synaptic correlations fixed.

Now, we are prepared to give the complete definition of the dynamic recall mode of a generalized sigma-pi Hopfield neural network. In this context, it is essential to note that the so-called sigma-pi weights $w_R \in \mathbb{R}$, $R \subset \{1, \ldots, n\}$, and the dilation and translation weights $\alpha_k, d_k \in \mathbb{R}$, $\alpha_k > 0$, $k \in \{1, \ldots, n\}$, appearing in the above definition are in some sense universal for the whole network and are not restricted to one fixed neuron. For example, as the following definition will show, a sigma-pi weight w_R will be used by all neurons which have a numbering index belonging to the set R. This means that all weights

$$w_R \,, \quad R \subset \{1, 2, \ \ldots \ , n\} \ \text{ and } \ i \in R \,, \tag{2.9}$$

are sigma-pi weights for the i-th neuron, and in the same sense

$$w_R \,, \quad R \subset \{1, 2, \ \ldots \ , n\} \ \text{ and } \ j \in R \,, \tag{2.10}$$

are sigma-pi weights for the j-th neuron. Especially, the weights

$$w_R \,, \quad R \subset \{1, 2, \ \ldots \ , n\} \ \text{ and } \ i, j \in R \,, \tag{2.11}$$

are sigma-pi weights which are used by the i-th and j-th neuron, simultaneously. In a similar fashion, $\alpha_k, d_k \in \mathbb{R}$, $\alpha_k > 0$, $k \in \{1, \ldots, n\}$ fixed, is a dilation resp. translation weight for all neurons of the network which have a numbering index different from k.
Before we now formulate our central definition let us briefly face the question on how to get the weights w_R, α_k and d_k in order to learn t different autoassociations

$$(\vec{x}^{(1)}, \vec{x}^{(1)}), (\vec{x}^{(2)}, \vec{x}^{(2)}), \ldots, (\vec{x}^{(t)}, \vec{x}^{(t)}) \in \{-1,1\}^n \times \{-1,1\}^n. \tag{2.12}$$

In general, this problem can be attacked by various so-called learning algorithms. For example, in the standard case $\alpha_k = 1$ and $d_k = 0$, $1 \leq k \leq n$, the most prominent one is the well-known Hebb learning rule which generates the sigma-pi weights via

$$w_R := \sum_{s=1}^{t} \prod_{k \in R} x_k^{(s)} , \quad R \subset \{1, 2, \ \ldots \ , n\} . \tag{2.13}$$

A sigma-pi Hopfield network using all 2^n weights defined by (2.13) reproduces all t autoassociations (2.12) perfectly, however, in practice the use of 2^n sigma-pi weights is often impossible because of complexity restrictions (cf. [1], [7], [11], or [15], for details). Therefore, we propose the learning rules

$$w_R := \sum_{s=1}^{t} \prod_{k \in R} x_k^{(s)} , \quad R \in \Gamma , \tag{2.14}$$

$$d_k := \lambda \sum_{s=1}^{t} x_k^{(s)} , \quad 1 \leq k \leq n , \tag{2.15}$$

and

$$\alpha_k := |d_k| + 1 , \quad 1 \leq k \leq n , \tag{2.16}$$

with Γ a reasonable small collection of subsets of $\{1, 2, \ldots, n\}$ and $\lambda \geq 0$ a learning parameter which can be tuned appropriately. Note that for $\lambda = 0$ resp. $d_k = 0$, $1 \leq k \leq n$, a standard Hopfield configuration with the usual Hebb learning scheme is reproduced. However, for $d_k \gg 0$ ($d_k \ll 0$) the corresponding activation factor $(\alpha_k x_k - d_k)$ becomes quite negative (positive) for $x_k = -1$ ($x_k = 1$) while it remains equal to 1 (-1) for $x_k = 1$ ($x_k = -1$). Since the sign of d_k indicates whether there are more 1 ($d_k > 0$) or more -1 ($d_k < 0$) entries in the k-th component of all training patterns the proposed learning scheme makes an x_k-value with low probability enter the network in an overemphasized way, therefore increasing the probability of suitable changes in network activity. On the other hand, a quite likely x_k-value remains uneffected by dilation and translation and is fed into the network as it is. It is quite clear that this should be a reasonable strategy. However, since a detailed discussion of the learning method is beyond the scope of this contribution and a recent paper with a more detailed experimental justification of (2.14)–(2.16) has just been submitted we finish the intuitive explanation of our learning rule, now. In the following, we simply assume that the relevant sigma-pi, dilation and translation weights for our Hopfield networks are generated by any reasonable learning algorithm (for example, the one given by (2.14)–(2.16)). Our main interest in this contribution is to take a look at the so-called recall mode of the networks, in other words, what do the networks do with an input vector handed over to them? An answer to this question will be given in the following definition.

Definition 2.3. *Let $n \in \mathbb{N}$ and Γ be a non-empty collection of subsets of $\{1, \ldots, n\}$. The generalized sigma-pi Hopfield neural network may have n neurons, $\#\Gamma$ sigma-pi weights w_R, $R \in \Gamma$, and n dilation and translation weights $\alpha_k, d_k \in \mathbb{R}$, $\alpha_k > 0$, $k \in \{1, \ldots, n\}$, which we assume to be generated by means of any reasonable learning algorithm. Now, in case that an input vector is entered,*

$$\vec{x} =: \vec{x}^{[0]} = (x_1^{[0]}, x_2^{[0]}, \ \ldots \ , x_n^{[0]}) \in \{-1, 1\}^n, \tag{2.17}$$

the generalized sigma-pi Hopfield neural network generates a sequence of vectors $(\vec{x}^{[u]})_{u \in \mathbb{N}_0}$ ($\mathbb{N}_0 := \mathbb{N} \backslash \{0\}$), via

$$a_j^{[u]} := \sum_{\substack{R \in \Gamma \\ j \in R}} w_R \prod_{\substack{k \in R \\ k < j}} (\alpha_k x_k^{[u+1]} - d_k) \prod_{\substack{k \in R \\ k > j}} (\alpha_k x_k^{[u]} - d_k), \tag{2.18}$$

$$x_j^{[u+1]} := \begin{cases} T_S\left(a_j^{[u]}\right), & \text{for} \quad a_j^{[u]} \neq 0 \, , \\ x_j^{[u]}, & \text{for} \quad a_j^{[u]} = 0 \, , \end{cases} \tag{2.19}$$

for $1 \le j \le n$ and $u \in \mathbb{N}_0$. As the output vector the so-defined Hopfield network yields the vector $\vec{x}^{[v]} \in \{-1, 1\}^n$, which for the first time satisfies

$$\vec{x}^{[v]} = \vec{x}^{[v+1]} \tag{2.20}$$

for some $v \in \mathbb{N}_0$. In analogy to the case of generalized sigma-pi neurons, the sigma-pi network is called complete of order r, $r \in \{1, 2, \ldots, n-1\}$, in case that Γ coincides with the set of all subsets of $\{1, \ldots, n\}$ with at most $r+1$ elements; if no such r exists, we say that it is incomplete.

Remark 2.3. *Note that the definition of order of a generalized sigma-pi Hopfield neural network as given above is compatible with the order definition for a single generalized sigma-pi Hopfield neuron (cf. Definition 2.2). To see this let us point to the fact that in the network context one index of the set R is always used to indicate the number of the neuron under consideration without implying an additional factor in its activation. In other words, an index set R implies $\#R$ activation factors for* *an isolated neuron while only $(\#R-1)$ factors appear in a network context. The deeper reason for this definition is that self-reflection (direct feedback) is usually not considered in Hopfield-type neural networks.*

Remark 2.4. *The above definition of the recall mode of a generalized sigma-pi Hopfield neural network implicitly states that the recall mode terminates, i.e., that there exists a natural number $v \in \mathbb{N}_0$ satisfying*

$$\vec{x}^{[v]} = \vec{x}^{[v+1]}, \tag{2.21}$$

which immediately implies

$$\vec{x}^{[v]} = \vec{x}^{[v+u]}, \quad u \in \mathbb{N}. \tag{2.22}$$

We will prove this in the following theorem.

Theorem 2.1. *Let $n \in \mathbb{N}$ and Γ be a non-empty collection of subsets of $\{1,\ldots,n\}$. The generalized sigma-pi Hopfield neural network may have n neurons, $\#\Gamma$ sigma-pi weights w_R, $R \in \Gamma$, and n dilation and translation weights $\alpha_k, d_k \in \mathbb{R}$, $\alpha_k > 0$, $k \in \{1,\ldots,n\}$, which we assume to be generated by means of any reasonable learning algorithm. Now, in case that an input vector is entered,*

$$\vec{x} =: \vec{x}^{[0]} = (x_1^{[0]}, x_2^{[0]}, \ldots, x_n^{[0]}) \in \{-1,1\}^n, \tag{2.23}$$

then there exists a natural number $v \in \mathbb{N}_0$, such that for the sequence of vectors $\vec{x}^{[u]} \in \{-1,1\}^n$, $u \in \mathbb{N}_0$, generated by means of the generalized sigma-pi Hopfield recall mode given in Definition 2.3 we have

$$\vec{x}^{[v]} = \vec{x}^{[v+u]}, \quad u \in \mathbb{N}. \tag{2.24}$$

Proof: Let $\vec{x} =: \vec{x}^{[0]} \in \{-1,1\}^n$ be given arbitrarily and

$$\vec{x}^{[u]} \in \{-1,1\}^n, \quad u \in \mathbb{N}_0, \tag{2.25}$$

be the sequence of vectors generated via

$$a_j^{[u]} := \sum_{\substack{R\in\Gamma \\ j\in R}} w_R \prod_{\substack{k\in R \\ k<j}} (\alpha_k x_k^{[u+1]} - d_k) \prod_{\substack{k\in R \\ k>j}} (\alpha_k x_k^{[u]} - d_k), \tag{2.26}$$

$$x_j^{[u+1]} := \begin{cases} T_S\left(a_j^{[u]}\right), & \text{for} \quad a_j^{[u]} \neq 0, \\ x_j^{[u]}, & \text{for} \quad a_j^{[u]} = 0, \end{cases} \tag{2.27}$$

for $1 \le j \le n$ and $u \in \mathbb{N}_0$. For any bipolar coded vector $\vec{x} \in \{-1,1\}^n$ we define the energy functional (or Hamilton function) E with respect to the sigma-pi weights w_R, $R \in \Gamma$, and the dilation and translation weights $\alpha_k, d_k \in \mathbb{R}$, $\alpha_k > 0$, $k \in \{1, \ldots, n\}$, as

$$E(\vec{x}) \;:=\; -\sum_{R\in\Gamma} w_R \prod_{k\in R} (\alpha_k x_k - d_k) \;. \tag{2.28}$$

Now, we consider any u-th recall cycle and take a detailed look at the j-th neuron in this cycle. Moreover, we assume that the neuron in this cycle indeed changes its activity, i.e., that we have $x_j^{[u+1]} \neq x_j^{[u]}$. Therefore, the two vectors

$$\vec{y} := (x_1^{[u+1]}, \ldots, x_{j-1}^{[u+1]}, x_j^{[u]}, x_{j+1}^{[u]}, \ldots, x_n^{[u]}) \tag{2.29}$$

and

$$\vec{z} := (x_1^{[u+1]}, \ldots, x_{j-1}^{[u+1]}, x_j^{[u+1]}, x_{j+1}^{[u]}, \ldots, x_n^{[u]}) \tag{2.30}$$

only differ in the j-th component; all other components are identical. Now, comparing the energies $E(\vec{y})$ and $E(\vec{z})$ we obtain

$$\begin{aligned}
&E(\vec{z}) - E(\vec{y}) \\
&= -\sum_{R\in\Gamma} w_R \left(\prod_{k\in R} (\alpha_k z_k - d_k) - \prod_{k\in R} (\alpha_k y_k - d_k) \right) \\
&= -\sum_{\substack{R\in\Gamma \\ j\in R}} w_R \left(\prod_{k\in R} (\alpha_k z_k - d_k) - \prod_{k\in R} (\alpha_k y_k - d_k) \right) \qquad (2.31) \\
&= -\sum_{\substack{R\in\Gamma \\ j\in R}} w_R \left(\prod_{\substack{k\in R \\ k<j}} (\alpha_k x_k^{[u+1]} - d_k) \prod_{\substack{k\in R \\ k>j}} (\alpha_k x_k^{[u]} - d_k) \right) \left(\alpha_j x_j^{[u+1]} - \alpha_j x_j^{[u]} \right) \\
&= -\alpha_j \left(x_j^{[u+1]} - x_j^{[u]} \right) \sum_{\substack{R\in\Gamma \\ j\in R}} w_R \prod_{\substack{k\in R \\ k<j}} (\alpha_k x_k^{[u+1]} - d_k) \prod_{\substack{k\in R \\ k>j}} (\alpha_k x_k^{[u]} - d_k) \\
&= -2\alpha_j \left| \sum_{\substack{R\in\Gamma \\ j\in R}} w_R \prod_{\substack{k\in R \\ k<j}} (\alpha_k x_k^{[u+1]} - d_k) \prod_{\substack{k\in R \\ k>j}} (\alpha_k x_k^{[u]} - d_k) \right| \\
&< 0 \,.
\end{aligned}$$

The two final steps hold because $\alpha_j > 0$ by presupposition, $x_j^{[u+1]} = -x_j^{[u]}$ since we assumed that in the cycle under consideration the j-th neuron indeed changed its state from 1 to -1 or vice versa and, finally,

$$x_j^{[u+1]} \;=\; T_S \left(\sum_{\substack{R\in\Gamma \\ j\in R}} w_R \prod_{\substack{k\in R \\ k<j}} (\alpha_k x_k^{[u+1]} - d_k) \prod_{\substack{k\in R \\ k>j}} (\alpha_k x_k^{[u]} - d_k) \right) \tag{2.32}$$

by means of (2.26) (note that $a_j^{[u]}$ could not be zero because of the proper change of state) and (2.27).
In conclusion, we have shown that

$$E(\vec{z}) \ < \ E(\vec{y}) \ . \tag{2.33}$$

Now, in case that we consider a complete update or recall cycle for all neurons with indices $j \in \{1, \ldots, n\}$, we can obviously conclude that for a given $u \in \mathbb{N}_0$ we have

$$E(\vec{x}^{[u+1]}) \ < \ E(\vec{x}^{[u]}) \ , \tag{2.34}$$

if there exists at least one index $j \in \{1, 2, \ldots, n\}$ satisfying

$$x_j^{[u+1]} \ \neq \ x_j^{[u]} \ . \tag{2.35}$$

Since the set $\{-1, 1\}^n$ is finite, the energy functional E can only decrease a finite number of times on the set of vectors $\vec{x}^{[u]} \in \{-1, 1\}^n$, $u \in \mathbb{N}_0$. Therefore, there exists a natural number $v \in \mathbb{N}_0$ with

$$E(\vec{x}^{[v+1]}) \ = \ E(\vec{x}^{[v]}) \ . \tag{2.36}$$

This necessarily implies

$$\vec{x}^{[v]} \ = \ \vec{x}^{[v+1]} \tag{2.37}$$

resp.

$$\vec{x}^{[v]} \ = \ \vec{x}^{[v+u]} \ , \quad u \in \mathbb{N} \ . \ \blacksquare \tag{2.38}$$

§3. Hyperbolic-Type Sigma-Pi Hopfield Neural Networks

Obviously, the idea to introduce dilation and translation weights only makes sense in case of incomplete sigma-pi networks and when additional helpful information concerning the structure of patters to be stored or recognized is available. In case of complete sigma-pi networks the dilation and translation weights do not increase the degree of freedom since calculating the products and reorganizing the terms a standard complete sigma-pi network without dilation and translation is reproduced (comp. the Example in §2). Therefore, only incomplete situations are of interest and often these situations appear because of theoretical or practical (complexity) limitations for the allowed multiplicative correlation of synaptic activities. In the following, we take a look at a problem where for these reasons essentially only the highest possible multiplicative interferences are of interest. This situation is usually called the hyperbolic case. Here, we have

$$\Gamma \ = \ \{\{1, 2, \ldots, n\}\}, \tag{3.1}$$

$$w_{\{1,2,\dots,n\}} = 1, \tag{3.2}$$

$$\alpha_k, d_k \in \mathbb{R},\ \alpha_k > 0,\quad k \in \{1,2,\dots,n\}, \tag{3.3}$$

and in case that an input vector is entered,

$$\vec{x} =: \vec{x}^{[0]} = (x_1^{[0]}, x_2^{[0]}, \ \dots\ , x_n^{[0]}) \in \{-1,1\}^n, \tag{3.4}$$

the hyperbolic sigma-pi Hopfield network generates a sequence of vectors $(\vec{x}^{[u]})_{u\in\mathbb{N}_0}$ via

$$a_j^{[u]} := \prod_{k=1}^{j-1} (\alpha_k x_k^{[u+1]} - d_k) \prod_{k=j+1}^{n} (\alpha_k x_k^{[u]} - d_k), \tag{3.5}$$

$$x_j^{[u+1]} := \begin{cases} T_S\left(a_j^{[u]}\right), & \text{for} \quad a_j^{[u]} \neq 0\ , \\ x_j^{[u]}, & \text{for} \quad a_j^{[u]} = 0\ , \end{cases} \tag{3.6}$$

for $1 \le j \le n$ and $u \in \mathbb{N}_0$. As the output vector the hyperbolic Hopfield network again yields the vector $\vec{x}^{[v]} \in \{-1,1\}^n$, which for the first time satisfies

$$\vec{x}^{[v]} = \vec{x}^{[v+1]} \tag{3.7}$$

for some $v \in \mathbb{N}_0$. This simply follows from Theorem 2.1 as a special case. Hyperbolic sigma-pi Hopfield neural networks are the canonical candidates to solve the so-called parity problem (comp. also [7], p. 281, in this context). To see this, we first of all have to reformulate the parity problem as an autoassociative classification problem. In case that we enter an arbitrary vector $\vec{x} \in \{-1,1\}^n$ into the network, then the first component x_1 of $\vec{x}$ should be set to -1, if

$$\prod_{k=2}^{n} x_k = -1, \tag{3.8}$$

and should be set to 1, if

$$\prod_{k=2}^{n} x_k = 1. \tag{3.9}$$

In the first case, $x_1 = -1$ indicates that the vector

$$(x_2, x_3, \ \dots\ , x_n) \in \{-1,1\}^{n-1} \tag{3.10}$$

has odd parity (odd number of -1 components), in the second case, $x_1 = 1$ indicates that vector (3.10) has even parity (even number of -1 components). As can be easily checked, a hyperbolic sigma-pi Hopfield network with all dilation weights set equal to one and all translation weights set equal to zero,

$$\alpha_1 = \alpha_2 = \ \dots\ = \alpha_n = 1, \tag{3.11}$$

$$d_1 = d_2 = \ \dots \ = d_n = 0, \tag{3.12}$$

solves this problem perfectly. For this type of problems hyperbolic neural networks are obviously superior to any classical Hopfield neural network (both, in view of complexity and in view of robustness). Translation enters the field, for example, in case that we have problems which are loosely connected to the general parity problem. We briefly explain the main ideas, now, but omit the simple explicit autoassociative encoding sketched above. Moreover, we will only consider one single neuron which will be tuned in order to fulfill a specific task. The reason for this restriction is that configuring a whole Hopfield network would imply to discuss each neuron separately which would make the examples quite unreadable. Our argument for justification is that since we show that one neuron can be configured appropriately for a special non-trivial task it is almost clear that also a whole network – i. e. a set of neurons with partly independent and partly overlapping weights and dilation and translation parameters – can be tuned as well. Such more sophisticated examples of complete networks which cannot be calculated by hand any longer are part of a recent experimental paper which has just been submitted. We start with the standard parity problem for $n = 3$ in the usual heteroassociative encoding which is suitable for being solved by only one single neuron:

$$\begin{array}{rclrcl}
\vec{x}^{(1)} = (-1,-1,-1) & \leftrightarrow & y^{(1)} = -1, & \vec{x}^{(2)} = (1,1,-1) & \leftrightarrow & y^{(2)} = -1,\\
\vec{x}^{(3)} = (1,-1,-1) & \leftrightarrow & y^{(3)} = 1, & \vec{x}^{(4)} = (1,-1,1) & \leftrightarrow & y^{(4)} = -1,\\
\vec{x}^{(5)} = (-1,1,-1) & \leftrightarrow & y^{(5)} = 1, & \vec{x}^{(6)} = (-1,1,1) & \leftrightarrow & y^{(6)} = -1,\\
\vec{x}^{(7)} = (-1,-1,1) & \leftrightarrow & y^{(7)} = 1, & \vec{x}^{(8)} = (1,1,1) & \leftrightarrow & y^{(8)} = 1.
\end{array}$$

Obviously, a neuron defined by

$$y = T_S(x_1 x_2 x_3) \tag{3.13}$$

solves the problem. The embedded twodimensional parity problem which is – let's say – independent of the first component

$$\begin{array}{rclrcl}
\vec{x}^{(1)} = (-1,-1,-1) & \leftrightarrow & y^{(1)} = 1, & \vec{x}^{(2)} = (1,1,-1) & \leftrightarrow & y^{(2)} = -1,\\
\vec{x}^{(3)} = (1,-1,-1) & \leftrightarrow & y^{(3)} = 1, & \vec{x}^{(4)} = (1,-1,1) & \leftrightarrow & y^{(4)} = -1,\\
\vec{x}^{(5)} = (-1,1,-1) & \leftrightarrow & y^{(5)} = -1, & \vec{x}^{(6)} = (-1,1,1) & \leftrightarrow & y^{(6)} = 1,\\
\vec{x}^{(7)} = (-1,-1,1) & \leftrightarrow & y^{(7)} = -1, & \vec{x}^{(8)} = (1,1,1) & \leftrightarrow & y^{(8)} = 1.
\end{array}$$

is fitted by

$$y = T_S((x_1 + 2)x_2 x_3). \tag{3.14}$$

Moreover, in the Hopfield situation in the case that, for example, the first neuron plays the role of presynaptic inhibition the choice

$$y = T_S((x_1 - 1)x_2 x_3) \tag{3.15}$$

is a proper one. In this case, $x_1 = 1$ gives total activation zero and leaves the Hopfield network uneffected.
Obviously, in case of pure hyperbolic situations positive dilation does not give any further degree of freedom, since only the signs of the activation and not its absolute values are relevant (this situation immediately changes when sigmoidal transfer functions are considered). In the Hopfield setting dilations become interesting in case that more than two summation terms are allowed as arguments in the signum transfer function. For example, a large dilation parameter makes the respective term dominant in sign compared to others. Consider the problem

$$\begin{array}{rclcrcl} \vec{x}^{(1)} = (-1,-1,-1) & \leftrightarrow & y^{(1)} = -1, & & \vec{x}^{(2)} = (1,1,-1) & \leftrightarrow & y^{(2)} = -1, \\ \vec{x}^{(3)} = (1,-1,-1) & \leftrightarrow & y^{(3)} = -1, & & \vec{x}^{(4)} = (1,-1,1) & \leftrightarrow & y^{(4)} = -1, \\ \vec{x}^{(5)} = (-1,1,-1) & \leftrightarrow & y^{(5)} = 1, & & \vec{x}^{(6)} = (-1,1,1) & \leftrightarrow & y^{(6)} = -1, \\ \vec{x}^{(7)} = (-1,-1,1) & \leftrightarrow & y^{(7)} = 1, & & \vec{x}^{(8)} = (1,1,1) & \leftrightarrow & y^{(8)} = 1. \end{array}$$

A brief analysis shows that this problem can be interpreted as a superposition of the threedimensional parity problem and two simple linear corrections (with respect to x_2 and x_3), however, in case that $x_1 = -1$ the threedimensional parity association should dominate. Therefore,

$$y = T_S((2x_1 - 1)x_2x_3 + x_2 + x_3) \tag{3.16}$$

is a perfect choice for the transfer dynamic as can be easily checked. For comparison, let us also calculate the sigma-pi weights for the transfer dynamic by means of a standard Hebb-type approach. Using the learning rule (2.13) adapted to the heteroassociative single neuron case (cf. [7], pp. 280f., or [15], pp. 155f., for details),

$$w_R := \sum_{s=1}^{8} y^{(s)} \prod_{k\in R} x_k^{(s)} \,, \quad R \subset \{1,2,3\} \,, \tag{3.17}$$

we obtain the weights $w_{\{1,2,3\}} = 6$, $w_{\{1,2\}} = 2$, $w_{\{1,3\}} = 2$, $w_{\{2,3\}} = -2$, $w_{\{1\}} = -2$, $w_{\{2\}} = 2$, $w_{\{3\}} = 2$, and $w_{\{\}} = -2$. These weights imply the activation

$$y = T_S(6x_1x_2x_3 + 2x_1x_2 + 2x_1x_3 - 2x_2x_3 - 2x_1 + 2x_2 + 2x_3 - 2), \tag{3.18}$$

resp.

$$y = T_S(3x_1x_2x_3 + x_1x_2 + x_1x_3 - x_2x_3 - x_1 + x_2 + x_3 - 1), \tag{3.19}$$

which also solves the problem, however, using much more operations than solution (3.16). Even this simple example shows that in special situations with some a priori knowledge about the problem to be simulated it is possible to reduce the network complexity by essentially restricting to higher order correlations and allowing dilations and translations.

§4. Concluding Remarks

In this paper, we introduced dilation and translation parameters for sigma-pi neurons in order to increase the flexibility of the neurons while keeping the number of allowed multiplicative interferences of synaptic activations fixed. The approach especially makes sense in case that for theoretical or practical reasons the number of non-zero sigma-pi weights is limited and an incomplete situation appears. We explained the idea in the Hopfield context, but it immediately applies to all neural networks where incomplete sigma-pi units can be used. Especially, networks for invariant pattern recognition problems and highly non-linear optimization or separation problems are of interest and further studies in these directions – using learning schemes like those sketched in (2.14)–(2.16) – are initiated.

References

1. Baldi, P., Neural networks, acyclic orientations of the hypercube, and sets of orthogonal vectors, SIAM J. Disc. Math. **1** (1988), 1–13.
2. Baldi, P., Neural networks, orientations of the hypercube, and algebraic threshold functions, IEEE Transactions on Information Theory **34** (1988), 523–530.
3. Baldi, P. and S. S. Venkatesh, Number of stable points for spin-glasses and neural networks of higher orders, Physical Review Letters **58** (1987), 913–916.
4. Chen, H. H., Y. C. Lee, T. Maxwell, G. Z. Sun, H. Y. Lee and C. L. Giles, High order correlation model for associative memory, in *AIP Conference Proceedings*, J. S. Denker (ed.), American Institute of Physics, New York, 1986, 86–99.
5. Giles, C. L. and T. Maxwell, Learning, invariance, and generalization in high-order neural networks, Applied Optics **26** (1987), 4972–4978.
6. Giles, C. L., R. D. Griffin and T. Maxwell, Encoding geometric invariances in higher-order neural networks, in *Neural Information Processing Systems – Natural and Synthetic –*, D. Z. Anderson (ed.), American Institute of Physics, New York, 1988, 301–309.
7. Fahner, G. and R. Eckmiller, Structural adaptation of parsimonious higher-order neural classifiers, Neural Networks **7** (1994), 279–289.
8. Hopfield, J. J., Neural networks and physical systems with emergent collective computational abilities, Proceedings of the National Academy of Sciences (USA) **79** (1982), 2554–2558.
9. Hopfield, J. J., Neurons with graded response have collective computational properties like those of two-state neurons, Proceedings of the National Academy of Sciences (USA) **81** (1984), 3088–3092.
10. Hopfield, J. J. and D. W. Tank, Neural computation of decisions in optimization problems, Biological Cybernetics **52** (1985), 141–152.
11. Kamp, Y. and M. Hasler, *Recursive Neural Networks for Associative Memory*, John Wiley & Sons, Chichester, 1990.

12. Lenze, B., Constructive multivariate approximation with sigmoidal functions and applications to neural networks, in *Numerical Methods of Approximation Theory*, D. Braess and L. L. Schumaker (eds.), ISNM **105**, Birkhäuser-Verlag, Basel, 1992, 155–175.
13. Lenze, B., Note on a density question for neural networks, Numerical Functional Analysis and Optimization **15** (1994), 909–913.
14. Lenze, B., How to make sigma-pi neural networks perform perfectly on regular training sets, Neural Networks **7** (1994), 1285–1293.
15. Psaltis, D., C. H. Park and J. Hong, Higher order associative memories and their optical implementations, Neural Networks **1** (1988), 149–163.
16. Rumelhart, D. E. and J. L. McClelland, *Parallel Distributed Processing: Explorations in the Microstructure of Cognition*, vols. I,II, MIT-Press, Cambridge, Massachusetts, 1986.

Burkhard Lenze
Fachbereich Informatik
Fachhochschule Dortmund
Postfach 105018
D-44047 Dortmund
Germany
lenze@fh-dortmund.de

Parametric Shape-Preserving Hermite Interpolation by Piecewise Quadratics

C. Manni

Abstract. We present a local scheme for C^1 Hermite interpolation. The constructed interpolant is a parametric curve with piecewise quadratic components which maintains the monotonicity and/or the convexity of the data. The approximation order is also investigated: the method has "optimal" accuracy.

§1. Introduction

We deal with the well–known problem of the construction of a C^1 Hermite interpolant which maintains the monotonicity and/or the convexity of the data. Specifically we assume that a set of points $x_0 < x_1 < \ldots < x_n$ and the values of a function $f(x_i)$ are given. In addition, we assume that the derivatives $f'(x_i)$ at the data points are known (given or computed). We search for a function, $s(x) \in C^1[x_0, x_n]$, interpolating f and its first derivative at the data points, which has in $[x_i, x_{i+1}]$ the same shape as the data.

A great deal of research has been done in recent years in constructing shape preserving interpolants (see for example [1], [3], [4], [7], [8], [9], [15] and references quoted therein) stimulated by the need of efficient practical methods and by theoretical problems.

In the case of Hermite interpolation it is known that the preservation of monotonicity cannot be achieved if we assume that $s(x)$ is a polynomial with a preassigned degree in each subinterval $[x_i, x_{i+1}]$, [5]. This is also true if we choose $s(x)$ as a piecewise polynomial with one additional knot in each subinterval, [5]. Then, following [11], in this paper we situate the problem in the general setting of parametric curves which has received a great attention in recent years especially for the interesting results concerning the accuracy of non constrained parametric polynomial

Advanced Topics in Multivariate Approximation
F. Fontanella, K. Jetter and P. J. Laurent (eds.) pp. 211–226.

ISBN 981-02-2852-X

interpolation [2], [13], [14]. We will construct our interpolant by considering the graph of the function $x \to s(x)$ as the support of a bidimensional parametric curve $(x(t), y(t))$ where $x(t)$ is strictly increasing (see also [8], [10]).

In [11] a method to construct a comonotone interpolant $s(x(t))$, $x \in [x_i, x_{i+1}]$, as a single valued parametric curve with cubic components is presented. The comonotonicity of the curve is controlled by the amplitude of the tangent vector at the extremities of each subinterval. A local algorithm for the construction of the interpolant, which yields at least a fourth order approximation to a C^4 function with no vanishing derivative, is provided. The mentioned method provides good performances when only the graph of the interpolant is needed, but it entails solving an unpleasant third degree equation if the value $s(\bar{x})$, where $\bar{x}$ is fixed, has to be computed.

Thus, from this point of view, if the problem is kept in the general setting of parametric curves, it is better to consider simpler structures for the components of the interpolant, even though this simpler structure may influence the approximation order of the curve.

In this paper we investigate this aspect and we present a scheme based on piecewise quadratics with one additional knot in each subinterval $[x_i, x_{i+1}]$.

Once again, the shape of the curve is controlled by the amplitude of the tangent vector at the extremities of each subinterval. Due to the simple structure of the space chosen for the components, the mentioned amplitude can be easily computed using conditions which are necessary and sufficient to retain both the monotonicity and/or the convexity of the data.

We propose simple local algorithms for choosing the free parameters in such a way that the comonotonicity and/or the convexity of the data is preserved. In addition "local shape parameters " can be provided as inputs to the algorithms, making it possible for the user to obtain visually pleasant interpolants without changes on the approximation order (see figures in section 6).

The approximation order is investigated: in both situations the method is at least third order accurate for C^3 functions.

The proposed scheme is purely local, easy to implement and produce visually pleasant plots. More important, it allows both "optimal" accuracy and local control of the shape. Therefore the scheme appears to be an attractive alternative to the existing ones.

From the accuracy point of view we recall that, while the third order approximation is optimal for quadratics in a non parametric setting, it can be increased if parametric approximation is considered. It has been shown in [13] that a planar curve can be approximated with fourth order accuracy near a point with non vanishing curvature by parametric quadratics. However it should be noticed that the approximating scheme of [13] is based on the Taylor expansion of the curve, up to the third derivatives, with respect to a suitable reparameterization and no shape constraint is imposed.

It is not simple to compare the proposed scheme with the earlier local parametric curves methods as the one in [8], where the authors are more interested in the geometric continuity than in the approximation order, and where no error estimates are provided. A more detailed comparison can be made with the parametric cubic scheme proposed in [11]. As previously said the main advantage of using piecewise quadratic components relies in the computational cost. In fact only quadratic expressions are considered and only a second order equation has to be solved in order to compute $s(\overline{x})$, where $\overline{x}$ is fixed. Therefore, even if parametric curves are used, the computational cost of the scheme is comparable with that of the existent non parametric methods, especially the ones based on rationals. In addition in [11] the preservation of convexity has not been considered while it can be easily obtained with the method proposed in this paper.

The simple "quadratic" structure provides that $s(x)$ is easily invertible whenever it is monotone. Then the presented scheme can be effective even in the context of curves reparametrization, [6], where the inversion of the parameterization is sometimes required.

As a final remark we observe that the parametric techniques considered here and in [11] seem to provide an useful tool for shape preserving interpolation in the bivariate case. They are in fact used in [12] for comonotone interpolation over rectangular grids and an extension to non gridded data is under study.

The remainder of this paper is divided into five sections. Section 2 defines the problem and states the hypotheses. In section 3 we describe the general piecewise quadratic parametric interpolation scheme. In section 4 we study how to choose the free parameters to determine the comonotonicity of the interpolant and we investigate the approximation order of the constructed curve. This procedure is repeated in section 5 for the preservation of convexity. Finally, section 6 outlines some graphical examples.

For the sake of simplicity we assume that the values of the function and of its derivative at the given points are exactly prescribed. In practice the latter are often not available but there exists a wide class of methods to approximate them with a good accuracy. It is trivial to generalize the results on the approximation order of the proposed scheme to the general case where the data are approximated with a given error.

For the same reason, we consider in the following only increasing and/or convex data as the decreasing and/or concave cases are analogous.

§2. The Problem

Let (x_i, f_i, f_i'), $f_i = f(x_i)$, $f_i' = f'(x_i)$, $i = 0, \dots, n$, be given. We put

$$h_i = x_{i+1} - x_i, \quad \Delta_i = f_{i+1} - f_i, \quad i = 0, \dots, n-1, \tag{2.1}$$

and we assume $h_i > 0$.

The data are *monotone increasing* in $[x_i, x_{i+1}]$ if

$$\Delta_i > 0, \quad f_i',\ f_{i+1}' \geq 0, \qquad \text{or} \quad \Delta_i = 0, \quad f_i' = f_{i+1}' = 0. \tag{2.2}$$

The data are *convex* in $[x_i, x_{i+1}]$ if

$$f_i' < \frac{\Delta_i}{h_i} < f_{i+1}', \qquad \text{or} \quad f_i' = \frac{\Delta_i}{h_i} = f_{i+1}'. \tag{2.3}$$

We want to construct a C^1 function, $s(x)$, interpolating the data:

$$\begin{array}{l} s(x_i) = f_i, \\ s'(x_i) = f_i', \end{array} \quad i = 0, \ldots, n, \tag{2.4}$$

which is *comonotone* and/or *coconvex* with the data, i.e. $s'(x) \geq 0$, and/or $s'(x)$ is increasing in $[x_i, x_{i+1}]$ if the data are monotone increasing and/or convex in the interval.

§3. The General Construction of the Interpolant

For each subinterval $[x_i, x_{i+1}]$ we want to construct a solution, $s(x)$, to the problem stated in the previous section by considering the graph of the function $x \to s(x)$ as the support of a bidimensional parametric curve with components belonging to the four dimensional linear space, V, of C^1 quadratic splines in $[0, 1]$ with one additional interior knot at θ i.e.

$$V = \{q(t) \in C^1([0,1]) : q(t)|_{[0,\theta]} \in \mathbb{P}_2, q(t)|_{[\theta,1]} \in \mathbb{P}_2\},$$

where $\mathbb{P}_2$ denotes the space of polynomials of degree less than or equal to 2. For the sake of simplicity we choose $\theta = 0.5$.

For each couple $(h_i^{(0)},\ h_i^{(1)}) \in \mathbb{R}^2$ there exists unique a parametric curve

$$\begin{cases} x = X_i(t; h_i^{(0)}, h_i^{(1)}) \\ y = Y_i(t; h_i^{(0)}, h_i^{(1)}) \end{cases}, \quad i = 0, \ldots, n-1, \quad 0 \leq t \leq 1, \tag{3.1}$$

with

$$X_i(t; h_i^{(0)}, h_i^{(1)}), \ Y_i(t; h_i^{(0)}, h_i^{(1)}) \in V,$$

such that the components satisfy the following Hermite interpolation conditions:

$$\begin{aligned} X_i(0; h_i^{(0)}, h_i^{(1)}) = x_i, \quad & X_i(1; h_i^{(0)}, h_i^{(1)}) = x_{i+1}, \\ \frac{d}{dt} X_i(t; h_i^{(0)}, h_i^{(1)})|_{t=0} = h_i^{(0)}, \quad & \frac{d}{dt} X_i(t; h_i^{(0)}, h_i^{(1)})|_{t=1} = h_i^{(1)}, \\ Y_i(0; h_i^{(0)}, h_i^{(1)}) = f_i, \quad & Y_i(1; h_i^{(0)}, h_i^{(1)}) = f_{i+1}, \\ \frac{d}{dt} Y_i(t; h_i^{(0)}, h_i^{(1)})|_{t=0} = h_i^{(0)} f_i', \quad & \frac{d}{dt} Y_i(t; h_i^{(0)}, h_i^{(1)})|_{t=1} = h_i^{(1)} f_{i+1}'. \end{aligned} \tag{3.2}$$

To be more specific let us consider the elements of the cardinal basis for Hermite interpolation in this space, i.e.

$$Q_0^{(0)}(t) = \begin{cases} 1-2t^2 & \text{if } 0 \le t < \frac{1}{2} \\ 2(1-t)^2 & \text{if } \frac{1}{2} \le t \le 1 \end{cases}, \quad Q_1^{(0)}(t) = \begin{cases} 2t^2 & \text{if } 0 \le t < \frac{1}{2} \\ 1-2(1-t)^2 & \text{if } \frac{1}{2} \le t \le 1 \end{cases},$$

$$Q_0^{(1)}(t) = \begin{cases} t-\frac{3}{2}t^2 & \text{if } 0 \le t < \frac{1}{2} \\ \frac{(1-t)^2}{2} & \text{if } \frac{1}{2} \le t \le 1 \end{cases}, \quad Q_1^{(1)}(t) = \begin{cases} -\frac{1}{2}t^2 & \text{if } 0 \le t < \frac{1}{2} \\ \frac{(1-t)(1-3t)}{2} & \text{if } \frac{1}{2} \le t \le 1 \end{cases}, \tag{3.3}$$

we have

$$\begin{aligned} &Q_r^{(0)}(s) = \delta_{rs}, \quad Q_r^{(1)}(s) = 0, \\ &\frac{d}{dt}Q_r^{(0)}(t)|_{t=s} = 0, \quad \frac{d}{dt}Q_r^{(1)}(t)|_{t=s} = \delta_{rs}, \end{aligned} \qquad r,s = 0,1.$$

then

$$\begin{aligned} X_i(t;h_i^{(0)},h_i^{(1)}) &= x_i Q_0^{(0)}(t) + x_{i+1}Q_1^{(0)}(t) + h_i^{(0)}Q_0^{(1)}(t) + h_i^{(1)}Q_1^{(1)}(t), \\ Y_i(t;h_i^{(0)},h_i^{(1)}) &= f_i Q_0^{(0)}(t) + f_{i+1}Q_1^{(0)}(t) + h_i^{(0)}f_i'Q_0^{(1)}(t) + h_i^{(1)}f_{i+1}'Q_1^{(1)}(t), \\ &0 \le t \le 1, \end{aligned}$$

and, from (2.1), (3.3), we obtain

$$\begin{aligned} X_i(t;h_i^{(0)},h_i^{(1)}) &= x_i + h_i Q_1^{(0)}(t) + h_i^{(0)}Q_0^{(1)}(t) + h_i^{(1)}Q_1^{(1)}(t), \\ Y_i(t;h_i^{(0)},h_i^{(1)}) &= f_i + \Delta_i Q_1^{(0)}(t) + h_i^{(0)}f_i'Q_0^{(1)}(t) + h_i^{(1)}f_{i+1}'Q_1^{(1)}(t), \end{aligned} \qquad 0 \le t \le 1.$$

For the sake of simplicity we will simply write $X_i(t)$, $Y_i(t)$ instead of $X_i(t;h_i^{(0)},h_i^{(1)})$, $Y_i(t;h_i^{(0)},h_i^{(1)})$ whenever is not crucial to know the specific values of the parameters $h_i^{(0)}$, $h_i^{(1)}$.

If it is possible to determine $h_i^{(0)}$, $h_i^{(1)}$, $i = 0,\ldots,n-1$, such that

$$\frac{d}{dt}X_i(t;h_i^{(0)},h_i^{(1)}) > 0, \tag{3.4}$$

the first component in (3.1) implicitly defines a function $t = t(x)$ and $Y_i(t;h_i^{(0)},h_i^{(1)})$ can be expressed as a function of x. In this case, putting

$$s(x) = Y_i(t(x);h_i^{(0)},h_i^{(1)}), \quad x_i \le x \le x_{i+1}, \; i = 0,\ldots,n-1, \tag{3.5}$$

since $\frac{dt}{dx} = \left(\frac{dX_i}{dt}\right)^{-1}$, from (3.1), (3.2), we have

$$\begin{aligned} s(x_i) &= f_i, \\ s'(x_i^+) &= \frac{d}{dt}Y_i(t;h_i^{(0)},h_i^{(1)})|_{t=0}\frac{dt}{dx}|_{x=x_i^+} = h_i^{(0)}f_i'\frac{1}{h_i^{(0)}} = f_i', \\ s'(x_i^-) &= \frac{d}{dt}Y_{i-1}(t;h_{i-1}^{(0)},h_{i-1}^{(1)})|_{t=1}\frac{dt}{dx}|_{x=x_i^-} = h_{i-1}^{(1)}f_i'\frac{1}{h_{i-1}^{(1)}} = f_i'. \end{aligned}$$

Thus (3.5) defines a function of class $C^1[x_0, x_n]$ which satisfies the interpolation conditions (2.4).

Since $\frac{dX_i}{dt}$ is a continuous piecewise linear function, from a simple manipulation we obtain that (3.4) holds if and only if

$$0 < h_i^{(0)},\ h_i^{(1)},\quad 0 < h_i^{(0)} + h_i^{(1)} < 4h_i. \tag{3.6}$$

Then whenever the parameters $h_i^{(0)}$, $h_i^{(1)}$ satisfy (3.6) it is possible to define $s(x)$ via (3.5) and $s(x)$ interpolates the data.

The parameters $h_i^{(0)}$, $h_i^{(1)}$ do not affect the interpolation conditions (2.4) for $s(x)$ but they determine the amplitude of the tangent vectors of the curve at the extremities of the interval (see (3.2)) and act as tension parameters. In fact setting $h_i^{(0)} = h_i^{(1)} = h_i$, we obtain

$$X_i(t; h_i, h_i) = x_i + h_i t, \tag{3.7}$$

thus $Y_i(t; h_i, h_i) = Y_i(t(x); h_i, h_i)$, $0 \le t \le 1$, is the piecewise quadratic spline, with a breakpoint at $x = x_i + \frac{h_i}{2}$, that interpolates the data (x_i, f_i, f_i'), $(x_{i+1}, f_{i+1}, f_{i+1}')$. On the contrary, considering the limit case $h_i^{(0)} = h_i^{(1)} = 0$, the calculated $(X_i(t; 0, 0),\ Y_i(t; 0, 0))$ is the straight line that interpolates (x_i, f_i), (x_{i+1}, f_{i+1}), in fact

$$Y_i(t; 0, 0) = Y_i(t(x); 0, 0) = f_i + \frac{\Delta_i}{h_i}(x - x_i), \quad x_i \le x \le x_{i+1}. \tag{3.8}$$

It turns out that the (tension) parameters $h_i^{(0)}$, $h_i^{(1)}$ can be chosen to control the shape of the curve as we do in the following sections. In the following, from (3.6), only positive values of the tension parameters will be considered.

§4. Preserving Monotonicity

Let $s(x)$ be defined by (3.5). Since

$$\frac{d}{dx} s(x) = \frac{d}{dt} Y_i(t) \left(\frac{d}{dt} X_i(t) \right)^{-1} \tag{4.1}$$

from (3.4) $s(x)$ is increasing in $[x_i, x_{i+1}]$ if and only if $\frac{d}{dt} Y_i(t; h_i^{(0)}, h_i^{(1)}) \ge 0, t \in [0, 1]$, that is if and only if

$$h_i^{(j)} f_{i+j}' \ge 0,\ j = 0, 1,\quad h_i^{(0)} f_i' + h_i^{(1)} f_{i+1}' \le 4\Delta_i. \tag{4.2}$$

It is then possible to choose an infinite number of couples $h_i^{(0)}$, $h_i^{(1)} > 0$, such that (3.6) and (4.2) are satisfied if the data satisfy (2.2). The choice of different parameters determines not only the shape of the interpolant but also its approximation order since $s(x)$ moves from the straight line (3.8) to the piecewise quadratic spline interpolating the data as $h_i^{(j)}$, $j = 0, 1$ increase from 0 to h_i. Our goal is to choose the parameters $h_i^{(j)}$ to combine good graphical performances and accuracy.

To determine the mentioned values of the tension parameters we prefer to write a simple algorithm involving a geometric interpretation of (4.2) (see Fig. 1) instead of the simple final expressions of the parameters.

Later we will investigate the approximation order of the constructed interpolant.

Algorithm 1.

0. *Given* $(f_i,\ f_i',\ i = 0, \dots, n)$ $(h_i, i = 0, \dots, n-1)$, $\frac{1}{2} \le \tau_i \le 1$, $i = 0, \dots, n-1$.
1. *Set* $h_i^{(0)} = h_i^{(1)} = h_i$, $\tau = \tau_i$.
2. *Set* $(a, b) = \left(h_i f_i', h_i f_{i+1}'\right)$, $\Delta = \Delta_i$.
3. *If* $a + b > 4\Delta\tau$:
    1. *Compute the intersection* (α, β) *of the straight lines* $ay - bx = 0$ *and* $x + y = 4\Delta\tau$.
    2. *If* $f_i' \ne 0$, *compute* $h_i^{(0)}$ *such that* $\alpha = h_i^{(0)} f_i'$.
    3. *If* $f_{i+1}' \ne 0$, *compute* $h_i^{(1)}$ *such that* $\beta = h_i^{(1)} f_{i+1}'$.
    4. *If* $f_i' = 0$, *set* $h_i^{(0)} = h_i^{(1)}$.
    5. *If* $f_{i+1}' = 0$, *set* $h_i^{(1)} = h_i^{(0)}$.

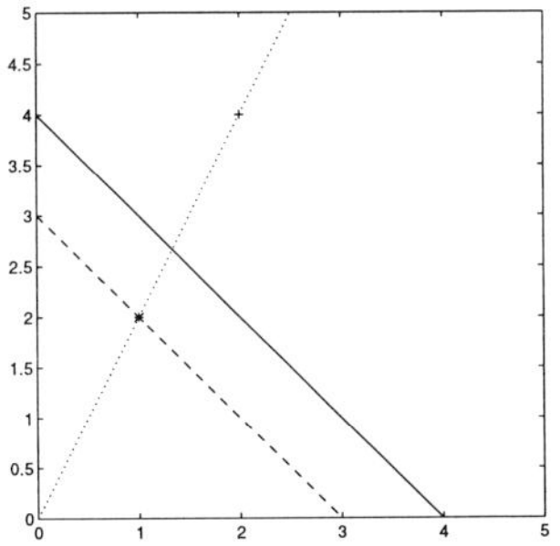

Fig. 1. Algorithm 1 with $\tau_i = \frac{3}{4}$, $(+) = (a, b)$, $(*) = (\alpha, \beta)$, $\Delta = 1$.

The following lemma holds.

Lemma 1. *Let* $h_i^{(0)}$, $h_i^{(1)}$, *be obtained from algorithm 1, then* $h_i^{(1)} = h_i^{(0)} \le h_i$ *so that (3.6) are satisfied.* ■

Remark 1. *The method is local, since a change in the data at* x_i *is only reflected in* $[x_{i-1}, x_{i+1}]$.

Remark 2. *The parameters τ_i in algorithm 1 act as local shape parameters. If $\tau_i = 1$ algorithm 1 determines parameters $h_i^{(0)} = h_i^{(1)}$ such that the y–component of the curve (3.1) has zero derivative at $t = 0.5$ (see (4.2)). The plot then shows a tangent parallel to the x-axis at $t = 0.5$ (i.e. at $x = x_i + \frac{h_i}{2}$). To avoid this we can consider $\tau_i \in [\frac{1}{2}, 1)$ so that the first derivative of the y–component does not vanishes in $[0,1]$ and, at the same time, the approximation order of the constructed interpolant is not reduced (see lemma 2).*

Using Taylor expansion, it is simple to show that the proposed scheme provides at least a second order approximation to a C^2 function if $h_i^{(0)}$, $h_i^{(1)}$ satisfy (3.6). Now we detail the case of C^3 functions.

Putting, as usual, $\|g\|_{[a,b]} = \max_{x\in[a,b]} |g(x)|$, $g \in C[a,b]$, the following theorem holds:

Theorem 1. *Assume f is a monotone increasing function in $C^3[x_i, x_{i+1}]$, and let $h_i^{(0)} = h_i^{(1)}$, $0 \le i \le n-1$, be obtained from algorithm 1. Then $s(x)$ constructed via (3.5) is such that*

$$\|s - f\|_{[x_i,x_{i+1}]} \le M_0 h_i^3. \tag{4.3}$$

Moreover, as $h_i \to 0$, if $|f'(x)| \ge m_i > 0$, $x \in [x_i, x_{i+1}]$,

$$\|s' - f'\|_{[x_i,x_{i+1}]} \le M_1 h_i^2, \tag{4.4}$$

where M_0, M_1 are constants not depending on h_i.

In order to prove theorem 1 we need the following lemma:

Lemma 2. *Assume f is a monotone increasing function in $C^3[x_i, x_{i+1}]$. If*

$$0 < h_i^{(0)}, h_i^{(1)} \le h_i, \tag{4.5}$$

and

$$h_i^{(0)} f_i' + h_i^{(1)} f_{i+1}' \ge 2\Delta_i, \tag{4.6}$$

then

$$0 \le (h_i - h_i^{(j)}) f_{i+j}' \le \frac{1}{6}\|f'''\| h_i^3, \; j = 0, 1. \tag{4.7}$$

Proof: Considering the divided differences of f, for some $\nu \in [x_i, x_{i+1}]$ we have:

$$2\Delta_i = h_i f_i' + h_i f_{i+1}' - h_i^3[x_i, x_i, x_{i+1}, x_{i+1}]f = h_i f_i' + h_i f_{i+1}' - \frac{h_i^3}{6} f'''(\nu),$$

then from (4.6)

$$0 \le h_i^{(0)} f_i' + h_i^{(1)} f_{i+1}' - 2\Delta_i = (h_i^{(0)} - h_i) f_i' + (h_i^{(1)} - h_i) f_{i+1}' + \frac{h_i^3}{6} f'''(\nu),$$

so that from (4.5)

$$0 \le (h_i - h_i^{(j)}) f'_{i+j} \le \frac{h_i^3}{6} f'''(\nu), \quad j = 0, 1. \quad \blacksquare$$

Proof of theorem 1.

During the proof D will denote a constant not depending on h_i.

Concerning (4.3), let $x \in [x_i, x_{i+1}]$ be fixed, then there exist t, $\hat{t} \in [0, 1]$, such that

$$x = X_i(t; h_i^{(0)}, h_i^{(1)}) = X_i(\hat{t}; h_i, h_i), \tag{4.8}$$

hence

$$|f(x) - s(x)| = |f(x) - Y_i(t; h_i^{(0)}, h_i^{(1)})| \le |f(x) - Y_i(\hat{t}; h_i, h_i)| +$$

$$|Y_i(\hat{t}; h_i, h_i) - Y_i(t; h_i, h_i)| + |Y_i(t; h_i, h_i) - Y_i(t; h_i^{(0)}, h_i^{(1)})| =: A + B + C.$$

Since $Y_i(\hat{t}(x); h_i, h_i)$ is the piecewise quadratic polynomial with a breakpoint at $x = x_i + \frac{h_i}{2}$ that interpolates the data, we obtain

$$A \le D h_i^3.$$

Let us now bound B. We have:

$$\frac{d}{dt} Y_i(t; h_i, h_i) = \begin{cases} 4t\Delta_i + (1 - 3t) h_i f'_i - t h_i f'_{i+1}, & 0 \le t < \frac{1}{2} \\ 4(1 - t)\Delta_i - (1 - t) h_i f'_i - (2 - 3t) h_i f'_{i+1}, & \frac{1}{2} \le t < 1 \end{cases}, \tag{4.9}$$

Since from (4.8) and (3.7)

$$h_i(\hat{t} - t) = \begin{cases} (h_i^{(0)} - h_i)(t - \frac{3}{2}t^2) - (h_i^{(1)} - h_i)(\frac{1}{2}t^2), & \text{if } 0 \le t < \frac{1}{2} \\ (h_i^{(0)} - h_i)\frac{(1-t)^2}{2} - (h_i^{(1)} - h_i)\frac{(1-3t)(1-t)}{2}, & \text{if } \frac{1}{2} \le t \le 1 \end{cases} \tag{4.10}$$

and (see proof of lemma 2)

$$\Delta_i = \frac{h_i}{2}(f'_i + f'_{i+1}) - \frac{h_i^3}{12} f'''(\nu), \ \nu \in [x_i, x_{i+1}],$$

then from (4.10)

$$|\Delta_i(\hat{t} - t)| \le \frac{h_i^3}{12} \|f'''\| \frac{1}{4} + |\frac{h_i}{2}(f'_i + f'_{i+1})(\hat{t} - t)|, \tag{4.11}$$

while from lemma 2, since $h_i^{(0)} = h_i^{(1)}$,

$$|h_i f'_i(\hat{t} - t)|, \ |h_i f'_{i+1}(\hat{t} - t)| \le \|f'''\| \frac{h_i^3}{6} \frac{1}{4}. \tag{4.12}$$

Thus from the mean value theorem, (4.9), (4.11) and (4.12)

$$B \le Dh_i^3.$$

Finally:

$$C = \begin{cases} \left|(h_i - h_i^{(0)})f_i'(t - \frac{3}{2}t^2) - (h_i - h_i^{(1)})f_{i+1}'(\frac{1}{2}t^2)\right|, & \text{if } 0 \le t < \frac{1}{2} \\ \left|(h_i - h_i^{(0)})f_i'\frac{(1-t)^2}{2} - (h_i - h_i^{(1)})f_{i+1}'\frac{(1-3t)(1-t)}{2}\right|, & \text{if } \frac{1}{2} \le t \le 1 \end{cases},$$

and (4.3) follows since from lemma 2

$$C \le Dh_i^3.$$

Concerning the first derivatives let us consider again (4.8), then

$$|f'(x) - s'(x)| \le \left|f'(x) - \frac{d}{dx}Y_i(\hat{t}(x); h_i, h_i)\right| + \left|\frac{d}{dx}Y_i(\hat{t}(x); h_i, h_i) - \frac{d}{dx}Y_i(t(x); h_i, h_i)\right|$$

$$+ \left|\frac{d}{dx}Y_i(t(x); h_i, h_i) - \frac{d}{dx}Y_i(t(x); h_i^{(0)}, h_i^{(1)})\right| =: A_1 + B_1 + C_1.$$

To be specific, let $x \in [x_i, x_i + \frac{h_i}{2})$; then from (3.4), t, $\hat{t} \in [0, 0.5)$ since $X_i(0.5; \delta, \delta) = x_i + \frac{h_i}{2}, \forall \delta$. Thus A_1, B_1 can be bounded similarly to A, B. So let us consider C_1. A straightforward calculation shows that

$$C_1 = \left|\left[(4h_i t + h_i^{(0)}(1 - 3t) - h_i^{(1)}t)h_i\right]^{-1}\left[4t(1 - 3t)(h_i^{(0)} - h_i)(\Delta_i - h_i f_i')\right.\right.$$

$$\left.\left.-t(1 - 3t)(h_i^{(1)} - h_i^{(0)})(h_i f_i' - h_i f_{i+1}') + 4t^2(h_i^{(1)} - h_i)(\Delta_i - h_i f_{i+1}')\right]\right|,$$

hence the thesis follows since from (4.7) and the hypotheses on the first derivative of f we have

$$0 \le h_i - h_i^{(j)} \le Dh_i^3, \quad j = 0, 1,$$

and

$$\frac{dX_i(t; h_i^{(0)}, h_i^{(1)})}{dt} = (4h_i t + h_i^{(0)}(1 - 3t) - h_i^{(1)}t) \ge h_i - Dh_i^3.$$

In a similar way we conclude when $x \in [x_i + \frac{h_i}{2}, x_{i+1}]$. ■

§5. Preserving Convexity

Let $s(x)$ be defined by (3.5) and assume the data satisfy (2.3). We want to show that it is possible to choose the parameters $h_i^{(0)}$, $h_i^{(1)} > 0$, such that $s'(x)$ is increasing, i.e. $s(x)$ is convex in $[x_i, x_{i+1}]$.

Let us put $\xi_i = X_i(0.5; h_i^{(0)}, h_i^{(1)})$: since $s'(x)$ is an absolutely continuous function in $[x_i, x_{i+1}]$, (see (4.1)) we can describe its behaviour via $s''(x), x \neq \xi_i$. We have

$$\frac{d^2}{dx^2}s(x) = \left(\frac{d^2Y_i}{dt^2}\frac{dX_i}{dt} - \frac{d^2X_i}{dt^2}\frac{dY_i}{dt}\right)\left(\frac{dX_i}{dt}\right)^{-3}, \quad x_i \leq x \leq x_{i+1}, \; x \neq \xi_i,$$

then, from (3.4), the convexity of $s(x)$ in $[x_i, x_{i+1}]$ is determined by the positivity of

$$\kappa_i(t; h_i^{(0)}, h_i^{(1)}) := \frac{d^2Y_i}{dt^2}\frac{dX_i}{dt} - \frac{d^2X_i}{dt^2}\frac{dY_i}{dt}, \quad t \in [0,1], \; t \neq 0.5.$$

Since V is the space of C^1 quadratic splines, $\kappa_i(t; h_i^{(0)}, h_i^{(1)})$ is simply a piecewise linear function in $[0,1]$ with a discontinuity at $t = 0.5$, so that it suffices to examine its sign at $t = 0$, $t = 1$ and the limits for $t \to 0.5^-, 0.5^+$. From (3.3) we obtain:

$$\begin{aligned}\kappa_i(0; h_i^{(0)}, h_i^{(1)}) =& (4\Delta_i - 3f_i'h_i^{(0)} - f_{i+1}'h_i^{(1)})h_i^{(0)} - (4h_i - 3h_i^{(0)} - h_i^{(1)})f_i'h_i^{(0)} \\ =& h_i^{(0)}[4\Delta_i - 4h_if_i' + h_i^{(1)}(f_i' - f_{i+1}')] = \kappa_i(0.5^-; h_i^{(0)}, h_i^{(1)}),\end{aligned}$$

and in a similar way:

$$\kappa_i(0.5^+; h_i^{(0)}, h_i^{(1)}) = \kappa_i(1; h_i^{(0)}, h_i^{(1)}) = h_i^{(1)}[4h_if_{i+1}' - 4\Delta_i + h_i^{(0)}(f_i' - f_{i+1}')].$$

Thus $s(x)$ defined by (3.5) is convex in $[x_i, x_{i+1}]$, if and only if

$$4\Delta_i - 4h_if_i' + h_i^{(1)}(f_i' - f_{i+1}') \geq 0, \; 4h_if_{i+1}' - 4\Delta_i + h_i^{(0)}(f_i' - f_{i+1}') \geq 0. \quad (5.1)$$

If the data satisfy (2.3) it is then possible to choose an infinite number of couples $h_i^{(0)}$, $h_i^{(1)} > 0$, such that (3.6) and (5.1) are satisfied. As shown in the previous section different values of the parameters determine the shape of the interpolant and its approximation order. Once again we will try to choose $h_i^{(j)}$, $j = 0, 1$ not too far from h_i to retain the approximation order of the piecewise quadratic spline interpolating the data.

Since algorithm 1, that we have considered for preserving monotonicity, produces equal values for both tension parameters related to $[x_i, x_{i+1}]$, in the following we will consider $h_i^{(0)} = h_i^{(1)}$.

As in the previous section, we prefer to write a simple algorithm to choose $h_i^{(0)}$ by using a geometric interpretation of (5.1) (see Fig. 2), instead of the simple final expression for the parameter.

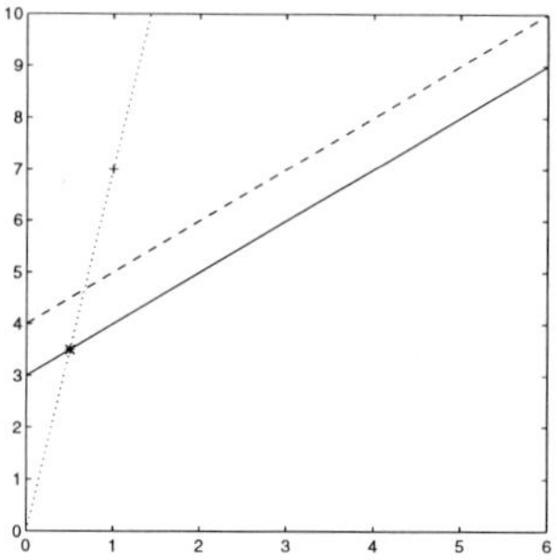

Fig. 2. Algorithm 2 with $\tau_i = \frac{3}{4}$, $(+) = (a, b)$, $(*) = (\alpha, \beta)$, $K = 1$.

Algorithm 2.

0. *Given* $(f_i,\ f'_i,\ i = 0, \ldots, n)$ $(h_i, i = 0, \ldots, n-1)$, $\frac{1}{2} \le \tau_i \le 1$, $i = 0, \ldots, n-1$.
1. *Set* $h_i^{(0)} = h_i^{(1)} = h_i,\ \tau = \tau_i$.
2. *Set* $(a, b) = \left(h_i f'_i, h_i f'_{i+1}\right)$, $K = min((f_{i+1} - f_i) - h_i f'_i; h_i f'_{i+1} - (f_{i+1} - f_i))$.
4. *If* $b - a > 4K\tau$
   1. *Compute the intersection* (α, β) *of the straight lines* $ay - bx = 0$ *and* $y - x = 4K\tau$.
   2. *Compute* $h_i^{(0)}$ *such that* $(\alpha, \beta) = \left(h_i^{(0)} f'_i, h_i^{(0)} f'_{i+1}\right)$.

We observe that remark 1 and lemma 1 apply even for algorithm 2 and the parameters τ_i play a similar role as in algorithm 1.

Remark 3. *If the data are neither convex or concave, that is for example* $f'_i < \frac{\Delta_i}{h_i}$, $\frac{\Delta_i}{h_i} > f_{i+1}$, *it is simple to see that* $\kappa_i(0; h_i^{(0)}, h_i^{(1)}) > 0$, $\kappa_i(1; h_i^{(0)}, h_i^{(1)}) < 0$ *for all* $h_i^{(0)}, h_i^{(1)} \le h_i$. *Hence* $s(x)$ *has exactly an inflection point at* $x = \xi_i$.

Let us now study the approximation order of the proposed method. The following theorem holds:

Theorem 2. *Let* f *be a convex function in* $C^3[x_i, x_{i+1}]$, *and let* $h_i^{(0)}$, $0 \le i \le n-1$, *be obtained from algorithm 2. Then* $s(x)$ *constructed via (3.5) is such that*

$$\|s - f\|_{[x_i, x_{i+1}]} \le K_0 h_i^3. \tag{5.2}$$

Moreover, as $h_i \to 0$, *if* $|f''(x)| \ge d_i > 0$, $x \in [x_i, x_{i+1}]$,

$$\|s' - f'\|_{[x_i, x_{i+1}]} \le K_1 h_i^2, \tag{5.3}$$

where K_0, K_1 *are constants not depending on* h_i.

In order to prove theorem 2 we need the following lemma. The proof is analogous to lemma 2 so that we omit the details.

In the following, F will denote a constant not depending on h_i.

Lemma 3. *Let f be a convex function in $C^3[x_i, x_{i+1}]$. Let K be as in algorithm 2. If*

$$0 < h_i^{(0)} \le h_i,$$

and

$$h_i^{(0)}(f'_{i+1} - f'_i) \ge 2K$$

then

$$0 \le (h_i - h_i^{(0)})h_i f''(x) \le F h_i^3, \; x \in [x_i, x_{i+1}]. \tag{5.4}$$

Proof of theorem 2.

Let $\hat{x} \in [x_i, x_{i+1}]$ be fixed; then from (3.4) there exists unique $\hat{t} \in [0,1]$ such that $X_i(\hat{t}; h_i^{(0)}, h_i^{(0)}) = \hat{x}$. To be specific let us assume $\hat{t} \in [0, 0.5)$, then from the interpolation conditions at $t = 0$ and the mean value theorem there exists $\theta \in (0, 0.5)$ such that

$$|f(\hat{x}) - s(\hat{x})| = \left| f(X_i(\hat{t}; h_i^{(0)}, h_i^{(0)})) - Y_i(\hat{t}; h_i^{(0)}, h_i^{(0)}) \right| \le \frac{1}{4} \left| \frac{d^2}{dt^2} [f(X_i(t)) - Y_i(t)]_{|t=\theta} \right|$$

From (3.3), if $t \in (0, 0.5)$, we have:

$$\begin{aligned} \left| \frac{d^2}{dt^2}[f(X_i(t)) - Y_i(t)] \right| &= \left| \frac{d^2 f}{dx^2} \left(\frac{dX_i}{dt} \right)^2 + \frac{df}{dx}\frac{d^2 X_i}{dt^2} - \frac{d^2 Y_i}{dt^2} \right| \\ &\quad + \Big| f''(X_i(t))[4h_i t + h_i^{(0)} - 4t h_i^{(0)}]^2 \\ &\quad + f'(X_i(t))4(h_i - h_i^{(0)}) - [4\Delta_i - 3h_i^{(0)} f'_i - h_i^{(0)} f'_{i+1}] \Big|, \end{aligned}$$

then, from Taylor expansion, since $0 < \theta \le \hat{t} < 0.5$,

$$\begin{aligned} |f(\hat{x}) - s(\hat{x})| \le f''(x_i) \Big[&(h_i - h_i^{(0)})(h_i^{(0)} + 2h_i) + h_i^{(0)}(h_i - h_i^{(0)}) \\ &+ (h_i^{(0)} - h_i)^2 + h_i(h_i - h_i^{(0)}) \Big] + F h_i^3, \end{aligned}$$

hence (5.2) follows from lemma 3.

Concerning (5.3) we just observe that from (4.1) there exists $\eta \in (0, 0.5)$, such that

$$\left| \frac{d}{dx}(f(x) - s(x))_{|x=\hat{x}} \right| \le \frac{1}{2} \left[\frac{d}{dt} X_i(t)_{|t=\hat{t}} \right]^{-1} \left| \frac{d^2}{dt^2}(f(X_i(t)) - Y_i(t))_{|t=\eta} \right|,$$

and that, from (5.4) and the hypotheses on f'',

$$\frac{d}{dt} X_i(t) \ge h_i - F h_i^2.$$

In a similar way we conclude if $\hat{t} \in [0.5, 1]$. ■

§6. Graphical Examples

In this section we present the performance of the proposed schemes in few classical tests.

In the first example we consider the same data as in [4] where the derivatives at each point are computed according to the Bessel scheme:

x_i :	0	2	3	5	6	8	9	11	12	14	15
f_i :	10	10	10	10	10	10	10.5	15	50	60	85 .
f_i' :	0	0	0	0	0	0	1.083	24.083	25	18.333	25

Figure 3 shows the parametric piecewise quadratic interpolants obtained from algorithm 1 for different values of the parameters τ_i. Other fittings for this test can be found in [3], [4], [10], [11].

To show the performances of the two algorithms in the second example we consider the data obtained from the function $f(x) = \sqrt{|x|}$ at the abscissas $x_0 = -0.5,\ x_1 = -0.01,\ x_2 = 0.01\ x_3 = 1$. Figure 4 shows the given function and the interpolants constructed using algorithm 1 or both algorithms 1 and 2 for $\tau_i = 0.8,\ 1,\ 0.8$.

Finally, in our last example, we consider the data coming from the following function, [7],

$$f(x) = \frac{1 + (x+\alpha)^{\frac{1}{3}}}{2 - (x+\alpha)^{\frac{1}{3}}}, \quad \alpha = 10^{-4},$$

at the abscissas:

$$x_i: \ -1 \quad -0.8 \quad -0.6 \quad -0.4 \quad -0.2 \quad -0.04 \quad 0.04 \quad 0.2 \quad 0.4 \quad 0.6 \quad 0.8 \quad 1\,.$$

Figure 5 shows the given function and the constructed interpolant obtained using algorithms 1 and 2 with $\tau_i = 0.6,\ i = 0, \ldots, 10$. Other fittings for this test can be found in [7].

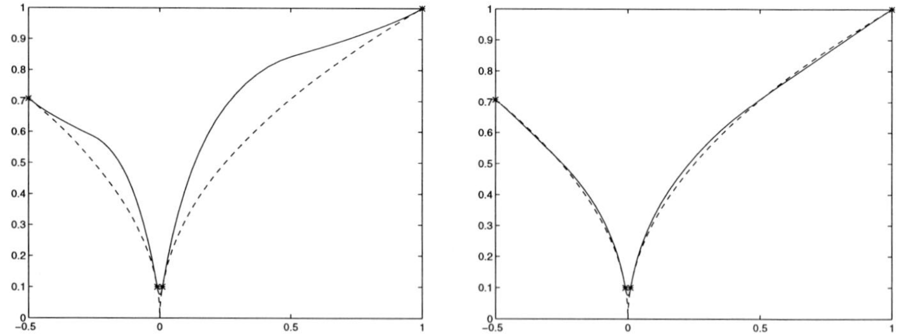

Fig. 4. Example 2. $\tau_i = 0.8, 1, 0.8$.
Left: $' - -'$ the given function, $'-'$ the interpolant obtained from algorithm 1.
Right: $' - -'$ the given function, $'-'$ the interpolant obtained using algorithms 1 and 2.

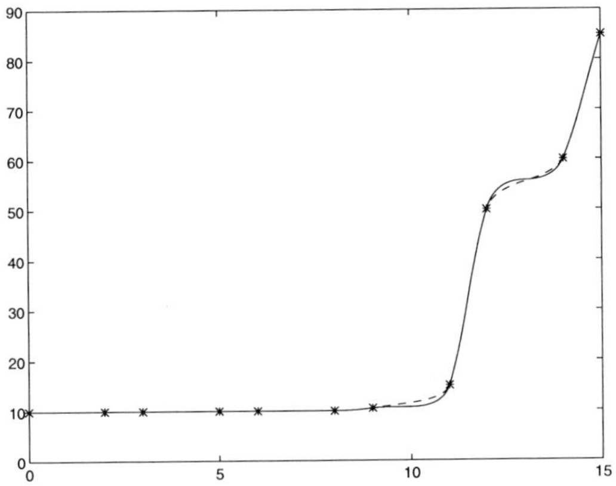

Fig. 3. Example 1. The interpolant obtained from algorithm 1. $'-'$: $\tau_i = 1,\ i = 0,\dots,9,\ '--'$: $\tau_i = 0.6,\ i = 0,\dots,9.$

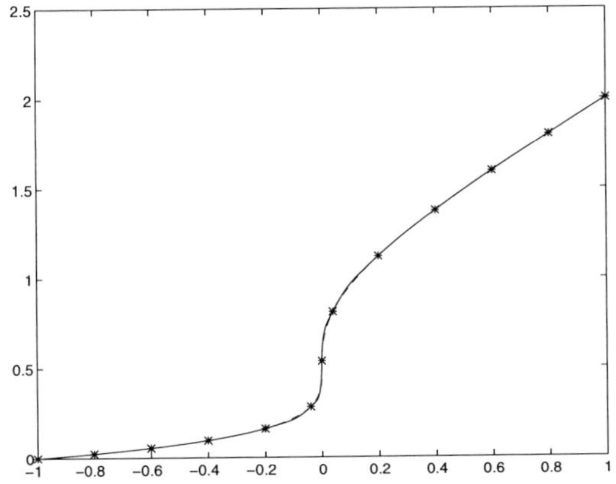

Fig. 5. Example 3. $'--'$ the given function, $'-'$ the interpolant; $\tau_i = 0.6$.

Acknowledgements. The author wishes to thank Prof. P. Sablonnière for helpful remarks.

References

1. Archer, J. C. and E. Le Gruyer, Two shape preserving Lagrange C^2– interpolants, Numer. Math. **64** (1993), 1–11.
2. de Boor, C., K. Höllig and M. Sabin, High accuracy geometric Hermite interpolation, Computer-Aided Geom. Design **4** (1987), 269–278.

3. Carlson, R. E. and F. N. Fritsch, Monotone piecewise cubic interpolation, SIAM J. Numer. Anal. **17** (1980), 238–246.
4. Costantini, P., An algorithm for computing shape–preserving interpolating splines of arbitrary degree, J. Comput. Appl. Math. **22** (1988), 89-136.
5. Edelman, A. and C. A. Micchelli, Admissible slopes for monotone and convex interpolation, Numer. Math. **51** (1987), 441–458.
6. Fuhr, R. D. and M. Kallay, Monotone linear rational spline interpolation, Computer-Aided Geom. Design **9** (1992), 313–319.
7. Gasparo, M. G. and R. Morandi, Piecewise cubic monotone interpolation with assigned slopes, Computing **46** (1991), 355–365.
8. Goodman, T. N. T. and K. Unsworth, Shape preserving interpolation by curvature continuous parametric curves, Computer-Aided Geom. Design **5** (1988), 341–355.
9. Gregory, J. A. and R. Delbourgo, Piecewise rational quadratic interpolation to monotonic data, I.M.A. J. Num. Anal. **2** (1982), 123–130.
10. Lee, E. T. Y., Choosing nodes in parametric curve interpolation, Comput. Aided Design **21** (1989), 363–370.
11. Manni, C., C^1 comonotone Hermite interpolation via parametric cubics, to appear in J. Comput. Appl. Math. (1995).
12. Manni, C. and P. Sablonnière, C^1 comonotone Hermite interpolation via parametric surfaces, in *Mathematical Methods in Computer Aided Geometric Design III*, M. Dæhlen, T. Lyche and L. L. Schumaker (eds.), Vanderbilt University Press, Nashville, 1995, 333–342.
13. Mørken, K. and K. Scherer, A general framework for high accuracy parametric interpolation, preprint 1994–7, University of Oslo, 1994.
14. Rababah, A., High order approximation method for curves, Computer-Aided Geom. Design **12** (1995), 89–102.
15. Schmidt, J.W. and W. Heß, An always successful method in univariate convex C^2 interpolation, Numer. Math. **71** (1995), 237–252.

Carla Manni
Dip. di Energetica
Universitá di Firenze
Via C. Lombroso 6/17
50134 Firenze, Italy
manni@ingfi1.ing.unifi.it

Marsden Identity, Blossoming and de Boor-Fix Formula

Marie-Laurence Mazure and Pierre-Jean Laurent

Abstract. For a finite dimensional space of functions (with non vanishing Wronskian) and its dual space, there is a unique canonical bilinear form and a unique reproducing function. These notions provide Marsden type identities and a simple characterization of a contact of order k between two functions. For Chebyshev spaces, this leads to a de Boor-Fix formula for defining the blossom and to a de Casteljau algorithm for computing the function.

§1. Introduction

In 1970, M.J. Marsden [10] proved the following identity:

$$E(x,y) := \frac{(x-y)^n}{n!} \equiv \sum_i N_i(x)\ D_i(y)\ , \tag{1.1}$$

where N_i denotes the B-spline of degree n whose support is $[t_i, t_{i+n+1}]$ and $D_i(y) := (t_{i+1}-y)\dots(t_{i+n}-y)/n!$. More recently, R. Goldman [8] investigated the relations between this type of identity for polynomials and blossoming : given a basis $(B_0,\dots,B_n)$ of the polynomial space $\mathcal{P}_n$, the polynomials $(B_0^*,\dots,B_n^*)$ such that

$$E(x,y) := \frac{(x-y)^n}{n!} \equiv \sum_{i=0}^{n} B_i(x)\ B_i^*(y)\ . \tag{1.2}$$

are called in [8] *the dual basis* of $(B_0,\dots,B_n)$. For instance, if the B_i's are the Bernstein polynomials, the dual basis is defined by $B_i^*(y) = (0-y)^{n-i}(1-y)^i/n!$, $i = 0,\dots,n$.

Advanced Topics in Multivariate Approximation
F. Fontanella, K. Jetter and P. J. Laurent (eds.) pp. 227–242.

ISBN 981-02-2852-X

In another famous paper, C. de Boor and G. Fix [6] developed in 1973 a formula for obtaining the coefficients of the development of a spline in a B-spline basis. If $S(x) = \sum_i \lambda_i N_i(x)$ is a spline of degree n, the coefficients λ_i are given by

$$\lambda_i = \sum_{j=0}^{n} S^{(j)}(a)(-1)^{n-j} D_i^{(n-j)}(a) \ , \tag{1.3}$$

where a is an arbitrary point in $]t_i, t_{i+n+1}[$. The results given by (1.1) and (1.3) are of course related. Formula (1.3) leads in a natural way to the following bilinear form on $\mathcal{P}_n \times \mathcal{P}_n$:

$$[P, Q] := \sum_{j=0}^{n} P^{(j)}(a)\ (-1)^{n-j}\ Q^{(n-j)}(a) \ , \tag{1.4}$$

which is actually connected to function E by:

$$[E(.,y), Q] = Q(y), \text{ for all } Q \in \mathcal{P}_n \text{ and all } y \in \mathbb{R} \ . \tag{1.5}$$

Function E is thus a kind of reproducing function. The bilinear form (1.4) provides a simple expression of the blossom p of a polynomial $P \in \mathcal{P}_n$. Given an n-tuple $\mathcal{T} = (x_1, \ldots, x_n)$, $p(\mathcal{T}) = [P, Q_\mathcal{T}]$, where $Q_\mathcal{T}(t) := (x_1 - t) \ldots (x_n - t)$ is the unique element of $\mathcal{P}_n$ which vanishes on $\mathcal{T}$ and satisfies $Q_\mathcal{T}^{(n)}(t) \equiv (-1)^n$. More generally, given an n-tuple $\mathcal{T} = (x_1, \ldots, x_n)$, a function Q will be said *to vanish on* $\mathcal{T}$ if it vanishes on each point of $\mathcal{T}$ *with multiplicity*. In other words, if $\mathcal{T}^{ord} = (x_1, \ldots, x_n)^{ord}$ denotes the n-tuple obtained by arranging the x_i's in the increasing order and if $\mathcal{T}^{ord} = (\tau_1^{\mu_1} \ldots \tau_r^{\mu_r})$, where $\tau_i < \tau_{i+1}$ (τ^μ means τ repeated μ times), then $Q^{(j)}(\tau_i) = 0, j = 0, \ldots, \mu_i - 1,\ i = 1, \ldots, r$.

In the more general framework of W-spaces (spaces of real functions with non vanishing Wronskian on an interval I), we define the dual $\mathcal{U}^*$ of an $(n+1)$-dimensional W-space $\mathcal{U}$ (see L.L. Schumaker [15]) and we introduce an intrinsic bilinear form on $\mathcal{U} \times \mathcal{U}^*$ and a reproducing function on $I \times I$. There are many relations between these two notions. Some of the tools used in this paper have already been introduced by N. Dyn and A. Ron [7], in particular the reproducing function. Using the bilinear form, we can characterize a C^k-contact between two functions in $\mathcal{U}$. In case $\mathcal{U}$ contains the constant functions and $D\mathcal{U}$ is an extended Chebyshev space (where D denotes the differentiation operator), the bilinear form allows us to introduce the notion of *blossom* f for a function $F \in \mathcal{U}$ by:

$$f(\mathcal{T}) := [F, \psi_\mathcal{T}^*], \tag{1.6}$$

where $\psi_\mathcal{T}^*$ is the unique element of $\mathcal{U}^*$ vanishing on $\mathcal{T}$ such that $[\mathbb{1}, \psi_\mathcal{T}^*] = 1$. A de Casteljau algorithm can be implemented and a Bernstein-Chebyshev basis can be defined. This constitutes another but equivalent approach to the one developed by H. Pottmann [13] (see also [12] and [16]) using intersections of osculating flats.

In case $\mathcal{U}$ is the extended complete Chebyshev space associated with the weight functions $w_0, \ldots, w_n$, the bilinear form can be expressed as follows:

$$[F, F^*] = \sum_{i=0}^{n} L_i F(a)\ (-1)^{(n-i)}\ \widehat{L}_{n-i} F^*(a) \tag{1.7}$$

where the L_i are differential operators defined by $L_0 F = F/w_0$, $L_i F = (L_{i-1}F)'/w_i$, $i = 1, \ldots, n$ and the $\widehat{L}_i$ are similarly defined from $\widehat{w}_0 = 1/(w_0 w_1 \ldots w_n)$, $\widehat{w}_i = w_{n+1-i}$, $i = 1, \ldots, n$. Formulae (1.6) and (1.7) yield the expression given by P. J. Barry [2] for the blossom. Observe that, for such a given space $\mathcal{U}$, there may exist several possible sets of weight functions w_i and associated differential operators L_i and $\widehat{L}_i$. As the bilinear form is intrinsic (i.e., depends only on $\mathcal{U}$), the right hand side of (1.8) depends neither on the chosen system of w_i, nor on the point $a \in I$.

§2. Bilinear Form and Reproducing Function

Let us start with a few basic algebraic results which will be useful later on (see [5]). Throughout this section, $\mathcal{U}$ and $\mathcal{V}$ will denote two $(n+1)$-dimensional spaces of functions defined on two sets X and Y respectively.

Theorem 2.1. *Given a function $E : X \times Y \longrightarrow \mathbb{R}$, the three following statements are equivalent:*

(i) $(\text{i})_1$ *for all $x \in X$, $E(x, .) \in \mathcal{V}$,*
$(\text{i})_2$ *the functions $E(., y)$, $y \in Y$, span the space $\mathcal{U}$,*

(ii) *for all basis $(V_0, \ldots, V_n)$ of $\mathcal{V}$, there exists a (unique) basis $(U_0, \ldots, U_n)$ of $\mathcal{U}$, such that*

$$E(x, y) = \sum_{i=0}^{n} U_i(x) V_i(y) \quad \textit{for all } (x, y) \in X \times Y\ , \tag{2.1}$$

(iii) *there exist a basis $(U_0, \ldots, U_n)$ of $\mathcal{U}$ and a basis $(V_0, \ldots, V_n)$ of $\mathcal{V}$ satisfying (2.1).*

Proof:
•(ii) $\Longrightarrow$ (iii) is obvious.
•(i) $\Longrightarrow$ (ii)
Let us suppose that (i) is satisfied and consider a basis $(V_0, \ldots, V_n)$ of $\mathcal{V}$. Then, for a given $x \in X$, the statement $(\text{i})_1$ proves that $E(x, .)$ may be written (in a unique way) as a linear combination of $V_0, \ldots, V_n$. In other words, $E(x, .) = \sum_{i=0}^{n} U_i(x)\, V_i$, where $U_i(x)$, $i = 0, \ldots, n$, denote $n+1$ real numbers depending on x. Thus, we obtain $n+1$ functions $U_0, \ldots, U_n$ defined on X which satisfy (2.1).

Hence, for any $y \in Y$, $E(.,y) = \sum_{i=0}^{n} V_i(y)\, U_i$. Therefore, using assumption $(\mathrm{i})_2$, we can derive that the $(n+1)$-dimensional space $\mathcal{U}$ is contained in the space spanned by $U_0, \ldots, U_n$, which proves the equality $\mathcal{U} = \mathrm{span}\,(U_0, \ldots, U_n)$, and that $(U_0, \ldots, U_n)$ is a basis of $\mathcal{U}$.

•(iii) $\Longrightarrow$ (i)
Let us start with a basis $(U_0, \ldots, U_n)$ of $\mathcal{U}$ and a basis $(V_0, \ldots, V_n)$ of $\mathcal{V}$ assumed to satisfy (2.1). Clearly, $E(.,y) \in \mathcal{U}$ for all $y \in Y$, and $E(x,.) \in \mathcal{V}$ for all $x \in X$. Moreover, it is possible to find $n+1$ elements $y_0, \ldots, y_n \in Y$ such that the matrix $A := \left(V_i(y_j)\right)_{i,j=0,\ldots,n}$ is regular.

Then, the equality

$$\left(E(.,y_0), \ldots, E(.,y_n)\right)^T = A \,.\, (U_0, \ldots, U_n)^T ,$$

proves that $\mathcal{U} = \mathrm{span}\,\left(E(.,y_0), \ldots, E(.,y_n)\right)$, which yields $(\mathrm{i})_2$. ■

Theorem and Definition 2.2. *Assume that $E : X \times Y \longrightarrow \mathbb{R}$ satisfies condition (i) of Theorem 2.1. Then, there exists a unique bilinear form $[.,.] : \mathcal{U} \times \mathcal{V} \longrightarrow \mathbb{R}$ which confers a reproducing property on E, in the sense that*

$$[E(.,y), V] = V(y) \quad \text{for all } V \in \mathcal{V} \text{ and for all } y \in Y , \tag{2.2}$$

It is a nondegenerate bilinear form which will be called the bilinear form associated with function E.

Proof: By Theorem 2.1, we can choose a basis $(U_0, \ldots, U_n)$ of $\mathcal{U}$ and a basis $(V_0, \ldots, V_n)$ of $\mathcal{V}$ satisfying (2.1). Given $U = \sum_{i=0}^{n} \alpha_i\, U_i \in \mathcal{U}$ and $V = \sum_{i=0}^{n} \beta_i\, V_i \in \mathcal{V}$, let us introduce the following quantity:

$$[U, V] := \sum_{i=0}^{n} \alpha_i\, \beta_i \,. \tag{2.3}$$

In doing so, we define the nondegenerate bilinear form on $\mathcal{U} \times \mathcal{V}$ characterized by $[U_i, V_j] = \delta_{ij}$, $i,j = 0, \ldots, n$. As a direct consequence of this definition, let us observe that $[U, V_i] = \alpha_i$ and $[U_i, V] = \beta_i$. In particular, for all $y \in Y$, the equality $E(.,y) = \sum_{i=0}^{n} V_i(y)\; U_i$, deduced from (2.1), implies that $[E(.,y), V_i] = V_i(y)$, $i = 0, \ldots, n$. Relation (2.2) follows by linearity.

On the other hand, let $[.,.] : \mathcal{U} \times \mathcal{V} \longrightarrow \mathbb{R}$ be a bilinear form such that (2.2) holds. It is possible to select $y_0, \ldots, y_n \in Y$ and $x_0, \ldots, x_n \in X$ such that the functions $E(.,y_i)$, $i = 0, \ldots, n$, and $E(x_i,.)$, $i = 0, \ldots, n$, form a basis of $\mathcal{U}$ and $\mathcal{V}$ respectively. The reproducing property of E implies that $[E(.,y_i), E(x_j,.)] = E(x_j, y_i)$, $i,j = 0, \ldots, n$, which proves the uniqueness of $[.,.]$. ■

In particular, the bilinear form defined in (2.3) depends only on function E, not on the two chosen bases $(U_0, \ldots, U_n)$ and $(V_0, \ldots, V_n)$, provided that these bases satisfy (2.1).

§3. W-spaces

3.1. W-Spaces and Dual Spaces

From now on, we shall denote by I a real interval with a nonempty interior. In order to simplify, all the functions that we shall deal with will be assumed to be infinitely many times differentiable on I. If $U_0, \ldots, U_n$ are any such functions, we shall set $\vec{U} := (U_0, \ldots, U_n)^T$. Recall that the Wronskian of these functions is the function $W(U_0, \ldots, U_n)$ defined on I by $W(U_0, \ldots, U_n)(t) := \det \mathcal{W}_{\vec{U}}(t)$, where $\mathcal{W}_{\vec{U}}$ is the square matrix of order $n+1$ defined by

$$\left(\mathcal{W}_{\vec{U}}(t)\right)_{i,j} := U_i^{(j)}(t) \quad t \in I \, .$$

Definition 3.1. *If $\mathcal{U}$ is an $(n+1)$-dimensional subspace of $C^\infty(I)$, $(U_0, \ldots, U_n)$ a basis of $\mathcal{U}$, then $\mathcal{U}$ is said to be a W-space on I if the Wronskian $W(U_0, \ldots, U_n)$ never vanishes on I.*

Clearly, this definition does not depend on the chosen basis. Actually, it means that a function $U \in \mathcal{U}$ is uniquely determined by the knowledge of $D_n U(a) := \left((U(a), \ldots, U^{(n)}(a)\right)^T$ for any given point $a \in I$.

Suppose that $\mathcal{U}$ is an $(n+1)$-dimensional W-space on I and that $(U_0, \ldots, U_n)$ is a basis of $\mathcal{U}$. Observe that $\vec{U}^{(i)}(x) = (U_0^{(i)}(x), \ldots, U_n^{(i)}(x))^T$ is in fact the $(i+1)$-th column of $\mathcal{W}_{\vec{U}}(x)$. Then, for a given $x \in I$, the $n+1$ following relations

$$\langle \vec{U}^{(i)}(x), \vec{U}^*(x) \rangle = \begin{cases} 0 & \text{for } i = 0, \ldots, n-1, \\ 1 & \text{for } i = n. \end{cases} \tag{3.1}$$

(where $\langle .,. \rangle$ denotes the scalar product in $\mathbb{R}^{n+1}$) uniquely define a vector $\vec{U}^*(x) = (U_0^*(x), \ldots, U_n^*(x))^T$, obtained by solving the linear system

$$\mathcal{W}_{\vec{U}}(x)^T \, . \, \vec{U}^*(x) = (0, \ldots, 0, 1)^T .$$

The system $(U_0^*, \ldots, U_n^*)$ defined by (3.1) is called *the dual system* of $(U_0, \ldots, U_n)$ and the space $\mathcal{U}^* := \operatorname{span}(U_0^*, \ldots, U_n^*)$ *the dual space* of $\mathcal{U}$ (see [15,7]). One can verify that this space $\mathcal{U}^*$ does not depend on the basis $(U_0, \ldots, U_n)$.

By differentiating relations (3.1), it can be recursively proved that

$$\langle \vec{U}^{(i)}, \vec{U}^{*(j)} \rangle \equiv \begin{cases} 0 & \text{for } i+j \le n-1, \\ (-1)^j & \text{for } i+j = n. \end{cases} \tag{3.2}$$

Let us consider the square matrix $\mathcal{Z}(x)$ of order $n+1$ defined, for $x \in I$, by

$$\mathcal{Z}(x) := \mathcal{W}_{\vec{U}}(x)^T \, . \, \mathcal{W}_{\vec{U}^*}(x) = \Big(\langle \vec{U}^{(i)}(x), \vec{U}^{*(j)}(x) \rangle \Big)_{i,j=0,\ldots,n} \, . \tag{3.3}$$

Then, relations (3.2) mean that $\mathcal{Z}(x)$ has the following structure:

$$\mathcal{Z}(x) \equiv \begin{pmatrix} 0 & \dots & \dots & 0 & (-1)^n \\ \vdots & \vdots & & (-1)^{n-1} & \cdot \\ & & \cdot^{\cdot^{\cdot}} & \cdot & \cdot \\ 0 & -1 & & & \\ 1 & \cdot & & \cdot & \cdot \end{pmatrix} . \tag{3.4}$$

Since $\mathcal{Z}(x)$ is regular for any $x \in I$, the space $\mathcal{U}^*$ is also an $(n+1)$-dimensional W-space on I. Actually, the case $i = 0$ in (3.2) shows that the dual system of $(U_0^*, \dots, U_n^*)$ is $(-1)^n(U_0, \dots, U_n)$. As an immediate consequence, the dual space of the dual space $\mathcal{U}^*$ is the initial space $\mathcal{U}$.

3.2. W-spaces and Reproducing Functions

If $(U_0, \dots, U_n)$ is a basis of $\mathcal{U}$ and $(U_0^*, \dots, U_n^*)$ its dual system, let us consider the function $E : X \times Y \longrightarrow \mathbb{R}$ defined by (see [7])

$$E(x,y) := \langle\, U(x), U^*(y) \,\rangle = \sum_{i=0}^{n} U_i(x)\, U_i^*(y) \quad x, y \in I \, . \tag{3.5}$$

According to Section 2, the bilinear form associated to E is given by

$$[U, U^*] = \sum_{i=0}^{n} \alpha_i \alpha_i^* \, , \tag{3.6}$$

for any $U = \sum_{i=0}^{n} \alpha_i U_i \in \mathcal{U}$ and any $U^* = \sum_{i=0}^{n} \alpha_i^* U_i^* \in \mathcal{U}^*$.

Theorem 3.2. *Let $x, y \in I$. Then, $E(.,y)$ is the only element $U \in \mathcal{U}$ such that $D_n U(y) = (0, \dots, 0, 1)^T$, while $E(x,.)$ is the only element $U^* \in \mathcal{U}^*$ such that $D_n U^*(x) = (0, \dots, 0, (-1)^n)^T$.*

Proof: From (3.5), we can calculate the partial derivatives of E as follows:

$$\partial_{1^i 2^j} E(x,y) = \langle\, \overrightarrow{U}^{(i)}(x), \overrightarrow{U^*}^{(j)}(y) \,\rangle \quad x, y \in I \, .$$

Consequently, relations (3.1) mean that, for a given $y \in I$, the function $U := E(.,y)$ satisfies $D_n U(y) = (0, \dots, 0, 1)^T$. Since U is an element of $\mathcal{U}$ which is a W-space on I, this completely determines U. A similar argument can be developed from (3.2) in order to prove the second part of Theorem 3.2. ■

Corollary and Definition 3.3. *The function E defined in (3.5) (hence also its associated bilinear form $[.,.]$), and the matrix $\mathcal{Z}$ defined in (3.4) are intrinsic, in the* *sense that they depend only on the space $\mathcal{U}$, not on the chosen basis $(U_0, \ldots, U_n)$. The function E will be called the reproducing function and* $[.,.]$ *the canonical bilinear form associated with the W-space $\mathcal{U}$.*

Proof: The fact that E is intrinsic is a direct consequence of the previous theorem. The same holds for $\mathcal{Z}$, since $(\mathcal{Z}(x))_{i,j} = \partial_{1^i 2^j} E(x,x)$ for all $x \in I$. ■

Examples 3.4.

1)- For $\mathcal{U} = \mathcal{P}_n$, Theorem 3.2 directly leads to $E(x,y) = (x-y)^n/n!$ for all $x, y \in \mathbb{R}$. Since $\big(E(x,.)\big)_{x\in\mathbf{R}}$ span the dual space $\mathcal{U}^*$, it follows that $\mathcal{U}^* \subset \mathcal{P}_n$. The space $\mathcal{U}^*$ being $(n+1)$-dimensional, this ensures that $\mathcal{U}^* = \mathcal{P}_n = \mathcal{U}$.

2)- The space $\mathcal{U}$ spanned by functions $U_0(x) = 1, U_1(x) = \cos x, U_2(x) = \sin x$ is a W-space on the whole real line, since $W(U_0, U_1, U_2)(x) = 1$ everywhere on $\mathbb{R}$. Relations (3.1) lead to $\vec{U}^*(x) = (1, -\cos x, -\sin x)^T$, which shows both that $\mathcal{U}^* = \mathcal{U}$ and that $E(x,y) = 1 - \cos(x-y)$ for all $x, y \in \mathbb{R}$.

4)- For $U_0(x) = 1$, $U_1(x) = x$ and $U_2(x) = x^3$, we obtain $W(U_0, U_1, U_2)(x) = 6x$. So, the space $\mathcal{U}$ spanned by these three functions is a W-space on $I =]0, +\infty[$. In that case, $\vec{U}^*(x) = (x^2/3, -x/2, 1/6x)^T$ for all $x > 0$ and the corresponding function E is defined for $x, y > 0$ by $E(x,y) = (x-y)^2\,(x+2y)/6y$.

3.3. The Canonical Bilinear Form

Theorem 3.7. *For any $U \in \mathcal{U}$ and any $U^* \in \mathcal{U}^*$, we have*

$$[U, U^*] = D_n U(a)^T \ . \ \mathcal{Z}(a)^{-T} \ . \ D_n U^*(a) \ , \tag{3.7}$$

where a denotes any point of I.

Proof: Given $a \in I$, let us choose the basis $(U_0, \ldots, U_n)$ of $\mathcal{U}$ characterized by ${U_i}^{(j)}(a) = \delta_{ij}$, $i, j = 0, \ldots, n$, that is to say, by

$$\mathcal{W}_{\vec{U}}(a) = \mathcal{I}_{n+1} \ . \tag{3.8}$$

Then, formula (3.3) proves that the dual system of this basis satisfies

$$\mathcal{W}_{\vec{U}^*}(a) = \mathcal{Z}(a) \ . \tag{3.9}$$

If $U = \sum_{i=0}^n \alpha_i U_i$ and $U^* = \sum_{i=0}^n \alpha_i^* U_i^*$, (3.8) and (3.9) lead to

$$(\alpha_0, \ldots, \alpha_n)^T = D_n U(a) \quad , \quad (\alpha_0^*, \ldots, \alpha_n^*)^T = \mathcal{Z}(a)^{-T} \ . \ D_n U^*(a) \ .$$

It follows that $[U, U^*] := (\alpha_0, \ldots, \alpha_n) . (\alpha_0^*, \ldots, \alpha_n^*)^T$ is given by (3.7). ■

For any subset $\mathcal{A}$ of $\mathcal{U}$, we shall denote by $\mathcal{A}^\circ$ the subspace of $\mathcal{U}^*$ which is orthogonal to $\mathcal{A}$ with respect to the canonical bilinear form, namely

$$\mathcal{A}^\circ := \{U^* \in \mathcal{U}^* \ \ | \ \ [U, U^*] = 0 \quad \forall\, U \in \mathcal{A}\} \ .$$

Of course, one can similarly define $\mathcal{A}^{*\circ}$ for any subset $\mathcal{A}^*$ of $\mathcal{U}^*$. According to the notations already defined in the introduction, a function U will be said "to vanish on a^{r+1}" if $U^{(j)}(a) = 0$, $j = 0, \ldots, r$. Then we can state:

Corollary 3.8. *If $a \in I$ and if r is a given integer, $-1 \leq r \leq n$, then*

$$\{U \in \mathcal{U} \mid U \text{ vanishes on } (a^{r+1})\}^{\circ} = \{U^* \in \mathcal{U}^* \mid U^* \text{ vanishes on } (a^{n-r})\} . \quad (3.10)$$

Proof: Let us introduce the following subspace of $\mathcal{U}$:

$$\mathcal{A} := \{U \in \mathcal{U} \mid U \text{ vanishes on } (a^{r+1})\} . \quad (3.11)$$

A function $U \in \mathcal{U}$ belongs to $\mathcal{A}$ iff the $r+1$ first components of $D_n U(a)$ are equal to 0. Consequently, on account of (3.7), a function $U^* \in \mathcal{U}^*$ belongs to $\mathcal{A}^{\circ}$ iff the $(n-r)$ last components of $\mathcal{Z}(a)^{-T} . D_n U^*(a)$ are equal to 0. Taking into account the structure of $\mathcal{Z}(a)^{-T}$, this occurs iff the $(n-r)$ first components of $D_n U^*(a)$ are equal to 0, i.e., iff U^* vanishes on (a^{n-r}). ■

Corollary 3.9. *Given an integer $r \in \{-1, \ldots, n\}$, $a \in I$ and $F, G \in \mathcal{U}$, the following statements are equivalent.*

(i) $F^{(i)}(a) = G^{(i)}(a)$ *for all* $i = 0, \ldots, r$,

(ii) $[F, U^*] = [G, U^*]$ *for all functions* U^* *which vanish on* (a^{n-r}).

Proof: Condition (i) means that $U := F - G$ belongs to the subspace $\mathcal{A}$ defined in (3.11). According to the previous theorem, (ii) can be replaced by $U \in \mathcal{A}^{\circ\circ}$. Thus, the equivalence between (i) and (ii) results from the equality $\mathcal{A}^{\circ\circ} = \mathcal{A}$. ■

§4. Extended Chebyshev Spaces

4.1. EC Spaces and Blossoming Principle

Definition 4.1. *A system $(U_0, \ldots, U_n)$ of $n+1$ functions defined on I is said to be an extended Chebyshev system (EC) on I if*

$$\begin{vmatrix} U_0(\tau_0) & \cdots & U_0{}^{(\mu_0-1)}(\tau_0) & U_0(\tau_1) & \cdots & U_0{}^{(\mu_r-1)}(\tau_r) \\ U_1(\tau_0) & \cdots & U_1{}^{(\mu_0-1)}(\tau_0) & U_1(\tau_1) & \cdots & U_1{}^{(\mu_r-1)}(\tau_r) \\ \vdots & \ddots & \vdots & \vdots & \ddots & \vdots \\ U_n(\tau_0) & \cdots & U_n{}^{(\mu_0-1)}(\tau_0) & U_n(\tau_1) & \cdots & U_n{}^{(\mu_r-1)}(\tau_r) \end{vmatrix} \neq 0 \quad (4.1)$$

for all distinct $\tau_0, \ldots, \tau_r \in I$, and all positive integers $\mu_0, \ldots, \mu_r$ summing to $n+1$.

Clearly, $(U_0, \ldots, U_n)$ is an EC system on I iff any non zero element of the space $\mathcal{U} := \text{span}\,(U_0, \ldots, U_n)$ has at most n zeros (counted with multiplicities) in I, which shows that this property does not depend on the particular basis $(U_0, \ldots, U_n)$ defining $\mathcal{U}$. Equivalently, each function $U \in \mathcal{U}$ is completely determined by the knowledge of $U^{(j)}(\tau_i)$, $i = 0, \ldots, r$, $j = 0, \ldots, \mu_i - 1$, where the τ_i's and the μ_i's are chosen as in Definition 4.1.

Such a space $\mathcal{U}$ will be said to be an $(n+1)$-*dimensional EC space on* I. Let us observe that, if $\mathcal{U}$ is an EC space on I, then it is a W-space on I so that we can consider the reproducing function E and the canonical bilinear form $[.,.]$ associated with $\mathcal{U}$ as in the previous section.

Theorem 4.2. *The dual space of an EC space is an EC space on the same interval* (see [11]).

Throughout this section, we shall deal with a space $\mathcal{U} = \text{span}\,(U_0, \ldots, U_n)$. Consider the two following assumptions :

$$(H_1) \qquad U_0 = 1\!\!1$$
$$(H_2) \qquad ({U_1}', \ldots, {U_n}') \text{ is an EC system on } I.$$

Let us observe that, when (H_1) holds, (H_2) implies that $\mathcal{U}$ is also an EC space on I, but the converse is not true as pointed out by the following example: if $U_0(x) = 1$, $U_1(x) = \cos x$, $U_2(x) = \sin x$, (U_0, U_1, U_2) is an EC system on $[0, 2\pi[$, while $({U_1}', {U_2}')$ is an EC system on $[0, \pi[$ only.

Theorem 4.3. *Let $(U_0, \ldots, U_n)$ be $n+1$ functions such that $W(U_0, \ldots, U_n)$ never vanishes on I and let $(U_0^*, \ldots, U_n^*)$ be the dual system of $(U_0, \ldots, U_n)$. If $U_0 = 1\!\!1$, then*

$$(H_2) \quad \Leftrightarrow \quad (U_1^*, \ldots, U_n^*) \text{ is an EC system on } I \ . \tag{4.2}$$

Proof: As soon as $U_0 = 1\!\!1$, one can verify that $(U_1^*, \ldots, U_n^*)$ is the dual system of $({U_1}', \ldots, {U_n}')$. Thus, Theorem 4.2 implies equivalence (4.2). ■

From now on, (H_1) and (H_2) will always be assumed to hold. As a consequence, we obtain the following fundamental result.

Theorem 4.4. *For all n-tuple $\mathcal{T} \in I^n$, there exists a unique function $\Psi_{\mathcal{T}}^* \in \mathcal{U}^*$ such that*

$$\Psi_{\mathcal{T}}^* \textit{ vanishes on } \mathcal{T}, \qquad [1\!\!1, \Psi_{\mathcal{T}}^*] = 1 \ . \tag{4.3}$$

Proof: For any $\Psi^* = \sum_{i=0}^n \alpha_i U_i^* \in \mathcal{U}^*$, $[1\!\!1, \Psi^*] = [U_0, \Psi^*] = \alpha_0^*$. Then, Ψ^* satisfies the normalization condition $[1\!\!1, \Psi^*] = 1$ (i.e., $\alpha_0^* = 1$), iff the function $\Phi^* := \Psi^* - U_0^*$ belongs to the n-dimensional EC space spanned by $U_1^*, \ldots, U_n^*$. Suppose that $\mathcal{T}^{ord} = ({\tau_1}^{\mu_1} \ldots {\tau_r}^{\mu_r})$ with $\tau_i < \tau_{i+1}$ and $\mu_i > 0$. Then, in this space, there exists exactly one element $\Phi_{\mathcal{T}}^*$ which coincides with $-U_0^*$ on $\mathcal{T}$, i.e., such that

$${\Phi_{\mathcal{T}}^*}^{(j)}(\tau_i) = -{U_0^*}^{(j)}(\tau_i) \quad 1 \le i \le r \ , \ 0 \le j \le \mu_i - 1 \ .$$

Hence, $\Psi_{\mathcal{T}}^* := U_0^* + \Phi_{\mathcal{T}}^*$ is the only element of $\mathcal{U}^*$ satisfying (4.3). ■

Definition 4.5. *For all $F \in \mathcal{U}$ and all n-tuple $\mathcal{T} \in I^n$, let us introduce*

$$f(\mathcal{T}) := [F, \Psi_{\mathcal{T}}^*] \ , \tag{4.4}$$

where $\Psi_{\mathcal{T}}^*$ is the only element of $\mathcal{U}^*$ which satisfies (4.3). With this, we define a function $f : I^n \to \mathbb{R}$ which will be called the blossom of F.

As a first obvious example, observe that, according both to the normalization condition required on each function $\Psi_{\mathcal{T}}^*$ and to the previous definition, the blossom of function $1\!\!1$ is equal to 1 everywhere on I^n.

Theorem 4.6. *The blossom f of any $F \in \mathcal{U}$ is a symmetric function such that*

$$f(x^n) = F(x) \quad \text{for all } x \in I \ . \tag{4.5}$$

Proof: The symmetry of f results from its definition. Now, by Theorem 3.2, we know that, if E is the reproducing function associated with $\mathcal{U}$, function $E(x,.)$ belongs to $\mathcal{U}^*$ and vanishes on (x^n). Moreover, $[\mathbb{1}, E(x,.)] = \mathbb{1}(x) = 1$, which finally proves that

$$E(x,.) = \Psi^*_{(x^n)} \ . \tag{4.6}$$

So, by (4.4), $f(x^n) = [F, E(x,.)]$. The reproducing property of E leads to (4.5). ■

Theorem 4.7. *For all $y \in I$, let us denote by $e(.;y) : I^n \to \mathbb{R}$ the blossom of $E(.,y)$. Then, for all $\mathcal{T} \in I^n$,*

$$e(\mathcal{T};y) = \Psi^*_{\mathcal{T}}(y) \ . \tag{4.7}$$

Proof: Relation (4.7) results straightforwardly from the reproducing property of E, since, according to (4.4), $e(\mathcal{T};y) := [E(.,y), \Psi^*_{\mathcal{T}}]$. ■

4.2. A de Casteljau Type Algorithm

Theorem 4.8. *Given $x_1, \dots, x_{n-1} \in I$ and $a < b$ two distinct points in I, there exists a C^∞ strictly increasing function β (depending on a, b and the x_i's), with $\beta(a) = 0$, $\beta(b) = 1$, such that, for all $F \in \mathcal{U}$, the blossom f of F satisfies*

$$f(x_1, \dots, x_{n-1}, x) = \big(1-\beta(x)\big)\, f(x_1, \dots, x_{n-1}, a) + \beta(x)\, f(x_1, \dots, x_{n-1}, b) \ . \tag{4.8}$$

The proof of this theorem relies on several lemmas.

Lemma 4.9. *Given an $(n-1)$-tuple $\mathcal{T} = ({\tau_1}^{\mu_1} \dots {\tau_r}^{\mu_r}) \in I^{n-1}$ and a, b two distinct points in I, let $\mathcal{T}_1 := (\mathcal{T}, a)$ and $\mathcal{T}_2 := (\mathcal{T}, b)$ be the two n-tuples obtained by adding a and b to $\mathcal{T}$. Then, the function $\Psi^* := \Psi^*_{\mathcal{T}_1} - \Psi^*_{\mathcal{T}_2}$ vanishes exactly on $\mathcal{T}$, which means that Ψ^* vanishes (with multiplicities) on $\mathcal{T}$, and that*

$$\Psi^*(x) \neq 0 \quad \text{if } x \notin \{\tau_1, \dots, \tau_r\}, \qquad \Psi^{*(\mu_i)}(\tau_i) \neq 0 \quad i = 1, \dots, r \ . \tag{4.9}$$

Proof: Thanks to the two normalization conditions $[\mathbb{1}, \Psi^*_{\mathcal{T}_1}] = 1$ and $[\mathbb{1}, \Psi^*_{\mathcal{T}_2}] = 1$, Ψ^* is a linear combination of $U_1^*, \dots, U_n^*$. On the other hand, since $a \neq b$, $\Psi^* \not\equiv 0$.

In other words, Ψ^* is a non zero element of the n-dimensional EC space spanned by $U_1^*, \dots, U_n^*$. Consequently, Ψ^* has at most $(n-1)$ zeros in I. Now, since $\Psi^*_{\mathcal{T}_1}$ and $\Psi^*_{\mathcal{T}_2}$ both vanish on the $(n-1)$-tuple $\mathcal{T}$, so does Ψ^*. Thus Ψ^* cannot vanish elsewhere (with multiplicity), in the sense of (4.9). ■

Lemma 4.10. *With the same assumptions and notations as in Lemma 4.9, for all $x \in I$, we have*

$$\Psi^*_{(\mathcal{T},x)} = \alpha(x)\ \Psi^*_{\mathcal{T}_1} + \beta(x)\ \Psi^*_{\mathcal{T}_2}\ , \tag{4.10}$$

where

$$\alpha(x) := \frac{\Psi^*_{\mathcal{T}_2}(x)}{\Psi^*_{\mathcal{T}_2}(x) - \Psi^*_{\mathcal{T}_1}(x)}\ , \quad \beta(x) := \frac{\Psi^*_{\mathcal{T}_1}(x)}{\Psi^*_{\mathcal{T}_1}(x) - \Psi^*_{\mathcal{T}_2}(x)} \quad \text{if } x \notin \{\tau_1, \ldots, \tau_r\}, \tag{4.11}$$

$$\alpha(\tau_i) := \frac{{\Psi^*_{\mathcal{T}_2}}^{(\mu_i)}(\tau_i)}{{\Psi^*_{\mathcal{T}_2}}^{(\mu_i)}(\tau_i) - {\Psi^*_{\mathcal{T}_1}}^{(\mu_i)}(\tau_i)}\ , \quad \beta(\tau_i) := \frac{{\Psi^*_{\mathcal{T}_1}}^{(\mu_i)}(\tau_i)}{{\Psi^*_{\mathcal{T}_1}}^{(\mu_i)}(\tau_i) - {\Psi^*_{\mathcal{T}_2}}^{(\mu_i)}(\tau_i)}\ . \tag{4.12}$$

Proof: Let us first notice that, if $x = a$, $(\mathcal{T}, x) = \mathcal{T}_1$, and clearly equality (4.10) is valid provided that we choose $\alpha(x) = 1, \beta(x) = 0$. These values can also be expressed either by (4.11) or by (4.12) depending on whether $a \notin \{\tau_1, \ldots, \tau_r\}$ (in which case, $\Psi^*_{\mathcal{T}_1}(a) = 0$, $\Psi^*_{\mathcal{T}_2}(a) \neq 0$), or $a = \tau_i$ (in which case, ${\Psi^*_{\mathcal{T}_1}}^{(\mu_i)}(a) = 0$, ${\Psi^*_{\mathcal{T}_2}}^{(\mu_i)}(a) \neq 0$). A similar argument can be developed for the case $x = b$. So that, in the rest of the proof, we shall assume that $x \neq a, b$.

The space of all elements of $\mathcal{U}^*$ which vanish on $\mathcal{T}$ is clearly 2-dimensional. Therefore, since $a \neq b$, a function $\Psi^* \in \mathcal{U}^*$ vanishes on $\mathcal{T}$ iff there exist $A, B \in \mathbb{R}$ such that $\Phi^* = A\ \Psi^*_{\mathcal{T}_1} + B\ \Psi^*_{\mathcal{T}_2}$. Let us determine coefficients A, B so that $\Phi^* = \Psi^*_{(\mathcal{T},x)}$, that is to say, so that Φ^* vanishes on $(\mathcal{T}, x)$ and satisfies the normalization condition $[\mathbb{1}, \Phi^*] = 1$. Since $\Psi^*_{\mathcal{T}_1}$ and $\Psi^*_{\mathcal{T}_2}$ are both normalized, condition $[\mathbb{1}, \Phi^*] = 1$ will hold iff

$$A + B = 1\ . \tag{4.13}$$

When $x \notin \{\tau_1, \ldots, \tau_r\}$, Φ^* vanishes on $(\mathcal{T}, x)$ iff $\Phi^*(x) = 0$, which means

$$A\, \Psi^*_{\mathcal{T}_1}(x) + B\, \Psi^*_{\mathcal{T}_2}(x) = 0\ . \tag{4.14}$$

By (4.9), $\Psi^*_{\mathcal{T}_2}(x) - \Psi^*_{\mathcal{T}_1}(x) \neq 0$, which proves that the linear system composed of (4.13) and (4.14) has a unique solution, namely $A = \alpha(x), B = \beta(x)$, where $\alpha(x)$ and $\beta(x)$ are defined by (4.11).

Suppose now that $x = \tau_i$ (hence, $a \neq \tau_i$). Then, the function Φ^* vanishes on the n-tuple $(\mathcal{T}, x) = (\tau_1^{\mu_1} \ldots \tau_i^{\mu_i+1} \ldots \tau_r^{\mu_r})$ iff $\Phi^{*(\mu_i)}(\tau_i) = 0$, i.e., iff

$$A\, {\Psi^*_{\mathcal{T}_1}}^{(\mu_i)}(\tau_i) + B\, {\Psi^*_{\mathcal{T}_2}}^{(\mu_i)}(\tau_i) = 0\ . \tag{4.15}$$

Again by (4.9), ${\Psi^*_{\mathcal{T}_2}}^{(\mu_i)}(\tau_i) - {\Psi^*_{\mathcal{T}_1}}^{(\mu_i)}(\tau_i) \neq 0$ and the two equations (4.13) and (4.15) have a unique solution, $A = \alpha(x), B = \beta(x)$, where $\alpha(x)$ and $\beta(x)$ are now defined by (4.12). ∎

Lemma 4.11. *Let f, g be two C^∞ functions defined on a real interval J containing a. If f and g both vanish on (a^k), with $g^{(k)}(a) \neq 0$ and $g(x) \neq 0$ for $x \in J \setminus \{a\}$, then the function h defined on J by*

$$h(x) := \frac{f(x)}{g(x)} \quad \text{if } x \neq a \quad , \quad h(a) := \frac{f^{(k)}(a)}{g^{(k)}(a)} \quad \text{if } x = a \ , \tag{4.16}$$

is C^∞ on J.

Proof: The assumptions on f and g imply that $f(x) = (x-a)^k f_1(x)$ and $g(x) = (x-a)^k g_1(x)$, where f_1 and g_1 are C^∞ on J and satisfy $f_1(a) = f^{(k)}(a)/k!$, $g_1(a) = g^{(k)}(a)/k!$ (see[12]). One can verify that $h(x) = f_1(x)/g_1(x)$ for all $x \in J$, which provides the announced result since g_1 does not vanish on J. ■

Proof of Theorem 4.8. By the very definition of the blossom, formula (4.8) is just an immediate consequence of (4.10). Moreover, from Lemma 4.11, we can derive that functions α and β defined by formulae (4.11) and (4.12) are C^∞on I.

Now, if x_1 and x_2 are two distinct points in I, $\beta(x_1) \neq \beta(x_2)$. Otherwise, we would have $\Psi^*_{(\mathcal{T},x_1)} = \Psi^*_{(\mathcal{T},x_2)}$ and this non zero function would have $n+1$ zeros in I. Hence, β being C^∞ and one-to-one on I, it is a strictly monotone function. Since $\beta(a) = 0$ and $\beta(b) = 1$, β is strictly increasing on I. ■

Theorem 4.8 may be interpreted by the following diagram

$$\begin{array}{ccc} f(\mathcal{T}, a) & & f(\mathcal{T}, b) \\ \searrow & & \swarrow \\ & f(\mathcal{T}, x) & \end{array}$$

in which the two arrows stand for the affine combination involved in (4.8). Then, starting from the $(n+1)$ values $f(a^{n-i}b^i)$, $i = 0, \ldots, n$, for a given $x \in I$, we can compute $F(x)$ in n steps as follows:

- first iteration : compute $f(a^{n-1-i}b^i x)$, $i = 0, \ldots, n-1$,
- second iteration : compute $f(a^{n-2-i}b^i x^2)$, $i = 0, \ldots, n-2$,

and so forth up to

- next to last iteration : compute $f(ax^{n-1})$, $f(bx^{n-1})$,
- last iteration : compute $f(x^n) = F(x)$.

Each computation is obtained by means of an affine combination the coefficients of which are positive as soon as $x \in]a, b[$.

This algorithm will be called *the de Casteljau-Chebyshev algorithm with respect to* (a, b), and, for any $F \in \mathcal{U}$, the real numbers $f(a^{n-i}b^i)$, $i = 0, \ldots, n$, will be called *the Bézier-Chebyshev points of F with respect to (a, b).*

4.3. Bernstein-Chebyshev Basis

As an immediate consequence of the de Casteljau-Chebyshev algorithm, we can derive that the linear mapping $\Phi : F \in \mathcal{U} \to \left(f(a^{n-i}b^i), 0 \leq i \leq n\right) \in \mathbb{R}^{n+1}$ is one-to-one, so that the linear functionals $\ell_i : F \in \mathcal{U} \to f(a^{n-i}b^i)$, $i = 0, \ldots, n$, are linearly independent. Therefore, all $F \in \mathcal{U}$ can be decomposed as

$$F = \sum_{i=0}^{n} f(a^{n-i}b^i)\ B_i\ , \tag{4.17}$$

where $(B_0, \ldots, B_n)$ is the dual basis of these functionals, that is to say, the elements of $\mathcal{U}$ which are characterized by

$$b_j(a^{n-i}b^i) = [B_j, \Psi^*_{(a^{n-i}b^i)}] = \delta_{i,j} \quad i = 0, \ldots, n, \tag{4.18}$$

where b_i stands for the blossom of B_i. In particular, on account of (4.7), applying (4.17) to $E(., y)$, $y \in I$, gives

$$E(x, y) = \sum_{i=0}^{n} B_i(x)\ \Psi^*_{(a^{n-i}b^i)}(y)\ . \tag{4.19}$$

Theorem and Definition 4.15. *The basis $(B_0, \ldots, B_n)$ defined in (4.18) is called the Bernstein-Chebyshev basis of $\mathcal{U}$ with respect to (a, b). It satisfies*

$$\sum_{i=0}^{n} B_i = 1\!\!1 \quad , \quad 0 < B_j(t) < 1 \quad \textit{for all } t \in]a, b[\textit{ and for all } j = 0, \ldots, n\ . \tag{4.20}$$

The function B_j is the only element of $\mathcal{U}$ which vanishes on $(a^j b^{n-j})$ and such that $[B_j, \Psi^*_{(a^{n-j}b^j)}] = 1$.

Proof: The first part of (4.20) is a consequence of (4.17). The second one is obtained by computing the values of B_j by means of the de Casteljau-Chebyshev algorithm starting with from its Bézier-Chebyshev points given in (4.18).

Consider the set $\mathcal{A} = \{U \in \mathcal{U} \mid U \text{ vanishes on } (a^{r+1})\}$ already introduced in (3.11). Recall that $\mathcal{A}^\circ = \{U^* \in \mathcal{U}^* \mid U^* \text{ vanishes on } (a^{n-r})\}$ (Corollary 3.8).

If $\mathcal{T}_i := (a^{n-i}b^i)$, $i = 0, \ldots, n$, we can deduce from (4.19) that $\left(\Psi^*_{\mathcal{T}_0}, \ldots, \Psi^*_{\mathcal{T}_n}\right)$ is the dual system of the Bernstein basis $(B_0, \ldots, B_n)$, hence, is a basis of $\mathcal{U}^*$. Therefore, $\left(\Psi^*_{\mathcal{T}_0}, \ldots, \Psi^*_{\mathcal{T}_r}\right)$ is a basis of the $(r+1)$-dimensional subspace $\mathcal{A}^\circ$.

Consequently, relations (4.18) ensure that $\mathcal{A}$ is spanned by $B_{r+1}, \ldots, B_n$. Thus, B_{r+1} vanishes on (a^{r+1}). A similar argument may be developed for b instead of a, which finally proves that B_i vanishes on $(a^i b^{n-i})$ for $i = 0, \ldots, n$. ∎

Theorem 4.16. *Let F, G be two elements of $\mathcal{U}$, and f, g, their blossoms. Then, given a, b two distinct points in I and given an integer r, $0 \leq r \leq n$, the three following statements are equivalent:*

(i) $F^{(i)}(a) = G^{(i)}(a)$ *for* $i = 0, \ldots, r$,

(ii) F *and* G *have the same first* $r + 1$ *Bézier-Chebyshev points w. r. to* (a, b),

(iii) $f(\mathcal{T}) = g(\mathcal{T})$ *for all n-tuple* $\mathcal{T}$ *containing* (a^{n-r}).

Proof: As an immediate consequence of Corollary 3.9, condition (i) holds iff $[F, \Psi^*] = [G, \Psi^*]$ for all Ψ^* belonging to a set $\mathcal{B}^* \subset \mathcal{U}^*$ which spans $\mathcal{A}^\circ$. The announced result is obtained by choosing either $\mathcal{B}^* = \{\Psi^*_{(a^{n-i}b^i)},\ i = 0, \ldots, r\}$ or $\mathcal{B}^* = \{\Psi^*_{\mathcal{T}},\ \mathcal{T} \text{ containing } (a^{n-r})\}$. ∎

§5. Complete EC Spaces and Blossoming

5.1. Some Preliminary Results on ECC Spaces

Definition 5.1. *An $(n + 1)$-dimensional EC space $\mathcal{U}$ is said to be complete (ECC space) on I if there exists a basis $(U_0, \ldots, U_n)$ of $\mathcal{U}$ such that, for $i = 0, \ldots, n$, the system $(U_0, \ldots, U_i)$ is an EC system on I.*

We shall need the two following theorems on ECC spaces, the proof of which can be found in [11].

Theorem 5.3. *If $\mathcal{U}$ is an $(n+1)$-dimensional ECC space on I, there exist $n+1$ never vanishing functions $w_0, \ldots, w_n \in C^\infty(I)$ (which we shall refer to as weight functions) such that $\mathcal{U} = \operatorname{Ker}(D \circ L_n)$, where D stands for the ordinary differentiation and where L_i, $i = 0, \ldots, n$, denote the differential operators defined on $C^\infty(I)$ by*

$$L_0 U = \frac{1}{w_0} U \quad , \quad L_i U = \frac{1}{w_i} \Big(L_{i-1} U \Big)' \quad i = 1, \ldots, n. \tag{5.1}$$

Theorem 5.4. *Let $\mathcal{U}$ be an ECC space on I, and $(w_0, \ldots, w_n)$ a system of weight functions defining $\mathcal{U}$. Then, its dual space $\mathcal{U}^*$ is the ECC space associated with the weight functions $\widehat{w}_0, \ldots, \widehat{w}_n$ given by*

$$\widehat{w}_0 := \frac{1}{w_0 \ldots w_n} \quad , \quad \widehat{w}_i := w_{n+1-i} \quad i = 1, \ldots, n. \tag{5.2}$$

Moreover, any basis $(U_0, \ldots, U_n)$ of $\mathcal{U}$, and its dual system $(U_0^*, \ldots, U_n^*)$ satisfy

$$\mathcal{L}_{\vec{U}}(t)^T \ . \ \widehat{\mathcal{L}}_{\vec{U}^*}(t) \equiv \mathcal{R} \quad t \in I \ , \tag{5.3}$$

where $\mathcal{L}_{\vec{U}}(t)$ denotes the $(n+1) \times (n+1)$ matrix $(\, L_0 \vec{U}(t), \ldots, L_n \vec{U}(t)\,)$ and similarly $\widehat{\mathcal{L}}_{\vec{U}^*}(t) := (\, \widehat{L}_0 \vec{U}(t), \ldots, \widehat{L}_n \vec{U}(t)\,)$, and where $\mathcal{R}$ is the $(n + 1) \times (n + 1)$ antidiagonal matrix such that $\mathcal{R}_{n-j,j} = (-1)^j$.

5.2. Blossoming

Throughout this section, $\mathcal{U}$ will denote an $(n+1)$-dimensional ECC space with associated weight functions $\omega_0 = \mathbb{I}, \omega_1, \ldots, \omega_n$. Then, the assumption (H_2) introduced in Section 4 is automatically satisfied.

Theorem 5.5. *Let $a \in I$. Then, for all $U \in \mathcal{U}$ and all $U^* \in \mathcal{U}^*$,*

$$[U, U^*] = \sum_{i=0}^{n} L_i U(a)\ (-1)^{n-i}\ \widehat{L}_{n-i} U^*(a)\ . \tag{5.4}$$

Proof: Since the quantity $[U, U^*]$ does not depend on the basis, we can express it in a relevant one, for instance the basis $(\overline{U}_0, \ldots, \overline{U}_n)$ characterized by

$$\mathcal{L}_{\vec{\overline{U}}}(a) = \mathcal{I}_{n+1}\ . \tag{5.5}$$

Then, by (5.3), the dual system $(\overline{U}_0^*, \ldots, \overline{U}_n^*)$ satisfies $\widehat{\mathcal{L}}_{\vec{\overline{U}^*}}(a) = \mathcal{R}$, or, equivalently, since $\mathcal{R}^T . \mathcal{R} = \mathcal{I}_{n+1}$,

$$\widehat{\mathcal{L}}_{\vec{\widehat{U}^*}}(a) = \mathcal{I}_{n+1}\ , \tag{5.6}$$

where $\widehat{U} = (\widehat{U}_0, \ldots, \widehat{U}_n)^T$ is defined by $\vec{\widehat{U}} := \mathcal{R}^T . \vec{\overline{U}}^*$. According to (5.5) and (5.6), the two functions U and U^* can be decomposed as follows

$$U = \sum_{i=0}^{n} L_i U(a)\ \overline{U}_i \quad , \quad U^* = \sum_{i=0}^{n} \widehat{L}_i U^*(a)\ \widehat{U}_i\ . \tag{5.7}$$

Since $\vec{\widehat{U}} := \mathcal{R}^T . \vec{\overline{U}}^*$, (5.7) leads to $U^* = \sum_{i=0}^{n} (-1)^{n-i}\ \widehat{L}_{n-i} U^*(a)\ \overline{U}_i^*$ and then, (3.6) becomes (5.4). ■

Corollary 5.6. (cf [2]) *For all $F \in \mathcal{U}$, the value of the blossom of F at any n-tuple $\mathcal{T} \in I^n$ is given by*

$$f(\mathcal{T}) = \sum_{i=0}^{n} L_i F(a)\ (-1)^{n-i}\ \widehat{L}_{n-i} \Psi_{\mathcal{T}}^*(a)\ . \tag{5.8}$$

Let us finally observe that the right hand side of (5.4) (hence, also that of (5.8)) is independent not only of the point $a \in I$, but also of the weight functions $(w_0, \ldots, w_n)$ defining $\mathcal{U}$.

References

1. Barry, P. J., de Boor-Fix functionals and polar forms, Computer-Aided Geom. Design **7** (1990), 425–430.
2. Barry, P. J., De Boor-Fix dual functionals and algorithms for Tchebycheffian B-splines curves, TR 94-28, University of Minnesota, to appear in Constr. Approx.
3. Barry, P. J., N. Dyn, R. N. Goldman and C. A. Micchelli, Identities for piecewise polynomials spaces determined by connection matrices, Aequationes Mathematicae **42** (1991), 123–136.
4. Barry, P. J., R. N. Goldman and C. A. Micchelli, Knot insertion algorithms for piecewise polynomial spaces determined by connection matrices, Adv. Comput. Math. **1** (1993), 139–171.
5. Brézinski C., *Biorthogonality and its Applications*, M. Dekker, 1992.
6. de Boor, C. and G. Fix, Spline approximation by quasi-interpolants, J. Approx. Th. **8** (1973), 19–45.
7. Dyn, N. and A. Ron, Recurrence relations for Tchebycheffian B-splines, Journal d'Analyse Mathématique **51** (1988), 118–138.
8. Goldman, R. N., Dual polynomial bases, J. Approx. Th. **79** (1994), 311–346.
9. Karlin, S., *Total Positivity*, Stanford University Press, Stanford, 1968.
10. Marsden, M. J., An identity for spline functions with applications to variation-diminishing spline approximation, J. Approx. Th. **3** (1970), 7–49.
11. Mazure, M. L., Chebyshev spaces, RR 952M IMAG, Université Joseph Fourier, Grenoble, 1996.
12. Mazure, M. L. and H. Pottmann, Tchebycheff curves, in *Total Positivity and its Applications*, M. Gasca and C. A. Micchelli (eds.), Kluwer Acad. Pub., Doordrecht, 1996, 187–218.
13. Pottmann, H., The geometry of Tchebycheffian splines, Computer-Aided Geom. Design **10** (1993), 181–210.
14. Ramshaw, L., Blossoming: A connect-the-dots approach to splines, Digital Equipment Corporation, SRC technical report **#19** (1987).
15. Schumaker, L. L., *Spline Functions: Basic Theory*, Wiley Interscience, 1981.
16. Seidel, H. P., New algorithms and techniques for computing with geometrically continuous spline curves of arbitrary degree, Math. Modelling and Num. Anal. **26** (1992), 149–176.
17. Stefanus, Y. and R. N. Goldman, Blossoming Marsden's identity, Computer-Aided Geom. Design **9** (1992), 73–84.

Marie-Laurence Mazure and Pierre-Jean Laurent
LMC-IMAG, Université Joseph Fourier
BP 53, 38041 Grenoble (cedex 09) France
mazure@imag.fr
pjl@imag.fr

Precise Controls in Numerical Grid Generation

R. Morandi and A. Sestini

Abstract. The Control Point Form method for numerical grid generation is presented for the two-dimensional case. This method allows the construction of grids which precisely conform to the boundaries of the domain by transfinite interpolation methods. It is also able to manipulate the coordinate curves through a set of control points used to define directions to be interpolated. A set of efficient C^1 piecewise quadratic interpolation functions is carried out to obtain a C^2 transformation from the computational domain into the physical domain. Furthermore, a criterion to control the grid-spacing by means of the parameter-spacing is proposed for the two-dimensional case.

§1. Introduction

The numerical solution of boundary value problems over arbitrarily shaped domains often requires the application of a grid generation algorithm. To generate a grid with specified properties, i.e. boundary conformity, concentrating or widening of the coordinate curves, control of their direction and position, suitable transformations $\mathbf{B}(s,t)$ from the computational domain into the physical domain must be considered ([1],[6],[7],[8],[10],[11]).

In [4],[5] P.R. Eiseman has studied the Control Point Form method which is an efficient and flexible method to generate computational grids on "quadrilateral" domains. In fact, it allows the construction of grids which precisely conform to all the boundaries through a transfinite interpolation method [1]. At the same time, it allows the control of the shape and the position of the coordinate curves by manipulation of control points used to define the directions to be interpolated. The efficiency of the method arises from the choice of local functions to assemble the data (boundaries and control points).

In [2] and [3] two sets of local interpolation functions have been introduced to generate C^1 and C^2 transformations, respectively. The necessity of C^2 transformations arises when three-dimensional boundary value problems are considered or

Advanced Topics in Multivariate Approximation
F. Fontanella, K. Jetter and P. J. Laurent (eds.) pp. 243–258.

ISBN 981-02-2852-X

when second derivatives of geometric quantities have to be computed. In any case the set of functions proposed in [3] for the C^2 case is not completely satisfactory. In fact the control of the position of the coordinate curves is more difficult because the convex hull property does not hold, unlike the C^1 case. In addition, the number of control points defining the directions to be interpolated increases, thus complicating control-point manipulation.

In this paper an analogous set of local interpolation functions to generate a C^2 transformation is presented. Under suitable assumptions, it is possible to prove that the convex hull property is recovered if these functions are used. In addition, the directions to be interpolated depend on the same number of control points as for the C^1 transformation [2].

Furthermore, the problem of controlling the grid-spacing by means of the parameter-spacing is investigated, in order to have a uniform/non-uniform distribution of the coordinate curves in the physical domain regulated by a uniform/non-uniform distribution of the coordinate lines in the computational domain ("uniformity" property). This behaviour is obtained assuming a "pseudo-rectangular" shape of the physical domain which, as explained is section 4., is a more restrictive hypothesis than the "quadrilateral" shape previously assumed. The criterion used for this purpose is analogous to that presented in [2] for the one-dimensional case. In this paper it is completed and extended to the two-dimensional case. As confirmed by the presented results, a parametrization technique very useful to control the grid-spacing by the parameter-spacing is obtained.

In section 2. the method is introduced and the properties of the transformation are studied. In section 3. the new set of C^1 interpolation functions is presented and, in section 4., conditions to obtain the "uniformity" property are given for "pseudo-rectangular" domains. Finally, in section 5. some numerical examples are illustrated.

§2. Control Point Form Method.

In this section the Control Point Form method is recalled to define a two-dimensional grid on a "quadrilateral" domain $\Omega \subset \mathbb{R}^2$, [4],[5], where, a "quadrilateral" domain is a domain such that

$$\partial\Omega \equiv \cup_{i=1}^{4} \partial\Omega_i,$$

being $\partial\Omega_1$, $\partial\Omega_3$, and $\partial\Omega_2$, $\partial\Omega_4$, two pairs of opposite regular curves with

$$\partial\Omega_1 \cap \partial\Omega_3 = \emptyset, \quad \partial\Omega_2 \cap \partial\Omega_4 = \emptyset,$$

and where each $\partial\Omega_i, i = 1, \dots, 4$, intersects $\partial\Omega_j, j = 1, \dots, 4$, $j \neq i + 2 - 4[\frac{i+1}{4}]$ and $j \neq i$ (where $[x]$ denotes the integer part of x), only in an extreme point.

Furthermore some properties of this method, useful in numerical grid generation, are analyzed.

The Control Point Form method is an algebraic method [10] which, besides $\partial\Omega$, uses a set of control points in $\mathbb{R}^2$ to construct a transformation $\mathbf{B} : R = [0,1] \times [0,1] \to \Omega$ to define the grid. This permits the control of the shape and the position of the coordinate curves of the grid, while the grid conforms precisely to all the boundaries. These features are obtained by defining the transformation $\mathbf{B}(s,t)$ as follows

$$\mathbf{B}(s,t) = \mathbf{B}_{\partial\Omega}(s,t) + \mathbf{Tp}_0(s,t) \tag{2.1}$$

where $\mathbf{B}_{\partial\Omega}(s,t) : R \to \mathbb{R}^2$ is a function so that $\mathbf{B}_{\partial\Omega}|_{\partial R} = \partial\Omega$ and $\mathbf{Tp}_0(s,t) : R \to \mathbb{R}^2$ is a control function vanishing on the boundaries of R and defined by using the control points.

First, let us consider the function $\mathbf{B}_{\partial\Omega}(s,t)$ and let us suppose that there exists a function $\mathbf{F}(s,t) : R \to \mathbb{R}^2$, $\mathbf{F} \in C^{p+1}(R)$, $p \geq 0$, so that

$$Im(\mathbf{F}|_{t=0}) = \partial\Omega_1; Im(\mathbf{F}|_{t=1}) = \partial\Omega_3;$$

$$Im(\mathbf{F}|_{s=0}) = \partial\Omega_4; Im(\mathbf{F}|_{s=1}) = \partial\Omega_2.$$

Then $\mathbf{B}_{\partial\Omega}(s,t)$ is defined as the Boolean sum

$$\mathbf{B}_{\partial\Omega}(s,t) = (P_1 \oplus P_2)[\mathbf{F}(s,t)] = (P_1 + P_2 - P_1P_2)[\mathbf{F}(s,t)], \tag{2.2}$$

where

$$P_1[\mathbf{F}(s,t)] = \alpha_0(s)\mathbf{F}(0,t) + \alpha_1(s)\mathbf{F}(1,t),$$

$$P_2[\mathbf{F}(s,t)] = \beta_0(t)\mathbf{F}(s,0) + \beta_1(t)\mathbf{F}(s,1),$$

and the blending functions $\alpha_0(s), \alpha_1(s), \beta_0(t), \beta_1(t)$ are $C^{p+1}([0,1])$, cardinal and local functions, that is

$$\begin{aligned} \alpha_0(0) &= 1, \ \alpha_0(1) = 0; \\ \alpha_1(0) &= 0, \ \alpha_1(1) = 1; \\ \beta_0(0) &= 1, \ \beta_0(1) = 0; \\ \beta_1(0) &= 0, \ \beta_1(1) = 1; \end{aligned} \tag{2.3}$$

$$\begin{aligned} \alpha_0(s) &= 0, \forall s \geq s_{u+1}, \ u \geq 1, \\ \alpha_1(s) &= 0, \forall s \leq s_{n-1-u}, \ u \geq 1, \\ \beta_0(t) &= 0, \forall t \geq t_{u+1}, \ u \geq 1, \\ \beta_1(t) &= 0, \forall t \leq t_{m-1-u}, \ u \geq 1, \end{aligned} \tag{2.4}$$

with $0 = s_1 < s_2 < \ldots < s_{n-1} = 1$ and $0 = t_1 < t_2 < \ldots < t_{m-1} = 1$, and with $m \geq 2u+1$, $n \geq 2u+1$, where u is a positive integer related to the local support of the functions $\alpha_0(s)$, $\alpha_1(s)$, $\beta_0(t)$, $\beta_1(t)$ and to the local support of the basic interpolation functions defined below. In practice it is $u = 1$ if $p = 0$ [2] and $u = 2$ if $p = 1$, as shown in the following section.

Now, let us consider the control function $\mathbf{Tp}_0(s,t)$. It is defined by a set of control points $\{\mathbf{Q}_{i,j},\ i=1,\dots,m,\ j=1,\dots,n\}$ as

$$\mathbf{Tp}_0(s,t) = \mathbf{Tp}(s,t) - (P_1 \oplus P_2)[\mathbf{Tp}(s,t)], \tag{2.5}$$

where

$$\mathbf{Tp}(s,t) = \sum_{i=1}^{m}\sum_{j=1}^{n} \mathbf{Q}_{i,j} G_j(s) H_i(t), \tag{2.6}$$

with

$$G_1(s) = 1 - \frac{g_1(s)}{g_1(s_{n-1})},\quad G_n(s) = \frac{g_{n-1}(s)}{g_{n-1}(s_{n-1})},$$

$$G_j(s) = \frac{g_{j-1}(s)}{g_{j-1}(s_{n-1})} - \frac{g_j(s)}{g_j(s_{n-1})},\quad j=2,\dots,n-1,$$

and

$$H_1(t) = 1 - \frac{h_1(t)}{h_1(t_{m-1})},\quad H_m(t) = \frac{h_{m-1}(t)}{h_{m-1}(t_{m-1})},$$

$$H_i(t) = \frac{h_{i-1}(t)}{h_{i-1}(t_{m-1})} - \frac{h_i(t)}{h_i(t_{m-1})},\quad i=2,\dots,m-1.$$

The functions $g_j(s), j=1,\dots,n-1$, and $h_i(t), i=1,\dots,m-1$, are defined as

$$g_j(s) = \int_0^s \phi_j(\sigma)\, d\sigma,\ j=1,\dots,n-1, \tag{2.7}$$

$$h_i(t) = \int_0^t \psi_i(\tau)\, d\tau,\ i=1,\dots,m-1,$$

where the basic interpolation functions $\phi_j(s), j = 1,\dots,n-1$, are $C^p([0,1])$, $p \geq 0$ functions with the following properties

$$\phi_j(s_k) = 0,\ \text{if } k \neq j, k=1,\dots,n-1,\ j=1,\dots,n-1, \tag{2.8}$$

$$\phi_j(s) = 0 \begin{cases} \forall s \notin [s_1, s_{j+u}], \text{ if } j=1,\dots,u, \\ \forall s \notin [s_{j-u}, s_{j+u}], \text{ if } j=u+1,\dots,n-u-1, \\ \forall s \notin [s_{j-u}, s_{n-1}], \text{ if } j=n-u,\dots,n-1, \end{cases} \tag{2.9}$$

$$\sum_{j=1}^{n-1} \frac{\phi_j(s)}{\phi_j(s_j)} = 1,\quad \forall s \in [0,1]. \tag{2.10}$$

The same properties are assumed for the functions $\psi_i(t), i=1,\dots,m-1$.

The hypotheses (2.4) and (2.9) guarantee the local influence of data on the grid. Furthermore, the hypothesis (2.8) allows the calculation of $\frac{\partial \mathbf{Tp}}{\partial s}(s_k, t_l)$ by using the $\mathbf{Q}_{\cdot,k+1}$ and the $\mathbf{Q}_{\cdot,k}$ control points only. In fact, it is easy to verify that the following formula holds (and obviously the analogous formula for the partial derivative with respect to t),

$$\frac{\partial \mathbf{Tp}}{\partial s}(s_k, t_l) = \frac{\phi_k(s_k)}{g_k(s_{n-1})} \sum_{i=l-u+1}^{l+u} (\mathbf{Q}_{i,k+1} - \mathbf{Q}_{i,k}) H_i(t_l), \tag{2.11}$$

$$k = 1, \ldots, n-1, \; l = 2, \ldots, m-2.$$

Then, the shape of the coordinate curves can be completely controlled in the internal region $[s_{u+1}, s_{n-1-u}] \times [t_{u+1}, t_{m-1-u}]$ using the control points since (2.4) implies $\mathbf{B}(s,t) \equiv \mathbf{Tp}(s,t)$ in this region. Finally, as it will be clear after reading section 4., hypothesis (2.10) and the analogous one related to the functions $\psi_i(t), i = 1, \ldots, m-1$, when combined with a suitable parametrization technique, guarantee the so called "uniformity" property [2], that is the capability of controlling the grid-spacing by the parameter-spacing. Other useful properties of the transformation (2.1) are the affine invariance property and the convex hull property which are important in numerical grid generation to control the position of the coordinate curves of the grid through the control points and the boundaries. In particular, the convex hull property is related to the value of the integer u. In fact, if $s \in [s_i, s_{i+1}]$, $i = u+1, \ldots, n-u-2$, $t \in [t_k, t_{k+1}], k = u+1, \ldots, m-u-2$, $\mathbf{B}(s,t)$ is in the convex hull of the following control points

$$\mathbf{Q}_{i-u+1,k-u+1}, \ldots \mathbf{Q}_{i+u+1,k-u+1}, \ldots \mathbf{Q}_{i-u+1,k+u+1}, \ldots \mathbf{Q}_{i+u+1,k+u+1}.$$

It is clear that, near the boundaries, the contribution of the boundaries themselves has to be considered. So the greater u is, the greater the number of control points involved is and so the convex hull property is less effective.

Concerning the affine invariance property, it is easy to verify that it holds for each choice of the blending functions $\alpha_0(s), \alpha_1(s), \beta_0(t), \beta_1(t)$ satisfying (2.3). In fact it is a consequence of (2.1),(2.6) and of the following unity partition property

$$\sum_{j=1}^{n} G_j(s) = 1, \; \forall s \in [0,1], \; \sum_{i=1}^{m} H_i(t) = 1, \; \forall t \in [0,1]. \tag{2.12}$$

On the other hand, in order to obtain the convex hull property for $\mathbf{B}(s,t)$, it is necessary to suppose

$$\begin{aligned} &G_j(s) \geq 0, \forall s \in [0,1], j = 1, \ldots, n, \\ &H_i(t) \geq 0, \forall t \in [0,1], i = 1, \ldots, m, \end{aligned} \tag{2.13}$$

and

$$\begin{aligned} \alpha_0(s) &= G_1(s), \quad \alpha_1(s) = G_n(s), \\ \beta_0(t) &= H_1(t), \quad \beta_1(t) = H_m(t). \end{aligned} \tag{2.14}$$

It is also remarkable that (2.14) insures the control on the shape of the coordinate curves throughout the interior of the domain because it is

$$\frac{\partial \mathbf{B}}{\partial s}(s_k, t_l) = \frac{\partial \mathbf{Tp}}{\partial s}(s_k, t_l), k = 2, \ldots, n-2, l = 2, \ldots m-2, \tag{2.15}$$

$$\frac{\partial \mathbf{B}}{\partial t}(s_k, t_l) = \frac{\partial \mathbf{Tp}}{\partial t}(s_k, t_l), k = 2, \ldots, n-2, l = 2, \ldots m-2. \tag{2.16}$$

§3. C^1 Interpolation Functions.

A set of suitable C^0 piecewise linear interpolation functions to construct a C^1 transformation $\mathbf{B}(s,t)$ has been introduced in [2]. For this set of basic functions $u = 1$ and it is easy to see that all the properties discussed in section 2. hold. But, when it is necessary to have continuous second derivatives of geometric quantities for the numerical solution of a boundary value problem on the domain Ω, or when the possibility of an extension to the three-dimensional case is considered, C^1 interpolation functions have to be looked for, in order to obtain a C^2 transformation. A proposal of piecewise quadratic C^1 interpolation functions is given in [3]. If C^1 basic interpolation functions satisfying all the hypotheses (2.8)- (2.10) are searched for and if unspecified flat spots have to be avoided along the coordinate curves, it is also shown that the lowest value obtainable for u is two ([3],Theorem 1). In any case, by using these functions, the convex hull property is lost, as outlined in [3]. Furthermore there can be no simplification of (2.11) with $u = 2$.
In this section a different set of C^1 piecewise quadratic interpolation functions is introduced. Even though it is always $u = 2$, these functions preserve the convex hull property and, at the same time, allow the modification of (2.11) in

$$\frac{\partial \mathbf{Tp}}{\partial s}(s_k, t_l) = \frac{\phi_k(s_k)}{g_k(s_{n-1})} \sum_{i=l}^{l+1} (\mathbf{Q}_{i,k+1} - \mathbf{Q}_{i,k}) H_i(t_l), \tag{3.1}$$

$$k = 1, \ldots, n-1, \quad l = 2, \ldots, m-2.$$

Without loss of generality, in the following it will be referred only to the s-parameter. To obtain a set of cardinal C^1 interpolation functions centered about a local maximum, with the exception of the functions $\phi_1(s)$ and $\phi_{n-1}(s)$, the following interpolation conditions are imposed:

$$\begin{cases} \phi_k(s_{k-2}) & = 0, k = 3, \ldots, n-1, \\ \phi_k'(s_{k-2}) & = 0, k = 3, \ldots, n-1, \\ \phi_k(s_{k-1}) & = 0, k = 2, \ldots, n-1, \\ \phi_k(s_k) & = A, k = 1, \ldots, n-1, \\ \phi_k'(s_k) & = 0, k = 2, \ldots, n-2, \\ \phi_k(s_{k+1}) & = 0, k = 1, \ldots, n-2, \\ \phi_k(s_{k+2}) & = 0, k = 1, \ldots, n-3, \\ \phi_k'(s_{k+2}) & = 0, k = 1, \ldots, n-3, \end{cases}$$

where A is a suitable constant value. The following new conditions are also imposed

$$\begin{aligned} &\int_{s_{k-2}}^{s_{k-1}} \phi_k(s)ds = 0, \;\; k = 3, \ldots, n-1, \\ &\int_{s_{k+1}}^{s_{k+2}} \phi_k(s)ds = 0, \;\; k = 1, \ldots, n-3, \end{aligned} \tag{3.2}$$

to obtain (3.1). In fact, assuming the same properties for the interpolation functions $\psi_i(t), i = 1, \ldots, m-1$, it is easy to verify that

$$H_1(t_l) = \begin{cases} 1, & \text{if } l = 1, \\ 0, & \text{otherwise}, \end{cases}$$

$$H_i(t_l) = \begin{cases} \neq 0, & \text{if } l = i, i-1, \\ 0, & \text{otherwise}, \end{cases} \qquad i = 2, \ldots, m-1,$$

$$H_m(t_l) = \begin{cases} 1, & \text{if } l = m-1, \\ 0, & \text{otherwise}, \end{cases}$$

and thus (3.1) holds.

Then, to obtain such a set of interpolation functions, a generic interval $[s_k, s_{k+1}]$ is decomposed in three subintervals

$$[s_k, s_{k+1}] = I_k^+ \cup I_k^{k+1} \cup I_{k+1}^-,$$

where $I_k^+ = [s_k, w_k^{(1)}]$, $I_k^{k+1} = [w_k^{(1)}, w_k^{(2)}]$, and $I_{k+1}^- = [w_k^{(2)}, s_{k+1}]$, with

$$w_k^{(1)} = s_k + \alpha_1 \delta_k, w_k^{(2)} = s_k + \alpha_2 \delta_k,$$

$$0 < \alpha_1 < \alpha_2 < 1, \;\; \delta_k = s_{k+1} - s_k,$$

and, on each of these subintervals the interpolation functions are assumed polynomial functions of degree 2. If, for simplicity, it is assumed that $\alpha_1 = \frac{1}{3}$ and $\alpha_2 = \frac{2}{3}$, the definition of the function $\phi_k(s), k = 3, \ldots, n-3$, is given by the following formula

$$\phi_k(s) = \begin{cases} \frac{9d_{k-2}}{\delta_{k-2}^2}(s-s_{k-2})^2, & \text{if } s \in I^+_{k-2}, \\ -\frac{63d_{k-2}}{2\delta_{k-2}^2}(s-w^{(1)}_{k-2})^2 + \frac{6d_{k-2}}{\delta_{k-2}}(s-w^{(1)}_{k-2}) + d_{k-2}, & \text{if } s \in I^{k-1}_{k-2}, \\ \frac{99d_{k-2}}{2\delta_{k-2}^2}(s-w^{(2)}_{k-2})^2 - \frac{15d_{k-2}}{\delta_{k-2}}(s-w^{(2)}_{k-2}) - \frac{1}{2}d_{k-2}, & \text{if } s \in I^-_{k-1}, \\ (\frac{9}{\delta_{k-1}^2}c_{k-1} - \frac{54}{\delta_{k-2}\delta_{k-1}}d_{k-2})(s-s_{k-1})^2 + & \\ +\frac{18d_{k-2}}{\delta_{k-2}}(s-s_{k-1}), & \text{if } s \in I^+_{k-1}, \\ (\frac{9A-36c_{k-1}}{2\delta_{k-1}^2} + \frac{81d_{k-2}}{2\delta_{k-2}\delta_{k-1}})(s-w^{(1)}_{k-1})^2 + (\frac{6c_{k-1}}{\delta_{k-1}} + & \\ -\frac{18d_{k-2}}{\delta_{k-2}})(s-w^{(1)}_{k-1}) + c_{k-1}, & \text{if } s \in I^k_{k-1}, \\ (\frac{-9A+18c_{k-1}}{2\delta_{k-1}^2} - \frac{27d_{k-2}}{2\delta_{k-2}\delta_{k-1}})(s-w^{(2)}_{k-1})^2 + & \\ (\frac{3A-6c_{k-1}}{\delta_{k-1}} + \frac{9d_{k-2}}{\delta_{k-2}})(s-w^{(2)}_{k-1}) + & \\ +(\frac{A+2c_{k-1}}{2} - \frac{3\delta_{k-1}d_{k-2}}{2\delta_{k-2}}), & \text{if } s \in I^-_k, \\ (\frac{-9A+18b_k}{2\delta_k^2} - \frac{27d_{k+1}}{2\delta_k\delta_{k+1}})(w^{(1)}_k - s)^2 + (\frac{3A-6b_k}{\delta_k} + & \\ +\frac{9a_{k+1}}{\delta_{k+1}})(w^{(1)}_k - s) + (\frac{A+2b_k}{2} - \frac{3\delta_k a_{k+1}}{2\delta_{k+1}}), & \text{if } s \in I^+_k, \\ (\frac{9A-36b_k}{2\delta_k^2} + \frac{81a_{k+1}}{2\delta_{k+1}\delta_k})(w^{(2)}_k - s)^2 + (\frac{6b_k}{\delta_k} + & \\ -\frac{18a_{k+1}}{\delta_{k+1}})(w^{(2)}_k - s) + b_k, & \text{if } s \in I^{k+1}_k, \\ (\frac{9b_k}{\delta_k^2} - \frac{54a_{k+1}}{\delta_k\delta_{k+1}})(s_{k+1}-s)^2 + & \\ +\frac{18a_{k+1}}{\delta_{k+1}}(s_{k+1}-s), & \text{if } s \in I^-_{k+1}, \\ \frac{99a_{k+1}}{2\delta_{k+1}^2}(w^{(1)}_{k+1}-s)^2 - \frac{15a_{k+1}}{\delta_{k+1}}(w^{(1)}_{k+1}-s) - \frac{1}{2}a_{k+1}, & \text{if } s \in I^+_{k+1}, \\ -\frac{63a_{k+1}}{2\delta_{k+1}^2}(w^{(2)}_{k+1}-s)^2 + \frac{6a_{k+1}}{\delta_{k+1}}(w^{(2)}_{k+1}-s) + a_{k+1}, & \text{if } s \in I^{k+2}_{k+1}, \\ \frac{9a_{k+1}}{\delta_{k+1}^2}(s_{k+2}-s)^2, & \text{if } s \in I^-_{k+2}, \\ 0, & \text{otherwise}, \end{cases} \tag{3.3}$$

where,

$$a_k = \phi_{k-1}(w^{(2)}_k),\ b_k = \phi_k(w^{(2)}_k),\ c_k = \phi_{k+1}(w^{(1)}_k),\ d_k = \phi_{k+2}(w^{(1)}_k).$$

The functions $\phi_2(s)$ and $\phi_{n-2}(s)$ are suitable truncated functions of the general function given above. In this case it is necessary to consider the intervals $[s_0, s_1]$

and $[s_{n-1}, s_n]$ to define δ_0, d_0 and δ_{n-1}, a_{n-1}, respectively. In the case where $k=1$, assuming $\phi_1'(s_1) = -\phi_2'(s_1) = \frac{-18d_0}{\delta_0}$, it is

$$\phi_1(s) = \begin{cases} (\frac{-9A+18b_1}{2\delta_1^2} + \frac{81d_0}{2\delta_0\delta_1} - \frac{27a_2}{2\delta_1\delta_2})(w_1^{(1)} - s)^2 - (\frac{-3A+6b_1}{\delta_1} + \\ +\frac{9d_0}{\delta_0} - \frac{9a_2}{\delta_2})(w_1^{(1)} - s) + (\frac{A+2b_1}{2} - \frac{3\delta_1 d_0}{2\delta_0} - \frac{3\delta_1 a_2}{2\delta_2}), & \text{if } s \in I_1^+, \\ (\frac{9A-36b_1}{2\delta_1^2} - \frac{27d_0}{2\delta_0\delta_1} + \frac{81a_2}{2\delta_1\delta_2})(w_1^{(2)} - s)^2 + (\frac{6b_1}{\delta_1} + \\ -\frac{18a_2}{\delta_2})(w_1^{(2)} - s) + b_1, & \text{if } s \in I_1^2, \\ (\frac{9b_1}{\delta_1^2} - \frac{54a_2}{\delta_1\delta_2})(s_2 - s)^2 + \frac{18a_2}{\delta_2}(s_2 - s), & \text{if } s \in I_2^-, \end{cases} \tag{3.4}$$

and the general definition for $k=1$ in the subintervals I_2^+, I_2^3, I_3^- holds. Analogously, in the case where $k = n-1$, assuming $\phi_{n-1}'(s_{n-1}) = -\phi_{n-2}'(s_{n-1}) = \frac{18a_{n-1}}{\delta_{n-1}}$, it is

$$\phi_{n-1}(s) = \begin{cases} (\frac{9c_{n-2}}{\delta_{n-2}^2} - \frac{54d_{n-3}}{\delta_{n-2}\delta_{n-3}})(s - s_{n-2})^2 + \\ +\frac{18d_{n-3}}{\delta_{n-3}}(s - s_{n-2}), & \text{if } s \in I_{n-2}^+, \\ (\frac{9A-36c_{n-2}}{2\delta_{n-2}^2} - \frac{27a_{n-1}}{2\delta_{n-2}\delta_{n-1}} + \frac{81d_{n-3}}{2\delta_{n-3}\delta_{n-2}})(s + \\ -w_{n-2}^{(1)})^2 + (\frac{6c_{n-2}}{\delta_{n-2}} - \frac{18d_{n-3}}{\delta_{n-3}})(s - w_{n-2}^{(1)}) + c_{n-2}, & \text{if } s \in I_{n-2}^{n-1}, \\ (\frac{-9A+18c_{n-2}}{2\delta_{n-2}^2} + \frac{81a_{n-1}}{2\delta_{n-2}\delta_{n-1}} - \frac{27d_{n-3}}{2\delta_{n-3}\delta_{n-2}})(s + \\ -w_{n-2}^{(2)})^2 - (\frac{-3A+6c_{n-2}}{\delta_{n-2}} + \frac{9a_{n-1}}{\delta_{n-1}} - \frac{9d_{n-3}}{\delta_{n-3}})(s + \\ -w_{n-2}^{(2)}) + (\frac{A+2c_{n-2}}{2} - \frac{3\delta_{n-2}a_{n-1}}{2\delta_{n-1}} - \frac{3\delta_{n-2}d_{n-3}}{2\delta_{n-3}}), & \text{if } s \in I_{n-1}^-, \end{cases} \tag{3.5}$$

and the general definition for $k = n-1$ in the subintervals $I_{n-3}^+, I_{n-3}^{n-2}, I_{n-2}^-$ holds.

Then the following symmetry conditions

$$\begin{aligned} &a_k = d_k, \quad c_k = b_k, \quad k = 2, \dots, n-3, \\ &b_1 = c_1 + d_1, \quad c_{n-2} = b_{n-2} + a_{n-2}, \\ &\frac{d_0}{\delta_0} = \frac{a_{n-1}}{\delta_{n-1}}, \end{aligned}$$

are imposed to obtain a symmetric behaviour of the non-vanishing interpolation functions in each interval $[s_k, s_{k+1}], k = 1, \dots, n-2$, and it is assumed $\delta_0 = \delta_1$, $\delta_{n-1} = \delta_{n-2}$.

Finally the condition (2.10) has to be imposed. Since $\phi_j(s_j) = A, \ j = 1\dots, n-1$, (2.10) can be equivalently written as

$$\sum_{j=1}^{n-1} \phi_j'(s) = 0. \tag{3.6}$$

Therefore, taking the locality of the interpolation functions into account and considering that they are C^1 piecewise quadratic polynomial functions, (3.6) becomes

$$\begin{aligned}
&\phi'_{k-1}(s_k) + \phi'_{k+1}(s_k) = 0, \ k = 2, \ldots, n-2, \\
&\phi'_{k-1}(w_k^{(1)}) + \phi'_k(w_k^{(1)}) + \phi'_{k+1}(w_k^{(1)}) + \phi'_{k+2}(w_k^{(1)}) = 0, \ k = 2, \ldots, n-3, \\
&\phi'_{k-1}(w_k^{(2)}) + \phi'_k(w_k^{(2)}) + \phi'_{k+1}(w_k^{(2)}) + \phi'_{k+2}(w_k^{(2)}) = 0, \ k = 2, \ldots, n-3, \\
&\phi'_1(w_1^{(1)}) + \phi'_2(w_1^{(1)}) + \phi'_3(w_1^{(1)}) = 0, \\
&\phi'_1(w_1^{(2)}) + \phi'_2(w_1^{(2)}) + \phi'_3(w_1^{(2)}) = 0, \\
&\phi'_{n-3}(w_{n-2}^{(1)}) + \phi'_{n-2}(w_{n-2}^{(1)}) + \phi'_{n-1}(w_{n-2}^{(1)}) = 0, \\
&\phi'_{n-3}(w_{n-2}^{(2)}) + \phi'_{n-2}(w_{n-2}^{(2)}) + \phi'_{n-1}(w_{n-2}^{(2)}) = 0.
\end{aligned}$$

In fact, the values previously assigned to $\phi'_1(s_1)$ and to $\phi'_{n-1}(s_{n-1})$ guarantee the validity of (3.6) in the values s_1, s_{n-1} as well. After some algebra it is possible to verify that only the parameter d_0 is really free. Figures 3.1a-3.1c show the C^1 interpolation functions for a uniform parametrization and for $d_0 = \frac{A}{30}, A = 1$. When d_0 is a positive value so that

$$\int_{s_{k-2}}^{s_{k-1}} |\phi_k(s)| ds << \int_{s_{k-1}}^{s_k} |\phi_k(s)| ds, \ k = 3, \ldots, n-1, \tag{3.7}$$

$$\int_{s_{k+1}}^{s_{k+2}} |\phi_k(s)| ds << \int_{s_k}^{s_{k+1}} |\phi_k(s)| ds, \ k = 1, \ldots, n-3, \tag{3.8}$$

it is also possible to prove that (2.13) holds [9] and thus the convex hull property is preserved.

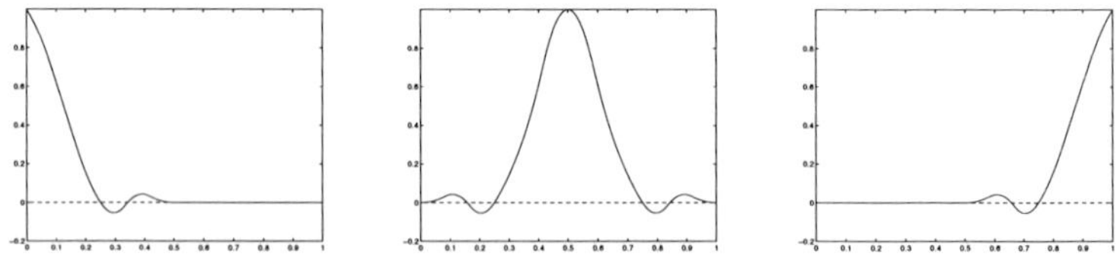

Fig. 3.1. Interpolation functions with $n = 6$.
Fig. 3.1a (on the left), 3.1b (in the center), 3.1c (on the right) show $\phi_1(s), \phi_3(s), \phi_{n-1}(s)$, respectively.

§4. Uniformity Property

In this section the uniformity property introduced in [2] for the one-dimensional case is extended to the two-dimensional case. With this property, it is possible to control the grid-spacing through the parameter-spacing. In particular, this control is obtained in the sense of the projections on assigned directions $\boldsymbol{\eta}$ and $\boldsymbol{\theta}$ of the vectors $(\mathbf{B}(s,t)-\mathbf{B}(0,t))$ and $(\mathbf{B}(s,t)-\mathbf{B}(s,0))$, respectively. In fact the goal is to look for these projections to be linearly increasing with s and t respectively, so that both the projected grid-spacings exactly conform to the related parameter-spacings. Obviously this behaviour is reasonably looked for when it is possible to associate each family of coordinate curves with a favorable direction, so that the control of the projected grid-spacings implies a good control of the grid-spacings themselves. This request can be reasonably formulated if the "quadrilateral" domain Ω is "pseudo-rectangular", that is if, for each pair of opposite boundaries, it is possible to find a favourite direction so that the related projections of the tangent vectors to both the opposite boundaries are positive everywhere. So in this section it is assumed that it is possible to find two non parallel unit vectors $\boldsymbol{\eta}$ and $\boldsymbol{\theta}$ so that,

$$(\mathbf{F}(s,0)-\mathbf{F}(0,0))\cdot\boldsymbol{\eta}=K_\eta(0)s,\ \ \forall s\in[0,1],$$

$$(\mathbf{F}(s,1)-\mathbf{F}(0,1))\cdot\boldsymbol{\eta}=K_\eta(1)s,\ \ \forall s\in[0,1],$$

where $K_\eta(0)=(\mathbf{F}(1,0)-\mathbf{F}(0,0))\cdot\boldsymbol{\eta},\ K_\eta(1)=(\mathbf{F}(1,1)-\mathbf{F}(0,1))\cdot\boldsymbol{\eta}$,

$$(\mathbf{F}(0,t)-\mathbf{F}(0,0))\cdot\boldsymbol{\theta}=K_\theta(0)t,\ \ \forall t\in[0,1],$$

$$(\mathbf{F}(1,t)-\mathbf{F}(1,0))\cdot\boldsymbol{\theta}=K_\theta(1)t,\ \ \forall t\in[0,1],$$

where $K_\theta(0)=(\mathbf{F}(0,1)-\mathbf{F}(0,0))\cdot\boldsymbol{\theta},\ K_\theta(1)=(\mathbf{F}(1,1)-\mathbf{F}(1,0))\cdot\boldsymbol{\theta}$, or, equivalently,

$$\frac{\partial\mathbf{F}}{\partial s}(s,0)\cdot\boldsymbol{\eta}=K_\eta(0),\forall s\in[0,1],\ \ \frac{\partial\mathbf{F}}{\partial s}(s,1)\cdot\boldsymbol{\eta}=K_\eta(1),\forall s\in[0,1],$$

$$\frac{\partial\mathbf{F}}{\partial t}(0,t)\cdot\boldsymbol{\theta}=K_\theta(0),\forall t\in[0,1],\ \ \frac{\partial\mathbf{F}}{\partial t}(1,t)\cdot\boldsymbol{\theta}=K_\theta(1),\forall t\in[0,1],$$

where " $\cdot$ " denotes the dot product. It has to be noted that this hypothesis is not restrictive with respect to the previous assumption of a "pseudo-rectangular" domain, because, eventually, it implies only a reparametrization of the boundaries. On the other hand, it is clear that not all "quadrilateral" domains are "pseudo-rectangular" and so the class of admissible geometries is reduced. In any case, as shown in the results presented in the following section, for this class of admissible geometries, a good control of the grid-spacing by the parameter-spacing is obtained quite easily, combining (2.10) with an adequate parametrization technique.

Then, with these assumptions on the boundaries, it is reasonable to look for the same behaviour for the coordinate curves defined by (2.1), that is for

$$(\mathbf{B}(s,t) - \mathbf{B}(0,t)) \cdot \boldsymbol{\eta} = K_\eta(t)s, \ \forall s \in [0,1], \ \forall t \in [0,1],$$

$$(\mathbf{B}(s,t) - \mathbf{B}(s,0)) \cdot \boldsymbol{\theta} = K_\theta(s)t, \ \forall t \in [0,1], \ \forall s \in [0,1],$$

or, equivalently,

$$\frac{\partial \mathbf{B}}{\partial s}(s,t) \cdot \boldsymbol{\eta} = K_\eta(t), \ \forall s \in [0,1], \forall t \in [0,1], \tag{4.1}$$

$$\frac{\partial \mathbf{B}}{\partial t}(s,t) \cdot \boldsymbol{\theta} = K_\theta(s), \ \forall t \in [0,1], \forall s \in [0,1].$$

As in the previous section, only the s-parameter is discussed.
Let us define the function

$$Sp(s) = \frac{1}{m-3} \sum_{i=2}^{m-2} (\mathbf{B}(s,t_i) - \mathbf{B}(0,t_i)) \cdot \boldsymbol{\eta}, \tag{4.2}$$

where, $t = t_1 = 0$ and $t = t_{m-1} = 1$ are not included in the sum, because of the hypothesis on the boundary. For each s, $Sp(s)$ is a discrete mean value with respect to the t-parameter of $(\mathbf{B}(s,t) - \mathbf{B}(0,t)) \cdot \boldsymbol{\eta}$. So the aim is to have

$$\frac{dSp}{ds}(s) = \overline{K}_\eta, \ \forall s \in [0,1], \tag{4.3}$$

where

$$\overline{K}_\eta = \overline{K}_\eta(t_2, \ldots, t_{m-2}) = \frac{1}{m-3} \sum_{i=2}^{m-2} K_\eta(t_i).$$

After a little algebra, if (2.15) holds, through the properties of the functions $H_i(t)$, $i = 1, \ldots, m$, the following formula can be written

$$\frac{dSp}{ds}(s) = \begin{cases} \left(\frac{1}{m-3}\sum_{i=2}^{m-2}(\mathbf{P}_2(t_i) - \mathbf{F}(0,t_i)) \cdot \boldsymbol{\eta}\right)\frac{\phi_1(s)}{g_1(s_{n-1})} \\ +\sum_{j=2}^{k+u}\left(\frac{1}{m-3}\sum_{i=2}^{m-2}(\mathbf{P}_{j+1}(t_i) - \mathbf{P}_j(t_i)) \cdot \boldsymbol{\eta}\right)\frac{\phi_j(s)}{g_j(s_{n-1})}, \\ s \in [s_k, s_{k+1}], \ 1 \le k \le u, \\ \\ \sum_{j=k-u+1}^{k+u}\left(\frac{1}{m-3}\sum_{i=2}^{m-2}(\mathbf{P}_{j+1}(t_i) - \mathbf{P}_j(t_i)) \cdot \boldsymbol{\eta}\right)\frac{\phi_j(s)}{g_j(s_{n-1})}, \\ s \in [s_k, s_{k+1}], \ u+1 \le k \le n-u-2, \\ \\ \sum_{j=k-u}^{n-2}\left(\frac{1}{m-3}\sum_{i=2}^{m-2}(\mathbf{P}_{j+1}(t_i) - \mathbf{P}_j(t_i)) \cdot \boldsymbol{\eta}\right)\frac{\phi_j(s)}{g_j(s_{n-1})} \\ +\left(\frac{1}{m-3}\sum_{i=2}^{m-2}(\mathbf{F}(1,t_i) - \mathbf{P}_{n-1}(t_i)) \cdot \boldsymbol{\eta}\right)\frac{\phi_{n-1}(s)}{g_{n-1}(s_{n-1})}, \\ s \in [s_k, s_{k+1}], \ n-u-1 \le k \le n-2, \end{cases}$$

where

$$\mathbf{P}_j(t) = \sum_{i=1}^{m} \mathbf{Q}_{i,j} H_i(t), \quad j = 1, \dots, n.$$

Then, by imposing

$$\frac{dSp}{ds}(s_j) = \overline{K}_\eta, j = 1, \dots, n-1,$$

the following set of equations

$$\begin{aligned}
&\frac{1}{m-3}\sum_{i=2}^{m-2}(\mathbf{P}_2(t_i) - \mathbf{F}(0,t_i)) \cdot \boldsymbol{\eta}\frac{\phi_1(s_1)}{g_1(s_{n-1})} = \overline{K}_\eta, \\
&\frac{1}{m-3}\sum_{i=2}^{m-2}(\mathbf{P}_{j+1}(t_i) - \mathbf{P}_j(t_i)) \cdot \boldsymbol{\eta}\frac{\phi_j(s_j)}{g_j(s_{n-1})} = \overline{K}_\eta, \; j = 2, \dots, n-2 \qquad (4.4)\\
&\frac{1}{m-3}\sum_{i=2}^{m-2}(\mathbf{F}(1,t_i) - \mathbf{P}_{n-1}(t_i)) \cdot \boldsymbol{\eta}\frac{\phi_{n-1}(s_{n-1})}{g_{n-1}(s_{n-1})} = \overline{K}_\eta
\end{aligned}$$

is obtained.
If (4.4) holds, (4.3) can be easily simplified in the condition (2.10). Considering as known quantities the t-parameter values $t_1, \dots, t_{m-1}$, the unknowns of the system (4.4) are the $(n-2)$ s-parameter steps $\delta_i = (s_{i+1} - s_i), i = 1, \dots, n-2$, which are constrained in $(0,1)$ and sum up to 1. The system is a linear system, being always $g_1(s_{n-1}) = \frac{A}{2}\delta_1$, $g_j(s_{n-1}) = \frac{A}{2}(\delta_{j-1} + \delta_j), j = 2, \dots, n-2$, $g_{n-1}(s_{n-1}) = \frac{A}{2}\delta_{n-2}$.

§5. Numerical Examples

In this section some examples of computational grids obtained by the described technique are presented just to confirm and illustrate the properties of the transformation $\mathbf{B}(s,t)$ discussed in the previous sections. All the grids shown in the figures have been constructed by using the C^1 interpolation functions presented in section 3. With the exception of the Figure 5.2a, the parametrization has been obtained through the approximated solution of the system (4.4) and of the analogous one in the t-parameter. In particular, it has been obtained solving the constrained least-squares problem associated to the system (4.4). In practice, an iterative strategy has been used, assuming $t_1, t_2, \dots, t_{m-1}$ as known quantities and $\delta_1, \delta_2, \dots \delta_{n-2}$ as unknowns in (4.4) and exactly the opposite assumption holding for the analogous system associated to the t-parameter.

Figure 5.1a and the enlargement 5.1b of its lower part relate to a vertical section of a "chalice" shaped domain and they are obtained by using a uniform parameter tabulation for both s and t.
Looking at these figures, the locality of the proposed scheme is clear. In fact, the influence of the control points, denoted with the symbol "o", diminishes quickly near

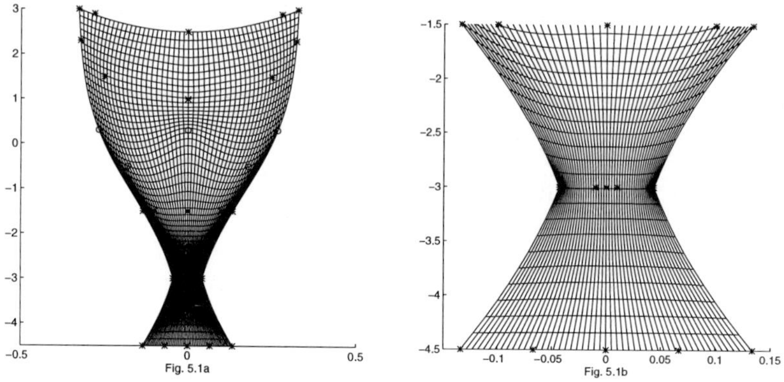

Fig. 5.1. The computational grid and its enlargement with $n = 5, m = 6$.

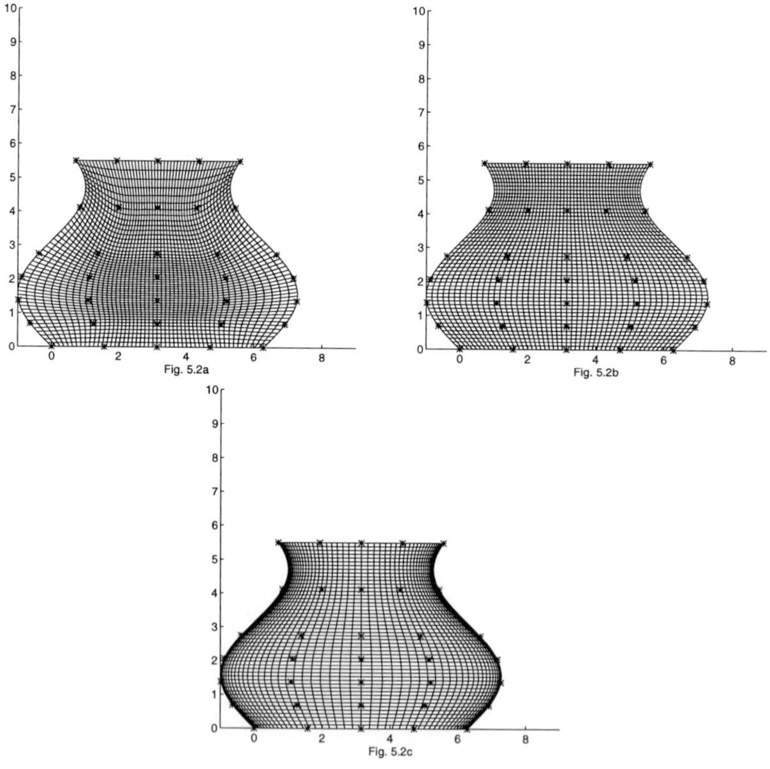

Fig. 5.2. The computational grid with $n = 5, m = 7$.

the bottom of the domain and the s-parameter grid curves become quite straight, as shown in the enlargement.

Figures 5.2a, 5.2b and 5.2c relate to a vertical section of an "amphora" shaped domain and are used to check the "uniformity" property. In fact the control points are unchanged in all the three figures but the parameter values $s_1, \ldots, s_{n-1}$, $t_1, \ldots, t_{m-1}$, and/or the s-parameter and t-parameter tabulations used to obtain the grids change. In Figures 5.2a and 5.2b a uniform parameter tabulation has been used for both s and t. In Figure 5.2a also a uniform parametrization has been used, so no approximation of the solution of the system (4.4) has been computed. The figure shows a very poor grid quality because, although there was no particular request, the vertical coordinate curves are more closely spaced in the center and in addition the horizontal ones are curved even if the control points are collinear. This behaviour is a consequence of a poor parametrization. In fact, the improved grid depicted in Figure 5.2b is obtained using the parametrization technique described at the beginning of this section. Figure 5.2c confirms the possibility of regulating the grid-spacing through the parameter-spacing. In this case, the s-parameter is tabulated by means of a stretching function. So, as shown in the figure, when the s-parameter is more closely spaced at the extremes of the interval $[0, 1]$, the vertical curves also concentrate at the lateral boundaries of the domain.

Acknowledgements. Work partially supported by M.U.R.S.T. The authors would like also to express their gratitude to the Referee for the useful suggestions and remarks.

References

1. Gordon, W. J. and C. A. Hall, Construction of curvilinear co-ordinate systems and applications to mesh generation, Int. J. Numer. Meth. Eng. **7** (1973), 461-477.
2. Eiseman, P. R., Coordinate generation with precise controls over mesh properties, J. Comput. Phys. **47** (1981), 331–351.
3. Eiseman, P. R., High level continuity for coordinate generation with precise controls, J. Comput. Phys **47** (1982), 352–374.
4. Eiseman, P. R., A control point form of algebraic grid generation, Int. J. Numer. Meth. Fluids **8** (1988), 1165–1181.
5. Eiseman, P. R., Control point grid generation, Comput. Math. Applic. **24** (1992), 57–67.
6. Khamayseh, A. and C. W. Mastin, Surface grid generation based on elliptic PDE models, Appl. Math. Comput. **65** (1994), 253-264.
7. Manke, J. W., A tensor product B-spline method for numerical grid generation, J. Comput. Phys. **108** (1993), 15-26.
8. Saunders, B. V. and P. W. Smith, Grid generation and optimization using tensor-product B-splines, Approx. Theory Appl. **3:4** (1987), 120-152.

9. Sestini A., Univalency conditions for the control point form method, preprint, University of Florence, 1996.
10. Thompson, J. F., Z. U. A. Warsi and C. W. Mastin, *Numerical Grid Generation. Foundations and Applications*, Elsevier Science Publishers, 1985.
11. Yang, J. C. and B. K. Soni, Structured adaptive grid generation, Appl. Math. Comput. **65** (1994), 265-278.

Rossana Morandi
Dipartimento di Matematica
Universitá di Perugia
Via Vanvitelli 1
06123 Perugia (Italy)
morandi@gauss.dipmat.unipg.it

Alessandra Sestini
Dipartimento di Energetica
Universitá di Firenze
Via C.Lombroso 17/6
50134 Firenze (Italy)
sestini@ingfi1.ing.unifi.it

A C^1 Positivity Preserving Scattered Data Interpolation Scheme

B. H. Ong and H. C. Wong

Abstract. Given a set of positive data on scattered points in the plane, a local non-parametric C^1 scheme is presented for generating a surface which interpolates the given set of data and preserves the positivity of the data. The domain is triangulated and the side-vertex method for interpolation in triangles is used. Rational cubics are used for univariate interpolation along line segments joining a vertex to the opposite edge of a triangle. The weights of these rational cubics are chosen to ensure that the blending surfaces are positivity preserving. The interpolating surface is obtained piecewise as a convex combination of three triangular surface patches.

§1. Introduction

The problem of positivity preserving interpolation, i.e. interpolation to nonnegative data by a nonnegative function, can be of interest. When the data arise from a physical experiment where negative values are not physically meaningful, it is vital that the interpolant preserves positivity. Positivity preserving or more generally range restricted interpolation has been considered in a number of papers. For example, in [3,5,7,14,15,17] polynomials, rationals or exponential splines are used in a direct approach for the univariate case of this problem while in [4,6,13] a variational approach is used. In [2,9,10,12,16,18] the multivariate case has been considered as an extension to the univariate case.

Brodlie et al in [2] and Mulansky and Schmidt in [10] have presented a local positivity preserving bivariate C^1 interpolation method for data on a rectangular grid by deriving sufficient conditions on the first partial derivatives and the mixed second partial derivatives at each data site. The former extend results from the positive univariate piecewise interpolation in [3] to ensure a piecewise bicubic interpolant lies between two bounding planes as the data. The latter extend the results

Advanced Topics in Multivariate Approximation
F. Fontanella, K. Jetter and P. J. Laurent (eds.) pp. 259–274.

ISBN 981-02-2852-X

from the univariate quadratic splines with additional knots in [9] and establish a system of linear inequalities with the first partial derivatives and second mixed partial derivatives as parameter ce and these estimates are modified as little as possible in order to fulfill the sufficient conditions. This leads to independent quadratic programs which can be solved efficiently by standard methods of quadratic opimization. We describe in [12,18] a C^1 positivity preserving local scheme for interpolating positive data given on a rectangular grid by using rational cubics and by extending the results in [7]. Utreras in [16] develops with the variational approach the positive thin plate splines by extending previous results in one variable to two variables. In [9] Mulansky and Schmidt construct a range restricted bivariate C^1 interpolant to scattered data using quadratic splines on a Powell-Sabin refinement of a triangulation of the data sites. The derived sufficient conditions result in a solvable system of linear inequalities with the gradients as parameters, which is separated with respect to the data sites. The selection of a suitable interpolant is based on the minimum norm modification of an initial interpolant which reduces to the solution of independent local quadratic programs, or on the minimization of the thin plate functional which results in a global optimization problem.

Here we consider a set of N given positive data $\{z_i\}$ on scattered points $\{(x_i, y_i)\}$ in the plane and describe the construction of a positivity preserving functional surface $F(x, y)$ with continuous first order partial derivatives and $F(x_i, y_i) = z_i, i = 1, \ldots, N$. The points $V_i = (x_i, y_i), i = 1, \ldots, N$ are used as the vertices of a triangulation of the domain D which is the convex hull of $\{V_i | \, i = 1, \ldots, N\}$. The interpolating surface depends on the triangulation method and on the adopted formulas for estimating the gradients.These aspects are not discussed here, and we concentrate our effort on the positivity preserving properties. We suppose that the domain have been triangulated with the points $V_i, i = 1, 2, \ldots, N$ as vertices by the Delaunay triangulation method. The estimation of the first order x and y partial derivatives F_x and F_y at each (x_i, y_i) for the surface F is obtained by using the method in [8]. Other methods of triangulation and derivative estimation may be used.The side-vertex method described by Nielson in [11] for interpolation in triangles is used to generate triangular patches. The first method of this type of blending for triangular patches was presented by Barnhill, Birkhoff and Gordon [1]. Their method was based upon the combination of univariate interpolation operators along lines parallel to the sides of the triangles. The univariate operators for the side-vertex method interpolate along lines joining a vertex and its opposite side. Among the two methods of combining the univariate operators described in [11], we adopt the method based upon a technique due to J. H. Brown and F. F. Little which uses some convex combination of the three operators O_i of a triangle, where each O_i is the Hermite operator along line segments joining a vertex V_i to its opposing edge. We have not used the method where the interpolant is based upon the Boolean sum of the operators O_i as the Boolean sum does not necessarily inherit the positivity preserving property of O_i but the convex combination does. We shall use rational cubics of the form:

$$R(t) = \frac{a(1-t)^3 + (a\omega + D_1)(1-t)^2t + (d\omega - D_2)(1-t)t^2 + dt^3}{1 + t(1-t)(\omega - 3)}, \tag{1.1}$$

where the weight $\omega > 0, 0 \le t \le 1$, and $a > 0, d > 0$ are its values while D_1, D_2 its derivatives at $t = 0$ and $t = 1$ respectively.The weights of these rational cubics have to be chosen suitably to ensure that the resulting curves are non-negative. As $R(t)$ reduces to a cubic polynomial when $\omega = 3$, we shall always start with a default value of 3 for ω and increase it when necessary. The interpolating surface is obtained piecewise as a convex combination of three triangular surface patches. The following theorem, quoted from [7], will be used to determine whether a rational cubic curve segment crosses a given constraint line by applying criterion (1.3) of the theorem to the numerator of the rational cubic. If the rational cubic does cross the constraint line, then its weight is increased so that equality in criterion (1.3) is achieved, in which case the modified curve segment just touches the constraint line.

Theorem 1. *Let*

$$p(x) = A(1-x)^3 + 3Bx(1-x)^2 + 3Cx^2(1-x) + Dx^3, \quad 0 \le x \le 1,$$

where $A, D > 0$, and $B < 0$ and/or $C < 0$.

If $p(x) < 0$ for some $x \in (0,1)$ [resp. $p(x) = 0$ for only one point in $(0,1)$], then

$$C^2 > BD, \quad B^2 > AC, \tag{1.2}$$

and

$$3B^2C^2 + 6ABCD - 4(AC^3 + B^3D) - A^2D^2 > 0 \quad [\text{resp.} = 0]. \tag{1.3}$$

Moreover if (1.3) holds, then (1.2) also holds and $p(x) < 0$ for some $x \in (0,1)$ [resp. $p(x) = 0$ for only one point in $(0,1)$].

In section 2, we describe the generation of a curve network on the edges of the triangulation of the domain. The blending in each triangle by the side-vertex method and the estimation of the weights are discussed in section 3. We conclude with some numerical examples in the last section.

§2. Generation of Curve Network

To localize the scheme, it is convenient to use the barycentric coordinates (b_i, b_j, b_k) for points of an arbitrary triangle T with vertices V_i, V_j and V_k. The vertices for any triangle will be labelled in the anti-clockwise order. Let e_{jk} represent the edge from vertex V_j to vertex V_k as shown in figure 2.1.

The derivative values $\frac{\partial F}{\partial e_{jk}}(V_\ell), \ell = j$ or k, in the direction of e_{jk} are given by :

$$\frac{\partial F}{\partial e_{jk}}(V_\ell) = \frac{\partial F}{\partial x}(V_\ell)\frac{(x_k - x_j)}{\|e_{jk}\|} + \frac{\partial F}{\partial y}(V_\ell)\frac{(y_k - y_j)}{\|e_{jk}\|}, \tag{2.1}$$

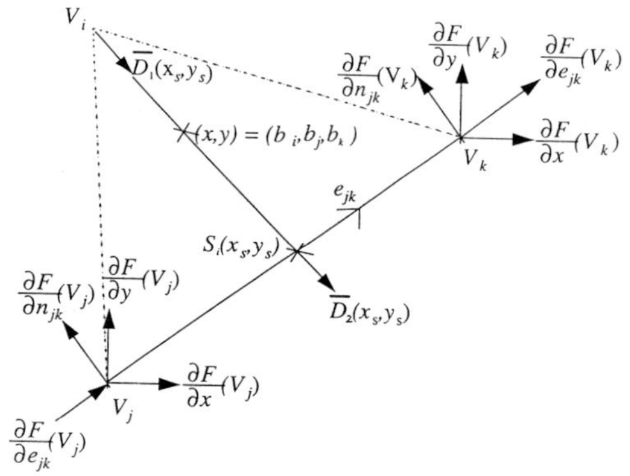

Fig. 2.1. Notations on a triangle.

where $\|e_{jk}\| = \sqrt{(x_k - x_j)^2 + (y_k - y_j)^2}$. A rational cubic of the form in (1.1) is constructed on every edge e_{jk} interpolating the data values and the partial derivatives $\|e_{jk}\| \frac{\partial F}{\partial e_{jk}}$ at the vertices V_j and V_k. The weight of the rational cubic is chosen such that the curve does not cross the plane $z = \delta$ for some $\delta > 0$. The choice of δ varies with the given data set. We suggest δ to be given as some fraction of the distance of the lowest data point from the xy plane. The resulting network of curves divides the surface which is yet to be constructed into an assembly of triangular patches. For convenience we shall refer to the curves in this network as boundary curves. This network of curves will be preserved as part of the final interpolating surface.

The derivative normal to the edge e_{jk} at the vertex $V_\ell, \ell = j$ or k, is given by

$$\frac{\partial F}{\partial n_{jk}}(V_\ell) = -\frac{\partial F}{\partial x}(V_\ell)\frac{(y_k - y_j)}{\|e_{jk}\|} + \frac{\partial F}{\partial y}(V_\ell)\frac{(x_k - x_j)}{\|e_{jk}\|}. \tag{2.2}$$

We shall now define the derivatives of the surface F on all the edges of the triangles in the domain. With the boundary curve defined on edge e_{jk}, the derivative $\frac{\partial F}{\partial e_{jk}}$ at each point on the edge e_{jk} is determined by the derivative of the boundary curve. To fully determine the derivative of F at each point on e_{jk}, the normal derivative $\frac{\partial F}{\partial n_{jk}}$ at each point on e_{jk} is defined as a linear combination of $\frac{\partial F}{\partial n_{jk}}(V_j)$ and $\frac{\partial F}{\partial n_{jk}}(V_k)$, so that the normal derivatives vary linearly and hence continuously along e_{jk}.

§3. Generation of Triangular Surface Patches

3.1 Blending

The interpolating surface F which is so far only defined on the boundaries of the triangles in domain D is extended to D by means of the side-vertex method for interpolation in triangles in [11]. Each triangular surface patch is generated as a convex combination of three surface patches, each of which is obtained by blending the data value at one vertex with its opposite boundary curve. The blending consists of univariate interpolation by rational cubics in the form of (1.1) along line segments joining a vertex and its opposite edge. The weights of these rational cubics are chosen to ensure that these curves do not cross the xy plane. The generation of each triangular patch is independent of one another. We shall just restrict our attention to the blending process in one triangle T with vertices V_i, V_j, V_k which blends the data value at vertex V_i with its opposite boundary curve on the edge e_{jk}, that is, the generation of the triangular surface patch which will be denoted by P_i. Similarly, the triangular surface patches obtained by blending the data values at vertices V_j, V_k with their respective opposite boundary curves will be labelled as P_j and P_k.

The patch P_i is obtained by univariate interpolation along line segments joining vertex V_i and its opposite edge e_{jk} as shown in figure 2.1. This interpolation preserves the values and the derivatives which are already defined at the vertex and on the edge e_{jk}. Let (x, y) be an arbitrary point of the triangle T and (b_i, b_j, b_k) denote its barycentric coordinates. Suppose that the line segment joining vertex $V_i(x_i, y_i)$ and the point (x, y) meets the edge e_{jk} at $S_i(x_s, y_s)$.

Let us denote the derivatives along V_iS_i at V_i and S_i by $\overline{D}_1$ and $\overline{D}_2$ respectively, the length of V_iS_i by $\|V_iS_i\|$ and the value of the boundary curve at S_i by $d(x_s, y_s)$. Then we have

$$\overline{D}_1(x_s, y_s) = \frac{(x_s - x_i)}{\|V_iS_i\|}\frac{\partial F}{\partial x}(V_i) + \frac{(y_s - y_i)}{\|V_iS_i\|}\frac{\partial F}{\partial y}(V_i),$$

and

$$\overline{D}_2(x_s, y_s) = \frac{(x_s - x_i)}{\|V_iS_i\|}\frac{\partial F}{\partial x}(S_i) + \frac{(y_s - y_i)}{\|V_iS_i\|}\frac{\partial F}{\partial y}(S_i).$$

The rational cubic $r(t; x_s, y_s)$ of the form in (1.1) is defined along V_iS_i interpolating z_i, $D_1(x_s, y_s)$ at V_i and $d(x_s, y_s)$, $D_2(x_s, y_s)$ at S_i, where for $\ell = 1, 2$,

$$D_\ell(x_s, y_s) = \overline{D}_\ell(x_s, y_s)\|V_iS_i\|. \tag{3.1}$$

Thus

$$r(t; x_s, y_s) = \frac{A(1-t)^3 + 3B(1-t)^2t + 3C(1-t)t^2 + Dt^3}{1 + t(1-t)(\omega(x_s, y_s) - 3)}, \quad 0 \le t \le 1, \tag{3.2}$$

where

$$A = z_i, \quad B = [z_i\,\omega(x_s, y_s) + D_1(x_s, y_s)]/3,$$

and

$$C = [d(x_s, y_s)\,\omega(x_s, y_s) - D_2(x_s, y_s)]/3, \quad D = d(x_s, y_s).$$

The weight $\omega(x_s, y_s)$ will be chosen so that $r(t; x_s, y_s)$ does not cross the xy plane and that $\omega(x_s, y_s)$ varies in a C^1 manner when S_i varies along the edge e_{jk}. The value of the patch P_i at $(x, y) \in T$ is then defined to be

$$P_i(x, y) = r(t; x_s, y_s), \quad \text{where } t = 1 - b_i.$$

The patch P_i thus obtained is C^1 and it preserves the values and derivatives which are already defined at V_i and on the edge e_{jk}.

3.2 Estimation of Weights

We shall discuss how the weight $\omega(x_s, y_s)$ is chosen. First observe that in (3.2) A is fixed while D varies with (x_s, y_s), B increases as $D_1(x_s, y_s)$ increases while C increases as $D_2(x_s, y_s)$ decreases. For fixed (x_s, y_s), $\omega(x_s, y_s)$ can be chosen so that $r(t; x_s, y_s)$ is non-negative. We shall find only one value for all the $\omega(x_s, y_s)$, but such that all the resulting curve segments $r(t; x_s, y_s)$ are non-negative.

Consider

$$r^*(t) = \frac{A^*(1-t)^3 + 3B^*(1-t)^2 t + 3C^*(1-t)t^2 + D^* t^3}{1 + t(1-t)(\omega^* - 3)}, \quad 0 \le t \le 1$$

where

$$A^* = z_i, \quad D^* = \min\{d(x_s, y_s) : (x_s, y_s) \in e_{jk}\},$$

$$B^* = [z_i\,\omega^* + D_1^*]/3, \quad C^* = [D^*\omega^* - D_2^*]/3,$$

with D_1^* and D_2^* being respectively a lower bound of $D_1(x_s, y_s)$ and an upper bound of $D_2(x_s, y_s)$ along e_{jk}.

With a default value of $\omega^* = 3.0$, theorem 1 is applied to test whether $r^*(t)$ is non-negative. If $r^*(t)$ is negative for some $t \in (0, 1)$, then theorem 1 determines a larger value for ω^* such that $r^*(t)$ just touches the xy plane.

Let us define

$$\omega(x_s, y_s) = \omega^*, \quad \forall (x_s, y_s) \in e_{jk}, \tag{3.3}$$

then by the following three propositions, theorem 1 and the choice of values for D^*, D_1^*, D_2^* used in determining ω^*, all the rational cubics $r(t; x_s, y_s)$ for the patch P_i will be non-negative. These propositions can be proved by simple algebra.

Proposition 3.1. *Suppose* $f(D) = 3B^2C^2 + 6ABCD - 4(AC^3 + B^3D) - A^2D^2$ *where* $A > 0$, *and* B *and/or* $C < 0$. *If there exists* $D^* > 0$ *such that* $f(D^*) \leq 0$, *then* $f(D) \leq 0, \forall D \geq D^*$.

Proposition 3.2. *Suppose* $g(B) = 3B^2C^2 + 6ABCD - 4(AC^3 + B^3D) - A^2D^2$ *where* A, C, D *are constants and* $A, D > 0$. *Let* $G = \{B \in (-\infty, \infty) \,|\, B < 0$ *or* $C < 0\}$. *If there exist* $B^* \in G$ *such that* $g(B^*) \leq 0$, *then* $\forall B \geq B^*$ *and* $B \in G$, $g(B) \leq 0$.

Proposition 3.3. *Suppose* $h(C) = 3B^2C^2 + 6ABCD - 4(AC^3 + B^3D) - A^2D^2$ *where* A, B, D *are constants and* $A, D > 0$. *Let* $H = \{C \in (-\infty, \infty) \,|\, C < 0$ *or* $B < 0\}$. *If there exist* $C^* \in H$ *such that* $h(C^*) \leq 0$, *then* $\forall C \geq C^*$ *and* $C \in H$, $h(C) \leq 0$.

We shall now try to find a lower bound for $D_1(x_s, y_s)$ and an upper bound for $D_2(x_s, y_s)$ which are used to determine ω^* in (3.3).

From (3.1), $D_1(x_s, y_s)$ varies linearly with the coordinates (x_s, y_s) of S_i along the side e_{jk}, hence its extrema can only occur at the two end points, V_j and V_k and we have

$$D_1(x_s, y_s) \geq \min\{D_1(x_j, y_j), D_1(x_k, y_k)\}.$$

By substituting $\frac{\partial F}{\partial x}(S_i)$ and $\frac{\partial F}{\partial y}(S_i)$ in (3.1) in terms of $\frac{\partial F}{\partial e_{jk}}(S_i)$ and $\frac{\partial F}{\partial n_{jk}}(S_i)$ and simplifying, $D_2(x_s, y_s)$ can be rewritten as

$$D_2(x_s, y_s) = D_n(x_s, y_s) + D_e(x_s, y_s),$$

where

$$D_n = K \frac{\partial F}{\partial n_{jk}}(S_i), \tag{3.4}$$

for some constant K which depends only on the three vertices $(x_\ell, y_\ell), \ell = i, j, k$, and

$$D_e = \left[(x_s - x_i)\frac{(x_k - x_j)}{\|e_{jk}\|} + (y_s - y_i)\frac{(y_k - y_j)}{\|e_{jk}\|}\right] \frac{\partial F}{\partial e_{jk}}(S_i). \tag{3.5}$$

Thus we could find an upper bound for $D_2(x_s, y_s)$ based upon the expressions (3.4) and (3.5) by summing up the upper bounds of the two expressions.

As the normal derivative $\frac{\partial F}{\partial n_{jk}}(S_i)$ has been defined to vary linearly along e_{jk} from $\frac{\partial F}{\partial n_{jk}}(V_j)$ to $\frac{\partial F}{\partial n_{jk}}(V_k)$, we have from (3.4)

$$D_n(x_s, y_s) \leq \max\left\{K \frac{\partial F}{\partial n_{jk}}(V_j),\, K \frac{\partial F}{\partial n_{jk}}(V_k)\right\}.$$

Consider now the expression (3.5). As the boundary curve along e_{jk} is a rational cubic $R(t)$ in the form of (1.1) where $(x_s, y_s) = t(x_k, y_k) + (1-t)(x_j, y_j)$, so D_e could be expressed as

$$D_e = \frac{Q(t)}{v(t)^2}, \tag{3.6}$$

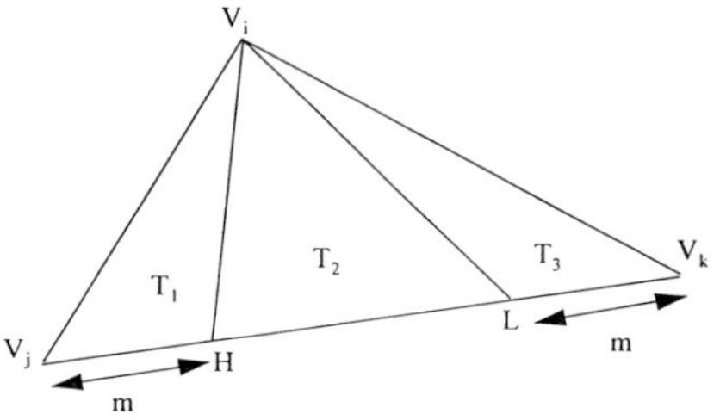

Fig. 3.1. Three sub blending surfaces.

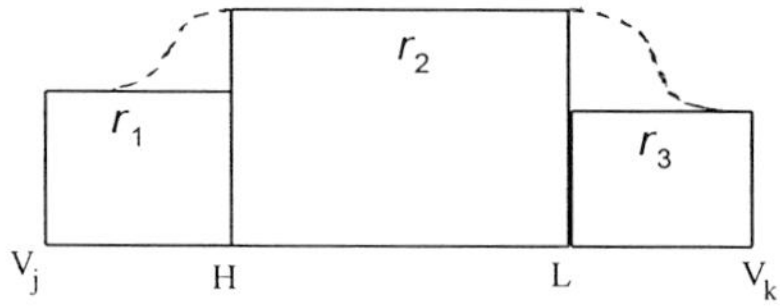

Fig. 3.2. Step function of weights for P_i.

where $Q(t)$ is a polynomial in t at most of fifth degree and $v(t) = 1+(\omega-3)t(1-t)$ for some weight $\omega \geq 3$. Clearly $1 \leq v(t) \leq 1+\frac{1}{4}(\omega-3)$. An upper bound for D_e in (3.6) can be estimated numerically.

When the weight for the boundary curve along the edge e_{jk} is 3.0, $v(t) = 1$ and thus D_e is a cubic polynomial. In this case we could find the maximum of D_e easily. We denote the upper bound for D_e obtained from the above discussion as M.

We thus have

$$D_2(x_s, y_s) \leq M + \max\Big\{K\,\frac{\partial F}{\partial n_{jk}}\,(V_j)\,,\;K\,\frac{\partial F}{\partial n_{jk}}\,(V_k)\Big\}.$$

Earlier we have defined $\omega(x_s, y_s)$ for the patch P_i to be the constant ω^* in (3.3). But in order not to over-estimate the weight $\omega(x_s, y_s)$ too much for the blending surface P_i, we subdivide the blending surface into three smaller surfaces as shown in figure 3.1 where m denotes a fraction of the length of e_{jk}. $\omega(x_s, y_s)$ is readjusted as described below.

For each smaller blending surface, we would estimate the weight $\omega(x_s, y_s)$ as before. The lower bound for $D_1(x_s, y_s)$, the upper bound for $D_2(x_s, y_s)$ and the smallest $d(x_s, y_s)$ are computed as before except that now the variation of (x_s, y_s) is not along the full edge e_{jk} but is restricted to the portion of the edge concerned.

Let the values of $\omega(x_s, y_s)$ obtained for the three sub blending surfaces of P_i on sub triangles T_1, T_2, T_3 be r_1, r_2, r_3 (figure 3.2). Then $\omega(x_s, y_s)$ is a step function along e_{jk} with values r_1, r_2 and r_3. For the blending surface P_i to be C^1, we require $\omega(x_s, y_s)$ to vary in a C^1 manner as (x_s, y_s) varies along e_{jk}.

By applying some simple transformation for cutting off the corners of functions, we can convert the step function into a smooth C^1 function $g(x_s, y_s)$ such that g is greater than or equal to the step function along e_{jk}. We then define $\omega(x_s, y_s)$ for patch P_i to be $g(x_s, y_s)$, thus ensuring a C^1 variation of the weight $\omega(x_s, y_s)$ along e_{jk} (dotted line in figure 3.2).

3.3 Convex Combination of Triangular Patches

The triangular patch P_i preserves the values and the derivatives which are already defined at the vertex V_i and on the edge e_{jk} before the blending process, but it does not necessarily preserve those at the other two vertices and edges. Similarly, P_j and P_k preserve the values and derivatives at the respective vertices and edges.

The interpolating surface P_T on the triangle T is generated as a convex combination of the above three blending patches, $P_\ell, \ell = i, j$ and k, that is, $\forall (x, y) = (b_i, b_j, b_k) \in T$,

$$P_T(b_i, b_j, b_k) = (C_i P_i + C_j P_j + C_k P_k)(b_i, b_j, b_k),$$

where the coefficients are given as

$$C_p = \frac{b_q^2 \, b_r^2}{b_q^2 \, b_r^2 + b_r^2 \, b_p^2 + b_p^2 \, b_q^2},$$

with distinct $p, q, r \in \{i, j, k\}$.

Since the coefficients C_i, C_j, C_k have the properties

$$\begin{aligned} C_i + C_j + C_k &= 1, \\ C_p \mid_{e_q} &= \delta_{pq}, \\ \partial C_p \mid_{e_q} &= 0, \end{aligned}$$

where e_q denotes the edge of triangle T opposite the vertex V_q, ∂ represents any first order differentiation and δ_{pq} is the Kronecker delta, P_T preserves all the values and derivatives which have been defined along the edges of the triangle T before the blending process.

The interpolating surface F which we are looking for is the composite surface defined as $F \mid_T = P_T$, for each triangle T in the domain. F is thus a positivity preserving C^1 surface interpolating the given scattered data.

§4. Numerical Examples and Concluding Remarks

We shall apply our scheme to the following three test functions:

$$G(x,y) = \exp\left((x-0.5)^2 + (y-0.5)^2\right) - 1.004,$$

$$H(x,y) = 1.025 - \Big[0.75\exp\Big(\frac{-(12x-2)^2-(12y-2)^2}{4}\Big)$$
$$+0.75\exp\Big(-\frac{(9x+1)^2}{49} - \frac{(9y+1)}{10}\Big)$$
$$+0.5\exp\Big(\frac{-(9x-7)^2-(9y-3)^2}{4}\Big) - 0.5\exp\Big(-(10x-4)^2-(10y-7)^2\Big)\Big]$$

$$J(x,y) = 5\Big(1+2\exp\Big(-3(|x^2+y^2|^{\frac{1}{2}}-6.7)\Big)\Big)^{-\frac{1}{2}}.$$

The domain used for the first two functions is the unit square $[0,1]\times[0,1]$ while for the last fuction, the sigmoidal function, the domain is $[0,9]\times[0,9]$. The given constraint plane for all the three examples is the xy plane.

The first example consists of 32 data points from the function G. The interpolating surface uses $m = 0.1$ and $\delta = \frac{3}{10}$ of the lowest data value. In figure 4.1.a, we have the triangulated domain. Figure 4.1.b shows the interpolating surface to all the data points, constructed with all the weights set to 3.0 irrespective of whether the curves concerned cross the constraint plane or not. This interpolating surface does cross the constraint plane. In figures 4.1.c and 4.1.d, with the same given data set and the same surface scheme, now we generate an interpolating surface where the weights are chosen, ranging from 3.0 to 8.0, to ensure that the surface does not cross the constraint plane. Although the test function $G(x,y)$ is symmetric about $x = 0.5$ and also symmetric about $y = 0.5$, due to the fact that the data points are chosen at random and the partial derivative at each data point is only an estimation, the generated interpolating surface is not symmetrical (figure 4.1.d).

We note that in (1.1) as $\omega\to\infty$, $R(t), t\in[0,1]$, tends to be linear. To show that a similar asymtotical property is maintained in the two dimensional triangular case, the above data set is reused for generating the interpolating surface where all the weights are set to be 15.0. The 'flattening' effect due to the increased weights is apparent in the resulting interpolating surface (figure 4.1.e).

The second example uses 34 data points from function $H(x,y)$ and the triangulated domain is shown in figure 4.2.a. $m = 0.3$ and $\delta = \frac{1}{10}$ of the lowest data value. In figure 4.2.b, we have the true surface of H where its minimum is approximately -0.1852. The interpolating surface constructed with all the weights set to 3.0 crosses the constraint plane. With chosen weights computed by the scheme, the interpolating surface generated does not cross the constraint plane (figure 4.2.c).

In the third example 35 data points from the function $J(x,y)$ are used, $m = 0.1$ and $\delta = \frac{1}{2}$ of the lowest data value. Figure 4.3.b shows the true surface of the function which is bounded between 0.0 and 5.0. The interpolating surface with all the weights set to 3.0 is shown in figure 4.3.c which has an approximate maximum and minimum of 5.289 and −0.179 respectively. The interpolating surface with the weights chosen by the scheme is shown in figure 4.3.d.

Although the interpolating surface P_T on the triangle T preserves all the boundary curves and derivatives which are defined on the edges of T before the blending process, the patches P_i, P_j and P_k do not necessarily agree with one another in values and derivatives along the edges of T. This could sometimes lead to some unpleasant effect on the visual smoothness of the interpolating surface F since $F\mid_T$ is a weighted average of P_i, P_j, P_k with the weights C_i, C_j, C_k varying with (b_i, b_j, b_k). It could sometimes happen, depending upon the given data set, that the interpolating surface F might not be visually smooth enough, for example undulatng a little near certain edges of the triangulation. The subdivision of the triangular patch P_i (similarly for P_j and P_k) on triangle T into three sub patches for the estimation of $\omega(x_s, y_s)$ is an attempt to reduce the weight $\omega(x_s, y_s)$ and hence reduce the difference (due to the reduction in the difference in weights) between the boundary curves along the edges e_{ij} and e_{ik} obtained before the blending process and the curves obtained along the same edges from the blending process for surface patch P_i. The variation of the parameters m and δ can be used to help in improving the visual smoothness of the interpolating surface.

This scattered data interpolation scheme can be easily extended to the case where the constraint plane is any horizontal plane. It could also be extended to the case where the interpolating surface is constrained to lie between two bounding horizontal planes as the given data. In this case, the set of weights would have to be calculated twice, once with respect to each constraint plane, and the larger of the two corresponding weights would be used. It would involve much computation as it does in the case of one constraint plane.

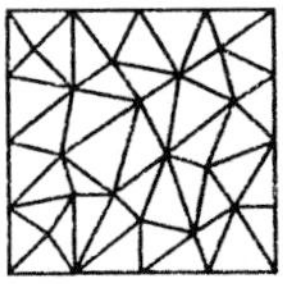

Fig. 4.1a. Triangulated domain for G.

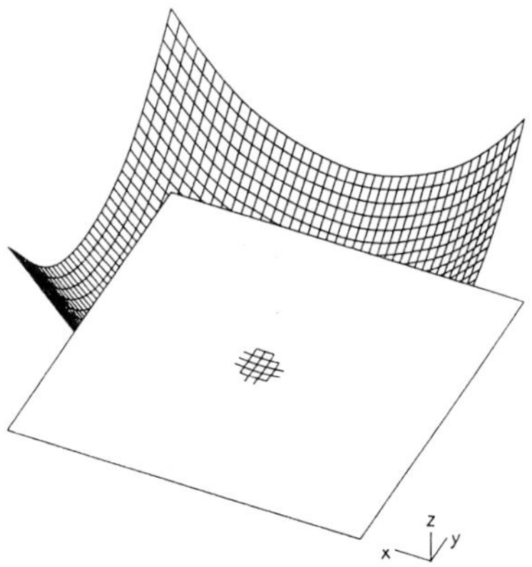

Fig. 4.1.b. Interpolating surface to the data from G where all the weights are 3.0

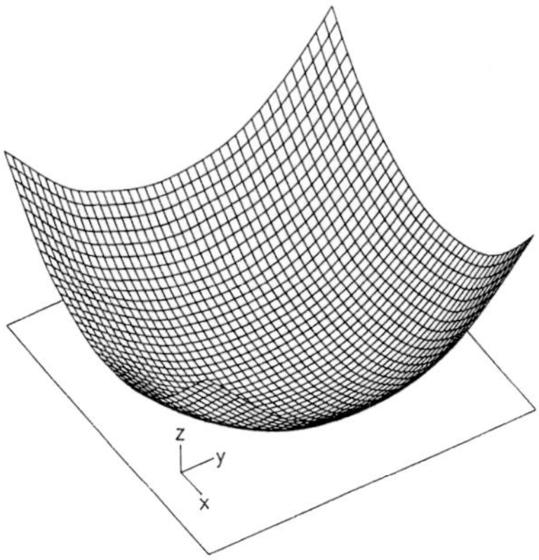

Fig. 4.1.c. Interpolating surface to the data from G.

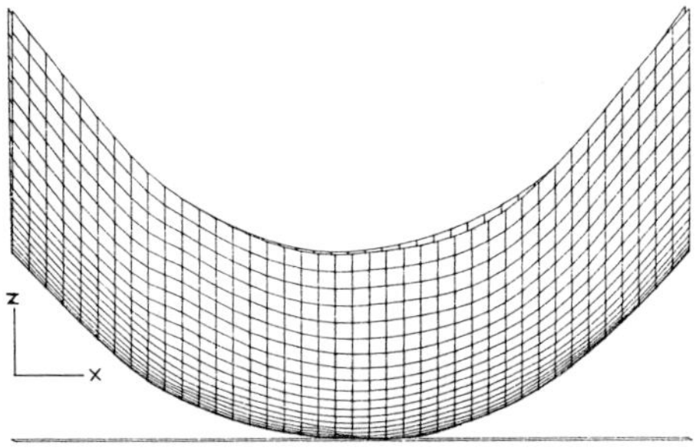

Fig. 4.1.d. Interpolating surface to the data from G.

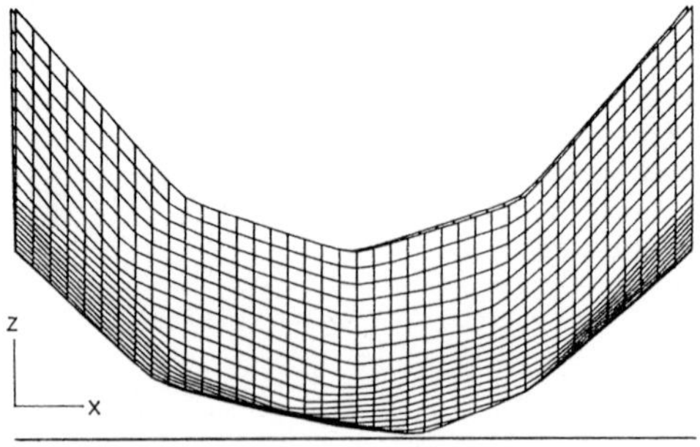

Fig.4.1.e. 'Flattened' interpolating surface to the data from G.

Fig. 4.2.a. Triangulated domain for H.

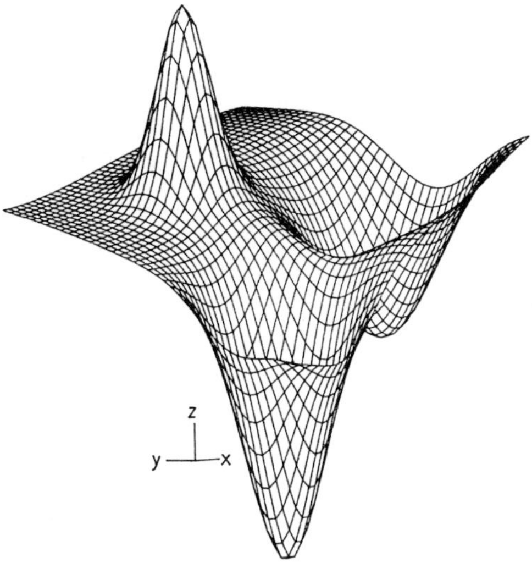

Fig. 4.2.b. True surface of H.

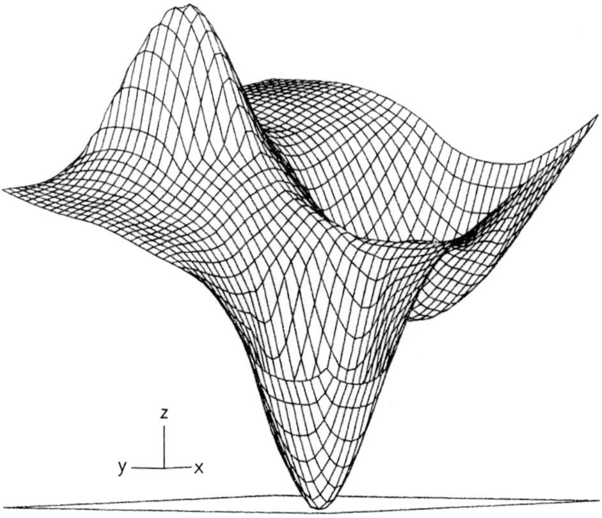

Fig. 4.2.c. Interpolating surface to the data from H.

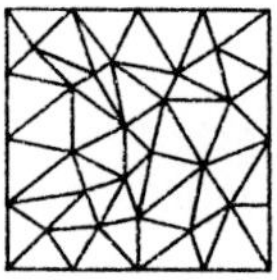

Fig. 4.3.a. Triangulated domain for J.

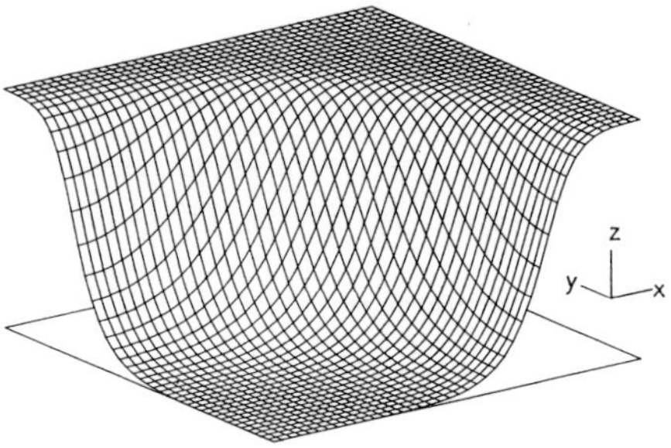

Fig. 4.3.b. True surface of J.

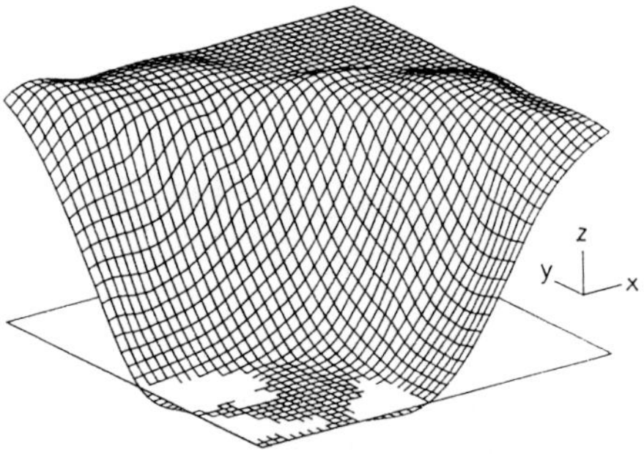

Fig. 4.3.c. Interpolating surface to the data from J where all the weights are 3.0.

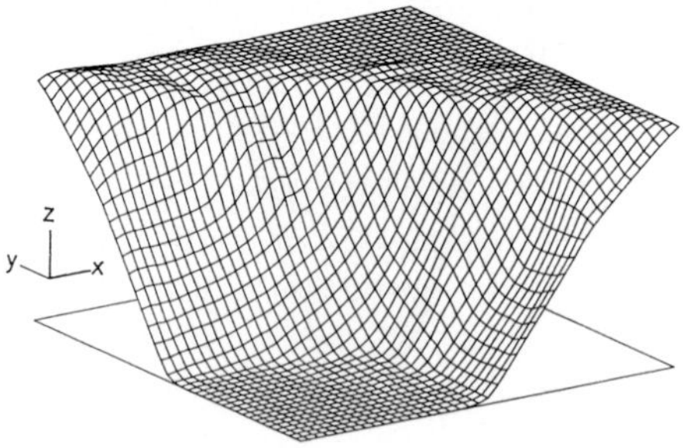

Fig. 4.3.d. Interpolating surface to the data from J.

Acknowledgements. We are very grateful to the referee for several valuable comments and suggestions.

References

1. Barnhill, R. E., Birkhoff, G. and Gordon, W. J., Smooth interpolation in triangles, J. Approx. Th. **8** (1973), 114–128.
2. Brodlie, K. W., S. Butt and P. Mashwama, Visualization of surface data to preserve positivity and other simple constraints, Computers and Graphics **19** (1995), 585–594.
3. Butt, S. and K. W. Brodlie, Preserving positivity using piecewise cubic interpolation, Computers and Graphics **17** (1993), 55–64.
4. Dauner, H. and C. H. Reinsch, An analysis of two algorithms for shape preserving cubic spline interpolation, IMA J. Numer. Anal. **9** (1989), 299–314.
5. Dougherty, R. L., A. Edelman and J. M. Hyman, Nonnegativity-, monotonicity- or convexity-preserving cubic and quintic Hermite interpolation, Math. Comp. **52** (1989), 471–494.
6. Fisher, B., G. Opfer and M. L. Puri, A local algorithm for constructing nonnegative cubic splines, J. Approx. Th. **64** (1991), 1–16.
7. Goodman, T. N. T., B. H. Ong and K. Unsworth, Constrained interpolation using rational cubic splines, in *Nurbs for Curve and Surface Design*, G. Farin (ed.), SIAM, Philadelphia, 1991, 59–74.
8. Goodman, T. N. T, H. B. Said and L. H. T. Chang, Local derivative estimation for scattered data interpolation, Applied Math. Comput. **68** (1995), 41–50.
9. Mulansky, B. and J. W. Schmidt, Powell-Sabin splines in range restricted interpolation of scattered data, Computing **53** (1994), 137–154.
10. Mulansky, B. and J. W. Schmidt, Nonnegative interpolation by biquadratic splines on refined rectangular grids, in *Wavelets, Images, and Surface Fitting*, P.-J. Laurent, A. Le Méhauté, and L. L. Schumaker (eds.), A K Peters, Wellesley, 1994, 379–386.
11. Nielson, G. M., The side-vertex method for interpolation in triangles, J. Approx. Th. **25** (1979), 318–336.
12. Ong, B. H. and H. C. Wong, A C^1 positivity preserving surface interpolation scheme, Technical Report No: M1/95, Universiti Sains Malaysia, 1995.
13. Opfer, G. and H. J. Oberle, The derivation of cubic splines with obstacles by methods of optimization and optimal control, Numer. Math. **52** (1988), 17–31.
14. Sakai, M. and J. W. Schmidt, Positive interpolation with rational splines, Bit **29** (1989), 140–147.
15. Schmidt, J. W. and W. Hess, Positivity of cubic polynomial on intervals and positive spline interpolation, Bit **28** (1988), 340–352.
16. Utreras, F. I., Positive thin plate splines, J. Approx. Theory Appl. **1** (1985), 77–108.

17. Wever, U., Non-negative exponential splines, Comput. Aided Design **20** (1988), 11–16.
18. Wong, H. C., Positivity preserving surface interpolation, M. Sc. thesis, Universiti Sains Malaysia, 1994.

B. H. Ong
School of Mathematical Sciences
Universiti Sains Malaysia
11800 Penang
Malaysia
bhong@cs.usm.my

H. C. Wong
Information Technology Centre
Universiti Sains Malaysia
11800 Penang
Malaysia
hcwong@rs1.itc.usm.my

Designing Rational Surfaces with Rational Offsets

Martin Peternell and Helmut Pottmann

Abstract. Classical Laguerre geometry provides an efficient approach to the design of rational surfaces with rational offsets. The geometric transformation which describes the change from the isotropic model of 3–dimensional Euclidean Laguerre space to the standard model, maps an arbitrary rational surface onto a rational surface all whose offsets are rational. It is shown how to apply this concept to surface modeling. In particular, we present a new surface design scheme which uses triangular patches on parabolic Dupin cyclides.

§1. Introduction

Offset curves and surfaces arise in various applications including NC milling, path planning for rapid prototyping, font design and geometric tolerancing. It is well–known that rational curves or surfaces do in general not possess rational offsets. In order to comply with current industry standards, offset curves or surfaces therefore have to be approximated in rational B–spline form [6,11].

A different approach to offsets takes its origin in the work of R. Farouki and T. Sakkalis [9] who proposed to use only those curves in the design process that do possess rational offsets. Farouki and Sakkalis introduced the so–called *Pythagorean–hodograph (PH) curves*, which are polynomial curves $x(t) = (x_1(t), x_2(t))$ with polynomial parametric speed $\sigma(t) = \sqrt{\dot{x}_1^2 + \dot{x}_2^2}$. Among other remarkable properties, these curves possess rational offsets $x_d(t) = x(t) + dn(t)$, where $n(t)$ is a field of unit normal vectors of the progenitor curve $x(t)$. Recent work on PH curves and their generalizations to the full class of rational curves with rational offsets showed that these curves are well suited for practical design purposes [1,2,7,8,12,19,20].

Using the dual representation, it has been shown in [19] that all rational surfaces with rational offsets (briefly called *PH surfaces* in the sequel) can be described explicitly. Based on the dual Bézier or B–spline representation, one can construct PH

Advanced Topics in Multivariate Approximation
F. Fontanella, K. Jetter and P. J. Laurent (eds.) pp. 275–286.

ISBN 981-02-2852-X

surfaces from the dual representation of spherical patches. Furthermore, unexpected results on offsets of special classes of surfaces could be derived [13,14,15,18,21].

We will now present a new geometric approach to PH surfaces based on Laguerre geometry. After a brief introduction into this classical type of geometry, we describe a geometric transformation that maps any rational surface onto a PH surface. Any PH surface may be obtained in this way. Graphs of polynomial or rational bivariate spline functions are mapped onto patchworks of PH surfaces. As an example, paraboloids are mapped onto parabolic Dupin cyclides and any modeling scheme with bivariate quadratic splines is transformed into a surface design scheme with cyclide patches. This new application of cyclides in free–form surface design offers more flexibility than existing techniques (see [10,16,23,24,25,26] and the references therein).

§2. A Brief Introduction to Laguerre Geometry

The fundamental elements of Laguerre geometry in Euclidean $\mathbb{R}^3$ are *oriented planes* and *cycles*. A cycle is an oriented sphere or a point (sphere with radius 0). The orientation is determined by a unit normal vector field or equivalently by a signed radius in the case of the sphere. If no ambiguity can arise, oriented planes will simply be called 'planes' in the sequel.

The basic relation in Laguerre geometry is that of *oriented contact of cycle and plane*. An oriented sphere and a plane are in oriented contact, if they are tangent and the unit normals coincide in the point of contact. For a point and a plane, oriented contact equals incidence.

Laguerre geometry studies properties which are invariant under Laguerre transformations. A *Laguerre transformation* consists of two bijective maps, one in the set of cycles and the other one in the set of planes. Additionally, a Laguerre transformation preserves oriented contact and non–contact between cycles and planes. A simple example of a Laguerre transformation is a *dilatation* which adds a constant $d \neq 0$ to the signed radius of each cycle and leaves its midpoint unchanged. Note that this map does not preserve points. Considering a surface as envelope of its oriented tangent planes, a dilatation maps the surface onto its offset at distance d. This already indicates the advantage of using Laguerre geometry in connection with offsets.

Using cartesian coordinates x_i in $\mathbb{R}^3$, each plane e is defined by a linear equation $e_0 + e_1x_1 + e_2x_2 + e_3x_3 = 0$. The coefficients e_i are called plane coordinates of e. The vector (e_1, e_2, e_3) defines the orientation of the plane and is always assumed to be normalized, $e_1^2+e_2^2+e_3^2 = 1$. Embedding $\mathbb{R}^3$ in $\mathbb{R}^4$ as hyperplane $x_4 = 0$, we map each plane e to the hyperplane $E = \zeta(e)$, defined by the homogeneous coordinates

$$E = (e_0, e_1, e_2, e_3, 1). \tag{2.1}$$

Note that the euclidean angle of E and $\mathbb{R}^3$ equals $\pi/4$. A sphere with midpoint m and signed radius r is represented by the point (m_1, m_2, m_3, r) in $\mathbb{R}^4$. This four–dimensional model of Laguerre geometry is called *cyclographic model*.

If more emphasis is on planes rather than cycles, one might be interested in a model where planes appear as points. This can easily be done by applying a duality $\delta : \mathbb{R}^{4*} \to \mathbb{R}^4$ (Blaschke map),

$$\delta(E) = (1, e_1, e_2, e_3, e_0). \tag{2.2}$$

The set of hyperplanes E is mapped to points contained in a hypercylinder $\Delta \subset \mathbb{R}^4$ with equation $\Delta : x_1^2+x_2^2+x_3^2 = 1$. In this *Blaschke model*, Laguerre transformations are seen as projective maps which preserve Δ.

Finally, we obtain an affine space (appropriately extended) as model of Laguerre space. For that, let w be the generator line of Δ containing the point $W(0,0,1,0)$. Furthermore, let $\bar{\mathbb{R}}^3$ be the hyperplane $x_3 = 0$ in $\mathbb{R}^4$; we use coordinate functions $y_1 = x_1, y_2 = x_2, y_3 = x_4$ in it. Applying the 'stereographic' projection $\sigma : \Delta - w \to \bar{\mathbb{R}}^3$ with center W to the cylindrical model, we get a map from planes e with unit normal $\neq (0,0,1)$ to points in $\bar{\mathbb{R}}^3$ via

$$\sigma \circ \delta \circ \zeta(e) = \frac{1}{1-e_3}(e_1, e_2, e_0). \tag{2.3}$$

Interpreting cycles as sets of tangent planes, we may state an important well-known result.

Lemma 2.1. *The set of tangent planes of a cycle Σ is mapped with $\sigma \circ \delta \circ \zeta$ to the set of points of a paraboloid of revolution or a plane Ψ satisfying*

$$\Psi : 2y_3 + (y_1^2 + y_2^2)(r + m_3) + 2y_1 m_1 + 2y_2 m_2 + r - m_3 = 0. \tag{2.4}$$

This can be proved by straightforward calculation.

So far, planes with unit normal $(0,0,1)$ do not have an image point in $\bar{\mathbb{R}}^3$. One therefore forms the so-called isotropic conformal closure $I^3 := \bar{\mathbb{R}}^3 \cup \mathbb{R}$ of $\bar{\mathbb{R}}^3$ and an extended map

$$\Lambda := \bar{\sigma} \circ \delta \circ \zeta,$$

which maps the plane $(e_0, 0, 0, 1)$ onto the real number e_0. To fix the problem of missing images of exceptional planes in Lemma 2.1, we have to extend the paraboloids Ψ by $r + m_3$, which equals 0 for a plane Ψ. The resulting model of Laguerre space, where planes are represented as points and cycles appear as paraboloids or planes, is called *isotropic model*. The transformation Λ describes the change from the standard model to the isotropic model.

In I^3, y_3-parallel lines and planes (called *isotropic lines and planes*) need a special treatment. The points of isotropic lines represent parallel planes. Non-isotropic lines as well as ellipses, whose projection onto $y_3 = 0$ are circles, and parabolas with isotropic axis are called isotropic Möbius circles. They may be obtained as intersection of 2 surfaces (2.4) and therefore are the Λ-image of the common tangent planes of 2 cycles, *i.e.*, the planes of a pencil or the tangent planes of a cone of revolution. This also shows that isotropic planes (as well as isotropic cylinders through the other isotropic Möbius circles) represent planes parallel to the tangent planes of a cone of revolution, that may degenerate to a pencil.

In the isotropic model, Laguerre transformations are seen as so–called isotropic Möbius transformations. We just mention two special cases. For a translation in the standard model, represented in plane coordinates by $(e_0, \dots, e_3) \mapsto (e_0 + ae_1 + be_2 + ce_3, e_1, e_2, e_3)$, we obtain in I^3

$$(y_1, y_2, y_3) \mapsto (y_1, y_2, y_3 + ay_1 + by_2 + \frac{c}{2}(y_1^2 + y_2^2 - 1)). \tag{2.5}$$

A dilatation $(e_0, \dots, e_3) \mapsto (e_0 + d, e_1, e_2, e_3)$ appears as isotropic Möbius transformation

$$(y_1, y_2, y_3) \mapsto (y_1, y_2, y_3 + \frac{d}{2}(y_1^2 + y_2^2 + 1)). \tag{2.6}$$

For more details on Laguerre geometry, we refer the reader to Benz [3], Blaschke [4] and Cecil [5].

§3. A Laguerre Geometric Approach to PH Surfaces

A surface Φ in $\mathbb{R}^3$ is a PH surface iff it possesses a rational parametric representation $s(u, v) = (s_1(u, v), s_2(u, v), s_3(u, v))$ such that the field of unit normal vectors $n(u, v)$ is rational in u and v. Then, all its offsets $s_d = s + dn$ are rational.

We now view Φ as set of its oriented tangent planes. They are oriented with n and their plane coordinates are rational functions,

$$(e_0, \dots, e_3) = (-n_1 s_1 - n_2 s_2 - n_3 s_3, n_1, n_2, n_3). \tag{3.1}$$

Applying the transformation $\delta \circ \zeta$, we obtain a rational 2–surface in the Blaschke–cylinder Δ. Conversely, if a rational 2–surface in Δ is given, $(\delta \circ \zeta)^{-1}$ maps it onto a PH surface Φ. Laguerre transformations appear as projective maps in the Blaschke model and projective maps preserve the rationality of a surface. This proves the following result.

Theorem 3.1. *Laguerre transformations map PH surfaces (viewed as sets of oriented tangent planes) onto PH surfaces.*

The stereographic projection σ defines a bijective map between rational surfaces in Δ and rational surfaces in the isotropic model I^3, which gives us a simple construction of PH surfaces.

Theorem 3.2. *The geometric transformation Λ^{-1}, which describes the change from the isotropic model of Laguerre space to the standard model, maps arbitrary* *rational surfaces onto PH surfaces. All non–developable PH surfaces may be obtained in this way.*

Note that surfaces Φ with a one–parameter set of tangent planes (*developable PH surfaces*) correspond to *rational curves* $\Psi \subset I^3$.

A further degeneracy should be pointed out. An isotropic line is mapped via Λ^{-1} onto a pencil of parallel planes. Hence, a rational cylinder Ψ with isotropic generators belongs to a 2–parameter set of planes that touch a rational curve at infinity. Moreover, if the rational surface Ψ possesses a real curve along which it is touched by isotropic lines (occluding contour for y_3–parallel projection), the corresponding surface Φ contains a real curve at infinity.

Let us assume that Ψ is the graph surface of a bivariate piecewise rational function defined over a triangulation in the y_1, y_2–plane. Then Φ is composed of triangular PH patches without points at infinity and it possesses an injective Gauss map $s(u,v) \mapsto n(u,v)$. If Φ is regular and C^2, it does not possess a change in the sign of Gaussian curvature. Note that parabolic surface points of Φ correspond to singularities in the Gauss map and therefore to singularities in the surface Ψ.

Conversely, suppose that Ψ is regular, then Φ is regular as a set of planes, but not necessarily as a point set. Note also that Laguerre geometry cannot distinguish between a surface and its offsets (those surfaces are related by Laguerre transformations!). But clearly, the offsets of a regular surface need not be regular.

Let Ψ_1, Ψ_2 be two rational surfaces that meet at a point p with a common non–isotropic tangent plane τ. Then Φ_1, Φ_2 are PH surfaces meeting with common tangent plane $\Lambda^{-1}(p)$ at a common point $x \in \mathbb{R}^3$ (the point of tangency between $\Lambda^{-1}(p)$ and the cycle $\Lambda^{-1}(\tau)$). Thus, two adjacent rational patches Ψ_1, Ψ_2 on a G^1 surface are mapped onto two PH patches Φ_1, Φ_2 meeting with common tangent planes along a common curve. This joint can either be smooth or like an edge of regression (see Fig. 3). If it is smooth and Ψ_1, Ψ_2 meet with G^k continuity, also Φ_1, Φ_2 form a G^k joint.

§4. Surface Design with Cyclide Patches

4.1 Parabolic Dupin Cyclides as Λ^{-1} Images of Paraboloids

We summarize some properties of parabolic Dupin cyclides which are important for surface design with cyclide patches. Let $\mathbb{R}^3$ be Euclidean 3-space, $\mathbb{P}^3$ its projective extension. $\mathbb{P}^3 - \mathbb{R}^3 = \Omega$ is the *ideal plane* at infinity. Using homogeneous coordinates $(x_0 : x_1 : x_2 : x_3)$, points in Ω are characterized by $x_0 = 0$.

A parabolic Dupin cyclide Φ is a special rational surface of algebraic order 3. Using an appropriate coordinate system, a homogeneous representation is

$$\phi(u,v) = (1 + u^2 + v^2, au + (a-b)uv^2, bv + (b-a)u^2v, au^2 + bv^2). \qquad (4.1)$$

An implicit equation corresponding to (4.1) is

$$x_0(bx_1^2 + ax_2^2 + (a+b)x_3^2) - x_3(abx_0^2 + x_1^2 + x_2^2 + x_3^2) = 0. \tag{4.2}$$

The set of tangent planes of Φ has the parametrization

$$\psi(u,v) = (-(au^2 + bv^2), 2u, 2v, u^2 + v^2 - 1). \tag{4.3}$$

The intersection of a parabolic Dupin cyclide Φ with the plane at infinity $\Omega : x_0 = 0$ consists of the absolute conic j, which is determined by $x_1^2 + x_2^2 + x_3^2 = 0$ and a real line $g : x_3 = 0$. Further, Φ contains two real lines $e, f \subset \mathbb{R}^3$ which are perpendicular to each other and determined by the equations $e : x_1 = 0, x_3 - ax_0 = 0$ and $f : x_2 = 0, x_3 - bx_0 = 0$.

For a plane ε passing through e or f, the intersection $\varepsilon \cap \Phi$ is a circle. These circles (parameter lines in (4.1)) are principal curvature lines. Along each of the lines e, f the tangent plane is constant, thus e, f are called *parabolic lines.*

There are three types of parabolic Dupin cyclides, determined by the choice of the constants a, b. We speak of a *horn cyclide* for $ab > 0$ (Fig.1) or a *ring cyclide* for $ab < 0$. The limit case ($a = 0$ or $b = 0$) is sometimes called *thorn cyclide.* The special choice $a = b$ yields a sphere. Offsetting needs not preserve the type of the cyclide, hence these types are not distinguishable within Laguerre geometry.

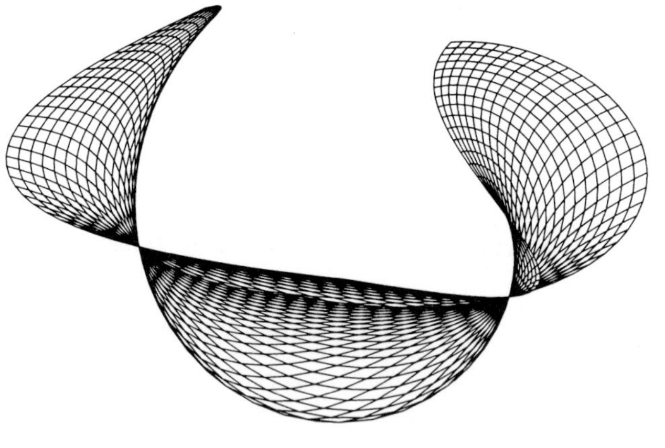

Fig. 1. A rational bicubic patch on a parabolic horn cyclide.

Parabolic Dupin cyclides form an eight parameter family of surfaces in $\mathbb{R}^3$. Let $g \in \Omega$ be the real line $x_0 = 0,\ x_3 = 0$. The family of parabolic Dupin cyclides passing through g has six real parameters, and will be denoted by G.

Using a linear parameter transformation, each parabolic Dupin cyclide $\Phi \in G$ admits a dual parametrization of the form

$$\psi(u,v) = (q(u,v), 2u, 2v, u^2 + v^2 - 1). \tag{4.4}$$

where $q(u,v)$ is a quadratic polynomial. Λ maps tangent planes of $\Phi \in G$ to points of the graph of a quadratic polynomial, denoted by Ψ. With $u = y_1$ and $v = y_2$ the equation of Ψ is

$$y_3 = \frac{1}{2} q(y_1, y_2).$$

Theorem 4.1. *The Λ^{-1} image of the graph Ψ of a quadratic function is a parabolic Dupin cyclide Φ or a sphere.*

Substituting $u = t$ and $v = \lambda t + \mu$ in (4.3), the parametrization defines a quadratic cone $\Gamma(\lambda, \mu)$. The same substitution applied to (4.1), we obtain in general cubic curves $c(\lambda, \mu)$ and $\Gamma(\lambda, \mu)$ is tangent to Φ at points of c. The intersection $c \cap \Omega$ consists of one real point in g and two conjugate complex points in j. Thus c is called *cubic circle*. Let $k = \Gamma \cap \Omega$. Then k is tangent to g and j in the points $c \cap \Omega$. This implies that $\Gamma(\lambda, \mu)$ is a cone of revolution, and leads us to the following

Theorem 4.2. *For a parabolic Dupin cyclide Φ there exists a two parameter family of cones of revolution each of which is tangent to the surface along a rational cubic or along a circle and a line.*

These properties can also be derived in I^3. But note that $\Lambda(\Phi)$ depends on the stereographic projection σ. By our special choice of the center, only parabolic Dupin cyclides $\Phi \in G$ are mapped to quadratic graph surfaces. Then, Λ–images of cones $\Gamma(\lambda, \mu)$ are non–isotropic lines or parabolas. If $\Phi \notin G$, $\Lambda(\Phi)$ is a rational surface of order 4. $\Lambda(\Gamma)$ can be any type of isotropic Möbius circle.

Remark 4.1. Along a cubic c the surface normals form a constant angle with the axis of $\Gamma(\lambda, \mu)$, *i.e.*, c is an *isophote* on the cyclide. More general isophotes correspond to circles in the parameter plane. Then, (4.1) shows that the general isophotes on parabolic cyclides are rational sextics. They are good candidates for boundaries of trimmed cyclide patches, since one may apply a circular spline algorithm to generate the boundary in the parameter plane.

A quadratic Bézier patch on the paraboloid Ψ over a triangle in the y_1, y_2–plane yields via Λ^{-1} a dual quadratic Bézier triangle on a cyclide Φ with control planes u_{ijk}. To perform the conversion from the dual to the point representation, we run one step of the de Casteljau algorithm. This results in 3 planes that intersect in the desired surface point,

$$x(r,s,t) = (ru_{200} + su_{110} + tu_{101}) \times (ru_{110} + su_{020} + tu_{011}) \times (ru_{101} + su_{011} + tu_{002}),$$

where r, s, t are the barycentric coordinates and $a \times b \times c$ denotes the exterior product of 3 vectors in $\mathbb{R}^4$. Hence, in the standard point representation the patch is representable as a *rational cubic Bézier triangle* on the cyclide Φ, whose control points follow immediately from the formula above. Analogously, parallelogram domains in the parameter plane yield *rational bicubic cyclide patches* (Fig. 1).

Remark 4.2. Consider the lines e and f on the cyclide (4.1) as axes of a *hyperbolic net* consisting of all lines that intersect both e and f. Then pick a point $\phi(u,v) \notin e, f$ on the surface, pass the unique line of the net through it and intersect it with the plane $x_3 - cx_0 = 0$. One gets the point

$$(x_1/x_0, x_2/x_0) = ((a-c)u, (b-c)v).$$

We may view the image plane as parameter plane and this shows that the map from the surface to the parameter plane is a net projection composed with an affine map (similarity for $c = (a+b)/2$). This standard way to parameterize a cubic surface (see also [26]) gives a better understanding of the geometry of the parametrization. It is now quite obvious that (4.1), $\mathbb{R}^2 \to \Phi$, does not reach the regular points on the parabolic lines e, f. We can generate rational curves and surface patches on the cyclide by projecting rational curves and surfaces with the net projection onto it. The image of a straight line, which does not intersect e, f, is a hyperbola of the form $uv + \alpha u + \beta v + \gamma = 0$ in the parameter plane and a rational quartic on the surface. We would like to mention that a detailed investigation of rational curves and surface patches on cyclides is currently performed by C. Mäurer [17] (for special patches, see also [26]).

Remark 4.3. *The transformation Λ preserves principal curvature lines in the following sense.* Take a principal curvature line c on a surface Φ and consider the developable surface Γ touching Φ along c. Then Λ maps the planes of Γ onto the points of an *isotropic principal curvature line* of the image surface Ψ of Φ. Isotropic principal curvature directions at a surface point are conjugate directions, whose orthogonal projections onto the plane $y_3 = 0$ are perpendicular. This result may be proved by considering an osculating paraboloid (second orderTaylor approximant) in I^3, which corresponds to an osculating parabolic cyclide in the standard model.

As an example, consider in I^3 the rational translational surface

$$\Psi(u,v) = (f_1(u), g_1(v), f_2(u) + g_2(v)),$$

with arbitrary rational functions f_1, f_2, g_1, g_2. Clearly, the parameter lines are isotropic curvature lines on Ψ. Thus, the transformation Λ^{-1} into the standard model yields a *rational surface with rational offsets and rational principal curvature lines.* The principal curvature lines are isophotes and therefore *planar* (see [4], p. 279). Any parameter rectangle $[a,b] \times [c,d]$ corresponds to a *rational principal patch* in the sense of R.R. Martin [16]. It is an interesting subject for future research to use Laguerre geometry for generating an even larger class of rational principal PH patches, in particular to study those of low degree. Applications in NC milling mainly motivated the study of PH surfaces. Since curvature lines have been proposed as good cutter paths, principal PH patches are also interesting from a practical point of view.

4.2 Surface Modeling

Proposition 4.3. *If two parabolic Dupin cyclides Φ_1, Φ_2 share a rational cubic c and the tangent cone of revolution Γ along it, then they possess the same line at infinity.*

Proof: The intersection of Φ_1 and Φ_2 is an algebraic curve of order 9, which contains the absolute conic j and the doubly counted cubic c. Therefore, the remaining component of the intersection is a line g. The complete intersection of a parabolic cyclide and the ideal plane Ω is the union of j and a line. By our asssumption, Φ_1 and Φ_2 possess a common tangent plane ($\neq \Omega$) at the real point at infinity U of c. Hence, the curves at infinity of the two cyclides share a common tangent at U and this must be the common ideal line g. ■

Φ_1 and Φ_2 belong to the six parameter family G of all parabolic Dupin cyclides passing through the line at infinity g. Hence, if we would like to construct a G^1 surface composed of parabolic Dupin cyclide patches, such that the patches meet smoothly along segments of cubic circles, we have to use patches on cyclides in the same family G. In the isotropic model (with an appropriate center of the stereographic projection), such a surface is the graph of a piecewise quadratic function. This suggests the following modeling technique.

Let (A_i, α_i) be coinciding vertices and oriented planes, considered as data elements in $\mathbb{R}^3$. The unit normals of α_i are denoted by a_i. The data must stem from a smooth surface with a constant sign of Gaussian curvature. We describe the construction of a G^1 surface, interpolating the given data. The surface is composed of rational cubic triangular patches in parabolic Dupin cyclides.

First, a coordinate transformation of the data (A_i, α_i) may be necessary. As the 'northpole' W is the center of the stereographic projection, the points a_i in the unit sphere S^2 should be situated suitably, *i.e.*, the Gaussian image of the interpolating surface should not contain W. This means, for example, that we are not able to model closed convex surfaces. Note that W determines the familiy G of cyclides we are using; they all contain the line at infinity $x_0 = 0, x_3 = 0$.

Λ maps the elements (A_i, α_i) to elements (B_i, β_i), where $B_i = \Lambda(\alpha_i)$ is a point in I^3. Interpreting A_i as cycle, Λ maps it to a paraboloid of revolution or plane (2.4). Then β_i is the tangent plane to $\Lambda(A_i)$ at B_i.

Let b_i be the normal projection of B_i onto $y_3 = 0$. We triangulate the vertices b_i, and then interpolate the data (B_i, β_i) with the graph of a piecewise quadratic function. For that, Powell–Sabin elements [11,22] can be used. Although the isotropic model depends on the choice of the origin, our result will be independent of it. This follows from (2.5), which shows that a translation in design space will 'add' a quadratic function to the surface Ψ in I^3. Translational invariance is now achieved by the quadratic precision of Powell–Sabin interpolants. Similarly, (2.6) shows that the method is *invariant under offsetting*: offsetting the data and

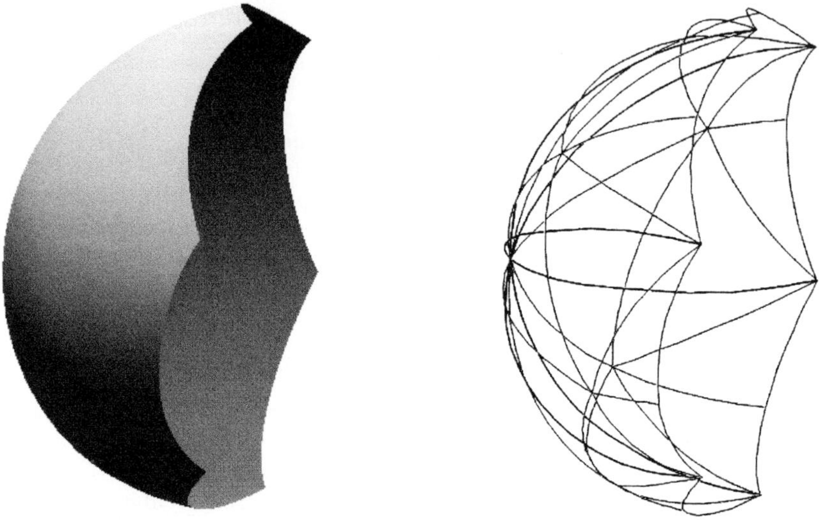

Fig. 2. Convex surface modeled with parabolic Dupin cyclide patches.

then constructing the interpolating surface yields the offset to the interpolant of the original data.

An example of the modeling technique based on Powell–Sabin 6–splits is shown in Figure 2. Here, the patches meet with G^1 continuity. As noted earlier, it is possible that although the patches share boundary curves with common tangent planes, the transition is not smooth, but forms a rigde (see Figure 3).

Fig. 3. The singularity problem.

Our modeling technique offers more flexibility than existing design schemes based on Dupin cyclides, although we use just cyclides of order 3. Of course, there are still the restrictions mentioned above. Some of them may be eliminated in future research. For example, the freedoms in constructing the microtriangles of a Powell–Sabin element may be used to apply a variational technique with an energy functional that serves to regularize the patches. However, smooth surfaces with vanishing Gaussian curvature along curves other than straight lines can never be modeled with parabolic Dupin cyclides.

In summary, the paper shows that appropriate geometric transformations can help to solve advanced geometric design problems with known approximation techniques. This has not yet been fully explored and seems to be a promising direction for future research.

Acknowledgements. This work has been supported by the Austrian Science Foundation through project P09790.

References

1. Ait Haddou, R. and L. Biard, G2 approximation of an offset curve by Tschirnhausen quartics, in *Mathematical Methods in Computer Aided Geometric Design III*, T. Lyche and L. L. Schumaker (eds.), Academic Press, New York, 1995, 1–10.
2. Albrecht G. and R. T. Farouki, Construction of C^2 Pythagorean–hodograph interpolating splines by the homotopy method, Adv. Comput. Math., to appear.
3. Benz, W., *Geometrische Transformationen*, BI-Wiss.-Verlag, Mannheim, 1992.
4. Blaschke, W., *Vorlesungen über Differentialgeometrie*, Band 3, Springer-Verlag, Berlin, 1929.
5. Cecil, T. E., *Lie Sphere Geometry*, Springer-Verlag, Berlin, 1992.
6. Farin, G., *NURBS for Rational Curve and Surface Design*, AK Peters, Wellesley, MA, 1994.
7. Farouki, R. T., Pythagorean–hodograph curves in practical use, in *Geometry Processing for Design and Manufacturing*, R.E. Barnhill (ed.), SIAM, Philadelphia, 1992, 3–33.
8. Farouki, R. T. and C. A. Neff, Hermite interpolation by Pythagorean–hodograph quintics, Math. Comp. **64** (1995), 1589–1609.
9. Farouki, R. T. and T. Sakkalis, Pythagorean hodographs, IBM J. Res. Develop. **34** (1990), 736–752.
10. Gallagher, T. and B. Piper, Convexity preserving surface interpolation, in *Designing Fair Curves and Surfaces*, N. S. Sapidis (ed.), SIAM, Philadelphia, 1994, 161–209.
11. Hoschek, J. and D. Lasser, *Fundamentals of Computer Aided Geometric Design*, A K Peters, Wellesley, MA, 1993.

12. Lü, W., Offset–rational parametric plane curves, Computer-Aided Geom. Design **12** (1995), 601–616.
13. Lü, W., Rational parameterization of quadrics and their offsets, Computing, to appear.
14. Lü, W., Rational canal surfaces, preprint.
15. Lü, W. and H. Pottmann, Pipe surfaces with rational spine curve are rational, Computer-Aided Geom. Design **13** (1996), to appear.
16. Martin, R. R., Principal patches for computational geometry, PhD thesis, Cambridge University, 1982.
17. Mäurer, C., Rational curves and surfaces on Dupin ring cyclides, preprint.
18. Peternell, M. and H. Pottmann, Computing rational parametrizations of canal surfaces, preprint.
19. Pottmann, H., Rational curves and surfaces with rational offsets, Computer-Aided Geom. Design **12** (1995), 175–192.
20. Pottmann, H., Curve design with rational Pythagorean–hodograph curves, Adv. Comput. Math. **3** (1995), 147–170.
21. Pottmann, H., W. Lü and B. Ravani, Rational ruled surfaces and their offsets, preprint.
22. Powell, M. J. D. and Sabin, M. A., Piecewise quadratic approximation on triangles, ACM Trans. Math. Software **3** (1977), 317–325.
23. Pratt, M. J., Cyclides in computer aided geometric design, Computer-Aided Geom. Design **7** (1990), 221–242.
24. Pratt, M. J., Cyclides in computer aided geometric design II, Computer-Aided Geom. Design **12** (1995), 131–152.
25. Srinivas, Y. L. and D. Dutta, An intuitive procedure for constructing geometrically complex objects using cyclides, Comput. Aided Design **26** (1994), 327–335.
26. Srinivas, Y. L. and D. Dutta, Rational parametric representation of parabolic cyclide: formulation and applications, Computer-Aided Geom. Design **12** (1995), 551–566.

Martin Peternell and Helmut Pottmann
Institut für Geometrie
Technische Universität Wien
Wiedner Hauptstrasse 8–10
A-1040 Wien
Austria
martin@geometrie.tuwien.ac.at
pottmann@geometrie.tuwien.ac.at

Kernels of Spherical Harmonics and Spherical Frames

Daniel Potts, Gabriele Steidl and Manfred Tasche

Abstract. Our concern is with the construction of a frame in $L^2(S)$ consisting of smooth functions based on kernels of spherical harmonics. The corresponding decomposition and reconstruction algorithms utilize discrete spherical Fourier transforms. Numerical examples confirm the theoretical expectations.

§1. Introduction

Traditionally, wavelets were tailored to problems on the Euclidean space $\mathbb{R}^d$. However, in most applications one has to analyze functions defined on compact domains. In particular, in geophysics wavelets on the unit sphere S of $\mathbb{R}^3$ are of interest. There exist different approaches to the constructions of spherical wavelets. Having spherical coordinates in mind, the idea of using tensor–products of periodic wavelets and wavelets on the interval was suggested in [7]. Applying tensor–products of periodic exponential spline–wavelets and spline–wavelets on the interval, wavelets on S were constructed in [3]. Unfortunately, tensor–product wavelets can possess singularities at the poles of S. To avoid these singularities, the coefficients of scaling functions and wavelets have to satisfy a linear system on each level, whose dimension grows very fast (see [3]).

A tensor–product method with a completely different kind of wavelets, namely trigonometric wavelets [2] and polynomial wavelets [11], was considered in [10]. But these wavelets cannot be used for the detection of singularities of a given function at the poles of S.

Advanced Topics in Multivariate Approximation
F. Fontanella, K. Jetter and P. J. Laurent (eds.) pp. 287–301.

ISBN 981-02-2852-X

Another simple and powerful technique for constructing biorthogonal wavelets on the sphere applies the so–called lifting scheme. This method was developed in [12]. One example are lifted versions of generalized Haar–wavelets on spherical triangles.

Finally, spherical wavelets were constructed by using spherical harmonics. This idea was realized in a different manner in [6] (for equidistributed nodes on S) and [9] (for scattered points on S).

Our paper deals with the construction of a frame in $L^2(S)$ consisting of smooth functions arising from kernels of spherical harmonics. The corresponding decomposition and reconstruction algorithms utilize discrete spherical Fourier transforms. Numerical examples emphasize that our method is suited for the detection of singularities of a given function at the poles of S, too.

This paper is organized as follows. Section 2 briefly recalls properties of spherical harmonics. Introducing the sampling spaces of band–limited functions in Section 3, we verify in Section 4 that these spaces form a multiresolution of $L^2(S)$. In Section 5 we construct the corresponding tight frame in $L^2(S)$. We describe decomposition and reconstruction algorithms in Section 6 and conclude with numerical tests in Section 7.

§2. Spherical Harmonics

Starting with the *Legendre polynomials*

$$P_k(x) := \frac{1}{2^k\, k!}\, \frac{\mathrm{d}^k}{\mathrm{d}x^k}(x^2-1)^k \quad (x \in [-1,1];\; k \in \mathbb{N}_0)\,,$$

we define the *associated Legendre functions* P_k^n $(n \in \mathbb{N}_0;\; k = n, n+1, \ldots)$ by

$$P_k^n(x) := \left(\frac{(k-n)!}{(k+n)!}\right)^{1/2} (1-x^2)^{n/2}\, \frac{\mathrm{d}^n}{\mathrm{d}x^n} P_k(x) \quad (x \in [-1,1])\,.$$

For any fixed $n \in \mathbb{N}_0$, the functions P_k^n $(k = n, n+1, \ldots)$ form a complete orthogonal system in $L^2[-1,1]$ with

$$\frac{1}{2}\int_{-1}^{1} P_k^n(x) P_l^n(x)\,\mathrm{d}x \;=\; \frac{1}{2k+1}\,\delta_{k,l} \quad (n \in \mathbb{N}_0;\; k,l = n, n+1, \ldots)\,.$$

Moreover, the associated Legendre functions fulfil the three–term recurrence relation

$$P_{n-1}^n(x) := 0, \qquad P_n^n(x) := \frac{((2n)!)^{1/2}}{2^n\, n!}(1-x^2)^{n/2}\,,$$

$$P_{k+1}^n(x) - v_k^n\, x P_k^n(x) + w_k^n\, P_{k-1}^n(x) = 0 \quad (k = n, n+1, \ldots) \tag{2.1}$$

with

$$v_k^n := \frac{2k+1}{((k-n+1)(k+n+1))^{1/2}}\,, \qquad w_k^n := \frac{((k-n)(k+n))^{1/2}}{((k-n+1)(k+n+1))^{1/2}}\,.$$

As usual, we parametrize the points of the unit sphere S in $\mathbb{R}^3$ by their spherical coordinates $(\theta,\varphi) \in [0,\pi] \times [0,2\pi)$. Then the *spherical distance* between the two points $p = (\theta,\varphi)$ and $p' = (\theta',\varphi')$ of S is determined by

$$pp' := \arccos\,(p \cdot p')$$

with the inner (cartesian) product

$$p \cdot p' = \cos\theta\ \cos\theta' \ + \ \sin\theta\ \sin\theta'\ \cos(\varphi-\varphi')\,.$$

We are interested in the Hilbert space $L^2(S)$ of all square integrable functions on S with the scalar product

$$\langle f, g\rangle := \frac{1}{4\pi}\int_0^\pi \int_0^{2\pi} f(\theta,\varphi)\,\overline{g(\theta,\varphi)}\sin\theta\,\mathrm{d}\varphi\,\mathrm{d}\theta \quad (f,\,g \in L^2(S))$$

and with the corresponding norm $\|\cdot\|$. Set

$$\hat{\mathrm{I}} := \{(k,n) : k \in \mathbb{N}_0,\ n = -k,\ldots,k\}\,.$$

An orthogonal basis of $L^2(S)$ is given by the set $\{Y_k^n : (k,n) \in \hat{\mathrm{I}}\}$ of *spherical harmonics*

$$Y_k^n(\theta,\varphi) := P_k^{|n|}(\cos\theta)\ \mathrm{e}^{in\varphi}\,.$$

It is easy to check that

$$\begin{aligned}\langle Y_k^n, Y_l^m\rangle &= \frac{1}{2}\int_{-1}^{1} P_k^{|n|}(x)P_l^{|m|}(x)\,\mathrm{d}x \cdot \frac{1}{2\pi}\int_0^{2\pi} \mathrm{e}^{\mathrm{i}(n-m)\varphi}\,\mathrm{d}\varphi \\ &= \frac{1}{2k+1}\,\delta_{k,l}\,\delta_{n,m} \quad ((k,n),\,(l,m) \in \hat{\mathrm{I}})\,. \qquad (2.2)\end{aligned}$$

Furthermore, it holds the following *addition theorem for spherical harmonics*

$$\sum_{n=-k}^{k} Y_k^n(\theta,\varphi)\ Y_k^{-n}(\theta',\varphi') \ = \ P_k(\cos\theta\ \cos\theta' \ + \ \sin\theta\ \sin\theta'\ \cos(\varphi-\varphi'))$$

or shortly with $p = (\theta,\varphi)$ and $p' = (\theta',\varphi')$,

$$\sum_{n=-k}^{k} Y_k^n(p)\ Y_k^{-n}(p') \ = \ P_k(p\cdot p')\,. \qquad (2.3)$$

We consider the Fourier expansion of $f \in L^2(S)$ with respect to the spherical harmonics

$$f = \sum_{(k,n)\in\hat{\mathrm{I}}} (2k+1)\, a_k^n(f)\, Y_k^n\,, \quad a_k^n(f) := \langle f, Y_k^n \rangle\,.$$

Let $l^2(\hat{\mathrm{I}})$ be the Hilbert space of all complex sequences $(a_k^n)_{(k,n)\in\hat{\mathrm{I}}}$ with

$$\sum_{(k,n)\in\hat{\mathrm{I}}} (2k+1)\, |a_k^n|^2 < \infty\,.$$

Then we refer to $F:\ L^2(S) \to l^2(\hat{\mathrm{I}})$ defined by

$$F\, f := (a_k^n(f))_{(k,n)\in\hat{\mathrm{I}}} \quad (f \in L^2(S))$$

as *spherical Fourier transform*. For more details on spherical harmonics, see e.g. [8].

§3. Sampling Spaces

For $j \in \mathbb{N}_0$ we set

$$\mathrm{I}^j := \{(s,t)\ :\ s = 0,\ldots,2^{j+1};\ \ t = 0,\ldots,2^{j+1}-1\}\,,$$
$$\hat{\mathrm{I}}^j := \{(k,n)\ :\ k = 0,\ldots,2^j-1;\ n = -k,\ldots,k\}\,.$$

We consider the *sampling spaces of level* j $(j \in \mathbb{N}_0)$

$$V^j := \operatorname{span}\{Y_k^n\ :\ (k,n) \in \hat{\mathrm{I}}^j\}$$

consisting of so–called *band–limited* functions. Clearly, $\dim V^j = 2^{2j}$. Moreover, for band–limited functions, it holds the following sampling theorem.

Theorem 3.1. *Let $f \in V^j$ $(j \in \mathbb{N}_0)$ be given . Then we have for $(k,n) \in \hat{\mathrm{I}}^j$*

$$a_k^n(f) = \frac{1}{2^{j+1}} \sum_{(s,t)\in\mathrm{I}^j} \varepsilon_s^{(j+1)} w_s^{(j+1)} f(p_{s,t}^j)\, Y_k^{-n}(p_{s,t}^j)$$

with

$$p_{s,t}^j := \left(\frac{\pi s}{2^{j+1}}, \frac{\pi t}{2^j}\right),$$

$$\varepsilon_0^{(j)} = \varepsilon_{2^j}^{(j)} := 2^{-1}, \quad \varepsilon_s^{(j)} := 1 \ (s = 1,\ldots,2^j-1)$$

and with the Clenshaw–Curtis weights

$$w_s^{(j+1)} := \frac{1}{2^{j+1}} \sum_{u=0}^{2^j} \varepsilon_u^{(j)} \frac{-2}{4u^2-1} \cos\frac{su\pi}{2^j} \quad (s = 0,\ldots,2^{j+1})\,. \tag{3.1}$$

Proof: By definition of V^j, it suffices to consider the functions $f(\theta,\varphi) = Y_l^m(\theta,\varphi)$ ($(l,m) \in \hat{\mathrm{I}}^j$). Their Fourier coefficients can be written as

$$a_k^n(f) = \frac{1}{2}\int_{-1}^{1} P_l^{|m|}(x) P_k^{|n|}(x)\,\mathrm{d}x \cdot \frac{1}{2\pi}\int_0^{2\pi} \mathrm{e}^{\mathrm{i}(m-n)\varphi}\,\mathrm{d}\varphi\,. \tag{3.2}$$

Now it holds for m, $n = 1-2^j, \ldots, 2^j-1$ that

$$\delta_{m,n} = \frac{1}{2\pi}\int_0^{2\pi} \mathrm{e}^{\mathrm{i}(m-n)\varphi}\,\mathrm{d}\varphi = \frac{1}{2^{j+1}} \sum_{t=0}^{2^{j+1}-1} \mathrm{e}^{\pi\mathrm{i}(m-n)t/2^j}\,. \tag{3.3}$$

Hence, for $m \neq n$ we are done. For $m = n$, we verify that $P_l^{|n|} P_k^{|n|}$ is an algebraic polynomial of degree at most $2^{j+1}-2$ such that Clenshaw–Curtis quadrature gives

$$\frac{1}{2}\int_{-1}^{1} P_l^{|n|}(x) P_k^{|n|}(x)\,\mathrm{d}x = \sum_{s=0}^{2^{j+1}} \varepsilon_s^{(j+1)}\, w_s^{(j+1)} P_l^{|n|}\left(\cos\frac{\pi s}{2^{j+1}}\right) P_k^{|n|}\left(\cos\frac{\pi s}{2^{j+1}}\right).$$

Together with (3.2) and (3.3), this completes the proof. ∎

Note that a similar sampling theorem for band–limited function on S was proved in [4]. From Theorem 3.1 it follows a discrete orthogonality relation for spherical harmonics.

Corollary 3.2. *Let $j \in \mathbb{N}_0$ be given. Then we have for all (k,n), $(l,m) \in \hat{\mathrm{I}}^j$*

$$\frac{1}{2^{j+1}} \sum_{(s,t)\in \mathrm{I}^j} \varepsilon_s^{(j+1)} w_s^{(j+1)}\, Y_l^m(p_{s,t}^j)\, Y_k^{-n}(p_{s,t}^j) \;=\; \frac{1}{2k+1}\,\delta_{k,l}\,\delta_{m,n}\,.$$

Proof: Using Theorem 3.1 for $f = Y_l^m$, we obtain the result by (2.2). ∎

Theorem 3.1 leads to the following algorithm for the computation of the spherical Fourier transform of a band–limited function $f \in V^j$:

Algorithm 3.3 (Discrete Spherical Fourier Transform)
Input: For fixed $j \in \mathbb{N}_0$, let $f(p_{s,t}^j) \in \mathbb{C}$ ($(s,t) \in \mathrm{I}^j$) be given.
1. For every $s = 0, \ldots, 2^{j+1}$ form by fast Fourier transform

$$\hat{f}_s^n := \frac{1}{2^{j+1}} \sum_{t=0}^{2^{j+1}-1} f(p_{s,t}^j)\, \mathrm{e}^{-\mathrm{i}n\pi t/2^j} \quad (n = 1-2^j, \ldots, 2^j-1)\,.$$

2. For every $n = 1-2^j, \ldots, 2^j-1$ compute

$$\hat{\alpha}_{k,n}^j := \sum_{s=0}^{2^{j+1}} \varepsilon_s^{(j+1)} w_s^{(j+1)}\; \hat{f}_s^n\; P_k^{|n|}\left(\cos\frac{\pi s}{2^{j+1}}\right) \quad (k = |n|, \ldots, 2^j-1)$$

by the three–term recurrence relation (2.1).
Output: $\hat{a}_k^n(f) := \hat{\alpha}_{k,n}^j \in \mathbb{C}$ ($(k,n) \in \hat{\mathrm{I}}^j$).

The size of the Clenshaw–Curtis weights can be estimated as follows:

Lemma 3.4. *For $j \in \mathbb{N}_0$ and $s = 0, \ldots, 2^{j+1}$, it holds*

$$(2^{2j+2}-1)^{-1} \le w_s^{(j+1)} < 2^{-j} - (2^{2j+2}-1)^{-1}\,.$$

Proof: From (3.1) it follows that

$$\begin{aligned} 2^{j+1} w_s^{(j+1)} &\ge 2^{j+1} w_0^{(j+1)} \\ &= 1 - \sum_{u=1}^{2^j-1} \left(\frac{1}{2u-1} - \frac{1}{2u+1}\right) - \frac{1}{2}\left(\frac{1}{2^{j+1}-1} - \frac{1}{2^{j+1}+1}\right) \\ &= \frac{1}{2}\left(\frac{1}{2^{j+1}-1} + \frac{1}{2^{j+1}+1}\right) = 2^{j+1}(2^{2j+2}-1)^{-1}\,. \end{aligned}$$

On the other hand, we have

$$\begin{aligned} 2^{j+1} w_s^{(j+1)} &< 1 + \sum_{u=1}^{2^j-1} \left(\frac{1}{2u-1} - \frac{1}{2u+1}\right) + \frac{1}{2}\left(\frac{1}{2^{j+1}-1} - \frac{1}{2^{j+1}+1}\right) \\ &= 2 - \frac{1}{2}\left(\frac{1}{2^{j+1}-1} + \frac{1}{2^{j+1}+1}\right) = 2 - 2^{j+1}(2^{2j+2}-1)^{-1}\,. \end{aligned}$$

This completes the proof. ■

§4. Multiresolution of $L^2(S)$

The following definition of scaling functions is motivated by the outstanding properties of so–called kernel polynomials associated with orthogonal polynomials in weighted L^2–spaces (see e.g. [1, pp. 35–38]). The constructions of polynomial scaling functions and corresponding wavelets in [2,5,10,11] are based on kernel polynomials. Clearly, the concept of kernel polynomials can be transferred to orthogonal systems in other Hilbert spaces. We are concerned with the *kernels* $K^j : S \times S \to \mathbb{C}$ *of level* j $(j \in \mathbb{N}_0)$

$$K^j(p, p') := \sum_{(k,n) \in \hat{I}^j} (2k+1)\, Y_k^n(p)\, Y_k^{-n}(p') \quad (p, p' \in S)$$

defined with respect to the spherical harmonics in V^j. The name "kernel" is explained by the *reproducing property*

$$\langle f, K^j(\cdot\,, p')\rangle = f(p') \quad (f \in V^j,\ p' \in S)\,, \tag{4.1}$$

which follows immediately from the Fourier expansion

$$f = \sum_{(k,n) \in \hat{I}^j} (2k+1)\, a_k^n(f)\, Y_k^n \tag{4.2}$$

of $f \in V^j$ and from (2.2).

We introduce the *scaling functions* ϕ^j *of level* j ($j \in \mathbb{N}_0$) by

$$2^j \phi^j := \sum_{(k,n)\in \hat{\mathrm{I}}^j} (2k+1)\, Y_k^n$$

and the *"rotated" scaling functions* $\phi^j_{s,t} \in V^j$ ($(s,t) \in \mathrm{I}^j$) by

$$2^j \phi^j_{s,t} := K^j(\,\cdot\,, p^j_{s,t}) = \sum_{(k,n)\in \hat{\mathrm{I}}^j} (2k+1)\, Y_k^n\, Y_k^{-n}(p^j_{s,t})\,.$$

Then we have by (4.1) that

$$a_k^n(\phi^j_{s,t}) = Y_k^{-n}(p^j_{s,t})\, a_k^n(\phi^j) = \begin{cases} 2^{-j}\, Y_k^{-n}(p^j_{s,t}) & (k,n) \in \hat{\mathrm{I}}^j, \\ 0 & \text{otherwise.} \end{cases} \tag{4.3}$$

The functions $\phi^j_{s,t} \in V^j$ ($(s,t) \in \mathrm{I}^j$) are smooth on the whole sphere S. Unfortunately, they are linearly dependent. By the addition theorem (2.3), the scaling functions read in terms of the spherical distance as

$$2^j \phi^j_{s,t}(p) = \sum_{k=0}^{2^j-1} (2k+1)\, P_k\left(\cos(pp^j_{s,t})\right) = \sum_{k=0}^{2^j-1} (2k+1)\, P_k(p \cdot p^j_{s,t})\,. \tag{4.4}$$

Theorem 4.1 collects further important properties of $\phi^j_{s,t}$.

Theorem 4.1. *For* $j \in \mathbb{N}_0$ *and* $(s,t) \in \mathrm{I}^j$ *it holds:*

(i) *Reproducing property:*

$$2^j \langle f, \phi^j_{s,t}\rangle = f(p^j_{s,t}) \quad (f \in V^j)\,.$$

(ii) $\|\phi^j_{s,t}\| = 1\,, \quad \phi^j_{s,t}(p^j_{s,t}) = 2^j$.

(iii) *Localization property:* $\phi^j_{s,t}$ *is localized around* $p^j_{s,t}$, i.e.

$$\frac{\|\phi^j_{s,t}\|}{\phi^j_{s,t}(p^j_{s,t})} = 2^{-j} = \min\left\{\|f\| : f \in V^j,\ f(p^j_{s,t}) = 1\right\}\,.$$

Proof: The reproducing property (i) follows immediately from (4.1). In particular, (i) yields

$$\|\phi^j_{s,t}\|^2 = \langle \phi^j_{s,t}, \phi^j_{s,t}\rangle = 2^{-j}\, \phi^j_{s,t}(p^j_{s,t})\,.$$

On the other hand, we have by (4.4) and $P_k(1) = 1$ $(k \in \mathbb{N}_0)$ that

$$\phi^j_{s,t}(p^j_{s,t}) = 2^{-j} \sum_{k=0}^{2^j-1} (2k+1)\, P_k(\cos 0) \;=\; 2^j\,,$$

which proves (ii).

The localization property (iii) can be seen by the following standard arguments: Let $f \in V^j$ with $f(p^j_{s,t}) = 1$, i.e.

$$1 = \sum_{(k,n)\in\hat{\mathrm{I}}} (2k+1)\, a^n_k(f)\, Y^n_k(p^j_{s,t})\,.$$

Then by applying the Cauchy–Schwarz inequality, the Parseval identity and (ii), we obtain

$$\begin{aligned} 1 &\le \left(\sum_{(k,n)\in\hat{\mathrm{I}}^j} (2k+1)\, |a^n_k(f)|^2 \right) \left(\sum_{(k,n)\in\hat{\mathrm{I}}^j} (2k+1)\, |Y^n_k(p^j_{s,t})|^2 \right) \\ &= \|f\|^2\, 2^j\, \phi^j_{s,t}(p^j_{s,t}) \;=\; 2^{2j}\, \|f\|^2\,, \end{aligned}$$

where the equality holds if and only if $a^n_k(f) = 2^{-j}\, Y^{-n}_k(p^j_{s,t})$ $((k,n) \in \hat{\mathrm{I}}^j)$, i.e. $f = 2^{-j}\, \phi^j_{s,t}$. ■

Theorem 4.2. *The subspaces V^j $(j \in \mathbb{N}_0)$ form a multiresolution of $L^2(S)$ with the following properties:*

(i) $V^j \subset V^{j+1}$ $(j \in \mathbb{N}_0)$, $\quad \operatorname{clos} \bigcup_{j=0}^{\infty} V^j = L^2(S)$.

(ii) $V^j = \operatorname{span}\{\phi^j_{s,t} : (s,t) \in \mathrm{I}^j\}$.

(iii) *The set* $\{(2^{j-1}\varepsilon^{(j+1)}_s w^{(j+1)}_s)^{1/2} \phi^j_{s,t} : (s,t) \in \mathrm{I}^j\}$ *is a tight frame in* V^j, *i.e., for all* $f \in V^j$ *we have*

$$2^{j-1} \sum_{(s,t)\in\mathrm{I}^j} \varepsilon^{(j+1)}_s w^{(j+1)}_s\, |\langle f, \phi^j_{s,t}\rangle|^2 = \|f\|^2\,.$$

Proof: 1. By definition of V^j, property (i) is straightforward.

2. Any $f \in V^j$ can be represented in the form (4.2), where the Fourier coefficients $a^n_k(f)$ $((k,n) \in \hat{\mathrm{I}}^j)$ can be computed by Theorem 3.1. Together with the definition of $\phi^j_{s,t}$, this results in

$$f = \frac{1}{2} \sum_{(s,t)\in\mathrm{I}^j} \varepsilon^{(j+1)}_s w^{(j+1)}_s\, f(p^j_{s,t})\, \phi^j_{s,t}\,.$$

Thus (ii) is clear.

3. By the reproducing property of $\phi^j_{s,t}$ the above expression can be replaced by

$$f = 2^{j-1} \sum_{(s,t)\in \mathrm{I}^j} \varepsilon_s^{(j+1)} w_s^{(j+1)} \langle f, \phi^j_{s,t}\rangle \, \phi^j_{s,t} \, .$$

Hence we obtain the final result

$$\|f\|^2 = \langle f, f\rangle = 2^{j-1} \sum_{(s,t)\in \mathrm{I}^j} \varepsilon_s^{(j+1)} w_s^{(j+1)} \, |\langle f, \phi^j_{s,t}\rangle|^2 \, . \quad \blacksquare$$

The Fourier–transformed *two–scale relation of* ϕ^j reads as follows

$$a_k^n(\phi^j) = A^{j+1}_{k,n} \, a_k^n(\phi^{j+1}) \quad ((k,n) \in \hat{\mathrm{I}}) \tag{4.5}$$

with

$$A^{j+1}_{k,n} := \begin{cases} 2 & (k,n) \in \hat{\mathrm{I}}^j, \\ 0 & (k,n) \in \hat{\mathrm{I}} \setminus \hat{\mathrm{I}}^j. \end{cases}$$

§5. Wavelet Spaces

We introduce the *wavelet space* W^j *of level* j ($j \in \mathbb{N}_0$) as the orthogonal complement of V^j in V^{j+1}, i.e. $V^{j+1} = V^j \oplus W^j$. This implies the orthogonal sum decomposition

$$L^2(S) = V^{j_0} \oplus \bigoplus_{j=j_0}^{\infty} W^j \quad (j_0 \in \mathbb{N}_0) \, .$$

Clearly,

$$\dim W^j = \dim V^{j+1} - \dim V^j = 3 \cdot 2^{2j} \, .$$

Let the function ψ^j ($j \in \mathbb{N}_0$) be defined by

$$2^j \, \psi^j := \sum_{(k,n)\in \hat{\mathrm{I}}^{j+1}\setminus \hat{\mathrm{I}}^j} (2k+1) \, Y_k^n$$

and the "rotated" functions $\psi^j_{s,t} \in W^j$ ($(s,t) \in \mathrm{I}^{j+1}$) by

$$2^j \, \psi^j_{s,t} := K^{j+1}(\,\cdot\,, p^{j+1}_{s,t}) - K^j(\,\cdot\,, p^{j+1}_{s,t}) = \sum_{(k,n)\in \hat{\mathrm{I}}^{j+1}\setminus \hat{\mathrm{I}}^j} (2k+1) \, Y_k^n \, Y_k^{-n}(p^{j+1}_{s,t}) \, .$$

By the addition theorem (2.3), the functions $\psi^j_{s,t}$ can be written alternatively as

$$2^j \psi^j_{s,t}(p) = \sum_{k=2^j}^{2^{j+1}-1} (2k+1) \, P_k \left(\cos(p \, p^{j+1}_{s,t})\right) .$$

Then we have

$$a_k^n(\psi^j_{s,t}) = Y_k^{-n}(p^{j+1}_{s,t}) \, a_k^n(\psi^j) = \begin{cases} 2^{-j} \, Y_k^{-n}(p^{j+1}_{s,t}) & (k,n) \in \hat{\mathrm{I}}^{j+1} \setminus \hat{\mathrm{I}}^j, \\ 0 & \text{otherwise.} \end{cases} \tag{5.1}$$

The functions $\psi^j_{s,t}$ ($(s,t) \in \mathrm{I}^{j+1}$) are smooth on the whole sphere S. Obviously, they are linearly dependent. Moreover, they possess the following important properties:

Theorem 5.1. *For $j \in \mathbb{N}_0$ and $(s,t) \in \mathrm{I}^{j+1}$ it holds:*

(i) *Reproducing property:*

$$2^j \langle f, \psi^j_{s,t}\rangle = f(p^{j+1}_{s,t}) \quad (f \in W^j)\,.$$

(ii) *Orthogonality property:*

$$\langle \phi^j_{u,v}, \psi^j_{s,t}\rangle = 0 \quad ((u,v) \in \mathrm{I}^j)\,.$$

(iii) $\|\psi^j_{s,t}\| = 3^{1/2}\,, \quad \psi^j_{s,t}(p^{j+1}_{s,t}) = 3 \cdot 2^j\,.$

(iv) *Localization property: $\psi^j_{s,t}$ is localized around $p^{j+1}_{s,t}$, i.e.*

$$\frac{\|\psi^j_{s,t}\|}{\psi^j_{s,t}(p^{j+1}_{s,t})} = 3^{-1/2}\,2^{-j} = \min\,\{\|f\| : f \in W^j,\ f(p^{j+1}_{s,t}) = 1\}\,.$$

Proof: The orthogonality property is a direct consequence of the definition of $\psi^j_{s,t}$. The proof of (i), (iii) and (iv) is omitted here, since it follows exactly the lines of the proof of Theorem 4.1. ■

Theorem 5.2. *We have*

$$W^j = \operatorname{span}\{\psi^j_{s,t} : (s,t) \in \mathrm{I}^{j+1}\}\,.$$

The functions $(2^{j-2}\varepsilon^{(j+2)}_s w^{(j+2)}_s)^{1/2}\psi^j_{s,t}$ $((s,t) \in \mathrm{I}^{j+1})$ constitute a tight frame in W^j, i.e.,

$$2^{j-2} \sum_{(s,t)\in\hat{\mathrm{I}}^{j+1}} \varepsilon^{(j+2)}_s w^{(j+2)}_s\, |\langle f, \psi^j_{s,t}\rangle|^2 = \|f\|^2 \tag{5.2}$$

for all $f \in W^j$.

Proof: Fourier expansion of $f \in W^j$ with Fourier coefficients $a^n_k(f)$ determined by Theorem 3.1 gives

$$\begin{aligned} f &= \sum_{(k,n)\in\hat{\mathrm{I}}^{j+1}\setminus\hat{\mathrm{I}}^j} (2k+1)\, a^n_k(f)\, Y^n_k \\ &= \frac{1}{4} \sum_{(s,t)\in\mathrm{I}^{j+1}} \varepsilon^{(j+2)}_s w^{(j+2)}_s f(p^{j+1}_{s,t})\, \psi^j_{s,t} \end{aligned} \tag{5.3}$$

such that $W^j = \operatorname{span}\,\{\psi^j_{s,t} : (s,t) \in \mathrm{I}^{j+1}\}$. Now (5.2) follows as in the proof of Theorem 4.2. ■

The Fourier–transformed *two–scale relation of ψ^j* reads

$$a^n_k(\psi^j) = B^{j+1}_{k,n}\, a^n_k(\phi^{j+1}) \quad ((k,n) \in \hat{\mathrm{I}}) \tag{5.4}$$

with

$$B^{j+1}_{k,n} := \begin{cases} 2 & (k,n) \in \hat{\mathrm{I}}^{j+1} \setminus \hat{\mathrm{I}}^j\,, \\ 0 & \text{otherwise.} \end{cases}$$

§6. Decomposition and Reconstruction Algorithms

In this section, we derive efficient decomposition and reconstruction algorithms. In order to *decompose* a given function $f^{j+1} \in V^{j+1}$ $(j \in \mathbb{N}_0)$ of the form

$$f^{j+1} = \frac{1}{2} \sum_{(s,t)\in \mathrm{I}^{j+1}} \varepsilon_s^{(j+2)} w_s^{(j+2)} \alpha_{s,t}^{j+1} \, \phi_{s,t}^{j+1} \tag{6.1}$$

with $\alpha_{s,t}^{j+1} := f^{j+1}(p_{s,t}^{j+1})$ $((s,t) \in \mathrm{I}^{j+1})$, the uniquely determined functions $f^j \in V^j$ and $g^j \in W^j$ have to be found such that

$$f^{j+1} = f^j + g^j \, . \tag{6.2}$$

Assume that the coefficients $\alpha_{s,t}^{j+1} \in \mathbb{C}$ $((s,t) \in \mathrm{I}^{j+1})$ or their *discrete spherical Fourier transform data*

$$\hat{\alpha}_{k,n}^{j+1} := \frac{1}{2^{j+2}} \sum_{(s,t)\in \mathrm{I}^{j+1}} \varepsilon_s^{(j+2)} w_s^{(j+2)} \alpha_{s,t}^{j+1} \, Y_k^{-n}(p_{s,t}^{j+1}) \quad ((k,n) \in \hat{\mathrm{I}}^{j+1}) \tag{6.3}$$

are known. The functions $f^j \in V^j$ and $g^j \in W^j$ can be represented by

$$f^j = \frac{1}{2} \sum_{(s,t)\in \mathrm{I}^{j}} \varepsilon_s^{(j+1)} w_s^{(j+1)} \alpha_{s,t}^{j} \, \phi_{s,t}^{j} \, , \tag{6.4}$$

$$g^j = \frac{1}{4} \sum_{(u,v)\in \mathrm{I}^{j+1}} \varepsilon_u^{(j+2)} w_u^{(j+2)} \beta_{u,v}^{j} \, \psi_{u,v}^{j} \tag{6.5}$$

with unknown coefficients $\alpha_{s,t}^j$, $\beta_{u,v}^j \in \mathbb{C}$. Let $\hat{\alpha}_{k,n}^j$, $\tilde{\beta}_{l,m}^j \in \mathbb{C}$ denote the following discrete spherical Fourier transform data

$$\hat{\alpha}_{k,n}^{j} := \frac{1}{2^{j+1}} \sum_{(s,t)\in \mathrm{I}^{j}} \varepsilon_s^{(j+1)} w_s^{(j+1)} \, \alpha_{s,t}^{j} \, Y_k^{-n}(p_{s,t}^{j}) \quad ((k,n) \in \hat{\mathrm{I}}^{j}) \, , \tag{6.6}$$

$$\tilde{\beta}_{l,m}^{j} := \frac{1}{2^{j+2}} \sum_{(u,v)\in \mathrm{I}^{j+1}} \varepsilon_u^{(j+2)} w_u^{(j+2)} \, \beta_{u,v}^{j} \, Y_l^{-m}(p_{u,v}^{j+1}) \quad ((l,m) \in \hat{\mathrm{I}}^{j+1} \setminus \hat{\mathrm{I}}^{j}) \, . \tag{6.7}$$

In order to *reconstruct* $f^{j+1} \in V^{j+1}$ $(j \in \mathbb{N}_0)$, we have to compute the coefficients $\alpha_{s,t}^{j+1}$ of the sum (6.2) with given functions $f^j \in V^j$ and $g^j \in W^j$ (see (6.1)). Assume that the coefficients $\alpha_{s,t}^j$, $\beta_{u,v}^j \in \mathbb{C}$ in (6.4) – (6.5) or their corresponding transformed data (6.6) – (6.7) are known. The decomposition and reconstruction algorithms are based on the following connection between (6.3) and (6.6) – (6.7):

Theorem 6.1. *For $j \in \mathbb{N}_0$, let $f^{j+1} \in V^{j+1}$, $f^j \in V^j$ and $g^j \in W^j$ with (6.1) – (6.7) be given. Then we have*

$$\hat{\alpha}^j_{k,n} = \hat{\alpha}^{j+1}_{k,n} \quad ((k,n) \in \hat{\mathrm{I}}^j),$$
$$\tilde{\beta}^j_{l,m} = \hat{\alpha}^{j+1}_{l,m} \quad ((l,m) \in \hat{\mathrm{I}}^{j+1} \setminus \hat{\mathrm{I}}^j).$$

Proof: From (6.1), it follows by (4.3) and (6.3) that for all $(k,n) \in \hat{\mathrm{I}}^{j+1}$

$$a^n_k(f^{j+1}) = 2^{j+1}\, \hat{\alpha}^{j+1}_{k,n}\, a^n_k(\phi^{j+1}).$$

Analogously, by (4.3), (5.1), (6.4) and (6.5) we have

$$a^n_k(f^j) = 2^j\, \hat{\alpha}^j_{k,n}\, a^n_k(\phi^j) \quad ((k,n) \in \hat{\mathrm{I}}^j), \tag{6.8}$$
$$a^m_l(g^j) = 2^j\, \tilde{\beta}^j_{l,m} a^m_l(\psi^j) \quad ((l,m) \in \hat{\mathrm{I}}^{j+1}). \tag{6.9}$$

From (6.2) it follows that for all $(k,n) \in \hat{\mathrm{I}}^{j+1}$,

$$a^n_k(f^{j+1}) = a^n_k(f^j) + a^n_k(g^j).$$

Using the Fourier–transformed two–scale relations (4.5) and (5.4), we obtain for all $(k,n) \in \hat{\mathrm{I}}^{j+1}$,

$$2\,\hat{\alpha}^{j+1}_{k,n}\, a^n_k(\phi^{j+1}) = \hat{\alpha}^j_{k,n}\, A^{j+1}_{k,n}\, a^n_k(\phi^{j+1}) + \tilde{\beta}^j_{k,n}\, B^{j+1}_{k,n}\, a^n_k(\phi^{j+1}).$$

By $a^n_k(\phi^{j+1}) \neq 0$ $((k,n) \in \hat{\mathrm{I}}^{j+1})$, this yields the assertion. ■

As consequence we obtain the following procedures:

Algorithm 6.2 (Decomposition Algorithm)
Input: For fixed $j \in \mathbb{N}_0$, let $\hat{\alpha}^{j+1}_{k,n} \in \mathbb{C}$ $((k,n) \in \hat{\mathrm{I}}^{j+1})$ be given.
Set

$$\hat{\alpha}^j_{k,n} := \hat{\alpha}^{j+1}_{k,n} \quad ((k,n) \in \hat{\mathrm{I}}^j),$$
$$\tilde{\beta}^j_{l,m} := \hat{\alpha}^{j+1}_{l,m} \quad ((l,m) \in \hat{\mathrm{I}}^{j+1} \setminus \hat{\mathrm{I}}^j).$$

Output: $\hat{\alpha}^j_{k,n} \in \mathbb{C}$ $((k,n) \in \hat{\mathrm{I}}^j)$, $\tilde{\beta}^j_{l,m} \in \mathbb{C}$ $((l,m) \in \hat{\mathrm{I}}^{j+1} \setminus \hat{\mathrm{I}}^j)$.

Algorithm 6.3 (Reconstruction Algorithm)
Input: For fixed $j \in \mathbb{N}_0$, let $\hat{\alpha}^j_{k,n} \in \mathbb{C}$ $((k,n) \in \hat{\mathrm{I}}^j)$ and $\tilde{\beta}^j_{l,m} \in \mathbb{C}$ $((l,m) \in \hat{\mathrm{I}}^{j+1} \setminus \hat{\mathrm{I}}^j)$ be given.
Set

$$\hat{\alpha}^{j+1}_{k,n} := \hat{\alpha}^j_{k,n} \quad ((k,n) \in \hat{\mathrm{I}}^j),$$
$$\hat{\alpha}^{j+1}_{l,m} := \tilde{\beta}^j_{l,m} \quad ((l,m) \in \hat{\mathrm{I}}^{j+1} \setminus \hat{\mathrm{I}}^j).$$

Output: $\hat{\alpha}^{j+1}_{k,n} \in \mathbb{C}$ $((k,n) \in \hat{\mathrm{I}}^{j+1})$.

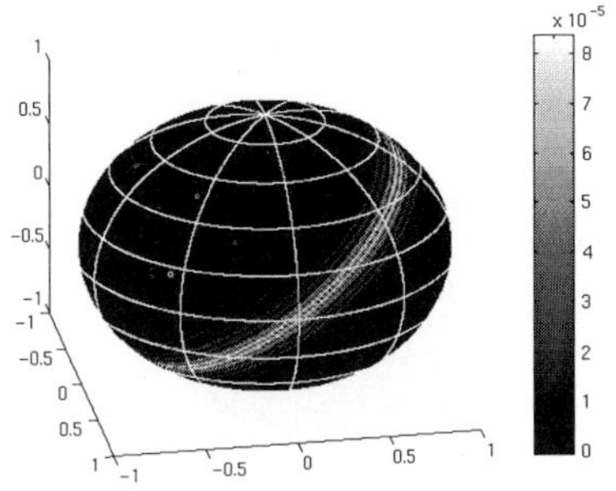

Fig. 1. Wavelet part of f_1 in W^6.

§7. Numerical Tests

Finally, we present numerical tests of the described decomposition algorithm. Consider the ellipsoid

$$x_1^2 + x_2^2 + 4\,x_3^2 = 1 \quad ((x_1, x_2, x_3) \in \mathbb{R}^3)\,.$$

Using spherical coordinates (θ, φ, r) of $\mathbb{R}^3$, we get $r^2\,(1 + 3\cos^2\theta) = 1$. Then we introduce the function $f : S \to \mathbb{R}$ defined by

$$f(\theta, \varphi) := \begin{cases} 1 & \theta \in [0, \pi/2], \\ (1 + 3\cos^2\theta)^{-1/2} & \theta \in (\pi/2, \pi]. \end{cases}$$

Hence f is the union of a half–sphere and a half–ellipsoid. This function f is smooth on S without the equator $\{(\pi/2, \varphi) : \varphi \in [0, 2\pi)\}$. If we rotate the function f around the x_1–axis by the angle $\pi/4$ and $\pi/2$, we obtain the functions f_1 and f_2, respectively.

By the decomposition algorithm 6.2, we analyze the local regularity of f_1 and f_2. In order to compute values of the related wavelet parts on a grid of S, we use the representation (5.3). As shown in Figures 1 and 2, the corresponding wavelet parts in W^6 describe the smoothness of f_1 and f_2 in the right way.

Acknowledgements. The authors would like to thank Günter Baszenski for his support in numerical questions.

References

1. Chihara, T. S., *An Introduction to Orthogonal Polynomials*, Gordon and Breach Science Publ., New York, 1978.
2. Chui, C. K. and H. N. Mhaskar, On trigonometric wavelets, Constr. Approx. **9** (1993), 167–190.

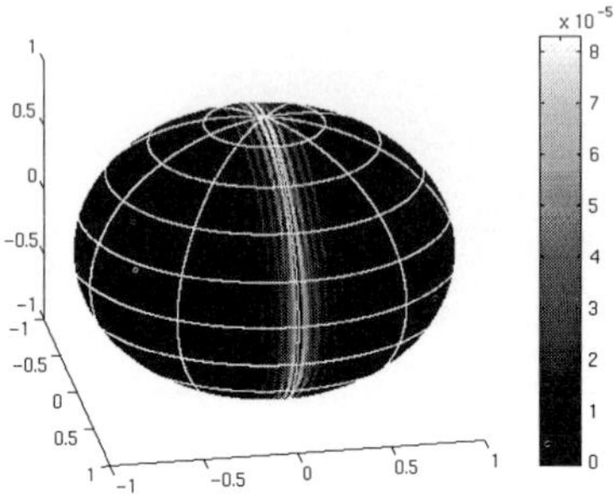

Fig. 2. Wavelet part of f_2 in W^6.

3. Dahlke, S., W. Dahmen, E. Schmitt and I. Weinreich, Multiresolution analysis and wavelets on S^2 and S^3, preprint, RWTH Aachen, 1994.
4. Driscoll, J. R. and D. M. Healy, Jr., Computing Fourier transforms and convolutions on the 2–sphere, Adv. Applied Math. **15** (1994), 202–250.
5. Fischer, B. and J. Prestin, Wavelets based on orthogonal polynomials, preprint, Medizinische Universität zu Lübeck, 1996.
6. Freeden, W. and U. Windheuser, Combined spherical harmonic and wavelet expansion, preprint, Universität Kaiserslautern, 1995.
7. Lyche, T. and L. L. Schumaker, L–spline wavelets, in *Wavelets: Theory, Algorithms and Applications*, C. K. Chui, L. Montefusco, and L. Puccio (eds.), Academic Press, San Diego, 1994, 197–212.
8. Müller, C., *Spherical Harmonics*, Springer-Verlag, Berlin, 1966.
9. Narcowich, F. J. and J. D. Ward, Nonstationary wavelets on the m–sphere for scattered data, preprint, Texas A&M University, 1995.
10. Potts, D. and M. Tasche, Interpolatory wavelets on the sphere, in *Approximation Theory VIII, Vol. 2: Wavelets and Multilevel Approximation*, C. K. Chui, and L. L. Schumaker (eds.), World Scientific Publ., Singapore, 1995, 335–342.
11. Tasche, M., Polynomial wavelets on $[-1, 1]$, in *Approximation Theory, Wavelets and Applications*, S. P. Singh (ed.), Kluwer Academic Publ., Dordrecht, 1995, 497–512.
12. Schröder, P. and W. Sweldens, Spherical wavelets: Efficiently representing functions on the sphere, preprint, University of South Carolina, 1995.

Daniel Potts
Fachbereich Mathematik, Universität Rostock
D – 18051 Rostock
Germany
daniel.potts@stud.uni-rostock.de

Gabriele Steidl
Fachbereich Mathematik, TH Darmstadt
Schloßgartenstr. 7
D – 64289 Darmstadt
Germany
steidl@mathematik.th-darmstadt.de

Manfred Tasche
Institut für Mathematik, Medizinische Universität zu Lübeck
Wallstr. 40
D – 23560 Lübeck
Germany
manfred.tasche@informatik.mu-luebeck.de

A Review of Algorithms for Thin Plate Spline Interpolation in Two Dimensions

M. J. D. Powell

Abstract. The thin plate spline interpolation problem can be expressed as the solution of an $(n+3)\times(n+3)$ system of linear equations, where n is the number of interpolation points. If the system is solved by a direct method, then $\mathcal{O}(n^3)$ computer operations are required. A particular direct method is recommended, that gives good results when this amount of work is tolerable. Conjugate gradient methods for positive definite or semi-definite symmetric systems are also applicable, and it is shown that the well-known variational formulation of the thin plate spline method assists the choice of pre-conditioners. Further, the work of each iteration can be reduced greatly by the use of Laurent expansion techniques when n is large. We also consider some iterative methods that are developed from the localization properties of the Lagrange functions of interpolation. Thus we identify a new algorithm that typically requires fewer than 15 iterations to achieve 10 decimals accuracy for large n.

§1. Some Useful Properties of the Thin Plate Spline Equations

The two-dimensional interpolation problem is the calculation of a function s from $\mathbb{R}^2$ to $\mathbb{R}$ that satisfies the conditions

$$s(\underline{x}_i) = f_i, \qquad i=1,2,\ldots,n, \tag{1.1}$$

where the points $\underline{x}_i \in \mathbb{R}^2$ and the right hand sides $f_i \in \mathbb{R}$, $i=1,2,\ldots,n$, are given. The number of possible choices of s is infinite, and it is very important in practice to pick a suitable one. In particular, the thin plate spline method takes up the freedom in s by minimizing the integral

$$I(s) = \iint_{\mathbb{R}^2} \left(\frac{\partial^2 s(\underline{x})}{\partial x^2}\right)^2 + 2\left(\frac{\partial^2 s(\underline{x})}{\partial x\,\partial y}\right)^2 + \left(\frac{\partial^2 s(\underline{x})}{\partial y^2}\right)^2 d\underline{x}, \tag{1.2}$$

Advanced Topics in Multivariate Approximation
F. Fontanella, K. Jetter and P. J. Laurent (eds.) pp. 303–322.

ISBN 981-02-2852-X

where x and y are the components of $\underline{x}$. Thus s is defined uniquely, provided that the points $\underline{x}_i$, $i = 1, 2, \ldots, n$, are not collinear, which is assumed throughout the paper. Further, the points must all be different, in order to avoid redundancies and inconsistencies.

Duchon [5] proves that the thin plate spline interpolant has the form

$$s(\underline{x}) = \sum_{j=1}^{n} \lambda_j \, \phi(\|\underline{x}-\underline{x}_j\|_2) + p(\underline{x}), \qquad \underline{x} \in \mathbb{R}^2, \tag{1.3}$$

where the numbers λ_j, $j=1,2,\ldots,n$, are real coefficients, where ϕ is the function

$$\phi(0) = 0 \quad \text{and} \quad \phi(r) = r^2 \log r, \qquad r>0, \tag{1.4}$$

and where p is a polynomial of degree at most one from $\mathbb{R}^2$ to $\mathbb{R}$. Further, the finiteness of the integral (1.2) implies the conditions

$$\sum_{j=1}^{n} \lambda_j = 0 \quad \text{and} \quad \sum_{j=1}^{n} \lambda_j \, \underline{x}_j = 0. \tag{1.5}$$

We take the view that s is a function of the form (1.3) whose coefficients satisfy the interpolation equations (1.1) and the constraints (1.5). Therefore, if $\underline{\lambda}$ and $\underline{f}$ are the vectors in $\mathbb{R}^n$ whose components are λ_j, $j=1,2,\ldots,n$, and f_i, $i=1,2,\ldots,n$, respectively, if $\underline{c}$ is the vector in $\mathbb{R}^3$ whose components are the coefficients of the polynomial

$$p(\underline{x}) = c_1 + c_2\, x + c_3\, y, \qquad \underline{x} \in \mathbb{R}^2, \tag{1.6}$$

if Φ is the $n\times n$ matrix that has the elements $\Phi_{ij} = \phi(\|\underline{x}_i - \underline{x}_j\|_2)$, and if P is the $n\times 3$ matrix whose i-th row is $(1 \; x_i \; y_i)$, where x_i and y_i are the components of the data point $\underline{x}_i$, then we require $\underline{\lambda}$ and $\underline{c}$ to be a solution of the $(n+3)\times(n+3)$ linear system

$$\begin{pmatrix} \Phi & P \\ P^T & 0 \end{pmatrix} \begin{pmatrix} \underline{\lambda} \\ \underline{c} \end{pmatrix} = \begin{pmatrix} \underline{f} \\ 0 \end{pmatrix}. \tag{1.7}$$

Several well-known properties of this system are established below, because they are useful to the methods that we will review. One of them is that the matrix of the system is nonsingular.

The reason for considering the thin plate spline interpolation method is that it is highly successful in many applications [7]. Of course the value of n in practice is restricted by the amount of computation that is tolerable when solving the equations (1.7). Direct methods become very laborious when n exceeds 1000, because there is hardly any sparsity in Φ. We will find, however, that iterative algorithms allow much larger values of n.

Some of the theory of this section employs the bilinear functional associated with the integral (1.2), namely the expression

$$\langle g, s\rangle = \iint_{\mathbb{R}^2} \frac{\partial^2 g(\underline{x})}{\partial x^2}\,\frac{\partial^2 s(\underline{x})}{\partial x^2} + 2\,\frac{\partial^2 g(\underline{x})}{\partial x\,\partial y}\,\frac{\partial^2 s(\underline{x})}{\partial x\,\partial y} + \frac{\partial^2 g(\underline{x})}{\partial y^2}\,\frac{\partial^2 s(\underline{x})}{\partial y^2}\, d\underline{x}, \tag{1.8}$$

where g and s are any functions from $\mathbb{R}^2$ to $\mathbb{R}$ that have square integrable second derivatives. Actually, s always has the form (1.3) subject to the conditions (1.5), and then, if we apply integration by parts to the derivatives of g in equation (1.8), the contributions from the boundaries at infinity are zero. Thus we deduce the identity

$$\langle g, s\rangle = \iint_{\mathbb{R}^2} g(\underline{x}) \left\{ \frac{\partial^4 s(\underline{x})}{\partial x^4} + 2\,\frac{\partial^4 s(\underline{x})}{\partial x^2\,\partial y^2} + \frac{\partial^4 s(\underline{x})}{\partial y^4} \right\} d\underline{x}. \tag{1.9}$$

Further, the term in braces is $\nabla^4 s(\underline{x})$, $\underline{x} \in \mathbb{R}^2$, and the biharmonic operator applied to $\phi(\|\underline{x}\|_2)$, $\underline{x} \in \mathbb{R}^2$, provides $8\pi\,\delta(\underline{x})$, $\underline{x} \in \mathbb{R}^2$, where $\delta(\cdot)$ is the Dirac delta function. It follows that expression (1.8) has the value

$$\langle g, s\rangle = 8\pi \iint_{\mathbb{R}^2} g(\underline{x}) \sum_{j=1}^{n} \lambda_j\, \delta(\underline{x} - \underline{x}_j)\, d\underline{x} = 8\pi \sum_{j=1}^{n} \lambda_j\, g(\underline{x}_j). \tag{1.10}$$

Thus $\langle g, s\rangle$ is zero if $g(\cdot)$ vanishes at all the interpolation points.

It is now straightforward to prove that the interpolant s minimizes the integral (1.2) subject to the constraints (1.1) if it has the form (1.3). Indeed, if $t(\underline{x})$, $\underline{x} \in \mathbb{R}^2$, is any other interpolant that has square integrable second derivatives, then the difference $g = t - s$ must satisfy $g(\underline{x}_i) = 0$, $i = 1, 2, \ldots, n$. Therefore it is elementary to derive the required inequality

$$I(t) = I(s+g,\, s+g) = I(s) + 2\,\langle g, s\rangle + I(g) = I(s) + I(g) \geq I(s), \tag{1.11}$$

where the last relation is a consequence of the nonnegativity of the integral (1.2).

The next theoretical result that is important to this review is a conditional positive definite property of the symmetric matrix Φ of the system (1.7). Specifically, if $\underline{\lambda}$ is any nonzero vector in $\mathbb{R}^n$ that satisfies the conditions (1.5), then $\underline{\lambda}^T \Phi \underline{\lambda}$ is strictly positive. In order to prove this assertion, we let s be the function (1.3) with $p \equiv 0$ for the moment, and we consider the identity

$$I(s) = \langle s, s\rangle = 8\pi \sum_{i=1}^{n} \lambda_i\, s(\underline{x}_i), \tag{1.12}$$

which is implied by the definitions (1.2) and (1.8) and by equation (1.10). The assumption $\underline{\lambda} \neq 0$ provides $I(s) > 0$. It follows from the definitions of s and Φ that we have the required relation

$$0 < \sum_{i=1}^{n} \lambda_i\, s(\underline{x}_i) = \sum_{i=1}^{n} \sum_{j=1}^{n} \lambda_i\, \lambda_j\, \phi(\|\underline{x}_i - \underline{x}_j\|_2) = \underline{\lambda}^T \Phi \underline{\lambda}. \tag{1.13}$$

We can now prove that the matrix of the system (1.7) is nonsingular. It is sufficient to show that, if $\underline{\lambda}$ and $\underline{c}$ satisfy $\Phi\underline{\lambda}+P\underline{c}=0$ and $P^T\underline{\lambda}=0$, then all of their components are zero. Because $P^T\underline{\lambda}=0$ implies $\underline{\lambda}^T P\underline{c}=0$, the first equation provides $\underline{\lambda}^T\Phi\underline{\lambda}=0$, but we found in the previous paragraph that $\underline{\lambda}^T\Phi\underline{\lambda}$ is positive if $\underline{\lambda}$ is a nonzero vector such that $P^T\underline{\lambda}=0$, this condition being identical to expression (1.5). Thus we deduce $\underline{\lambda}=0$ from the given equations. Hence the first equation becomes $P\underline{c}=0$, which is equivalent to the conditions $p(\underline{x}_i)=0$, $i=1,2,\ldots,n$, where p is the polynomial (1.6). In other words, all of the interpolation points are in the set $\{\underline{x} : p(\underline{x})=0\}$. We are assuming, however, that the interpolation points are not collinear. It follows that $p\equiv 0$, which implies the values $c_1=c_2=c_3=0$ as required.

Of course the nonsingularity of the system (1.7) ensures that the data define the parameter vectors $\underline{\lambda}$ and $\underline{c}$ uniquely. Thus the interpolating function (1.3) is also well-defined, and we have found that it minimizes the integral (1.2) subject to the interpolation conditions. Moreover, the equations (1.7) give the formula

$$\underline{\lambda} = A\,\underline{f}, \tag{1.14}$$

where A is the leading $n \times n$ submatrix of the $(n+3) \times (n+3)$ matrix on the right hand side of the relation

$$\begin{pmatrix} \underline{\lambda} \\ \underline{c} \end{pmatrix} = \begin{pmatrix} \Phi & P \\ P^T & 0 \end{pmatrix}^{-1} \begin{pmatrix} \underline{f} \\ 0 \end{pmatrix}. \tag{1.15}$$

The following properties of A are important to several of the algorithms that will be considered.

The symmetry of Φ and the definition of A imply that A is a symmetric matrix. Moreover, the integral (1.2) has the value

$$I(s) = 8\pi\,\underline{f}^T A\,\underline{f}, \tag{1.16}$$

which is proved as follows. Equations (1.2), (1.8), (1.10) and (1.1) provide the identity $I(s)=8\pi\,\underline{\lambda}^T\underline{f}$, and then we substitute the value of $\underline{\lambda}$ that occurs in formula (1.14). Expressions (1.2) and (1.16) show that $\underline{f}^T A\,\underline{f}$ is nonnegative for every vector $\underline{f}$ in $\mathbb{R}^n$. It follows that all the eigenvalues of the symmetric matrix A are nonnegative. Further, expression (1.16) is zero if and only if the interpolation conditions (1.1) allow s to be a polynomial of degree at most one. This happens when the data satisfy $f_i=p(\underline{x}_i)$, $i=1,2,\ldots,n$, where p is such a polynomial, so $\underline{f}$ is in a 3-dimensional linear subspace of $\mathbb{R}^n$. In this case the vector of coefficients (1.14) is zero, and therefore A has 3 eigenvalues of zero, the remaining $n-3$ eigenvalues being positive.

Section 2 considers algorithms for solving the system (1.7) that employ the conditional positive definiteness of Φ, much of the material being taken from Sibson and Stone [11] and from Powell [9]. The properties of A that have just been mentioned are put to good use by Dyn, Levin and Rippa [6], and their method is the subject of Section 3. Finally, Section 4 describes a new approach to the thin plate spline interpolation problem that is being developed by the author in collaboration with R.K. Beatson and G. Goodsell. It provides a new algorithm that requires far less iterations than the methods of Sections 2 and 3 in all the numerical experiments that have been tried.

§2. Algorithms that Employ the Conditional Positive Definiteness of Φ

The equations (1.7) imply that the required vector $\underline{\lambda} \in \mathbb{R}^n$ is in the null space of P^T, where P is the $n \times 3$ matrix whose i-th row is $(1\, x_i\, y_i)$. Further, because the data points $\underline{x}_i$, $i = 1, 2, \ldots, n$, are not collinear, the rank of P is 3. Therefore we can let Q be an $n \times (n-3)$ real matrix whose columns are linearly independent and span the null space of P^T. Many choices of Q are available, and, having picked one, the required vector $\underline{\lambda}$ is a unique linear combination of the columns of Q. In other words, we can express $\underline{\lambda}$ in the form

$$\underline{\lambda} = Q\, \underline{\mu}, \tag{2.1}$$

for some vector $\underline{\mu}$ in $\mathbb{R}^{n-3}$. The algorithms that make use of the conditional positive definiteness of Φ calculate $\underline{\mu}$ by solving an $(n-3)\times(n-3)$ system of linear equations. Then $\underline{\lambda}$ is given by formula (2.1). Having found $\underline{\lambda}$, the vector $\underline{c} \in \mathbb{R}^3$ can be generated by solving 3 of the n equations $P\underline{c} = \underline{f} - \Phi\underline{\lambda}$. Let α, β and γ be the indices of the 3 equations that are chosen. One can show that $\underline{c}$ is properly defined by this method if the data points $\underline{x}_\alpha$, $\underline{x}_\beta$ and $\underline{x}_\gamma$ are not collinear.

Because formula (2.1) provides $P^T\underline{\lambda}=0$ automatically, the calculation of $\underline{\mu}$ has to be derived from the first part of the system (1.7), namely the relation $\Phi\underline{\lambda}+P\underline{c}=\underline{f}$. We have to eliminate $\underline{c}$ from these n equations, which leaves $n-3$ conditions that are satisfied by $\underline{\lambda}=Q\underline{\mu}$. Pre-multiplication by Q^T achieves the elimination, because we have chosen the columns of Q to be orthogonal to the columns of P. Thus $\underline{\mu}$ is a solution of the $(n-3)\times(n-3)$ system

$$Q^T \Phi\, Q\, \underline{\mu} = Q^T \underline{f}. \tag{2.2}$$

Fortunately, the matrix $Q^T\Phi\, Q$ is not only symmetric, which is inherited from the symmetry of Φ, but also positive definite, which is proved in the next paragraph. Therefore $\underline{\mu}$ is defined uniquely by the equations (2.2).

We establish the positive definiteness of $Q^T\Phi\, Q$ by deducing that, if $\underline{\mu}$ is an arbitrary nonzero vector in $\mathbb{R}^{n-3}$, then $\underline{\mu}^T(Q^T\Phi\, Q)\,\underline{\mu}$ is positive. Equivalently, we require $\underline{\lambda}^T\Phi\underline{\lambda} > 0$, where $\underline{\lambda}$ is the vector (2.1). Now $\underline{\lambda} \in \mathbb{R}^n$ is nonzero, because the columns of Q are linearly independent, and the choice of Q provides the property $P^T\underline{\lambda} = 0$. Therefore the positive definiteness of $Q^T\Phi\, Q$ is equivalent to the conditional positive definiteness of Φ that has been proved already in Section 1.

When the thin plate spline equations are solved by a direct method, the author prefers to make a suitable choice of Q, to form the matrix $Q^T\Phi Q$ explicitly, and to solve the system (2.2) by calculating the Cholesky factorization of $Q^T\Phi Q$. A strong advantage of this approach is that the positive definiteness restricts the size of the off-diagonal elements of $Q^T\Phi Q$, but Φ can have very large elements away from its diagonal. It is convenient in this case to generate Q by deleting the first 3 columns of an $n\times n$ orthogonal matrix, Ω say, where Ω has the property that the $n\times 3$ matrix $\Omega^T P$ is upper triangular. Thus the columns of Q are linearly independent and orthogonal to the columns of P as required. Further, all of the elements of $Q^T\Phi Q$ can be calculated in only $\mathcal{O}(n^2)$ operations, because the condition that $\Omega^T P$ be upper triangular allows Ω to be a product of at most 3 Householder rotations or $3n-6$ Givens rotations. Therefore we can form $\Omega^T\Phi\Omega$ in $\mathcal{O}(n^2)$ operations, and $Q^T\Phi Q$ is just the bottom right $(n-3)\times(n-3)$ submatrix of $\Omega^T\Phi\Omega$. Moreover, the calculation of $\underline{\lambda}$ from equation (2.1) is straightforward, because this equation can be expressed as $\underline{\lambda}=\Omega\,\underline{\breve{\mu}}$, where $\underline{\breve{\mu}}$ is the vector in $\mathbb{R}^n$ whose components are 3 zeros followed by the components of $\underline{\mu}$.

When n is sufficiently large, however, it is more efficient to solve the positive definite system (2.2) by conjugate gradients instead of Cholesky, even if as many as n iterations are needed to achieve acceptable accuracy. The most expensive task of each conjugate gradient iteration is the calculation of the product $Q^T\Phi Q\underline{\mu}$ for a trial value of $\underline{\mu}\in\mathbb{R}^{n-3}$, and the assertion depends on the fact that this task can be done in fewer than $\mathcal{O}(n^2)$ computer operations, although Φ has hardly any zero elements. Many choices of Q, including the ones that are suggested in the previous paragraph, allow the vectors $\underline{\mu}$ and $\Phi Q\underline{\mu}$ to be premultiplied by Q and by Q^T, respectively, in only $\mathcal{O}(n)$ operations. We assume that Q has this property. Therefore we seek a fast method for generating the matrix-vector product $\Phi\underline{\nu}$, where $\underline{\nu}=Q\underline{\mu}$ need not satisfy any special conditions.

Such a method exists, because the components of $\Phi\underline{\nu}$ are just the values of the function

$$t(\underline{x}) = \sum_{j=1}^{n} \nu_j\,\phi(\|\underline{x}-\underline{x}_j\|_2), \qquad \underline{x}\in\mathbb{R}^2, \tag{2.3}$$

at the points $\underline{x}_i$, $i=1,2,\ldots,n$, where ν_j, $j=1,2,\ldots,n$, are the components of $\underline{\nu}$. The method employs Laurent expansions of the functions $\phi(\|\cdot-\underline{x}_j\|_2)$, $j=1,2,\ldots,n$, about suitable points in $\mathbb{R}^2$. For example, if we require the expansion of the j-th thin plate spline function about the origin, then, for $\|\underline{x}\|_2>\|\underline{x}_j\|_2$, we write $\phi(\|\underline{x}-\underline{x}_j\|_2)$ in the form

$$\phi(|z-z_j|) = |z-z_j|^2\log|z| - \Re\left\{(z-z_j)\,(\overline{z}-\overline{z}_j)\sum_{k=1}^{\infty}\frac{1}{k}\left(\frac{z_j}{z}\right)^k\right\}, \tag{2.4}$$

where z and z_j are the complex numbers whose real and imaginary parts are the components of $\underline{x}$ and $\underline{x}_j$, respectively. We form an approximation to $\phi(\|\underline{x}-\underline{x}_j\|_2)$

by truncating the infinite sum after about 20 terms, and the maximum error that is allowed is prescribed, so typically the truncated expansion is acceptable if $\underline{x}$ satisfies $\|\underline{x}\|_2 \geq 4\,\|\underline{x}_j\|_2$. The most important feature of this approximation is that it can be expressed as a linear combination of functions of z, where all of the dependencies on z_j are confined to the coefficients of the linear combination. Thus the sum $\sum_{j\in\mathcal{J}} \nu_j\, \phi(\|\underline{x}-\underline{x}_j\|_2)$ can be estimated by a linear combination of the same functions of z, where the set $\mathcal{J} \subset \{1,2,\ldots,n\}$ is chosen so that all of the points $\{\underline{x}_j : j \in \mathcal{J}\}$ are relatively close to the origin. The method constructs Laurent expansions of useful clusters of interpolation points, in order that expression (2.3) can be replaced by a sum of $\mathcal{O}(\log n)$ Laurent expansions, except perhaps for a few values of j such that $\|\underline{x}-\underline{x}_j\|_2$ is small. The number of expansions that are needed for every $\underline{x}$ in the set of interpolation points is of magnitude n, the ones that are required for each $\underline{x}$ being selected automatically. Details of the method can be found in Beatson and Newsam [3] and in Powell [8]. Typically, the number of computer operations to calculate all of the components of $\Phi\underline{\nu}$ to prescribed accuracy is bounded above by a large constant multiple of $n \log n$.

The other strong advantage of conjugate gradients is that, by choosing Q carefully, it is usually possible to terminate the calculation after far fewer than n iterations. It is well-known that one iteration is sufficient if the matrix $Q^T \Phi\, Q$ of the system (2.2) is a multiple of the unit matrix, and fortunately our knowledge of Φ can make this remark useful. Indeed, the identity $\nabla^4\phi(\|\underline{x}\|_2) = 8\pi\,\delta(\underline{x})$, $\underline{x} \in \mathbb{R}^2$, which we used in the derivation of equation (1.10), implies the relation

$$\nabla^2_{\underline{x}}\, \nabla^2_{\underline{y}}\phi(\|\underline{x}-\underline{y}\|_2) = 8\pi\,\delta(\underline{x}-\underline{y}), \qquad \underline{x},\, \underline{y}\in\mathbb{R}^2, \tag{2.5}$$

which we compare with the matrix elements

$$(Q^T\Phi\, Q)_{ij} = \sum_{k=1}^{n}\sum_{\ell=1}^{n} Q_{ki}\, Q_{\ell j}\, \phi(\|\underline{x}_k-\underline{x}_\ell\|_2), \qquad 1\leq i,j\leq n-3. \tag{2.6}$$

Thus, if the sums $\sum_{k=1}^{n} Q_{ki}\, \phi(\|\underline{x}_k-\underline{y}\|_2)$ and $\sum_{\ell=1}^{n} Q_{\ell j}\, \phi(\|\underline{x}-\underline{x}_\ell\|_2)$ are estimates of $\nabla^2_{\underline{x}}\phi(\|\underline{x}-\underline{y}\|_2)$ at $\underline{x}=\underline{x}_i$ and $\nabla^2_{\underline{y}}\phi(\|\underline{x}-\underline{y}\|_2)$ at $\underline{y}=\underline{x}_j$, respectively, it is possible that the conjugate gradient algorithm will perform very well. In other words, we require $\sum_{i=1}^{n} Q_{ij}\, g(\underline{x}_i)$ to be an estimate of the *second* derivative $\nabla^2 g(\underline{x})$ at $\underline{x}=\underline{x}_j$, where g is a general twice differentiable function from $\mathbb{R}^2$ to $\mathbb{R}$. Something has to be done about edge effects, however, partly because the dimensions of Q are only $n\times(n-3)$.

Following Powell [9], we test this idea in the easy case when the interpolation points form a finite, regular square grid in $\mathbb{R}^2$ that covers the unit square $[0,1]\times[0,1]$. We derive a column of Q for each interior grid point from the well-known 5-point formula

$$\nabla^2 g(\underline{x}) \approx h^{-2}\{g(\underline{x}-h\underline{e}_1) + g(\underline{x}+h\underline{e}_1) + g(\underline{x}-h\underline{e}_2) + g(\underline{x}+h\underline{e}_2) - 4\,g(\underline{x})\}, \tag{2.7}$$

where h is the grid size, and where $\underline{e}_1$ and $\underline{e}_2$ are the co-ordinate vectors in $\mathbb{R}^2$. Thus each of these columns of Q has 5 nonzero elements, and we introduce the scaling factor h^2 in order that they are 1, 1, 1, 1 and -4. At the interpolation points that are interior points of the sides of the unit square, we approximate only half of the operator $h^2\nabla^2 = h^2(\partial^2/\partial x^2 + \partial^2/\partial y^2)$, using the differences

$$g(\underline{x}-h\underline{e}_1) - 2g(\underline{x}) + g(\underline{x}+h\underline{e}_1) \text{ and } g(\underline{x}-h\underline{e}_2) - 2g(\underline{x}) + g(\underline{x}+h\underline{e}_2) \tag{2.8}$$

on the sides in the x- and y-directions, respectively, so these columns of Q have only 3 nonzero elements, specifically 1, -2 and 1. We require just one more column in the null space of P^T. We let it correspond to the corners of the unit square, using the remark that the functional $g(0)-g(\underline{e}_1)-g(\underline{e}_2)+g(\underline{e}_1+\underline{e}_2)$ is zero when g is any linear polynomial, so this column has the nonzero elements 1, -1, -1 and 1, which completes the specification of Q.

	Original Q		Diagonal scaling	
	$\varepsilon = 10^{-6}$	$\varepsilon = 10^{-10}$	$\varepsilon = 10^{-6}$	$\varepsilon = 10^{-10}$
$n = 49$	22	28	16	21
$n = 100$	29	37	20	26
$n = 196$	36	48	23	31
$n = 400$	44	60	29	37

Table 1: Numbers of iterations for the method of Section 2

The conjugate gradient algorithm was applied to the system (2.2) in this case, the components of $\underline{f}$ being $f_i = \sin(5\pi x_i y_i)$, $i = 1, 2, \ldots, n$, and $\underline{\mu}$ being set to the zero vector initially. This is one of the 3 vectors $\underline{f}$ that occur in the experiments of [9]. The numbers of iterations that were needed to achieve the condition

$$\| Q^T \Phi Q \underline{\mu} - Q^T \underline{f} \|_\infty \leq \varepsilon \tag{2.9}$$

for 2 choices of ε and 4 choices of n are given in the second and third columns of Table 1. We see that the iteration counts are much less than n, and that, when n is doubled, the numbers of iterations increase by only about 25%. Furthermore, it is easy to scale the columns of Q so that all the diagonal elements of $Q^T \Phi Q$ are one. Thus the iteration counts are reduced to the values that are given in the last two columns of the table. Therefore careful attention to the choice of Q before solving the system (2.2) by the conjugate gradient method can be very rewarding.

§3. Algorithms that Employ the Positive Semi-Definiteness of A

We recall from Section 1 that the $n \times n$ matrix A, introduced in equation (1.14), is symmetric and positive semi-definite, 3 of its eigenvalues being zero and the other $n-3$ being positive. Further, the system (1.7) causes $\underline{\lambda} = A\underline{f}$ to satisfy $P^T\underline{\lambda} = 0$ for every $\underline{f}$ in $\mathbb{R}^n$, which is equivalent to the condition $P^TA = 0$. Moreover, we have noted already that expression (1.14) is zero when $\underline{f}$ is in the column space of P, so A has the property $AP = 0$. Alternatively, one can deduce $AP = 0$ from $P^TA = 0$ and the symmetry of A.

The method of Dyn et al [6], which provides the material of this section, generates an estimate, W say, of A, that retains the properties of A that are given above. Indeed, W is also an $n \times n$ symmetric matrix with 3 zero eigenvalues and $n-3$ positive ones that satisfies $P^TW = 0$. Then W is used as a pre-conditioner in a conjugate gradient method that calculates the required vector of coefficients $\underline{\lambda} \in \mathbb{R}^n$. A simple description of that calculation would not show the close relation between the rôle of W and the use of Q in the previous section, so some preliminary ideas receive attention below before W is employed explicitly.

Let H be the positive semi-definite square root of W. Then H is a real symmetric matrix, whose eigenvectors are the eigenvectors of W, and whose eigenvalues are the nonnegative square roots of the corresponding eigenvalues of A. Thus, $P^TW = 0$ implies $P^TH = 0$, and the columns of H span the $(n-3)$-dimensional null space of P^T. It follows that, corresponding to equation (2.1), we can write

$$\underline{\lambda} = H\underline{\sigma}, \tag{3.1}$$

for some vector $\underline{\sigma} \in \mathbb{R}^n$, which is no longer unique, because equation (3.1) is preserved if we add any vector from the column space of P to $\underline{\sigma}$, but the required $\underline{\lambda}$ is still unique. We now pre-multiply the equation $\Phi\underline{\lambda} + P\underline{c} = \underline{f}$ by H, which eliminates $P\underline{c}$, so we find that $\underline{\sigma}$ has to be a solution of the $n \times n$ system

$$H\Phi H\underline{\sigma} = H\underline{f}, \tag{3.2}$$

which is analogous to equation (2.2). The condition $HP = 0$ implies that the symmetric matrix $H\Phi H$ has at least 3 eigenvalues of zero. The other eigenvalues must be nonzero, because the system (3.2) is consistent for every $\underline{f} \in \mathbb{R}^n$, and the dimension of the linear space $\{H\underline{f} : \underline{f} \in \mathbb{R}^n\} \subset \mathbb{R}^n$ is $n-3$. Furthermore, the following argument shows that these other eigenvalues are positive. The eigenvalue equation $H\Phi H\underline{\sigma} = \eta\,\underline{\sigma}$ implies $\eta = \underline{\sigma}^T H\Phi H\underline{\sigma}/\underline{\sigma}^T\underline{\sigma}$. Therefore, if η is any nonzero eigenvalue, it has the same sign as $\underline{\lambda}^T\Phi\,\underline{\lambda}$, where $\underline{\lambda} = H\underline{\sigma}$. Because the definition of H causes $\underline{\lambda}$ to be in the null space of P^T, the required condition $\eta > 0$ is a consequence of the conditional positive definiteness of Φ that is proved in Section 1. Thus the matrix $H\Phi H$ is positive semi-definite and $H\underline{f}$ is in its column space. It follows that one of the solutions $\underline{\sigma}$ of the system (3.2) can be calculated by the conjugate gradient method, the freedom in the solution being taken up by the initial choice of $\underline{\sigma}$, assuming exact arithmetic. Then the required $\underline{\lambda}$ always has the value (3.1). We see that this approach is comparable to the one that occurs in Section 2. Its details are as follows.

Let F be the quadratic function

$$F(\underline{\sigma}) = \frac{1}{2}\underline{\sigma}^T H \Phi H \underline{\sigma} - \underline{\sigma}^T H \underline{f}, \qquad \underline{\sigma} \in \mathbb{R}^n, \tag{3.3}$$

which is convex and bounded below due to the conditional positive definiteness of Φ. The conjugate gradient method calculates a minimizer of F, because condition (3.2) is equivalent to $\underline{\nabla} F(\underline{\sigma}) = 0$. We set $\underline{\sigma}_1 = 0$ for the first iteration, and $\underline{\sigma}_k$ is the estimate of the required $\underline{\sigma}$ at the beginning of the k-th iteration. Therefore termination occurs if the k-th iteration finds that the gradient vector

$$\underline{g}_k = \underline{\nabla} F(\underline{\sigma}_k) = H \Phi H \underline{\sigma}_k - H \underline{f} \tag{3.4}$$

is zero, and then $\underline{\sigma}_k$ is returned as the solution of the system (3.2). Otherwise, a search direction $\underline{s}_k$ is chosen that satisfies the strict descent condition $\underline{s}_k^T \underline{g}_k < 0$, the choice when $k=1$ being $\underline{s}_1 = -\underline{g}_1 = H\underline{f}$. Therefore we achieve the strict reduction $F(\underline{\sigma}_{k+1}) < F(\underline{\sigma}_k)$ by letting $\underline{\sigma}_{k+1}$ be the vector

$$\underline{\sigma}_{k+1} = \underline{\sigma}_k + \alpha_k \, \underline{s}_k, \tag{3.5}$$

where the steplength α_k is calculated so that $F(\underline{\sigma}_{k+1})$ is as small as possible. Thus the new gradient, which can be written in the form

$$\underline{g}_{k+1} = \underline{g}_k + \alpha_k \, H \Phi H \underline{s}_k, \tag{3.6}$$

is orthogonal to $\underline{s}_k$, so α_k has the value

$$\alpha_k = -\, \underline{s}_k^T \underline{g}_k \,/\, \underline{s}_k^T H \Phi H \underline{s}_k. \tag{3.7}$$

Moreover, the method causes the search directions to be conjugate with respect to the matrix $H\Phi H$ by applying the formula

$$\underline{s}_k = -\,\underline{g}_k + \frac{\underline{s}_{k-1}^T H \Phi H \underline{g}_k}{\underline{s}_{k-1}^T H \Phi H \underline{s}_{k-1}} \, \underline{s}_{k-1}, \tag{3.8}$$

when $k \geq 2$ and $\underline{g}_k \neq 0$. The theoretical description of the calculation of $\underline{\sigma}$ is now complete, but we will find that the definition $H = W^{1/2}$ allows an implementation that is much more convenient in practice.

Next, however, we deduce that it is ideal to let W be the matrix A. We ignore the trivial case when $H\underline{f}$ is zero, because then the method of the previous paragraph would terminate at the beginning of the first iteration. Indeed, we prove that $W=A$ and $H\underline{f}\neq 0$ imply termination at the beginning of the second iteration. Furthermore, the steplength $\alpha_1=1$ occurs, because we establish the property

$$\underline{\nabla} F(\underline{\sigma}_1+\underline{s}_1) = H\Phi H(\underline{\sigma}_1+\underline{s}_1) - H\underline{f} = 0. \tag{3.9}$$

The values $\underline{\sigma}_1 = 0$ and $\underline{s}_1 = H\underline{f}$ have been noted already, so it is sufficient to derive the equation $H\Phi H^2\underline{f} = H\underline{f}$. Let B^T be the bottom left $3\times n$ submatrix of the inverse matrix of expression (1.15). It follows from the definition of A that $\Phi A + PB^T = I$ occurs, where I is the $n\times n$ unit matrix. Hence $HP=0$ implies $H\Phi A = H$, which is equivalent to $H\Phi A\underline{f} = H\underline{f}$, $\underline{f}\in \mathbb{R}^n$. Therefore the required equation is a consequence of $W=A$ and $H^2=W$.

In practice, a matrix W is chosen that has the properties that are specified in the second paragraph of this section, and we are unwilling to calculate $H=W^{1/2}$. Fortunately, there is no need to do so, because it is shown below that only the square of H occurs if, instead of $\underline{\sigma}_k$, $\underline{s}_k$ and $\underline{g}_k$, we work with the vectors $\underline{\lambda}_k = H\underline{\sigma}_k$, $\underline{d}_k = H\underline{s}_k$ and $\underline{b}_k = H\underline{g}_k$, respectively. The previous choice $\underline{\sigma}_1 = 0$ gives the values $\underline{\lambda}_1 = 0$ and $\underline{b}_1 = H\underline{g}_1 = -W\underline{f}$ for the beginning of the first iteration. Further, the gradient (3.4) is zero if and only if $\underline{b}_k = H\underline{g}_k$ is zero, so it is usual to stop the iterations in practice when $\|\underline{b}_k\|$ is sufficiently small, in which case $\underline{\lambda}$ is set to the available vector $\underline{\lambda}_k$, because already the product (3.1) has been formed implicitly. Alternatively, when the search direction of the k-th iteration is required, the choice $\underline{d}_1 = -\underline{b}_1 = W\underline{f}$ is immediate, and we deduce from equation (3.8) that, for $k\geq 2$, we should pick the vector

$$\underline{d}_k = -\,\underline{b}_k + \frac{\underline{d}_{k-1}^T\Phi\,\underline{b}_k}{\underline{d}_{k-1}^T\Phi\,\underline{d}_{k-1}}\,\underline{d}_{k-1}. \tag{3.10}$$

The steplength of the k-th iteration is given the value

$$\alpha_k = (\underline{d}_k^T\underline{f} - \underline{d}_k^T\Phi\,\underline{\lambda}_k)\,/\,\underline{d}_k^T\Phi\,\underline{d}_k, \tag{3.11}$$

which is derived by substituting formula (3.4) into the numerator of expression (3.7). Finally, corresponding to equations (3.5) and (3.6), the iteration applies the recurrence relations

$$\underline{\lambda}_{k+1} = \underline{\lambda}_k + \alpha_k\,\underline{d}_k \quad\text{and}\quad \underline{b}_{k+1} = \underline{b}_k + \alpha_k\,W\Phi\,\underline{d}_k. \tag{3.12}$$

This algorithm not only avoids $H = W^{1/2}$ but also is parsimonious about matrix-vector products. Both W and Φ have to occur in such products on each iteration, because otherwise the influence of these matrices would be trivial. Therefore we welcome the usual property of the conjugate gradient method that only the vector $\underline{d}_k$ has to be multiplied by Φ during an iteration, because the symmetry of

Φ provides $\underline{d}_k^T \Phi = (\Phi\, \underline{d}_k)^T$, and $\Phi\, \underline{d}_{k-1}$ can be preserved from the previous iteration for use in formula (3.10). Moreover, W occurs only in $\underline{b}_1 = -W \underline{f}$ and in the second part of expression (3.12). Usually, W is so sparse that the work of pre-multiplying $\underline{f}$ and $\Phi\, \underline{d}_k$ by W is trivial in comparison with the calculation of $\Phi\, \underline{d}_k$. Furthermore, the Laurent expansion method of [3], that is considered briefly in Section 2, allows $\Phi\, \underline{d}_k$ to be formed to high accuracy in $\mathcal{O}(n \log n)$ computer operations.

A disadvantage of the above procedure, however, over the conjugate gradient method of Section 2, is that there tends to be more loss of accuracy from computer rounding errors, due partly to the fact that any deviations from the condition $P^T \underline{\lambda}_k = 0$ are inherited by $P^T \underline{\lambda}_{k+1}$ when the first of the formulae (3.12) is applied. This kind of deficiency occurs in the numerical example at the end of this section, and is studied in Chapter 6 of Baxter [1]. On the other hand, the choice of a suitable matrix W is easier than the construction of a successful pre-conditioner Q for the iterative calculation of Section 2. We have seen already that, from this point of view, it is unsatisfactory that Q has only $n-3$ columns. More importantly, one of the key ideas in [6] is that the relation (1.16), between A and the integral (1.2), is highly valuable for generating a W, that is not only a reasonable approximation to A, but also satisfies the conditions in the second paragraph of this section, even when the interpolation points $\underline{x}_i$, $i = 1, 2, \ldots, n$, are in general positions.

Specifically, the integral (1.2) is estimated by a quadrature rule with positive coefficients, so the value of the integrand is required at several discrete points. Then each of the second derivatives that occurs in the values of the integrand is approximated by a linear combination of the data $s(\underline{x}_i) = f_i$, $i = 1, 2, \ldots, n$. A routine feature of second derivative estimation is that such approximations vanish if the interpolation conditions (1.1) allow s to be a polynomial of degree one or less, and it is necessary to retain this feature. Further, we require at least one of the approximations to be nonzero when $\underline{f}$ is not in the column space of P. Let $J(s)$ be the estimate of $I(s)$ that is provided by this method. Due to the squares of second derivatives in the definition (1.2) and the positive coefficients of the quadrature rule, $J(s)$ is nonnegative, and its dependence on $\underline{f}$ has the form

$$J(s) = 8\pi\, \underline{f}^T W \underline{f}, \qquad \underline{f} \in \mathbb{R}^n, \tag{3.13}$$

for a particular $n \times n$ matrix W. Further, because $J(s)$ is zero if and only if the data can be interpolated by a polynomial of degree at most one, W enjoys all of the properties that are stated in the second paragraph of this section. Therefore this construction provides a suitable pre-conditioner for the given conjugate gradient method. Moreover, the element W_{ij} is nonzero only if at least one of the second derivative approximations employs both f_i and f_j, and it is usual for each approximation to be derived from a local cluster of interpolation points. Therefore, when n is large, it is routine for W to be very sparse.

We apply the above method to the numerical example of Table 1, where the interpolation points form a square grid on $[0,1] \times [0,1]$. Although the integral (1.2) is over the whole of $\mathbb{R}^2$, the quadrature points, which are not chosen explicitly, can be regarded as the data points $\underline{x}_i$, $i=1,2,\ldots,n$. Indeed, we include suitable second difference terms by letting the approximation to the integral be a constant multiple of the expression

$$\begin{aligned} &\textstyle\sum_{i\in\mathcal{I}_{xx}} [s(\underline{x}_i-h\underline{e}_1) - 2\,s(\underline{x}_i) + s(\underline{x}_i+h\underline{e}_1)]^2 \\ +2&\textstyle\sum_{i\in\mathcal{I}_{xy}} [s(\underline{x}_i) - s(\underline{x}_i+h\underline{e}_1) - s(\underline{x}_i+h\underline{e}_2) + s(\underline{x}_i+h\underline{e}_1+h\underline{e}_2)]^2 \\ +&\textstyle\sum_{i\in\mathcal{I}_{yy}} [s(\underline{x}_i-h\underline{e}_2) - 2\,s(\underline{x}_i) + s(\underline{x}_i+h\underline{e}_2)]^2, \end{aligned} \tag{3.14}$$

where $\underline{e}_1$ and $\underline{e}_2$ are still the co-ordinate vectors in $\mathbb{R}^2$, and where $\mathcal{I}_{xx}$, $\mathcal{I}_{xy}$ and $\mathcal{I}_{yy}$ are the largest possible subsets of $\{1,2,\ldots,n\}$ such that all of the required values of s occur in the interpolation equations (1.1). The overall scaling of W makes no difference to the conjugate gradient algorithm of this section, except that it may influence the termination condition. Therefore we define W by letting $\underline{f}^T W \underline{f}$ be the quadratic form (3.14), so its elements are integers whose moduli are at most 20. Further, we employ two termination conditions, one of them being the inequality

$$\|\underline{b}_k\|_\infty = \|W\Phi\underline{\lambda}_k - W\underline{f}\|_\infty \le \varepsilon, \tag{3.15}$$

instead of expression (2.9). The other one avoids unnecessary calculation when there is serious damage from computer rounding errors, there being an error return if the denominator $\underline{d}_k^T\,\Phi\,\underline{d}_k$ of formula (3.11) becomes nonpositive. As before, we let the components of $\underline{f}$ be $f_i=\sin(5\pi x_i y_i)$, $i=1,2,\ldots,n$, and $\underline{\lambda}_1$ is set to the zero vector.

	$\varepsilon=10^{-6}$	$\varepsilon=10^{-10}$	$\min_k \|\underline{b}_k\|_\infty$
$n=49$	13	22*	3.7×10^{-7}
$n=100$	15	26*	8.2×10^{-8}
$n=196$	16	29*	1.0×10^{-7}
$n=400$	16	29*	1.6×10^{-7}

Table 2: Numbers of iterations for the method of Section 3

The numbers of iterations that occur, for the values of n and ε that have been tried already, are presented in the second and third columns of Table 2. We see that the $\varepsilon=10^{-6}$ counts are much better than in Table 1, and that they show hardly any growth as n increases. The entries in the third column, however, are marked with asterisks, because in these cases termination was due to an error return, the least calculated values of $\|\underline{b}_k\|_\infty$ being given in the final column of the table. Of course the failures depend on the accuracy of the computer arithmetic, so we note that the calculations were run on a SPARC 2 workstation in double precision FORTRAN, which provides about 16 decimal places of accuracy. The error returns are inevitable for sufficiently small ε, because equation (3.3) and the definition $\underline{\lambda}_k = H\underline{\sigma}_k$ imply that the conjugate gradient method is minimizing the indefinite quadratic form

$$F_0(\underline{\lambda}) = \frac{1}{2}\underline{\lambda}^T\Phi\underline{\lambda} - \underline{\lambda}^T\underline{f}, \qquad \underline{\lambda}\in \mathbb{R}^n, \tag{3.16}$$

and computer rounding errors cause $\underline{\lambda}$ to be contaminated by eigenvectors of Φ that have negative eigenvalues. It was found when $n=100$, for example, that the calculated values of $\|\underline{b}_k\|_\infty$ decrease monotonically for 16 iterations, and then they increase for 10 iterations, reaching a final value of 4.7×10^{-1}. Nevertheless, the function values $\{F_0(\underline{\lambda}_k) : k=1,2,3,\ldots,\}$ decrease strictly monotonically, because the steplength (3.11) is just the value of α that minimizes $F_0(\underline{\lambda}_k+\alpha\,\underline{d}_k)$, $\alpha\in\mathbb{R}$. The undesirable behaviour of the sequence $\{\underline{b}_k : k=1,2,\ldots,26\}$ seems to be due to a component that is at the rounding error level initially, and that grows by about a factor of 5 on every iteration. In our experiments, the growing component has become so large when $\|\underline{b}_k\|_\infty \le 10^{-6}$ is achieved, that the condition $\|\underline{b}_k\|_\infty \le 10^{-10}$ cannot be satisfied at the beginning of a later iteration. Therefore, as suggested in [1], the method of this section should be extended in a way that forces the vectors $\{\underline{\lambda}_k : k=1,2,3,\ldots\}$ and $\{\underline{b}_k : k=1,2,3,\ldots\}$ to remain close to the null space of P^T in the presence of computer rounding errors.

§4. Algorithms that Employ Lagrange Functions of Subsets of the Data

Let χ_i, $i=1,2,\ldots,n$, be the Lagrange functions of the interpolation problem of this paper. Therefore χ_i is the required thin plate spline function when the right hand sides have the values $f_j=\delta_{ij}$, $j=1,2,\ldots,n$, where δ_{ij} is the Kronecker delta. Further, it follows from the linearity of the system (1.7) that, for general right hand sides, we seek the interpolant

$$s(\underline{x}) = \sum_{i=1}^{n} f_i\,\chi_i(\underline{x}), \qquad \underline{x}\in\mathbb{R}^2. \tag{4.1}$$

Beatson and Powell [4] propose the approximation of each χ_i by $\widehat{\chi}_i$, where $\widehat{\chi}_i$ is the thin plate spline solution of the Lagrange conditions

$$\widehat{\chi}_i(\underline{x}_j) = \delta_{ij}, \qquad j\in\mathcal{L}_i, \tag{4.2}$$

$\mathcal{L}_i$ being a suitable subset of $\{1,2,\ldots,n\}$ that contains far fewer than n elements when n is large. They use the estimates $\widehat{\chi}_i\approx\chi_i$, $i=1,2,\ldots,n$, in an iterative refinement procedure that is intended to provide a sufficiently accurate approximation to the function (4.1). We will consider some developments of this technique, that include the new algorithm that is mentioned at the end of Section 1.

Let $s(\underline{x})$, $\underline{x} \in \mathbb{R}^2$, be the approximation to the right hand side of equation (4.1) at the beginning of an iteration of the method of [4]. The iteration tries to drive the residuals $f_i - s(\underline{x}_i)$, $i=1,2,\ldots,n$, to zero by adding the term

$$\sum_{i=1}^{n} \{f_i - s(\underline{x}_i)\} \,\widehat{\chi}_i(\underline{x}), \qquad \underline{x} \in \mathbb{R}^2, \tag{4.3}$$

to s. Thus, for $j=1,2,\ldots,n$, the new j-th residual has the value

$$f_j - [s(\underline{x}_j) + \sum_{i=1}^{n} \{f_i - s(\underline{x}_i)\} \,\widehat{\chi}_i(\underline{x}_j)] = \sum_{i=1}^{n} \{\delta_{ij} - \widehat{\chi}_i(\underline{x}_j)\} \{f_i - s(\underline{x}_i)\}. \tag{4.4}$$

In other words, if $\underline{\eta}^{(k)}$ is the vector that has the components $f_j - s(\underline{x}_j)$, $j=1,2,\ldots,n$, at the beginning of the k-th iteration, then the method generates the sequence

$$\underline{\eta}^{(k+1)} = R\,\underline{\eta}^{(k)}, \qquad k=1,2,3,\ldots, \tag{4.5}$$

where R is the $n \times n$ matrix that has the elements

$$R_{ji} = \delta_{ij} - \widehat{\chi}_i(\underline{x}_j), \qquad 1 \le i,j \le n. \tag{4.6}$$

We pick $s \equiv 0$ initially, so $\underline{\eta}^{(1)}$ is just the data vector $\underline{f} \in \mathbb{R}^n$. It follows from equation (4.5) that, after k iterations, the residuals are the components of $\underline{\eta}^{(k+1)} = R^k \underline{f}$. Therefore s tends to satisfy the interpolation equations (1.1) in the limit as $k \to \infty$ if the spectral radius of R, namely $\rho(R)$, is less than one. Further, s tends to the required thin plate spline interpolant in this case, because each of the functions $\widehat{\chi}_i$, $i = 1,2,\ldots,n$, has the form (1.3), and the conditions (1.5) for square integrable second derivatives are satisfied, due to the definition of $\widehat{\chi}_i$.

Beatson, Goodsell and Powell [2] consider the choice of the sets $\mathcal{L}_i$, $i = 1, 2,\ldots,n$, for this method, when the interpolation points $\underline{x}_i$, $i = 1,2,\ldots,n$, form a square grid on the unit square $[0,1] \times [0,1]$, as in the numerical examples that have been given already. They take the view that, if the coefficient μ_{ij} of the true Lagrange function

$$\chi_i(\underline{x}) = \sum_{j=1}^{n} \mu_{ij}\, \phi(\|\underline{x} - \underline{x}_j\|_2) + p_i(\underline{x}), \qquad \underline{x} \in \mathbb{R}^2, \tag{4.7}$$

satisfies $|\mu_{ij}| \ge 0.5$, then the index j should be included in the set $\mathcal{L}_i$ of the approximate Lagrange function

$$\widehat{\chi}_i(\underline{x}) = \sum_{j \in \mathcal{L}_i} \widehat{\mu}_{ij}\, \phi(\|\underline{x} - \underline{x}_j\|_2) + \widehat{p}_i(\underline{x}), \qquad \underline{x} \in \mathbb{R}^2, \tag{4.8}$$

where p_i and $\widehat{p}_i$ are polynomials of degree at most one. Thus, if $\underline{x}_i$ is well away from the boundary of the grid, the set $\{\underline{x}_j : j \in \mathcal{L}_i\}$ contains about 61 points that are tightly clustered about $\underline{x}_i$. On the other hand, if $\underline{x}_i$ is one of the four corners of the unit square, then not only $\underline{x}_i$ but also the other three corners of $[0,1]\times[0,1]$ are in $\{\underline{x}_j : j \in \mathcal{L}_i\}$, while the number of indices in $\mathcal{L}_i$ is of magnitude $n^{1/2}$ for large n, because many of the coefficients μ_{ij} are substantial when $\underline{x}_j$ is on or adjacent to the grid boundary. Therefore an extension to the iteration of Beatson and Powell [4] is proposed in [2], which allows the largest of the sets $\mathcal{L}_i$ to be much smaller than before for square grids of many interpolation points.

The extension was motivated by the remark that, if all of the sets $\mathcal{L}_i$, $i = 1,2,\ldots,n$, have exactly q elements, where q is prescribed, and if we define each $\mathcal{L}_i$ by letting $\{\underline{x}_j : j \in \mathcal{L}_i\}$ contain $\underline{x}_i$ and the $q-1$ interpolation points that are closest to $\underline{x}_i$, then it is straightforward to generate each $\mathcal{L}_i$ when the interpolation points are in general positions. Let these choices be made, and let the current thin plate spline function s be updated by the addition of the term (4.3). Then the moduli of the residuals $f_i - s(\underline{x}_i)$, $i=1,2,\ldots,n$, may increase, but, when q is large enough, we expect the updating to remove any very high frequency components of the residuals. Further, the smoothness of the new residuals allows the construction of useful corrections to them using only about half of the interpolation conditions (1.1). Specifically, in the square grid case, Beatson et al [2] introduce the subset $\mathcal{S}^{(2)} \subset \mathcal{S}^{(1)} = \{\underline{x}_i : i=1,2,\ldots,n\}$, $\underline{x}_i$ being in $\mathcal{S}^{(2)}$ if and only if $\|\underline{x}_i\|_1$ is an integer multiple of $2h$, where h is the grid size and where $\|\underline{x}\|_1$ denotes the sum of the moduli of the components of $\underline{x} \in \mathbb{R}^2$. Furthermore, approximate Lagrange functions are constructed for the new set of interpolation points, by calculating the thin plate spline solution of the equations (4.2) for every $\underline{x}_i \in \mathcal{S}^{(2)}$, where now we define $\mathcal{L}_i$ by letting $\{\underline{x}_j : j \in \mathcal{L}_i\}$ contain $\underline{x}_i$ and the $q-1$ points of $\mathcal{S}^{(2)}$ that are closest to $\underline{x}_i$. Then, after introducing the original correction (4.3), the extended method applies the correction (4.3) that occurs when the set $\mathcal{S}^{(1)}$ of interpolation points is replaced by $\mathcal{S}^{(2)}$. Next, the analogous correction for the set $\mathcal{S}^{(3)}$ is made, where $\underline{x}_i$ is in $\mathcal{S}^{(3)}$ if and only if both components of $\underline{x}_i$ are even multiples of h. Then an analogous $\mathcal{S}^{(4)}$ correction may be applied, where $\mathcal{S}^{(4)}$ contains those points of $\mathcal{S}^{(3)}$ whose 1-norms are integer multiples of $4h$. Each iteration of the extended method continues in this way, until we reach a set, $\mathcal{S}^{(\ell)}$ say, that contains so few points that it is convenient to correct s by the exact thin plate spline interpolant to the current residuals $f_i - s(\underline{x}_i)$, $\underline{x}_i \in \mathcal{S}^{(\ell)}$, which completes the iteration. Thus an iteration can be regarded as a fine to coarse sweep of a multigrid algorithm. Moreover, the grid size h is chosen so that each of the sets $\mathcal{S}^{(k)}$, $k=1,2,\ldots,\ell$, includes all four corners of the unit square $[0,1]\times[0,1]$.

Again there exists an $n\times n$ matrix R such that each iteration of the extended method multiplies the vector of residuals by R, as in equation (4.5). Beatson et al [2] calculate the spectral radius of R for several choices of n, q and ℓ, finding that a good rate of convergence is achieved in many cases. On the other hand, the rate seems to deteriorate seriously as n increases, unless the correction to s by exact interpolation to the residuals $f_i - s(\underline{x}_i)$, $\underline{x}_i \in \mathcal{S}^{(\ell)}$, is made after every stage of the multigrid procedure. Therefore they also investigate the following idea.

Instead of correction terms of the form (4.3), they use the approximate Lagrange function (4.8) to revise the coefficient λ_i of the term $\phi(\|\underline{x}-\underline{x}_i\|_2)$, $\underline{x} \in \mathbb{R}^2$, in the current estimate (1.3) of the required interpolant. This can be done, because equation (1.14) implies that the exact correction to λ_i has the value $\sum_{j=1}^{n} A_{ij}\,[f_j - s(\underline{x}_j)]$. Moreover, by combining $\underline{\lambda} = A\,\underline{f}$ with the choice $f_j = \delta_{ij}$, $j=1,2,\ldots,n$, we find that the coefficients μ_{ij}, $j=1,2,\ldots,n$, of the true Lagrange function (4.7) are identical to the matrix elements A_{ji}, $j=1,2,\ldots,n$. It follows from the symmetry of A that $A_{ij} = A_{ji} = \mu_{ij}$, so the exact correction is the sum $\sum_{j=1}^{n} \mu_{ij}\,[f_j - s(\underline{x}_j)]$. Hence, assuming that the approximation $\widehat{\chi}_i \approx \chi_i$ is suitable, the idea is to change the coefficient λ_i by the amount

$$\tau_i = \sum_{j \in \mathcal{L}_i} \widehat{\mu}_{ij}\,[f_j - s(\underline{x}_j)], \qquad i=1,2,\ldots,n. \tag{4.9}$$

Indeed, Beatson et al [2] investigate the possibility of calculating the numbers (4.9), in order to replace the correction term (4.3) by the expression

$$\sum_{i=1}^{n} \tau_i\,\phi(\|\underline{x}-\underline{x}_i\|_2), \qquad \underline{x} \in \mathbb{R}^2. \tag{4.10}$$

Furthermore, they introduce this new correction procedure into the multigrid technique that has been described already, which provides another new algorithm, details being available in their paper. In particular, some extra work is needed to ensure that the conditions (1.5) are obtained by the thin plate spline coefficients of s at the end of each iteration. Successive vectors of residuals, however, still satisfy a recurrence relation of the form (4.5), where k is the iteration number and where R is another constant $n{\times}n$ matrix. Again they calculate $\rho(R)$ for each of the square grid numerical experiments, finding that these spectral radii are about the same as those that occurred in the first multigrid procedure of this section.

A disadvantage of the multigrid technique is that it may be difficult to choose suitable sets $\mathcal{S}^{(1)} \supset \mathcal{S}^{(2)} \supset \cdots \supset \mathcal{S}^{(\ell)}$ when the positions of $\underline{x}_i$, $i=1,2,\ldots,n$, are general. Moreover, one can update λ_i by τ_i immediately after each of the numbers (4.9) is calculated, instead of estimating all of these numbers before doing any updating. Further, if λ_i is revised successfully, then it should be possible to ignore the i-th interpolation condition (1.1) for the remainder of the iteration. Therefore we are going to consider an algorithm of the following form.

Before the first iteration, we order the interpolation equations (1.1) so that it is suitable to correct the coefficients λ_i, $i=1,2,\ldots,n$, in sequence, an example of this ordering being given later. We also pick an integer q, which is the number of elements in each of the sets $\mathcal{L}_i$ of the approximate Lagrange functions. Further, we let $\widehat{q}$ be an integer in the interval $[q,n]$ that is small enough to allow the exact calculation of the thin plate spline interpolant to the last $\widehat{q}$ of the residuals $f_i - s(\underline{x}_i)$, $i=1,2,\ldots,n$. Then, for $i=1,2,\ldots,n-\widehat{q}$, we define $\mathcal{L}_i$ by letting the set $\{\underline{x}_j : j\in\mathcal{L}_i\}$ be composed of $\underline{x}_i$ and the $q-1$ points in $\{\underline{x}_{i+1},\underline{x}_{i+2},\ldots,\underline{x}_n\}$ that are closest to $\underline{x}_i$, any ties being broken randomly. We also determine the coefficients $\{\widehat{\mu}_{ij} : j\in\mathcal{L}_i\}$ of the approximate Lagrange function (4.8) by using the equations (4.2) as before. We are now ready to begin the iterations, starting with $s\equiv 0$. Termination occurs at the beginning of an iteration if all of the residuals $f_i - s(\underline{x}_i)$, $i=1,2,\ldots,n$, are sufficiently small. Otherwise, the iteration revises the coefficients λ_i, $i=1,2,\ldots,n-\widehat{q}$, in sequence, where, for each i, a change is made to s that increases λ_i by τ_i without disturbing the earlier coefficients λ_j, $j<i$, and the most recently updated s is employed in formula (4.9). Finally, the iteration corrects the last $n-\widehat{q}$ coefficients and the linear polynomial part of s by adding to s the thin plate spline interpolant to the residuals

$$f_i - s(\underline{x}_i), \qquad i=n-\widehat{q}+1,\ n-\widehat{q}+2,\ \ldots,\ n. \tag{4.11}$$

The above modification of s for each integer i in $[1,n-\widehat{q}]$ is not necessarily the addition of the term $\tau_i\,\phi(\|\underline{x}-\underline{x}_i\|_2)$, $\underline{x}\in\mathbb{R}^2$. Instead, we prefer a correction that satisfies the constraints (1.5). Specifically, for each i, we let the set $\mathcal{I}_i$ contain just 3 of the integers $\{i+1,i+2,\ldots,n\}$, where the points $\{\underline{x}_j : j\in\mathcal{I}_i\}$ are relatively close to $\underline{x}_i$ and not collinear. Further, we seek the coefficients ν_{ij}, $j\in\mathcal{I}_i$, that are defined by the equations

$$\sum_{j\in\mathcal{I}_i} \nu_{ij} = 1 \quad \text{and} \quad \sum_{j\in\mathcal{I}_i} \nu_{ij}\,\underline{x}_j = \underline{x}_i, \tag{4.12}$$

these sets and coefficients being determined before the first iteration. Then the required updating of λ_i is achieved by the addition of the function

$$\tau_i\,[\,\phi(\|\underline{x}-\underline{x}_i\|_2) - \sum_{j\in\mathcal{I}_i} \nu_{ij}\,\phi(\|\underline{x}-\underline{x}_j\|_2)\,], \qquad \underline{x}\in\mathbb{R}^2, \tag{4.13}$$

to s. Another advantage of the correction (4.13) over $\tau_i\,\phi(\|\underline{x}-\underline{x}_i\|_2)$, $\underline{x}\in\mathbb{R}^2$, is that expression (4.13) and its derivatives are relatively small when $\underline{x}$ is far from $\underline{x}_i$, due to the conditions (4.12).

Some refinements have been omitted from the above description that are important in general, such as forcing each of the sets $\mathcal{L}_i$, $i=1,2,\ldots,n-\widehat{q}$, to have the property that the points $\{\underline{x}_j : j\in\mathcal{L}_i\}$ are not collinear. The reason is that the refinements are unnecessary for the square grid examples that have been considered already, and our main objective at present is to apply the new algorithm to such examples. Therefore we have to choose the order of the interpolation points $\underline{x}_i$, $i=1,2,\ldots,n$, which is guided by the sets $\mathcal{S}^{(1)}\supset\mathcal{S}^{(2)}\supset\cdots\supset\mathcal{S}^{(\ell)}$ that occur in the multigrid methods of this section. Specifically, we impose the constraints that, for $k=2,3,\ldots,\ell$, every point that is not in $\mathcal{S}^{(k)}$ must precede every point of $\mathcal{S}^{(k)}$, but all the remaining ordering decisions are taken randomly. Moreover, the sets $\mathcal{I}_i$,

$i=1,2,\ldots,n-\widehat{q}$, have the property that the 3 points $\{\underline{x}_j : j \in \mathcal{I}_i\}$ are as close as possible to $\underline{x}_i$, which leaves some freedom that is also taken up arbitrarily. We now let the data f_i, $i=1,2,\ldots,n$, be random numbers from the uniform distribution on $[-1,1]$, but the points $\underline{x}_i$, $i=1,2,\ldots,n$, are as before, except that different values of n were used in order to assist comparisons with the multigrid experiments of [2]. The multigrid investigations also influenced the value of $\widehat{q}$, which is $\max[q,16]$ when $n=169$ or 625 and $\max[q,25]$ when $n=289$ or 1089. The numbers of iterations that were needed to satisfy the termination condition

$$|f_i - s(\underline{x}_i)| \leq 10^{-10}, \qquad i=1,2,\ldots,n, \tag{4.14}$$

for these values of n and 4 values of q are reported in Table 3. There is a range of iteration counts, because each experiment was repeated 10 times using different random numbers.

n	$q=10$	$q=20$	$q=30$	$q=50$
169	21–25	9–10	7–7	6–6
289	22–25	10–12	7–8	6–6
625	28–34	12–14	8–8	6–7
1089	28–32	14–16	8–9	6–7

Table 3: Numbers of iterations for the method of Section 4

We see that the Table 3 results for $q \geq 20$ are much better than those of Tables 1 and 2, as it is appropriate to make comparisons with the $\varepsilon=10^{-10}$ columns of the earlier tables. The Table 3 experiments were also run after perturbing both coordinates of each of the data points $\underline{x}_i$, $i=1,2,\ldots,n$, by random numbers from the interval $[-0.4h,\ 0.4h]$, where h is still the grid size. It was highly encouraging that these perturbations caused hardly any changes to the numbers of iterations. Therefore a more general version of the algorithm has been developed recently. Numerical experiments show that it is very successful, even in cases when the interpolation points are chosen randomly without any grid structure. Indeed, the iteration counts in Table 3 are typical for interpolation points in general positions. The details of this work are available in [10].

Acknowledgements. I am very grateful to George Goodsell for much assistance with the numerical calculations that have guided the construction of the new algorithm of Section 4.

References

1. Baxter, B. J. C., The interpolation theory of radial basis functions, Ph.D. dissertation, University of Cambridge, 1992.
2. Beatson, R. K., G. Goodsell and M. J. D. Powell, On multigrid techniques for thin plate spline interpolation in two dimensions, report No. DAMTP 1995/NA08, University of Cambridge, 1995.
3. Beatson, R. K. and G. N. Newsam, Fast evaluation of radial basis functions: I, Comput. Math. Applic. **24** (1992), 7–19.
4. Beatson, R. K. and M. J. D. Powell, An iterative method for thin plate spline interpolation that employs approximations to Lagrange functions, in *Numerical Analysis 1993*, D. F. Griffiths and G. A. Watson (eds.), Longman Scientific & Technical, Burnt Mill, 1994, 17–39.
5. Duchon, J., Splines minimizing rotation-invariant seminorms in Sobolev spaces, in *Constructive Theory of Functions of Several Variables*, W. Schempp and K. Zeller (eds.), Lecture Notes in Mathematics 571, Springer-Verlag, Berlin, 1977, 85–100.
6. Dyn, N., D. Levin and S. Rippa, Numerical procedures for surface fitting of scattered data by radial functions, SIAM J. Sci. Statist. Comput. **7** (1986), 639–659.
7. Hardy, R. L., Theory and applications of the multiquadric-biharmonic method, Comput. Math. Applic. **19** (1990), 163–208.
8. Powell, M. J. D., Truncated Laurent expansions for the fast evaluation of thin plate splines, Numer. Algorithms **5** (1993), 99–120.
9. Powell, M. J. D., Some algorithms for thin plate spline interpolation to functions of two variables, in *Advances in Computational Mathematics: New Delhi, India*, H. P. Dikshit and C. A. Micchelli (eds.), World Scientific, Singapore, 1994, 303–319.
10. Powell, M. J. D., A new iterative method for thin plate spline interpolation in two dimensions, report No. DAMTP 1995/NA10, University of Cambridge, 1995.
11. Sibson, R. and G. Stone, Computation of thin plate splines, SIAM J. Sci. Statist. Comput. **12** (1991), 1304–1313.

M. J. D. Powell
Department of Applied Mathematics and Theoretical Physics
University of Cambridge
Silver Street
Cambridge CB3 9EW (UK)
mjdp@amtp.cam.ac.uk

B-splines on Uniform Meshes of the Plane

Paul Sablonnière

Abstract. We define new families of B-splines on the three and four direction meshes of the plane. Given an integer $k \geq 0$ and a simple polygon P consisting of a small number of triangles, we first study the existence of C^k splines with support P and of minimal degree. They are called simple B-splines. Then, we define composed B-splines by repeated convolution of simple B-splines with piecewise affine or quadratic box-splines. Finally, some properties of these functions are given: smoothness, degree and approximation order.

§1. Introduction

Let τ_1 denote the uniform triangulation of $\mathbb{R}^2$ having $\mathbb{Z}^2$ as set of vertices and whose edges are parallel to the three directions $e_1 = (1,0), e_2 = (0,1), e_3 = (1,1)$. Similarly, let τ_2 denote the uniform triangulation whose set of vertices is $\mathbb{Z}^2 \cup (\mathbb{Z} + 1/2)^2$ and whose edges are parallel to the four directions e_1, e_2, e_3 and $e_4 = (1,-1)$. We are interested in the construction of splines with bounded support (called B-splines in the following) obtained by repeated convolution of simple B-splines with the piecewise affine pyramid π on τ_1 and the C^1-quadratic box-spline ω on τ_2. Classical box-splines are important examples of such B-splines . Simple B-splines have simple supports for which there is a variety of possible shapes. On τ_1, we have restricted our study to those consisting of one triangle Δ_1 , or a pair of triangles Σ_1 (e.g. a square), or an hexagon H_1 formed by the six triangles surrounding a given vertex. The corresponding B-splines are called respectively Δ_1-splines, Σ_1-splines and H_1-splines. On τ_2, we consider simple supports consisting of one square Σ_1 decomposed into four triangles by its diagonals and one lozenge Λ_1 decomposed into eight triangles. The corresponding B-splines are called respectively Σ_1 and Λ_1-splines .

Once the support and the smoothness $r \geq 0$ are given, we try to determine the minimal degree $d = d(r)$ for which the construction of a simple B-spline of class C^r

Advanced Topics in Multivariate Approximation
F. Fontanella, K. Jetter and P. J. Laurent (eds.) pp. 323–338.

ISBN 981-02-2852-X

is possible. For sake of simplicity, we assume the invariance of the B-spline with respect to the isometries of its support. Otherwise it would be much more difficult to obtain satisfying results of existence and uniqueness. In general, the family of integer translates $B(\phi) = \{\phi(. - \alpha), \alpha \in \mathbb{Z}^2\}$ of a simple B-spline ϕ does not form a partition of unity. In this case, ϕ is called a *reduced* B-spline. Adding this property leads to new results of existence and unicity for a larger minimal degree, of course. In that case, ϕ is called a *complete* B-spline. In general, we have tried to prove results for all values of $r \in \mathbb{N}$. When it was not successfull, we give the largest values of r for which the construction of simple B-splines is possible.

Now, for a given composed B-spline ϕ, obtained by repeated convolution of one of the previous B-splines with π or ω, it is not hard either to compute its smoothness and its degree and to describe its support, or to give the space $\mathbb{P}(\phi) = \mathbb{P}_n$ of polynomials of maximal total degree included in the space $S(\phi)$ generated by the set $B(\phi)$ of its integer translates . This gives immediately the approximation order $n+1$ of a smooth function in the space $S(\phi)$. These properties and the construction of quasi-interpolants are direct consequences of the Strang-Fix theory applied to Fourier transforms of B-splines .

For B-splines which are not box-splines, we have not in general a refinement equation, therefore there is no subdivision algorithm. However, once the Bernstein-Bézier form of simple B-splines is known, the one of composed B-splines can be computed by standard convolution algorithms (similar to those proposed e.g. by Lai in [6]). The approximation order is in general proportional to the size of the support. Thus B-splines with small supports are easy to manipulate, but have a low approximation order while B-splines with larger supports are more difficult to construct, but have a higher approximation order. Moreover, the degrees are also decreasing when the support increases, for a given smoothness.

The first part of the paper is devoted to the trangulation τ_1. Complete proofs and constructions are given in the doctoral thesis by Sbibih [11]. The main results are given in [8] and some sketched proofs appear in [9]. Since the present paper is considered as a survey, we simply recall the main results, for sake of completeness. The second part of the paper is devoted to the triangulation τ_2. In this case, all the results are new and we give the complete proofs whenever they are not too lengthy. There are some related works, for example [4], which give some examples of simple B-splines (mainly H_1 splines) with a low order of smoothness and few general results. Concerning applications, we are now investigating some interpolation problems where B-splines with small supports are very convenient. We also hope that our B-splines are of interest for understanding the structure of spaces of piecewise polynomial functions on triangulations (uniform or not).

§ 2. B-splines on the Triangulation τ_1

For a given smoothness $r \geq 0$, we denote respectively β_r, χ_r and π_r the Δ_1, Σ_1 and (reduced) H_1-splines of minimal degree, when they exist (see figure 1 for the supports). We denote π_r the (complete) H_1-spline for which $B(\pi_r)$ is a partition of unity.

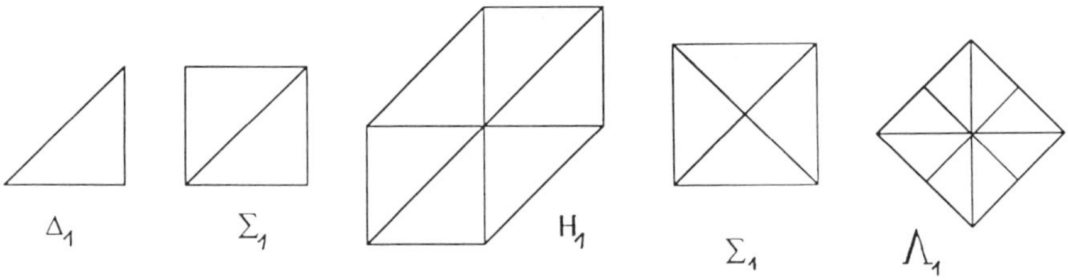

Fig. 1. Simple supports in τ_1 and τ_2.

Theorem 1.

(i) *for all $r \geq 0$, there exists a unique Δ_1-spline $\beta_r \geq 0$ of degree $3r+3$ (up to a multiplicative constant) and this degree is minimal.*

(ii) *for all $r \geq 0$, there exist Σ_1-splines $\chi_r \geq 0$ of minimal degree $3r+2$ for r even and $3r+3$ for r odd. They are not unique if we only assume the invariance of χ_r through the symmetries w.r.t. the two diagonals of the square Σ_1 .*

(iii) *for all $r \geq 0$, there exists a unique complete H_1-spline π_r of minimal degree $3r+1$ for r even and $3r+2$ for r odd . In particular, $\pi = \pi_0$.*

(iv) *there is no reduced H_1-spline π_r for $(r,d) = (1,3)$ and $(2,4)$, and for pairs $(2r-1, 4r)$ and $(2r, 4r+1)$ whenever $r \geq 2$. There exist reduced H_1-splines for $(r,d) = (1,4), (2,5), (3,9), (4,10), (5,13), (6,1)$ and the degrees obtained are minimal, but there is not yet a general result in this case .*

For composed B-splines , we use the notation $\pi^k := \pi * \pi * \ldots * \pi$ (k times): it is the box-spline of degree $3k+1$ and smoothness $2k$ (see e.g. [2]). By convolution with π^k of the above simple B-splines , we obtain the following families of composed B-splines . For $r \geq 0$, we set

$$\beta_r^0 := \beta_r \quad , \quad \chi_r^0 := \chi_r \quad , \quad \pi_r^0 := \pi_r \quad , \quad \pi_r^0 := \pi_r$$

and for $k \geq 1$, we define respectively

$$\beta_r^k := \beta_r * \pi^k \quad , \quad \chi_r^k := \chi_r * \pi^k$$

$$\pi_r^k := \pi_r * \pi^k \quad , \quad \pi_r^k := \pi_r * \pi^k$$

The supports of these B-splines are hexagons with slightly different shapes (see e.g.[9], definition 2). All of them are functions of smoothness $2k+r$ because of the convolution by π^k. For $k \geq 1$, their integer translates form a partition of unity. Moreover, they are non negative when the associated simple B-splines are non negative. Their main properties are gathered in the next theorem .

Theorem 2.

(i) *for any simple or composed B-spline ϕ, the family $B(\phi)$ is globally linearly independent, i.e., $\Sigma c(\alpha)\phi(. - \alpha) = 0$ implies $c(\alpha) = 0$ for all $\alpha \in \mathbb{Z}^2$.*

(ii) *β_r^k is of degree $3(r+k+1)$ and $\mathbb{P}(\beta_r^k) = \mathbb{P}_{2k-1}$ for $k \geq 1$, $r \geq 0$.*

(iii) *χ_r^k is of degree $3(r+k)+2$ for r even (resp. $3(r+k)+3$ for r odd), moreover $\mathbb{P}(\chi_r^k) = \mathbb{P}_{2k-1}$ for $k \geq 1$, $r \geq 0$.*

(iv) *π_r^k is of degree $3(r+k)+1$ for r even (resp. $3(r+k)+2$ for r odd), moreover $\mathbb{P}(\pi_r^k) = \mathbb{P}_{2k}$ for $k \geq 1$ and $r \geq 1$.*

Remarks

1) For $r = -1, \beta_{-1}$ is the characteristic function of Δ_1, and β_{-1}^k is the B-spline of degree $3k$ and class $2k-1$ with minimal support described e.g. in [3], chapter 4 . Similarly, χ_{-1}^k is the characteristic function of Σ_1 and χ_{-1}^k is the box-spline of degree $3k$ and class $2k-1$. Finally, $\pi_0^k = \pi^{k+1}$ is the box-spline of degree $3k+1$ and class $2k$. Therefore classical B-splines and box-splines belong to the families of B-splines described above .

2) Composed B-splines are not, in general, of minimal degree for given class and support . In [9], two different B-splines of class C^2 , with the same support and whose translates form a partition of unity are described : one of them is of degree 5 (which is the minimal degree) and the other one is β_0, which is of degree 6 .

3) Various discrete or integral quasi-interpolants can be constructed, which approximate a smooth function f in spaces of splines with the highest possible degree of approximation (see chapter 3 of [1] and section 5 of [9] and [10]).

§3. Simple B-splines on τ_2: Existence of Reduced Σ_1 and Λ_1 Splines

In the following sections, the results are new and more details are given than for B-splines on the triangulation τ_1. We first define reduced Σ_1 and Λ_1 splines (see figure 1).

Definition 1. *For $r \geq 0$, define $\tilde{\sigma}_r$ (resp. $\tilde{\lambda}_r$) as the simple B-splines of class C^r and of minimal degree whose support is the square Σ_1 (resp. the lozenge Λ_1) .*

Theorem 3. *There is no Σ_1-spline $\tilde{\sigma}_r$ (resp. Λ_1-spline $\tilde{\lambda}_r$) of class C^r , of degree $d \leq 2r$ for r even and $d \leq 2r+1$ for r odd.*

Proof: (i) Let us first prove the result for $d = 2r$ (r even) and Σ_1-splines. We assume that the support Σ_1 is centered at the origin (figure 2), that the axes Ox

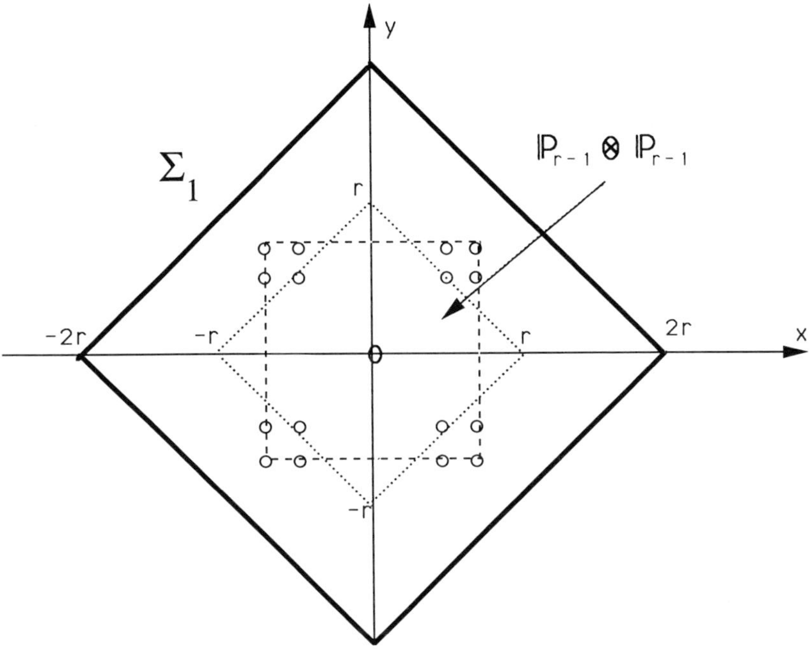

Fig. 2. Theorem 3 for Σ_1-splines.

and Oy are its two diagonals, and that its vertices are $(\pm 2r, 0)$ and $(0, \pm 2r)$. Because $\tilde{\sigma}_r$ is of class C^r, its Bernstein-Bézier coefficients (abbr. BB-coefficients) are zero on the parallel lines $|x| + |y| = m, r \leq m \leq 2r$. In particular, they are zero in the four small squares $\{r/2 \leq |x| \leq r-1, r/2 \leq |y| \leq r-1\}$. Since each of them contains $(r/2)^2$ BB-coefficients , there are r^2 coefficients in their union. These r^2 coefficients uniquely determine all the BB-coefficients inside the square $\{|x| \leq r-1, |y| \leq r-1\}$ which can be considered as those of one polynomial in the tensor-product space $\mathbb{P}_{r-1} \otimes \mathbb{P}_{r-1}$, but subdivided into four square pieces. Therefore, all the BB-coefficients of $\tilde{\sigma}_r$ are equal to zero.

(ii) For Λ_1-splines of degree $d = 2r$ (figure 3,left), the C^r continuity of $\tilde{\lambda}_r$ across the boundary of its support implies that the BB-coefficients are zero inside the annulus $\{r \leq |x| \leq 2r\} \cup \{r \leq |y| \leq 2r\}$. Therefore, the conditions of C^m continuity across the coordinate axes, for $r \leq m \leq 2r$, are automatically satisfied by these coefficients. This fact and the C^r continuity of $\tilde{\lambda}_r$ imply that this function is only composed of four polynomial pieces because the internal edges formed by axes are artificial. Thus $\tilde{\lambda}_r$ is simply another version of $\tilde{\sigma}_r$, with a support Λ_1

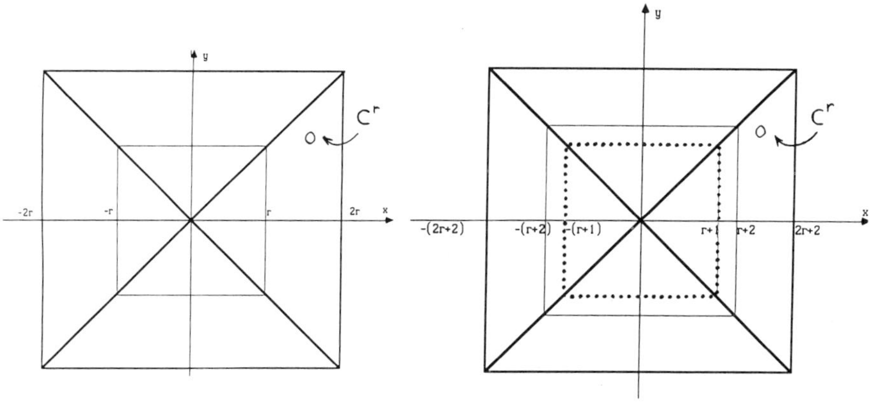

Fig. 3. Theorems 3 and 4 for Λ_1-splines.

deduced from Σ_1 by a rotation of angle $\pi/4$, therefore $\tilde{\lambda}_r$ must be equal to zero. The proofs are similar for r odd . Of course, the same result holds for lower degrees. ■

Theorem 4. *Up to a multiplicative constant, there exists a unique Σ_1-spline (resp Λ_1-spline) of class C^r, of degree $d = 2r+1$ for r even and $d = 2r+2$ for r odd. By theorem 3, these degrees are minimal.*

Proof: (i) For Σ_1-splines, r odd and $d = 2r+2$, the BB-coefficients are zero in the annulus $r+2 \le |x|+|y| \le 2r+2$. All of them are also zero, except one (the stars in figure 4) in each of the small squares $\{\frac{1}{2}(r+1) \le |x| \le r, \frac{1}{2}(r+1) \le |y| \le r\}$. As in the proof of theorem 3, these small squares uniquely determine the tensor-product polynomial in $\mathbb{P}_r \otimes \mathbb{P}_r$ whose BB-coefficients lie in the square $\{|x| \le r, |y| \le r\}$. Because of the symmetry properties of $\tilde{\sigma}_r$, the four star BB-coefficients are equal. Thus $\tilde{\sigma}_r$ is uniquely determined up to a multiplicative constant. The proof is similar for r even and $d = 2r+1$.

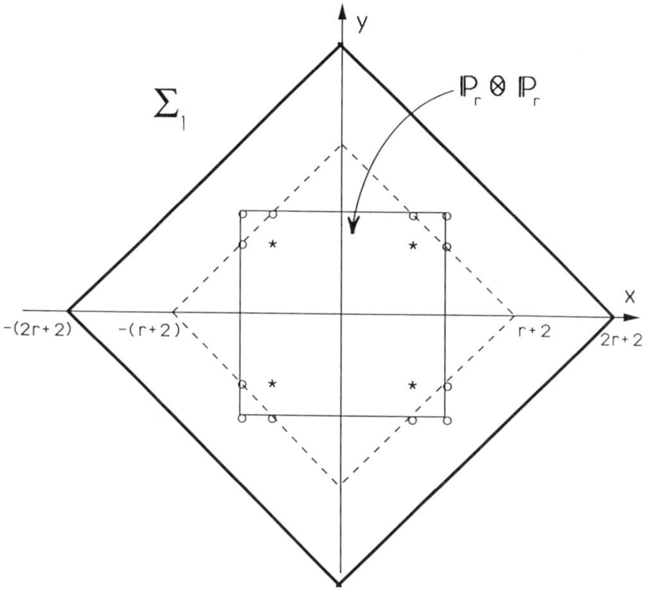

Fig. 4. Theorem 4 for Σ_1-splines.

(ii) For Λ_1-splines, r odd and $d = 2r+2$ (see figure 3,right), the BB-coefficients are zero on the rows $r+2 \le |x| \le 2r+2$ and $r+2 \le |y| \le 2r+2$. Therefore C^m continuity conditions across the axes hold for $r+2 \le m \le 2r+2$. Since $\tilde{\lambda}_r$ is of class C^r, they also hold for $0 \le m \le r$. It remains to consider the BB-coefficients on the rows $|x| = r+1$ and $|y| = r+1$. Since r is odd, $r+1$ is even, so the C^r continuity of $\tilde{\lambda}_r$ and its symmetry with respect to the axes imply that the C^{r+1} continuity conditions across these axes are automatically satisfied by the BB-coefficients lying on these particular rows. Therefore, $\tilde{\lambda}_r$ is in fact identical to $\tilde{\sigma}_r$ (up to a rotation of angle $\pi/4$) and we are brought back to the first part of the proof. For r even, the proof is quite similar to the one given in the second part of the proof of theorem 3. ■

Remark It is obvious that $B(\tilde{\sigma}_r)$ and $B(\tilde{\lambda}_r)$ do not form a partition of unity. For $\tilde{\sigma}_r$, this is due to the fact that all the translates are equal to zero on the horizontal and vertical lines of the mesh. In the second case, it is due to the fact that the translates are equal to zero at the centers of the squares. So, we are led naturally to consider the property of partition of unity for the family $B(\tilde{\sigma}_r) \cup B(\tilde{\lambda}_r)$.

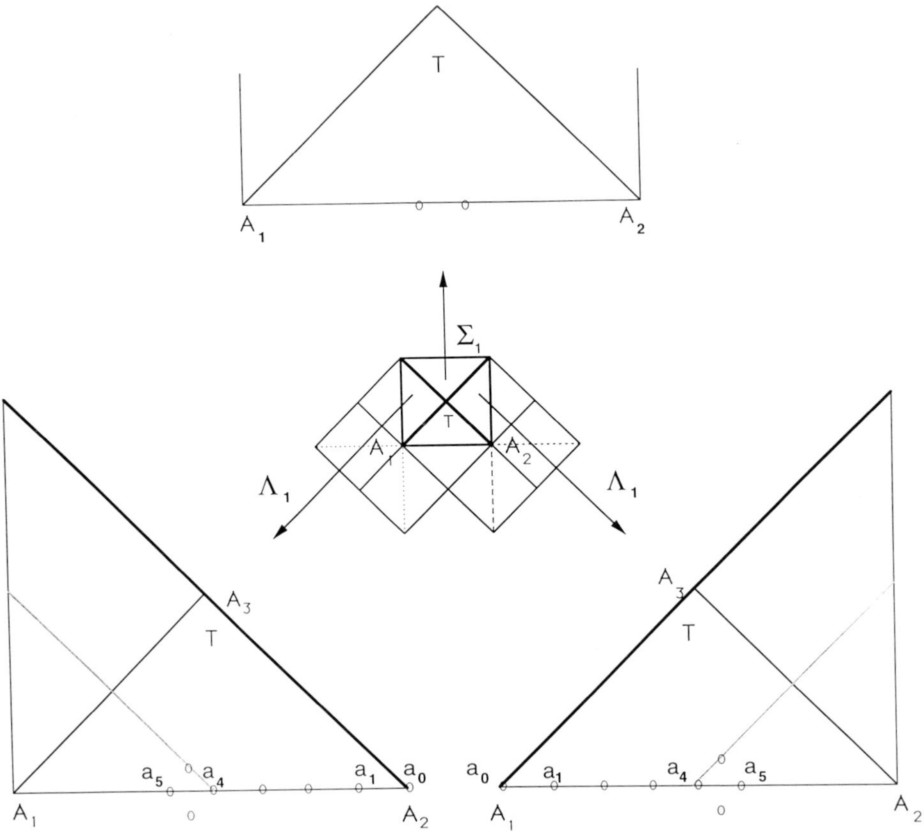

Fig. 5. Corollary of theorem 4 (r=4).

Corollary. *The integer translates of* $\tilde{\sigma}_r$ *and* $\tilde{\lambda}_r$ *do not form a partition of unity.*

Proof: Let us consider the case $r = 2s, d = 4s+1$ (see figure 5). There are three B-splines (one Σ_1-spline and two Λ_1-splines) whose supports contain one fixed triangle $T \in \tau_2$. We want to prove that each B-spline has at least two BB-coefficients that are zero on $\Gamma = A_1A_2$. On the south boundary Γ of Σ_1, the two central BB-coefficients are obviously equal to zero. Let us denote $\{a_{4s+1}, \ldots, a_1, a_0\}$ the set of BB-coefficients of the left Λ_1-spline $\tilde{\lambda}_r$ on Γ. By C^r continuity across the upper right boundary A_1A_3, we have $a_0 = a_1 = \ldots = a_{2s} = 0$. Now, the C^1 continuity of $\tilde{\lambda}_r$ across the internal edge Γ of Λ_1 implies $a_{2s+1} = 0$. Therefore the two central BB-coefficients a_{2s} and a_{2s+1} of the left Λ_1-spline are equal to zero. This holds also, by symmetry, for the right Λ_1-spline. Finally, the sum of the three B-splines cannot be equal to one on the triangle T. A similar proof holds for $r = 2s - 1$ and $d = 4s$: one can prove in this case that the central BB-coefficients on Γ of the three B-splines are zero. ■

§4. Simple B-splines on τ_2 : Existence of Complete Σ_1 and Λ_1- Splines

Definition 2. *(PU-property). For $r \geq 0$, we denote σ_r (resp λ_r) the complete Σ_1 -spline (resp complete Λ_1-spline) of class C^r and of minimal degree whose integer translates form a partition of unity :*

$$\sum\{\sigma_r(. - \alpha) + \lambda_r(. - \alpha)\} = 1$$

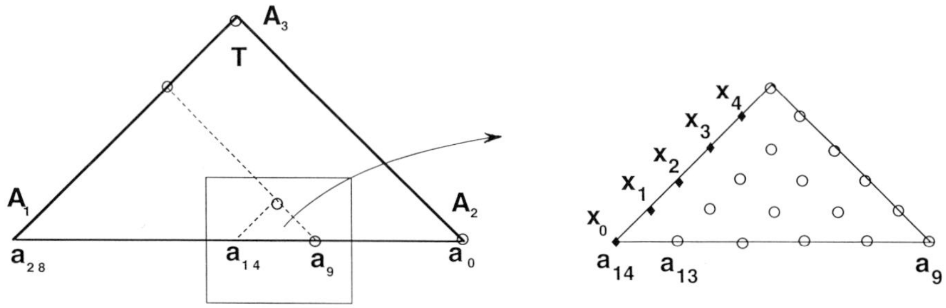

Fig. 6. Theorem 5 (s=5, r=9, d=28).

Theorem 5. *There do not exist complete Σ_1 and Λ_1-splines of class C^r, of degree $d \leq 3r$ for r even and $d \leq 3r + 1$ for r odd.*

Proof: For $r = 2s - 1, d = 6s - 2, s \geq 1$, consider the triangle $A_1A_2A_3$ (figure 6) which is covered by the south triangle in the support of the Σ_1-spline centered at

A_3 and by one triangle in each support of the Λ_1-splines centered at A_1 and A_2. We are interested in the BB-coefficients on the segment A_1A_2. Those of the Σ_1-splines are equal to zero because they lie on the boundary of its support. Denoting by $\{a_{6s-2}, a_{6s-3}, \ldots, a_1, a_0\}$ the set of BB-coefficients of the left Λ_1-spline, we obtain by symmetry the BB-coefficients of the right Λ_1-spline $\{a_0, a_1, \ldots, a_{6s-2}\}$. The PU-property implies that $a_i + a_{6s-2-i} = 1$ for $0 \le i \le 6s-2$, in particular $a_{3s-1} = 1/2$. Consider the left Λ_1-spline. By C^r continuity across the boundary edge A_2A_3, we get immediately $a_0 = a_1 = \ldots = a_{2s-1} = 0$. Now by C^r continuity across the internal edge A_1A_2, and the symmetry of upper and lower polynomials with respect to this edge, it is easy to prove that $a_{2s} = a_{2s+1} = \ldots = a_{3s-1} = 0$. In particular, this contradicts the fact that $a_{3s-1} = 1/2$. The proof can be done by induction. By C^1 continuity, we see immediately that $a_{2s} = 0$. Let us assume that we have already proved that $a_{2s} = \ldots = a_{3s-2} = 0$ and moreover that the BB-coefficients in the small square with diagonal $[a_{3s-2}, a_{2s-1}]$ are also equal to zero. Let us denote by $x_0 = a_{3s-1}, x_1, \ldots, x_{s-1}$ the BB-coefficients on the left upper edge of the small square with diagonal $[a_{3s-1}, a_{2s-1}]$. By C^k continuity, $k = 1, 3, \ldots, 2s-1$, we obtain s equations for the s unknowns $\{x_0, \ldots, x_{s-1}\}$:

$$x_0 - 2x_1 = 0$$

$$x_0 - 3x_1 + 3x_2 - 2x_3 = 0$$

$$\ldots$$

$$x_0 - \binom{2s-1}{1}x_1 + \binom{2s-1}{2}x_2 + \ldots + (-1)^{s-1}\binom{2s-1}{s-1}x_{s-1} = 0.$$

The matrix of this linear system being regular, we obtain $x_0 = \ldots = x_{s-1} = 0$. By symmetry, C^k continuity is automatically satisfied for k even . A similar proof has already been given for H_1-splines in [7]. ■

Theorem 6. *There exist complete Σ_1 and Λ_1-splines of degree $d = 3r+1$ for r even and $d = 3r+2$ for r odd (the same degrees as those of complete H_1-splines on τ_1). For $r = 2s-1$ or $2s, s \ge 1$, their construction depends on $\frac{1}{2}s(s+1)$ arbitrary parameters .*

Proof: Since the proof is rather technical, we give the details for $s = 3, r = 6, d = 19$. As in the proof of theorem 5, we compute the BB-coefficients of the Σ_1-spline σ_6 and of the two left and right Λ_1-splines λ_6 and λ'_6 in the triangle $T = A_1A_2A_3$ included in their supports. We decompose the sets of BB-coefficients into ten subregions as indicated in figure 7.

(i) By C^6-continuity and the PU-property of definition 2, it is clear that the BB-coefficients of σ_6 are equal to one in the subregion I and to zero in the subregions VI to X. Similarly, those of λ_6 (resp. λ'_6) are equal to one in the subregions VI and IX (resp VII and X) and to zero in the subregions I, III, V, VII, X (resp I, II, IV,

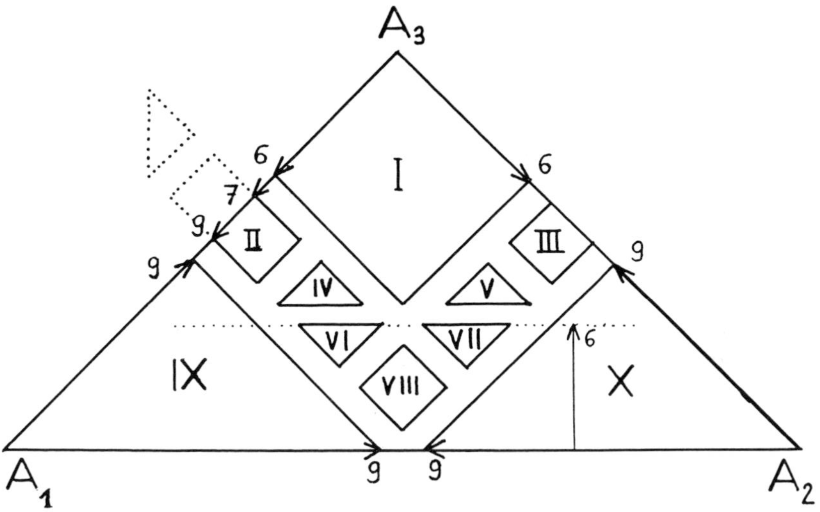

Fig. 7. Theorem 6, (i).

VI, IX). It remains to compute the BB-coefficients of σ_6 in subregions II and IV, which, by symmetry, are equal to those in subregions III and V respectively. For λ_6, there remain subregions II, IV and VIII and for λ'_6, subregions III, V and VIII.

(ii) Let $P = \{a_1, b_1, b_2, c_1, c_2, c_3\}$ (resp.$Q = \{\alpha_1, \beta_1, \beta_2, \gamma_1, \gamma_2, \gamma_3\}$) be the set of the six arbitrary BB-coefficients of σ_6 (resp.λ_6) lying in subregion IV of $T \in \Sigma_1$ (resp. $\in \Lambda_1$). Q is deduced from P by the PU-property: $a_1 + \alpha_1 = 1$, $b_i + \beta_i = 1$ for $i = 1, 2$, $c_i + \gamma_i = 1$ for $i = 1, 2, 3$. (the same result holds, by symmetry, for regions V in Σ_1 and Λ'_1). Now, by C^6 continuity and the symmetry of σ_6 with respect to A_1A_3, we can compute the other coefficients of σ_6 in subregion II (and by symmetry in subregion III). The corresponding equations are easy to solve in terms of parameters of P; they come from C^3 and C^5 continuity across A_1A_3 (see figure 8):

$$3a_2 - 2a_3 = a_1 \quad , \quad 10a_2 - 4a_3 = 10a_1,$$
$$3b_3 - 2b_4 = b_2 \quad , \quad 10b_3 - 4b_4 = 10b_2 - 5b_1,$$
$$3c_4 - 2c_5 = c_3 \quad , \quad 10c_4 - 4c_5 = 10c_3 - 5c_2 + c_1.$$

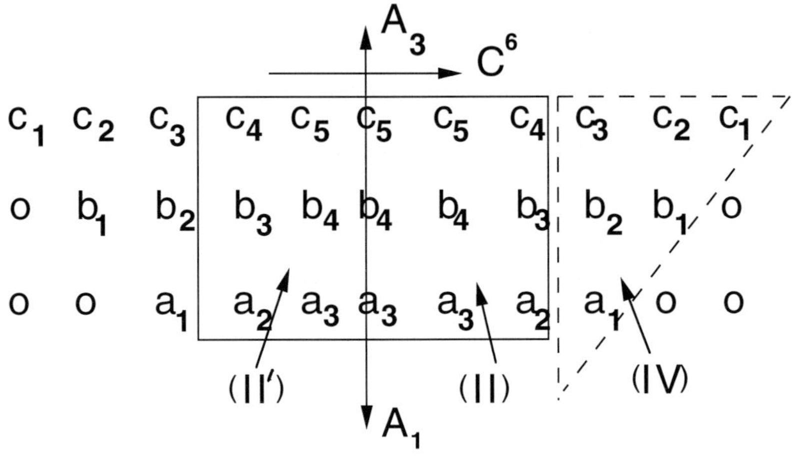

Fig. 8. Theorem 6, (ii).

In the same way, the BB-coefficients of λ_6 in subregion II of T (in Λ_1) are deduced from the above ones by using the PU property.

(iii) Finally, we have to compute the BB-coefficients of λ_6 and λ_6' in VIII (see figure 9). By symmetry and the PU-property, those of λ_6' satisfy the following equations:

$$u = x_2 = z = 1/2 \quad , \quad v_1 + v_2 = y_1 + y_2 = x_1 + x_3 = 1$$

By C^k continuity of λ_6' across $A_1A_2, 1 \le k \le 6$, we can compute all the remaining parameters. In fact, only $k = 3$ and $k = 5$ give new equations (for $k = 2, 4, 6$, they are identities because of the symmetry of λ_6' w.r.t. the edge). For C^3-continuity, we get

$$x_1 = 1 - 3u + 3v_1 - x_1 \quad , \quad x_2 = 4 - 3(2u+1) + 3(v_2 + v_1) - x_2.$$

(the second equation is an identity). For C^5-continuity, we obtain

$$0 = 1 - 5u + 10v_1 - 10x_1,$$
$$0 = 6 - 5(4u+1) + 10(3v_1 + v_2) - 10(2x_1 + x_2) + 5y_1,$$
$$z = 16 - 5(6u+5) + 10(3v_1 + 3v_2 + 1) - 10(x_1 + 2x_2 + x_3) + 5(y_1 + y_2) - z.$$

(the third one is an identity). After simplification, we obtain $3v_1 - 2x_1 = 1/2, 10v_1 - 10x_1 = 3/2$, which gives $v_1 = 1/5, v_2 = 4/5, x_1 = 1/20, x_3 = 19/20$, and finally $y_1 = 1/5, y_2 = 4/5$. ■

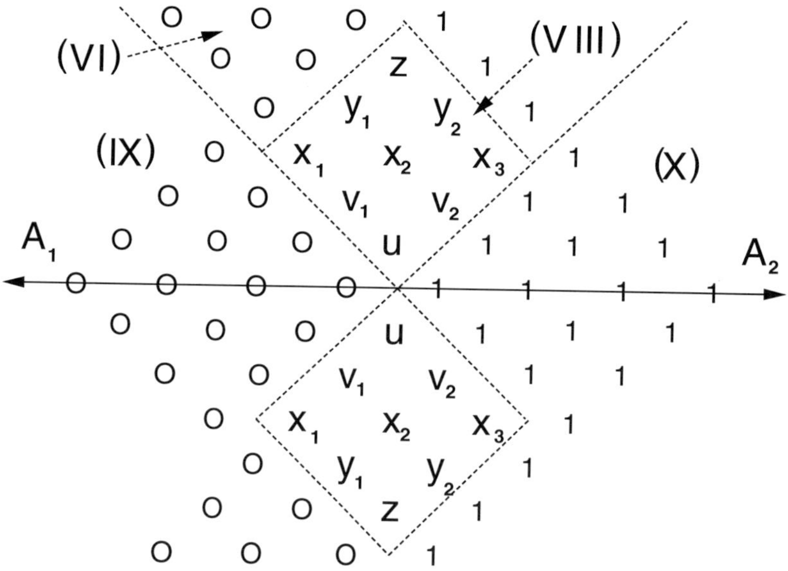

Fig. 9. Theorem 6, (iii).

§5. A Family of Composed B-splines on τ_2.

Definition 3. *Let ω be the C^1 quadratic box-spline and let $\tilde{\sigma}_r$ (resp $\tilde{\lambda}_r$) be the Σ_1-spline (resp Λ_1-spline) of class C^r and minimal degree defined in theorem 4. Let $\omega^k := \omega * \omega * \ldots * \omega$ (k times). For $r \geq 0$, we set $\tilde{\sigma}_r^0 = \tilde{\sigma}_r$, $\tilde{\lambda}_r^0 = \tilde{\lambda}_r$, and for $k \geq 1$, we define the composed B-splines*

$$\tilde{\sigma}_r^k := \tilde{\sigma}_r * \omega^k \quad , \quad \tilde{\lambda}_r^k := \tilde{\lambda}_r * \omega^k .$$

Definition 4.

(i) *Let $\Omega_{k+1,k}$ be the octagon which is the convex hull of the eight following vertices*

$$v_1 = (k+1)e_1, \ v_2 = v_1 + ke_3, \ v_3 = v_2 + (k+1)e_2, \ v_4 = v_3 + ke_4,$$
$$v_5 = v_4 - (k+1)e_1, \ v_6 = v_5 - ke_3, \ v_7 = v_6 - (k+1)e_2, \ v_8 = v_7 - ke_4.$$

(ii) *Let $\Omega_{k,k+1}$ be the octagon deduced from the preceding one by substituting (e_1, e_2) to (e_3, e_4) and vice-versa. (see figure 10 for $k = 2$).*

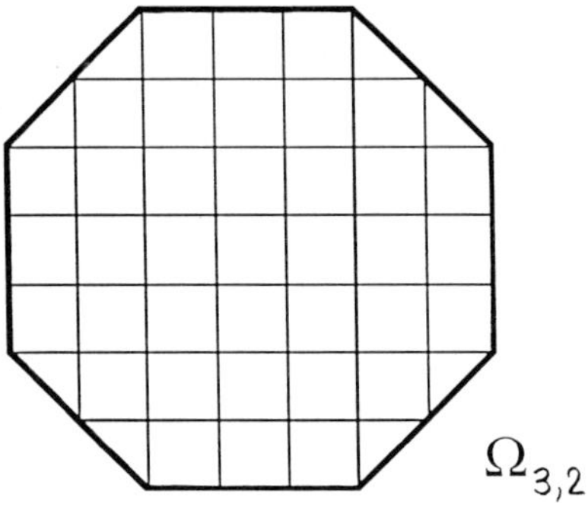

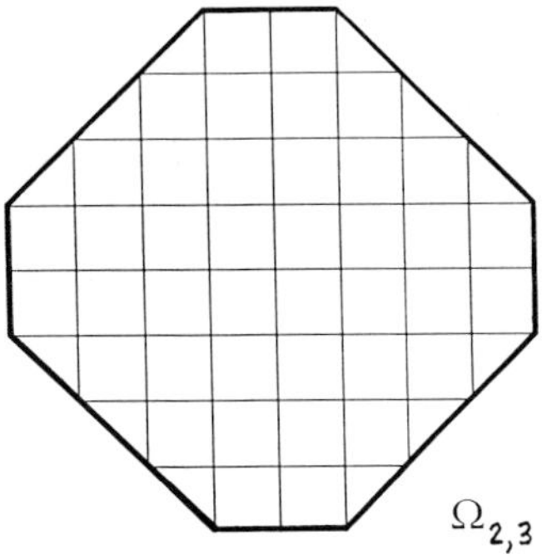

Fig. 10. Octagonal supports for $k = 2$.

Since $\mathbb{P}(\omega) = \mathbb{P}_2$, we easily prove the following theorem by using the Strang-Fix theory [12] connecting the maximal space $\mathbb{P}(\phi)$ to the zeros of derivatives of $\hat{\phi}$ on the mesh $(2\pi\mathbb{Z})^2$. We also need classical properties of convolution and Fourier transform in $\mathbb{R}^2$.

Theorem 7.

(i) *the support of* $\tilde{\sigma}_r^k$ *(resp.* $\tilde{\lambda}_r^k$*) is the octagon* $\Omega_{k+1,k}$ *(resp* $\Omega_{k,k+1}$*).*

(ii) $\tilde{\sigma}_r^k$ *and* $\tilde{\lambda}_r^k$ *are B-splines of class* C^{r+3k}*, of degree* $2r + 4k + 1$ *for* r *even and of degree* $2r + 4k + 2$ *for* r *odd.*

(iii) *for* $r \geq 0$ *and* $k \geq 1$, $\mathbb{P}(\tilde{\sigma}_r^k) = \mathbb{P}(\tilde{\lambda}_r^k) = \mathbb{P}_{3k-1}$.

Remarks

1) For $r = -1$, we can define $\tilde{\sigma}_{-1}$ (resp $\tilde{\lambda}_{-1}$) as the characteristic function of the square Σ_1 (resp. the lozenge Λ_1). Therefore $\tilde{\sigma}_{-1}^k$ and $\tilde{\lambda}_{-1}^k$ are box-splines of class C^{3k-1} and degree $4k$. In this case, $\mathbb{P}(\tilde{\sigma}_{-1}^k) = \mathbb{P}(\tilde{\lambda}_{-1}^k) = \mathbb{P}_{3k}$.

2) For $r = 0$, $\tilde{\sigma}_0$ and $\tilde{\lambda}_0$ are piecewise affine pyamids on Σ_1 and Λ_1 respectively. Therefore $\tilde{\sigma}_0^k$ and $\tilde{\lambda}_0^k$ are box-splines of class C^{3k} and degree $4k + 1$. In that case, $\mathbb{P}(\tilde{\sigma}_0^k) = \mathbb{P}(\tilde{\lambda}_0^k) = \mathbb{P}_{3k-1}$ (see [5] for details on these box-splines and their approximation properties). Thus, the family of B-splines described in this section is a natural extension of some families of classical box-splines .

3) It is possible, of course, to define a second family of B-splines by convolution of ω^k with simple B-splines σ_r and λ_r constructed in theorem 6. But since their

degrees are much higher than those of the first family, it is not yet clear wether they are more interesting in spite of their (slightly) better approximation properties.

4) There is a great variety of shapes for simple supports on τ_2. Moreover, they can be convolved separately by the characteristic functions of Σ_1 and Λ_1. This gives rise to new families of B-splines . Some of their properties together with applications to interpolation problems on rectangular domains will be discussed in two papers to be presented at the Program on Spline Functions and the Theory of Wavelets in Montréal (January 1996) and at the next conference on Curves and Surfaces in Chamonix (June 1996).

References

1. de Boor, C., K. Höllig and S. Riemenschneider, *Box-splines*, Applied Mathematical Sciences 98, Springer-Verlag, New York, 1993.
2. Bojanov, B. D., H. A. Hakopian and A. A. Sahakian, *Spline functions and Multivariate Interpolation*, Kluwer, Dordrecht, 1993.
3. Chui, C. K., *Multivariate Splines*, CBMS-NSF Series Applied Math.**54**, SIAM Publications, Philadelphia, 1988.
4. Chui, C. K., Vertex splines and their applications to interpolation of discrete data, in *Computation of Curves and Surfaces*, W. Dahmen et al. (eds), Kluwer, Dordrecht,1990.
5. Dahmen, W. and C. A. Micchelli, On the optimal approximation rates for criss-cross finite element spaces, J. Comput. Applied Math. **10** (1984), 255–273.
6. Lai, M. J., Fortran subroutines for B-nets of box-splines on three and four directional meshes, Numer. Algorithms **2** (1992), 33–38.
7. Sablonnière, P. and D. Sbibih, Existence and unicity of H_1-splines on an equilateral triangulation of the plane, Publication LANS **43**, INSA de Rennes, June 1992.
8. Sablonnière, P. and D. Sbibih, B-splines à supports hexagonaux sur un réseau tridirectionnel régulier du plan, Comptes Rendus Acad. Sci. Paris **319**, Série I, (1994), 277–282.
9. Sablonnière, P. and D. Sbibih, Some families of B-splines with hexagonal support on a three-direction mesh, in *Mathematical Methods in Computer Aided Geometric Design III*, M. Daehlen, T. Lyche and L. L. Schumaker (eds.), Vanderbilt University Press, Nashville, 1995, 467–475.
10. Sablonnière, P., Quasi-interpolants associated with H-splines on a three-direction mesh, Publication LANS **57**, INSA de Rennes (1994), to appear in J. Comput. Applied Math..
11. Sbibih, D., B-splines et quasi-interpolants sur un réseau tridirectionnel du plan, thèse de doctorat, Université d' Oujda, Maroc, 1995.
12. Strang, G. and G. Fix, A Fourier analysis of the finite element variational method, in *Constructive Aspects of Functional Analysis*, G. Geymonat et al. (eds.), CIME II Ciclo, 1971, 793–840.

Paul Sablonnière
Laboratoire LANS, INSA de Rennes
20, Avenue des Buttes de Coësmes
35043 Rennes cedex (France)
sablonni@perceval.univ-rennes1.fr

A Laurent Operator Technique for Multivariate Frames and Wavelet Bases

J. Stöckler

Abstract. A new approach to the study of multivariate frames and wavelet bases is presented. The product of the synthesis and the analysis operator defines a generalized Laurent operator. Standard methods of operator theory can be used in order to find exact upper and lower frame and Riesz bounds. For a class of generalized Meyer wavelets these bounds are determined by the singular values of a 2×2-matrix. A pyramidal algorithm for the decomposition by oversampled wavelet bases is described.

§1. Introduction

The construction of Riesz bases of $L_2(\mathbb{R})$ by dilation and shift of a single function has become an important tool in modern analysis, with applications in a wide range of areas as signal and image processing, quantum physics, numerical solution of differential and integral equations. These bases, commonly referred as *wavelet bases*, are usually combined with a pair of fast algorithms for the computation of the coefficients in the basis expansion of a given function (decomposition) and, vice versa, computation of the synthesized function from given wavelet coefficients (reconstruction).

Here we consider families of functions generated by dilates and shifts of a finite set $\Psi \subset L_2(\mathbb{R}^d)$. We follow the distinction of [3] into *Bessel families*, *wavelet bases* and *affine frames*. More specifically, let $M \in \mathbb{R}^{d\times d}$ be a *scaling matrix* with all its eigenvalues of modulus strictly larger than 1. Then the unitary operations of dilation and shift on the Hilbert space $L_2(\mathbb{R}^d)$ are defined by

$$Df := |\det M|^{1/2} f(M\,.), \qquad T_\alpha f := f(\,.-\alpha), \quad \alpha \in \mathbb{R}^d.$$

Advanced Topics in Multivariate Approximation
F. Fontanella, K. Jetter and P. J. Laurent (eds.) pp. 339–354.

ISBN 981-02-2852-X

Given a finite set $\Psi \subset L_2(\mathbb{R}^d)$ and a lattice $\mathcal{L} = L\mathbb{Z}^d$, where L is a non-singular matrix, we denote by X the family

$$X = (\Psi, M, \mathcal{L}) := \{D^j T_\alpha \psi;\ \psi \in \Psi,\ j \in \mathbb{Z},\ \alpha \in \mathcal{L}\}\ .$$

Definition 1. *Let X be defined as above.*

(i) X is a Bessel family, if there exists a constant $B > 0$ such that

$$\sum_{\eta \in X} |\langle f, \eta \rangle|^2 \le B \|f\|^2_{L_2(\mathbf{R}^d)}, \qquad f \in L_2(\mathbb{R}^d).$$

(ii) X is an affine frame, if there exist constants $A, B > 0$ such that

$$A\|f\|^2_{L_2(\mathbf{R}^d)} \le \sum_{\eta \in X} |\langle f, \eta \rangle|^2 \le B \|f\|^2_{L_2(\mathbf{R}^d)}, \qquad f \in L_2(\mathbb{R}^d).$$

(iii) X is a wavelet basis, if X is total in $L_2(\mathbb{R}^d)$ and there are constants $A, B > 0$ such that

$$A\|d\|^2_{\ell_2} \le \left\| \sum_{\eta \in X} d_\eta\, \eta \right\|^2 \le B\|d\|^2_{\ell_2}, \qquad d \in \ell_2(\Psi \times \mathbb{Z} \times \mathcal{L}).$$

If an affine frame satisfies (ii) with frame bounds $A = B$, then it is called *tight*. The frame is *inexact* if it stays a frame after removal of some of its elements. The estimates in (iii) are usually called Riesz-stability estimates, and the constants are the Riesz bounds. It is well-known that every wavelet basis is also an affine frame with the same bounds A, B. Let us mention several reasons for the study of such more general families. At first sight, the larger flexibility allows for better design results; e.g. smooth compactly supported (tight) frames with a number of symmetry relations were constructed in $L_2(\mathbb{R}^2)$ with scaling by $A = \begin{pmatrix} 1 & 1 \\ -1 & 1 \end{pmatrix}$, whereas no differentiable orthonormal wavelet basis with the same attractive properties is known at present [2, 14]. This was also the motivation for the study of univariate affine frames, see [7]. Secondly, the redundancy of a frame decomposition can increase the robustness of the representation which is subjected to quantization errors [12]. Furthermore, frame decompositions yield better results when translation invariance is an essential requirement.

Our main concern in this paper is the characterization of such families and the computation of sharp bounds A and B. Our treatment of affine frames is based on the definition of two operators, the *analysis operator*

$$T_X : L_2(\mathbb{R}^d) \longrightarrow \ell_2(\Psi \times \mathbb{Z} \times \mathcal{L}), \qquad f \mapsto (\langle f, \eta \rangle)_{\eta \in X},$$

and the *synthesis operator*

$$T_X^* : \ell_2(\Psi \times \mathbb{Z} \times \mathcal{L}) \longrightarrow L_2(\mathbb{R}^d), \qquad d = (d_\eta)_{\eta \in X} \mapsto \sum_{\eta \in X} d_\eta \, \eta \, .$$

An equivalent formulation of Definition 1 is saying that X is a Bessel sequence if T_X is bounded, or an affine frame it $T_X^* T_X$ is bounded and bounded below, and a wavelet basis if it is total and $T_X T_X^*$ is bounded and bounded below. It is clear from the definition that T_X^* is the adjoint of T_X. These and other general results can be found in [6, 9].

We devote much attention to operators of the form

$$S_{X,Y} = T_X \, T_Y^* \tag{1}$$

where $X = (\Psi, M, \mathcal{L}_X)$ and $Y = (\Theta, M, \mathcal{L}_Y)$ are Bessel families. In contrast to the results by Ron and Shen [13] there is a completely different sort of invariance inherent to these operators as will be explained in section 2 in terms of generalized Laurent operators. Standard methods of operator theory can then be applied in section 3 in order to find exact expressions for upper and lower frame and Riesz bounds. For a special class of univariate functions we compute these bounds in terms of the singular values of a 2×2 complex matrix in section 4. The last section is devoted to the special case of oversampled biorthogonal wavelet bases where the Laurent operator can be associated with a certain pyramidal algorithm.

Most of the material in this article is taken from [16], where the Laurent operator technique for affine frames was established. The results in section 4 are new and related to ongoing work of the author with C. K. Chui and X. Shi.

Let us fix some of our notations. We denote by $\mathbb{T}$ the unit circle in $\mathbb{C}$. The Fourier transform is defined as $\widehat{f}(\xi) = \int_{\mathbb{R}^d} f(x) e^{-ix\cdot\xi} \, dx$. The bracket product of $f, g \in L_2(\mathbb{R}^d)$ is the periodic function

$$[f \, , \, g] := \sum_{\alpha \in \mathbb{Z}^d} f(\,.\, + 2\pi\alpha) \, \overline{g}(\,.\, + 2\pi\alpha). \tag{2}$$

Note that this defines a function in $L_1(C)$ where $C = [-\pi, \pi)^d$. We frequently use the isomorphism of sequences and periodic functions

$$\ell_2(\mathbb{Z}^d) \ni (d_\alpha)_{\alpha \in \mathbb{Z}^d} \quad \sim \quad \tau(\xi) = \sum_{\alpha \in \mathbb{Z}^d} d_\alpha e^{-i\alpha\cdot\xi} \in L_2(C). \tag{3}$$

§2. The Laurent Operator Technique

It is convenient to explain the needed terminology in an abstract setting. Given a separable Hilbert space $\mathcal{H}$, we denote by $\ell_2(\mathcal{H})$ its direct product

$$\ell_2(\mathcal{H}) = \{(h_j)_{j\in\mathbb{Z}};\ h_j \in \mathcal{H},\ (\sum_{j\in\mathbb{Z}} \|h_j\|^2)^{1/2} < \infty\}\ .$$

This defines a separable Hilbert space whose norm is given above. The *bilateral shift* on $\ell_2(\mathcal{H})$ is the unitary operator

$$U(h_j)_{j\in\mathbb{Z}} := (h_{j-1})_{j\in\mathbb{Z}}, \qquad (h_j)_{j\in\mathbb{Z}} \in \ell_2(\mathcal{H})\ .$$

Any bounded linear operator on $\ell_2(\mathcal{H})$ which commutes with U is called a *generalized Laurent operator*, see [8]. The usefulness of this terminology lies in the following result which relates Laurent operators on $\ell_2(\mathcal{H})$ and certain operator-valued functions on the unit circle $\mathbb{T}$ with values in the operator algebra $\mathcal{B}(\mathcal{H})$. Before giving a precise formulation we recall the definition of Fourier transform of vector-valued functions, see [8: p. 33ff]. For any vector $(h_j)_{j\in\mathbb{Z}} \in \ell_2(\mathcal{H})$ there is a notion of Fourier series

$$h(z) := \sum_{j\in\mathbb{Z}} z^j h_j \in \mathcal{H} \quad \text{for a.e.} \quad z \in \mathbb{T}$$

in a weak sense; i.e., the complex function $\langle h(z), g\rangle_{\mathcal{H}}$ has Fourier coefficients $\langle h_j, g\rangle_{\mathcal{H}}$ in the usual sense for any $g \in \mathcal{H}$. With measurability and integration of vector-valued functions understood in the sense of Bochner's integral [17: chapter 5] we can then conclude that $h \in L_2(\mathbb{T}, \mathcal{H})$.

Proposition 1. [1: p.235f, 8: p.33f] *For any bounded linear operator V on $\ell_2(\mathcal{H})$ which commutes with the bilateral shift U there exists a unique essentially bounded operator-valued function*

$$F_V : \mathbb{T} \longrightarrow \mathcal{B}(\mathcal{H})$$

such that the following holds:

(i) For $(h_j)_{j\in\mathbb{Z}} \in \ell_2(\mathcal{H})$ let $h(z)$ be its Fourier series and $g(z)$ the corresponding Fourier series of $V(h_j)_{j\in\mathbb{Z}}$. Then

$$g(z) = F_V(z)\, h(z) \quad \text{for a.e.} \quad z \in \mathbb{T}.$$

(ii) The operator norm of V is given by

$$\|V\|_{\mathcal{B}(\ell_2(\mathcal{H}))} = \operatorname*{ess\,sup}_{z\in\mathbb{T}} \|F_V(z)\|_{\mathcal{B}(\mathcal{H})}.$$

(iii) V is bounded below by $\beta > 0$, i.e.

$$\|\, V(h_j)_{j\in\mathbb{Z}}\, \|^2_{\ell_2(\mathcal{H})} \ge \beta \|\, (h_j)_{j\in\mathbb{Z}}\, \|^2_{\ell_2(\mathcal{H})}, \qquad (h_j)_{j\in\mathbb{Z}} \in \ell_2(\mathcal{H}),$$

if and only if

$$\|F_V(z)\,g\|_{\mathcal{H}}^2 \geq \beta \|g\|_{\mathcal{H}}^2 \qquad \textit{for all } g \in \mathcal{H} \textit{ and a.e. } z \in \mathbb{T}.$$

Here we use the notion of weak measurability of operator-valued functions, see [17: chapter 5]. $F : \mathbb{T} \to \mathcal{B}(\mathcal{H})$ is essentially bounded by definition if $\|F(z)\|_{\mathcal{B}(\mathcal{H})}$ is an essentially bounded function on $\mathbb{T}$. Again, there exist Fourier transforms of such functions in the same weak sense as above. By this we can write

$$F(z) = \sum_{j \in \mathbb{Z}} z^j F_j \quad \text{for a.e.} \quad z \in \mathbb{T}, \tag{4}$$

where F_j are bounded operators on $\mathcal{H}$. In case of Proposition 1 the operators F_j, $j \in \mathbb{Z}$, can easily be determined by

$$(F_j\,g)_{j \in \mathbb{Z}} = V\,(\delta_{0,k}\,g)_{k \in \mathbb{Z}}\,, \qquad g \in \mathcal{H}, \tag{5}$$

see [1: p.235f]. Note that the argument of V in the previous expression has a constant vector-valued Fourier transform.

For pointing out the connection to the operator $S_{X,Y}$ in (1) we let $\mathcal{H} = \ell_2(\Psi \times \mathbb{Z}^d)$, or (in abuse of notation) its isomorphic image under (3)

$$\mathcal{H} = L_2(C)^I \quad \text{with} \quad I = \#\Psi.$$

Since the Bessel families X and Y are dilation invariant, the relation

$$T_X\,D\,f = U\,T_X\,f, \qquad f \in L_2(\mathbb{R}^d), \tag{6}$$

shows how dilation and the bilateral shift on $\ell_2(\mathcal{H})$ are connected to each other. Similarly, we also have

$$T_Y\,D^{-1}\,f = U^{-1}\,T_Y\,f, \qquad f \in L_2(\mathbb{R}^d).$$

By unitarity, this is equivalent to $DT_Y^* = T_Y^* U$, and combining this with (6) yields

$$S_{X,Y}\,U = T_X\,D\,T_Y^* = U\,S_{X,Y}.$$

Hence we have proved the following result.

Theorem 1. *Let $\Psi, \Theta \subset L_2(\mathbb{R}^d)$ be the generating sets of two Bessel families $X = (\Psi, M, \mathcal{L}_X)$ and $Y = (\Theta, M, \mathcal{L}_Y)$, and assume that Ψ and Θ have the same finite cardinality. Then*

$$S_{X,Y} := T_X\,T_Y^* : \ \ell_2(\mathcal{H}) \longrightarrow \ell_2(\mathcal{H})$$

is a generalized Laurent operator.

Our next aim is to find a precise description of the operator-valued function $F_{X,Y}$ which is associated with the Laurent operator $S_{X,Y}$ as in Proposition 1. With the notations at the end of section 1 we can give the following preparatory result.

Lemma 1. *Let X and Y be Bessel families as in Theorem 1, and L_X, L_Y be the non-singular matrices which generate the lattices $\mathcal{L}_X$ and $\mathcal{L}_Y$, respectively. For any $\psi \in \Psi$, $\theta \in \Theta$ the operator*

$$S_j^{\psi,\theta} : L_2(C) \longrightarrow L_2(C), \qquad j \in \mathbb{Z},$$

with

$$S_j^{\psi,\theta}\ \tau := \frac{|\det M|^{j/2}}{|\det L_X|} \left[\ \tau(L_Y^* \, M^{*j} \, L_X^{*\,-1}\,.)\ \widehat{\theta}(M^{*j} \, L_X^{*\,-1}\,.)\,,\ \widehat{\psi}(L_X^{*\,-1}\,.)\ \right]$$

is well-defined. Here the bracket product is used as in (2).

Proof: Let $\tau \in L_2(C)$ be given, and $(d_\alpha)_{\alpha\in\mathbb{Z}^d} \in \ell_2(\mathbb{Z}^d)$ the corresponding sequence in (3). Then the function

$$f = \sum_{\alpha\in\mathbb{Z}^d} d_\alpha\ \theta(\,.\ - L_Y\alpha)$$

is in $L_2(\mathbb{R}^d)$, since Y is a Bessel sequence which implies that T_Y^* is bounded. Its Fourier transform is given by $\widehat{f}(\xi) = \tau(L_Y^*\xi)\,\widehat{\theta}(\xi)$. Next, direct computations yield that

$$\big(\langle\ f\,,\ D^j\, T_{L_X\alpha}\,\psi\ \rangle\big)_{\alpha\in\mathbb{Z}^d} \sim \sigma_j(f) := \frac{|\det M|^{j/2}}{|\det L_X|}\ \left[\ \widehat{f}(M^{*j}\, L_X^{*\,-1}\,.)\,,\ \widehat{\psi}(L_X^{*\,-1}\,.)\ \right]$$

holds for any $f \in L_2(\mathbb{R}^d)$, where we use the isomorphism (3) once more. As mentioned before the bracket product converges in $L_1(C)$. It even defines a function in $L_2(C)$, since the sequence of its Fourier coefficients on the left is part of $T_X\, f$ and X is a Bessel sequence. By inserting the previous expression for $\widehat{f}$ we have shown that the family of operators is well-defined. ■

The above operators can be combined to a square operator matrix

$$S_j = \left(S_j^{\psi,\theta}\right)_{\psi\in\Psi,\theta\in\Theta}\,, \qquad j \in \mathbb{Z},$$

with domain $L_2(C)^I$ and I the cardinality of the sets Ψ and Θ. Then the operator-valued function

$$F_{X,Y}(z) := \sum_{j\in\mathbb{Z}} z^j\, S_j\,, \qquad z \in \mathbb{T}, \tag{7}$$

is connected with the Laurent operator $S_{X,Y}$ in the following way.

Theorem 2. *Let X and Y be Bessel families as in Theorem 1. Then the generalized Laurent operator $S_{X,Y}$ and the operator-valued function $F_{X,Y}$ in (7) are related as in Proposition 1, modulo the componentwise isomorphism (3) of $\ell_2(\Psi \times \mathbb{Z}^d)$ to $L_2(C)^I$. In particular, $F_{X,Y}$ is a well-defined essentially bounded operator-valued function.*

Proof: Let G be the operator-valued function in Propostion 1 defined by the operator $S_{X,Y}$. We have to show that the operators G_j on $\mathcal{H}$ which appear in the Fourier series (4) of G coincide with S_j modulo the isomorphic transformation (3). To this end let $d \in \mathcal{H}$ and $\tau \in L_2(C)^I$ be its isomorphic image under (3). Then we have

$$(G_j d)_{j\in\mathbb{Z}} = T_X\, T_Y^* \quad (\delta_{0,k}\, d)_{k\in\mathbb{Z}}$$

as in (5). Note that the function

$$f = T_Y^* \,(\delta_{0,k}\, d)_{k\in\mathbb{Z}} = \sum_{\theta\in\Theta} \sum_{\alpha\in\mathbb{Z}^d} d_{\theta,\alpha}\, \theta(\,.\, - L_Y\alpha)$$

has the same form as in the proof of Lemma 1 with an extra summation over θ. By a simple matrix calculation we can also see from that proof that $G_j d$, $j \in \mathbb{Z}$, has the isomorphic image $S_j\tau \in L_2(C)^I$ with S_j as above. This completes the proof of the theorem. ■

§3. Frame Bounds and Riesz Bounds

The results of section 2 provide a technique for affine frame operators which can be compared to the z-transform of ordinary Laurent operators. While the algebraic structure of $S_{X,Y}$ is usually quite intricate, the operator matrices S_j whose components are defined in Lemma 1 have a relatively simple form. For example, if the Bessel families $X = (\Psi, M, \mathbb{Z}^d)$, $Y = (\Theta, M, \mathbb{Z}^d)$ are both defined with respect to the standard lattice $\mathcal{L}_X = \mathcal{L}_Y = \mathbb{Z}^d$ and an integer matrix M, then the representation in Lemma 1 further simplifies to

$$S_j^{\psi,\theta}\tau(\xi) = |\det M|^{j/2} \left[\tau(M^{*j}\,.\,)\,\widehat{\theta}(M^{*j}\,.\,)\,,\,\widehat{\psi}\right]\,, \qquad j \in \mathbb{Z}.$$

By periodicity we observe that, for $j \geq 0$, this operator consists of dilation of τ followed by multiplication with $[\widehat{\theta}(M^{*j}\,.\,)\,,\,\widehat{\psi}]$. The analogue of this operator on the corresponding sequence space $\ell_2(\mathbb{Z}^d)$ is upsampling and convolution.

Let us further mention that for $X = Y$ the operator $S_X = S_{X,X}$ is self-adjoint. This is reflected by

$$S_j = S_{-j}^*, \qquad j \in \mathbb{Z},$$

which easily follows from the definition in Lemma 1. In particular, S_0 is a self-adjoint and positive operator. Hence the associated operator-valued function $F_X(z)$ is also self-adjoint for a.e. $z \in \mathbb{T}$.

Our next result contains several applications of Proposition 1 to Bessel families.

Theorem 3. *Let* $X = (\Psi, M, \mathcal{L}_X)$ *and* $Y = (\Theta, M, \mathcal{L}_Y)$ *be Bessel families of* $L_2(\mathbb{R}^d)$ *as in Theorem 1, and let* F_X *(resp.* $F_{X,Y}$*) be the operator-valued function associated with the Laurent operator* S_X *(resp.* $S_{X,Y}$*).*

(i) The exact bound of the Bessel family X *is given by*

$$B_X = \operatorname*{ess\,sup}_{z \in \mathbb{T}} \|F_X(z)\|_{\mathcal{B}(\mathcal{H})}.$$

(ii) If X *is total in* $L_2(\mathbb{R}^d)$*, then it is a wavelet basis if and only if the quadratic forms*

$$\langle F_X(z)\,\tau\,,\,\tau\rangle, \qquad \tau \in L_2(C)^I, \quad z \in \mathbb{T},$$

are bounded and bounded below for a.e. $z \in \mathbb{T}$*. The essential lower and upper bounds* A_X, B_X *of the quadratic form are the exact Riesz bounds of* X*.*

(iii) If Y *is an orthonormal basis of* $L_2(\mathbb{R}^d)$*, then* X *is an affine frame with bounds* A_X, $B_X > 0$ *if and only if*

$$A_X\|\tau\|^2 \le \|F_{X,Y}(z)\,\tau\|^2 \le B_X\|\tau\|^2, \qquad \tau \in L_2(C)^I, \tag{8}$$

holds for a.e. $z \in \mathbb{T}$*.*

(iv) If Y *is a wavelet basis of* $L_2(\mathbb{R}^d)$ *with bounds* A_Y, B_Y*, then* X *is an affine frame if and only if (8) holds with constants* $\widetilde{A}$, $\widetilde{B} > 0$*. Some frame bounds of* X *are then given by* $\widetilde{A}/B_Y$ *and* $\widetilde{B}/A_Y$*.*

Proof: The best possible upper bound for the Bessel family X in Definition 1(i) has the form

$$B_X = \|T_X\|^2 = \|T_X^*\|^2 = \|S_X\|.$$

Hence (i) directly follows from Proposition 1. For (ii) we only have to establish the relation between Riesz bounds and the essential upper and lower bounds of the quadratic form, since the totality of X is assumed. These bounds are also the bounds of the self-adjoint operator S_X on $\ell_2(\mathcal{H})$ by Proposition 1. The Riesz bounds in Definition 1(iii) tie in with this operator in an obvious way. For (iii) we note that T_Y^* defines an isometric isomorphism. Hence the self-adjoint operators $T_X^* T_X$ and $S_{X,Y}^* S_{X,Y}$ have the same lower and upper bounds. While the former gives the exact frame bounds of X by Definition 1(ii), the latter is related to $F_{X,Y}^* F_{X,Y}$ in the sense of Proposition 1. This proves (iii). A similar derivation for (iv) holds with isometry of T_Y^* replaced by the Riesz stability condition for Y. We do not go into further details, since this generalization of (iii) is straightforward. ∎

We wish to mention that the set Y in the previous theorem plays an auxiliary role only for finding the frame bounds of X. This was one motivation for the study of the general operators $S_{X,Y}$ with allowing different families X and Y. As noticed in [13] the operator $S_X = T_X T_X^*$ is not useful for finding lower frame bounds of X. A proper choice for Y can considerably simplify the analysis of $S_{X,Y}$, see section 4.

The assertions of Theorem 3 can be used in different ways. In one possible approach we use rough estimates of the form

$$B_X \le \sum_{j\in\mathbb{Z}} \|S_j\|, \qquad A_X \ge \lambda_0 - \sum_{j\neq 0} \|S_j\|,$$

where $\lambda_0 \ge 0$ is a lower bound of the self-adjoint operator S_0 of the Fourier series of S_X. Such estimates have been developed in [16]. A new result is the exact computation of bounds for a special class of functions which we will consider in the next section.

§4. A Class Related to Meyer's Wavelet

Based on the Laurent operator approach we develop exact Riesz bounds for a class of univariate functions ψ; namely we assume that

$$\operatorname{supp} \widehat{\psi} \subset [-8\pi/3, -2\pi/3] \cup [2\pi/3, 8\pi/3]. \tag{9}$$

This class has also been investigated with a different approach by Dai and Larson [5: chapter 5] where orthonormal wavelet bases are considered. The above set plays a central role in Meyer's construction of orthonormal wavelets [6: p. 117], too.

Theorem 4. *Let $\psi \in L_2(\mathbb{R})$ satisfy (9) and $X = (\psi, 2, \mathbb{Z})$ be the family obtained by dyadic scaling and shift of ψ.*

(i) X is a wavelet basis of $L_2(\mathbb{R})$ if and only if the 2×2 complex matrix

$$Z(\xi) = \begin{pmatrix} \widehat{\psi}(\xi) & \widehat{\psi}(\xi - 2\pi) \\ \widehat{\psi}(2\xi) & \widehat{\psi}(2\xi - 4\pi) \end{pmatrix} \tag{10}$$

and its inverse are essentially bounded on the interval $(2\pi/3, 4\pi/3)$. The exact Riesz bounds A_X, B_X are then given by

$$B_X = \operatorname*{ess\,sup}_{\xi\in(2\pi/3,4\pi/3)} |Z(\xi)|^2, \qquad A_X^{-1} = \operatorname*{ess\,sup}_{\xi\in(2\pi/3,4\pi/3)} |Z(\xi)^{-1}|^2.$$

(ii) There exists no inexact frame X of this type.

Proof: Our argument proceeds as follows. We first show that X is a frame of $L_2(\mathbb{R})$ if and only if the conditions on the matrix $Z(\xi)$ are satisfied. This shows that X is total if these conditions hold true. Next we prove that the Riesz stability estimates in Definition 1(iii) are equivalent to the same pair of matrix conditions. By combining both results we obtain (i) and (ii).

Let us begin with the analysis of the frame properties of X. We choose our second family $Y = (\theta, 2, \mathbb{Z})$ for an application of Theorem 3(iii) by defining

$$\widehat{\theta}(\xi) = \begin{cases} 1, & \text{if } \xi \in [-2\pi, -\pi) \cup [\pi, 2\pi), \\ 0, & \text{otherwise.} \end{cases}$$

This is the Littlewood-Paley orthonormal wavelet, see [6: p. 115]. The operator-valued function

$$F(z) := F_{X,Y}(z) = \sum_{j \in \mathbb{Z}} z^j \, S_j$$

is defined with components $S_j = S_j^{\psi,\theta}$ as in Lemma 1. It is easily seen that $S_j = 0$ for all $|j| \geq 2$ since the supports of the functions $\widehat{\theta}(2^j\,.)$ and $\widehat{\psi}$ are disjoint. For the remaining values of j we obtain

$$S_0\,\tau(\xi) = \begin{cases} \overline{\widehat{\psi}}(\xi - 2\pi)\,\tau(\xi), & \text{for } 2\pi/3 \leq \xi < \pi, \\ \overline{\widehat{\psi}}(\xi)\,\tau(\xi), & \text{for } \pi \leq \xi < 2\pi, \\ \overline{\widehat{\psi}}(\xi - 4\pi)\,\tau(\xi), & \text{for } 2\pi \leq \xi \leq 8\pi/3, \end{cases}$$

$$S_1\,\tau(\xi) = \begin{cases} 2^{1/2}\,\overline{\widehat{\psi}}(\xi)\,\tau(2\xi), & \text{for } 2\pi/3 \leq \xi < \pi, \\ 2^{1/2}\,\overline{\widehat{\psi}}(\xi - 2\pi)\,\tau(2\xi), & \text{for } \pi \leq \xi < 4\pi/3, \\ 0, & \text{for } 4\pi/3 \leq \xi \leq 8\pi/3, \end{cases}$$

$$S_{-1}\,\tau(\xi) = \begin{cases} 0, & \text{for } 2\pi/3 \leq \xi < 4\pi/3, \\ 2^{-1/2}\,\overline{\widehat{\psi}}(\xi - 4\pi)\,\tau(\xi/2), & \text{for } 4\pi/3 \leq \xi < 2\pi, \\ 2^{-1/2}\,\overline{\widehat{\psi}}(\xi)\,\tau(\xi/2), & \text{for } 2\pi \leq \xi \leq 8\pi/3. \end{cases}$$

Note that we specified all functions on the interval $[2\pi/3, 8\pi/3]$ which can be done by periodicity. We split the interval into two parts $I_1 = [2\pi/3, 4\pi/3]$ and $I_2 = 2I_1$. Then

$$F(z)\,\tau(\xi) = S_0\tau(\xi) + zS_1\tau(\xi) + \overline{z}S_{-1}\tau(\xi), \qquad \xi \in I_1 \cup I_2,$$

can also be written as a 2-vector

$$\begin{pmatrix} F(z)\,\tau(\xi) \\ 2^{1/2}\,z\,F(z)\,\tau(2\xi) \end{pmatrix} = \widetilde{Z}(\xi) \cdot \begin{pmatrix} \tau(\xi) \\ 2^{1/2} z\tau(2\xi) \end{pmatrix}, \qquad \xi \in I_1.$$

Here the matrix $\widetilde{Z}(\xi)$ is defined as

$$\widetilde{Z}(\xi) = \begin{cases} \overline{Z(\xi)} \cdot \begin{pmatrix} 0 & 1 \\ 1 & 0 \end{pmatrix}, & \text{for } 2\pi/3 \leq \xi < \pi, \\ \overline{Z(\xi)}, & \text{for } \pi \leq \xi \leq 4\pi/3. \end{cases}$$

Note that the components of the vectors are normalized so as to give

$$\int_{2\pi/3}^{4\pi/3} \left| \begin{pmatrix} \tau(\xi) \\ 2^{1/2} z\tau(2\xi) \end{pmatrix} \right|^2 d\xi = \int_{2\pi/3}^{8\pi/3} |\tau(\xi)|^2 \, d\xi = \|\tau\|_{L_2(C)}^2.$$

By using standard integral estimates we thus find that the inequalities of Theorem 3(iii) hold with the values of A_X and B_X in Theorem 4. This finishes the first part of the proof.

In the second part we look at the Riesz stability of X in terms of the operator bounds of $S_X = T_X T_X^*$. Let

$$G(z) := G_X(z) = \sum_{j \in \mathbb{Z}} z^j R_j, \qquad z \in \mathbb{T},$$

be its operator-valued function. Then $R_j = S_j^{\psi,\psi}$ in Lemma 1 vanishes for $|j| \geq 2$ by the support assumption. The quadratic form in Theorem 3(ii) simplifies to

$$\langle G(z)\tau \, , \, \tau \rangle = \langle R_0 \tau \, , \, \tau \rangle + 2 \, \mathrm{Re} \, (z \langle R_1 \tau \, , \, \tau \rangle) \; , \qquad \tau \in L_2(C),$$

since $R_{-1} = R_1^*$. Performing the same type of operations as before, we obtain

$$\langle G(z)\tau \, , \, \tau \rangle = \int_{2\pi/3}^{4\pi/3} \left| Z(\xi) \cdot \begin{pmatrix} \tau(\xi) \\ 2^{1/2} z\tau(2\xi) \end{pmatrix} \right|^2 d\xi.$$

The standard integral estimates give the upper and lower bounds A_X and B_X as in the theorem. As explained at the beginning of the proof this is combined with our first step and yields both assertions of Theorem 4. ∎

As an immediate consequence of the previous result we obtain a characterization of orthonormal wavelet bases which is also considered in [5].

Corollary 1. *A family X as in Theorem 4 defines an orthonormal basis of $L_2(\mathbb{R})$ if and only if the matrix $Z(\xi)$ in (10) is unitary for a.e. $\xi \in (2\pi/3, 4\pi/3)$.*

Proof: Any orthonormal basis of $L_2(\mathbb{R})$ is also a Riesz basis with bounds $A_X = B_X = 1$. In view of Theorem 4 this happens if and only if $Z(\xi)^* \, Z(\xi)$ has largest and smallest eigenvalue 1 for a.e. ξ in the given interval. Hence this is equivalent to the unitarity of $Z(\xi)$. ∎

§5. Oversampling of Wavelet Bases

We give a short account of the methods which were developed in [16] for obtaining fast decomposition algorithms using affine frames. A more complete treatment will be given elsewhere.

Let us first introduce some special assumptions which are used here. We let $X = (\Psi, M, \mathbb{Z}^d)$ and $Y = (\Theta, M, \mathbb{Z}^d)$ be a pair of biorthogonal wavelet bases of $L_2(\mathbb{R}^d)$; i.e. each X and Y are Riesz bases of $L_2(\mathbb{R}^d)$ and the biorthogonality relations

$$\left\langle \psi_{j,\alpha}^{(i)}, \theta_{k,\beta}^{(m)} \right\rangle = \delta_{i,m}\, \delta_{j,k}\, \delta_{\alpha,\beta}, \qquad j,k \in \mathbb{Z},\ \alpha,\beta \in \mathbb{Z}^d,\ 1 \le i,m \le I,$$

are satisfied. Here we used an upperscript index for numbering the elements of Ψ and Θ. We also pose some mild decay conditions on the Fourier transforms of $\psi \in \Psi$ and $\theta \in \Theta$. Typical assumptions (though there exist milder ones, see [16: section 3.3]) are that

$$|\widehat{\psi}(\xi)| \le c_1 |\xi|^{\sigma_1} \quad \text{and} \quad \sum_{\alpha \in \mathbb{Z}^d} |\widehat{\psi}(\xi + 2\pi\alpha)|^{2-\sigma_2} \le c_2$$

hold for a.e. $\xi \in C$ with constants c_k, $\sigma_k > 0$. A similar pair of conditions is assumed to be satisfied for all $\theta \in \Theta$.

Furthermore, we require that the scaling matrix M has *integer* entries. We are mostly interested in the families

$$X_n := (\Psi, M, \mathcal{L}_n) \quad \text{with} \quad \mathcal{L}_n = M^{-n}\mathbb{Z}^d, \quad n \ge 0. \tag{11}$$

The results in [4] can directly be applied in order to show that the above decay assumptions are sufficient for X_n being a Bessel family for any $n \ge 0$. Furthermore, in [16: chapter 3] we showed that the generalized Laurent operator

$$S_{X_n,Y} = T_{X_n} T_Y^*$$

has a Fourier series $\sum_{j\in\mathbb{Z}} z^j S_j$ which is strongly convergent, or even more

$$\sum_{j\in\mathbb{Z}} \|S_j\| < \infty.$$

The biorthogonality implies that T_X and T_Y^* are inverse to each other, which means that

$$f = \sum_{i=1}^{I} \sum_{j\in\mathbb{Z}} \sum_{\alpha\in\mathbb{Z}^d} \left\langle f, \psi_{j,\alpha}^{(i)} \right\rangle \theta_{j,\alpha}^{(i)}, \qquad f \in L_2(\mathbb{R}^d),$$

holds. Hence $S_{X,Y}$ is the identity operator. For $n > 0$ there is still some orthogonality left which is expressed in the following result.

Proposition 2. *Let $n \in \mathbb{N}$, X_n the family (11) and S_j the operators in the Fourier series of $T_{X_n} T_Y^*$. Then $S_j = 0$ for all $j \leq -n$.*

Proof: Recall that S_j is the operator matrix on $L_2(C)^I$ with entries given in Lemma 1. Here we are working with $L_X = M^{-n}$ and L_Y the identity matrix. For $j \in \mathbb{Z}$ we find that a typical entry of S_j has the form

$$S_j^{\psi,\theta}\tau(\xi) = |\det M|^{j/2+n}\left[(\tau\widehat{\theta})(M^{*j+n}\,.)\,,\,\widehat{\psi}(M^{*n}\,.)\right], \qquad \tau \in L_2(C).$$

If $j \leq -n$ we let $K = M^{*j+n}$ and take a complete set Γ of representatives of $2\pi\mathbb{Z}^d/(2\pi K^{-1}\mathbb{Z}^d)$. By the periodicity of τ the above bracket product decomposes into

$$\begin{aligned}
&\left[(\tau\widehat{\theta})(K\,.)\,,\,\widehat{\psi}(M^{*n}\,.)\right](\xi)\\
&\quad= \sum_{\gamma\in\Gamma}\tau(K(\xi+\gamma))\sum_{\alpha\in\mathbb{Z}^d}\widehat{\theta}(K(\xi+\gamma)+2\pi\alpha)\,\overline{\widehat{\psi}}(M^{*|j|}\,(K(\xi+\gamma)+2\pi\alpha))\\
&\quad= \sum_{\gamma\in\Gamma}\tau(K(\xi+\gamma))\,\left[\widehat{\theta}\,,\,\widehat{\psi}(M^{*|j|}\,.)\right]\,(K(\xi+\gamma)).
\end{aligned}$$

The last bracket product vanishes a.e. in $\mathbb{R}^d$ by the orthogonality relation, since $j \neq 0$. Hence the assertion of the proposition follows. ■

Another observation made from the representation in Lemma 1 is that

$$S_j\tau(\xi) = |\det M|^{j/2+n}\;E_j(\xi)\cdot\tau(M^{*j+n}\xi), \qquad \tau \in L_2(C)^I,$$

holds for all $j > -n$. Here the factor $E_j(\xi)$ is a matrix-valued function on C of the form

$$E_j = \left(\left[\widehat{\theta}(M^{*j+n}\,.)\,,\,\widehat{\psi}(M^{*n}\,.)\right]\right)_{\psi\in\Psi,\,\theta\in\Theta}. \tag{12}$$

Hence S_j is a multiplication operator on $L_2(C)^I$ combined with a dilation. Equivalently, S_j can be understood as convolution and upsampling on the space $\ell_2(\mathbb{Z}^d)^I$.

This simple structure of all non-vanishing S_j has a very useful consequence in view of the self-adjoint operator

$$S_{X_n,Y}^*\,S_{X_n,Y} =: \sum_{j\in\mathbb{Z}} z^j\,R_j.$$

The summability of $\|S_j\|$ gives

$$R_j = \sum_{k\in\mathbb{Z}} S_k^* S_{k+j} = \sum_{k\geq 1} S_{k-n}^* S_{k+j-n}, \qquad j \geq 0, \tag{13}$$

and $R_j = R_{-j}^*$ for all $j < 0$. It now turns out that the products $S_k^* S_{k+j}$, $k > -n$, are all of the same form, namely multiplication combined with dilation by M^{*j}.

Proposition 3. *With the assumptions of Proposition 2 the self-adjoint operator* $S^*_{X_n,Y} S_{X_n,Y}$ *is a generalized Laurent operator. The entries* R_j, $j \geq 0$, *of its Fourier series are given by*

$$R_j \tau(\xi) = |\det M|^{j/2+n}\ F_j(\xi) \cdot \tau(M^{*j}\xi), \qquad \tau \in L_2(C)^I,$$

where F_j *is a matrix-valued function on* C. *With* Γ_k, $k \geq 1$, *a complete set of representatives of* $2\pi\mathbb{Z}^d/(2\pi M^{*k}\mathbb{Z}^d)$ *it has the form*

$$F_j(\xi) = \sum_{k\geq 1} \sum_{\gamma\in\Gamma_k} (E^*_{k-n} \cdot E_{k+j-n})\,(M^{*-k}(\xi+\gamma)), \qquad \xi \in C.$$

The main part of the **proof** is to analyse the operator products $S^*_{k-n} S_{k+j-n}$ in (13). This is done by first computing the adjoint

$$S^*_{k-n}\tau(\xi) = |\det M|^{(n-k)/2} \sum_{\gamma\in\Gamma_k} (E^*_{k-n} \cdot \tau)\,(M^{*-k}(\xi+\gamma)), \qquad \tau \in L_2(C)^I.$$

Remark. The simple structure of the operator product $S^*_{X_n,Y} S_{X_n,Y}$ can be used for the purpose of preconditioning. When developing algorithms for the computation of frame decompositions one uses iterative methods for the inverse of this self-adjoint positive operator. In [16: section 5.4] we construct an approximate Cholesky decomposition where the Cholesky factor is a banded lower triangular Laurent operator. Its entries have the same type as the operators S_j, hence are composed of multiplication and dilation. They can be defined in such a way that inversion of the Cholesky factor (which is needed for the preconditioned iteration procedure) is done by a fast algorithm.

For finding the preconditioner one first needs to compute the operators R_j, $j \geq 0$, explicitly. Instead of using the above representation for the matrix-valued function F_j, one can also obtain its ith column, $1 \leq i \leq I$, numerically by an application of $S^*_{X_n,Y} S_{X_n,Y}$ to the element

$$(\tau_j)_{j\in\mathbb{Z}} \in \ell_2(\mathcal{H}) \quad \text{with} \quad \tau_j = 0,\ j \neq 0, \quad \text{and} \quad \tau_0 = (\delta_{i,m})_{1\leq m\leq I}.$$

Then the kth component of the image in $\ell_2(\mathcal{H})$ defines the ith column of F_k. Hence if there is a fast algorithm for $S_{X_n,Y}$ and its adjoint, the operators R_j are numerically accessible.

We further wish to explain the possibility to have a fast algorithm for $S_{X_n,Y}$. Note that we have not appealed to a multiresolution analysis in our previous results. From now on we assume that our wavelet bases X and Y are generated by a multiresolution analysis. This notion has been introduced in [11] and is generalized to our setting in [10], see also the references therein.

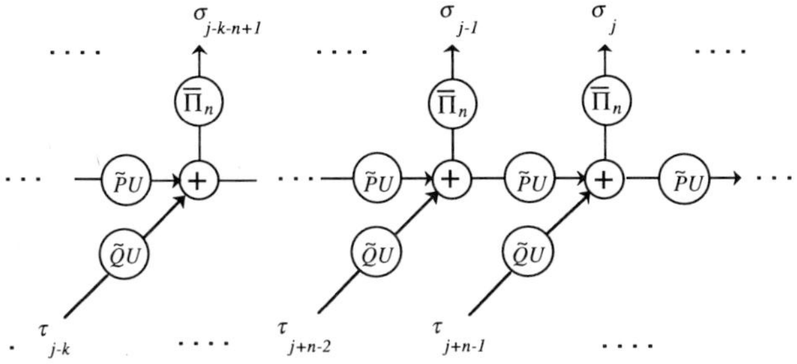

Fig. 1. The pyramidal algorithm for oversampled wavelet bases.

Let the mask coefficients of ϕ, η be defined by the periodic functions P, $\widetilde{P}$, hence

$$\widehat{\phi}(M^*\xi) = P(\xi)\widehat{\phi}(\xi), \quad \widehat{\eta}(M^*\xi) = \widetilde{P}(\xi)\widehat{\eta}(\xi), \qquad \xi \in \mathbb{R}^d.$$

Furthermore, we use the symbols of the wavelet masks Q_ψ, $\widetilde{Q}_\theta$, which are given by

$$\widehat{\psi}(M^*\xi) = Q_\psi(\xi)\widehat{\phi}(\xi), \quad \widehat{\theta}(M^*\xi) = \widetilde{Q}_\theta(\xi)\widehat{\eta}(\xi), \qquad \xi \in \mathbb{R}^d,$$

for all $\psi \in \Psi$ and $\theta \in \Theta$. The essential identity which is used for the representation of S_j, $j > -n$, is the biorthogonality relation

$$[\widehat{\phi}\,, \widehat{\eta}] \equiv 1 \qquad \text{a.e. in } \mathbb{R}^d,$$

of the scaling functions ϕ and η. This enables us to simplify the matrix-valued functions E_j in (12). By an inductive argument we obtain that

$$E_{j-n}(\xi) = \prod_{k=0}^{j-2} \widetilde{P}(M^{*k}\xi) \prod_{k=0}^{n-2} \overline{P}(M^{*k}\xi) \left(\widetilde{Q}_\theta(M^{*j-1}\xi)\, \overline{Q_\psi}(M^{*n-1}\xi)\right)_{\psi\in\Psi,\,\theta\in\Theta}, \tag{14}$$

which holds for all $j \geq 1$. Hence the entries of E_{j-n} are obtained by successive multiplication of the above periodic functions, or equivalently by convolution of the mask coefficients and appropriate upsampling.

This gives rise to a pyramidal algorithm which is a realization of the operator $S_{X_n,Y}$ as described in the diagram of Figure 1. There we use the symbol U for the upsampling operator $U\tau = |\det M|^{1/2}\tau(M^*\,.)$ (in distinction to the bilateral shift in section 2). The symbol $\overline{\Pi_n}$ stands for multiplication with the second product in (14) times $\overline{Q_\psi}(M^{*n-1}\xi)$. We only present the scheme for the case of singleton sets $\Psi = \{\psi\}$, $\Theta = \{\theta\}$. The general case is obtained by having I branches at the lower and upper part of the diagram, but keeping the horizontal path unchanged. The similarity to the wavelet reconstruction scheme of $\widetilde{\phi}$ is also apparent in the diagram. It shows that, in contrast to the 'algorithme à trous' [15] , decimation between different levels occurs for these frames.

References

1. Bercovici, H., *Operator Theory and Arithmetic in H^∞*, American Math. Soc., Providence, 1988.
2. Chui, C. K., K. Jetter and J. Stöckler, Wavelets and frames on the four-directional mesh, in *Wavelets: Theory, Algorithms and Applications*, C. K. Chui, L. Montefusco, and L. Puccio (eds.), Academic Press, San Diego, 1994, 215–230.
3. Chui, C. K. and X. Shi, Bessel sequences and affine frames, Appl. Comput. Harmonic Anal. **1** (1993), 29–49.
4. Cohen, A. and I. Daubechies, Non-separable bidimensional wavelet bases, Rev. Mat. Iberoamer. **9** (1993), 51–137.
5. Dai, X. and D. R. Larson, Wandering vectors for unitary systems and orthogonal wavelets, to appear in Memoirs of the American Mathematical Society.
6. Daubechies, I., *Ten Lectures on Wavelets*, CBMS-NSF Reg. Conf. Series in Appl. Math. 61, SIAM, Philadelphia, 1992.
7. Daubechies, I., A. Grossmann and Y. Meyer, Painless nonorthogonal expansions, J. Math. Phys. **27** (1986), 1271–1283.
8. Fillmore, P. A., *Notes on Operator Theory*, Van Nostrand Reinhold Co., New York, 1970.
9. Heil, C. and D. Walnut, Continuous and discrete wavelet transforms, SIAM Review **31** (1989), 628–666.
10. Jia, R. Q. and Z. Shen, Multiresolution and wavelets, Proc. Edinburgh Math. Soc. **37** (1994), 271–300.
11. Mallat, S., Multiresolution approximations and wavelet orthonormal bases of $L_2(\mathbb{R})$, Trans. Amer. Math. Soc. **315** (1989), 69–87.
12. Munch, N. J., Noise reduction in tight Weyl-Heisenberg frames, IEEE Trans. Inform. Theory **38** (1992), 608–616.
13. Ron, A. and Z. Shen, Affine systems in $L_2(\mathbb{R}^d)$: the analysis of the analysis operator, preprint, CMS Report 96-02, University of Wisconsin-Madison, 1996.
14. Ron, A. and Z. Shen, Compactly supported tight affine spline frames in $L_2(\mathbb{R}^d)$, preprint, CMS Report 96-05, University of Wisconsin-Madison, 1996.
15. Shensa, M. J., The discrete wavelet transform: wedding the à trous and Mallat algorithms, IEEE Trans. Signal Processing **40** (1992), 2464–2482.
16. Stöckler, J., Multivariate affine frames, Habilitationsschrift, Universität Duisburg, 1995.
17. Yosida, K., *Functional Analysis*, 6th ed., Springer-Verlag, Berlin, 1980.

Joachim Stöckler
Institut für Angewandte Mathematik und Statistik
Universität Hohenheim
Germany
stockler@uni-hohenheim.de

On Piecewise Algebraic Curves

Ren-Hong Wang

Abstract. In this paper, we will explain what piecewise algebraic curves are, why we need to study them, and how difficult is to derive piecewise algebraic curves. A survey will also be presented on the recent results in piecewise algebraic curves and related topics. Especially we will introduce some results which are concerning Bezout–type theorem on piecewise algebraic curves, real local branch of piecewise algebraic curves, and irreducibility of the so-called piecewise algebraic sets,etc.

Let D be a domain in $\mathbb{R}^2$, Δ be a partition consisting of a finite irreducible algebraic curves. Denote by $\Delta_i, i = 1, \dots, N$ the cells of Δ,

$$P_k := \{\sum_{i=0}^{k}\sum_{j=0}^{k-i} c_{ij}x^i y^j : c_{ij} \in R\},$$

and

$$S_k^r(\Delta) := \{s \in C^r(D) : s|_{\Delta_i} \in P_k, i = 1, \dots, N\}.$$

A piecewise algebraic curve is defined by

$$\Gamma := \{(x, y) \in D : s(x, y) = 0, s \in S_k^r(\Delta)\}; \tag{1}$$

usually we denote

$$\Gamma : \quad s(x, y) = 0.$$

Why we have to study the piecewise algebraic curve? Let us recall the interpolation problem by bivariate splines. The interpolation problem is how to choose a knot set $K := \{(x_i, y_i)\}_{i=1}^d$ such that for any given values $z_1, \dots, z_d$, there exists a unique $s \in S_k^r(\Delta)$ satisfying

$$s(x_i, y_i) = z_i, \qquad i = 1, \dots, d, \tag{2}$$

Advanced Topics in Multivariate Approximation
F. Fontanella, K. Jetter and P. J. Laurent (eds.) pp. 355–361.

ISBN 981-02-2852-X

where $d = \dim S_k^r(\Delta)$.

By using the representation of a bivariate spline ([6, 9]), (2) is a linear algebraic equation, therefore (2) has a unique solution if and only if the following linear homogeneous equations

$$s(x_i, y_i) = 0, \qquad i = 1, \ldots, d,$$

have only a trivial solution.

It means that (2) has a unique solution if and only if the knot set K does not lie on any piecewise algebraic curve as follows

$$\Gamma : \qquad s(x, y) = 0, \quad s \in S_k^r(\Delta).$$

So it will be important to study the piecewise algebraic curve defined as above.

It will be difficult, however, to derive the properties of piecewise algebraic curves. For instance, let $s(x, y) \in S_k^r(\Delta)$, and denote by $s_i(x, y)$ the polynomial determined by $s(x, y)$ on the i-th cell of Δ, then the relation $\{(x, y) \in \Delta_i : s_i(x, y) = 0\} \neq \emptyset$ can not be guaranteed.

Definitions. *Let $s_i(x, y) \in S_{k_i}^{r_i}(\Delta), i = 1, 2$ be two bivariate spline functions, and $s_{1,i}(x, y)$ and $s_{2,i}(x, y)$ the polynomials defined by $s_1(x, y)$ and $s_2(x, y)$ on the cell Δ_i respectively. If there exist some cells of Δ such that for each of those cells, say Δ_i, the polynomials $s_{1,i}(x, y)$ and $s_{2,i}(x, y)$ have a common factor $t_i(x, y)$, then we say that $s_1(x, y)$ and $s_2(x, y)$ have local common factors. In general, these common factors may be not the same.*

As we know that an algebraic curve $\gamma : l(x, y) = 0$ is called a branch of the algebraic curve $\Gamma : L(x, y) = 0$, if $l(x, y)$ is a factor of $L(x, y)$.

A piecewise algebraic curve $\gamma : t(x, y) = 0$ is called a local branch of the piecewise algebraic curve $\Gamma : s(x, y) = 0$, if there exists a union U of cells of Δ, such that γ is a branch of Γ on U.

It is clear that the statement "The bivariate splines $s_1(x, y)$ and $s_2(x, y)$ have a local common factor" is not equivalent to the statement "The piecewise algebraic curves $\Gamma_1 : s_1(x, y) = 0$ and $\Gamma_2 : s_2(x, y) = 0$ have a local branch".

Example 1. *Let $D = \mathbb{R}^2$, the partition Δ of D consist of $x = 0$. Denote by $R_-^2 = \{(x, y) : x < 0\}$, $R_+^2 = \{(x, y) : x \geq 0\}$. $s_1(x, y), s_2(x, y)$, and $s_3(x, y) \in S_1^0(\Delta)$ are defined by*

$$s_1(x, y) = \begin{cases} x - 1, & \text{on } R_-^2, \\ -x - 1, & \text{on } R_+^2, \end{cases}$$

$$s_2(x, y) = \begin{cases} x - 1, & \text{on } R_-^2, \\ 3x - 1, & \text{on } R_+^2, \end{cases}$$

and

$$s_3(x, y) = \begin{cases} x - y, & \text{on } R_-^2, \\ 2x - y, & \text{on } R_+^2 \end{cases}$$

respectively. Then $\Gamma_i : s_i(x, y) = 0$ are

$$\Gamma_1 = \emptyset,$$

$$\Gamma_2 = \{(x, y) : \emptyset \text{ on } R^2_-, x = \frac{1}{3} \text{ on } R^2_+\},$$

and

$$\Gamma_3 = \{(x, y) : x = y \text{ on } R^2_-, 2x = y \text{ on } R^2_+\}$$

respectively.

Example 2. *Let* $D = \mathbb{R}^2$, *the partition* Δ *of* D *consist of* $x = 0$ *and* $y = 0$. *Denote by* $D_1 = \{(x, y) : x > 0, y \geq 0\}, D_2 = \{(x, y) : x \leq 0, y \geq 0\}, D_3 = \{(x, y) : x \leq 0, y < 0\}$, *and* $D_4 = \{(x, y) : x > 0, y < 0\}$. $s_4(x, y) \in S^1_2(\Delta)$ *is defined as follows*

$$s_4(x, y) = \begin{cases} 3x^2 + 3y^2 - 1, & \text{on } D_1, \\ x^2 + 3y^2 - 1, & \text{on } D_2, \\ x^2 - 1, & \text{on } D_3, \\ 3x^2 - 1, & \text{on } D_4. \end{cases}$$

The piecewise algebraic curve $\Gamma_4 : s_4(x, y) = 0$ *is*

$$\Gamma_4 : \begin{cases} x^2 + y^2 = \frac{1}{3}, & \text{on } D_1, \\ x^2 + 3y^2 = 1, & \text{on } D_2, \\ x = -1, & \text{on } D_3, \\ x = \frac{\sqrt{3}}{3}, & \text{on } D_4. \end{cases}$$

Example 3. *Let* $D = \mathbb{R}^2$, *the partition* Δ *of* D *consist of* $x = 0$. *The piecewise algebraic curve*

$$\gamma = \{(x, y) : x - y = 0 \text{ on } R^2_-, \emptyset \text{ on } R^2_+\}$$

is a local common branch (on R^2_-*) of the following piecewise algebraic curves:*

$$\Gamma_1 = \{(x, y) : xy - y^2 - yx_+ = 0 \text{ on } \mathbb{R}^2\}$$

and

$$\Gamma_2 = \{(x, y) : x - y = 0 \text{ on } \mathbb{R}^2\}.$$

Suppose that a partition Δ is consisted by a finite straight segments, and v is an interior vertex of Δ. Denote by $st(v)$ the union of cells of sharing v.

Let $s_i(x, y) \in S^{r_i}_{k_i}(\Delta), i = 1, 2$. The intersection set of the piecewise algebraic curves $\Gamma_1 : s_1(x, y) = 0$, and $\Gamma_2 : s_2(x, y) = 0$ is denoted by $\text{Inter}(\Gamma_1, \Gamma_2)$. The following number

$$\begin{aligned} BN(k_1, r_1; k_2, r_2) = \max\{&\text{Card Inter}(\Gamma_1, \Gamma_2) < \infty : \\ &\Gamma_i : s_i(x, y) = 0, s_i(x, y) \in S^{r_i}_{k_i}(\Delta), i = 1, 2\} \end{aligned} \tag{3}$$

is called the Bezout's number.

It is obvious that

$$BN(k_1, r_1; k_2, r_2) \leq N k_1 k_2, \tag{4}$$

where N is the number of cells of Δ ([5]).

By using the resultant on the polar coordinates, R. H. Wang and G. H. Zhao proved ([10])

Theorem 1. *Let $\Gamma_i : s_i(x, y) = 0, s_i(x, y) \in S_{k_i}^{r_i}(\Delta), i = 1, 2$ be piecewise algebraic curves. For any given interior vertex v_i, the Bezout's number on $st(v_i)$ is satisfied the following inequality*

$$BN(k_1, r_1; k_2, r_2) \leq n_i(k_1 k_2 - 1) + 1, \tag{5}$$

where n_i is the number of edges of passing through v_i.

X.Q.Shi and R.H.Wang ([3]) showed that the Bezout's number $BN(k_1, 0; k_2, 0)$ depends on some properties of a triangulation Δ.

A triangulation Δ is called to be 2–signs if one can mark -1, or $+1$ on each of vertices of Δ such that the numbers marked on 3 vertices on any cell of Δ are not the same.

A triangulation Δ is called to be 3–signs if one can mark $-1, 0$, or $+1$ on each of vertices of Δ such that the numbers marked on 3 vertices on any cell of Δ are totally different. It is obvious that any 3-sign triangulation is also a 2-sign triangulation.

Conjecture. ([3]) *Any triangulation is of 2–signs.*

Let v be an interior vertex of Δ. The number of boundary vertices of the star $st(v)$ is called the degree of the interior vertex v. Denote by $d(v)$ the degree of v.

An interior vertex v is called to be even or odd, if $d(v)$ is even or odd respectively. A triangulation Δ is called to be even, if all of its interior vertices are even. We have to show that there exist some triangulations that are neither odd nor even.

It is clear that the even triangulation of a simple connected domain is of 3–signs. However, the odd triangulation of a simple connected domain can not be of 3-sign.

X. Q. Shi and R. H. Wang ([3]) proved the following two theorems.

Theorem 2. *If Δ is a triangulation of a simple connected domain in $\mathbb{R}^2$, then*

$$\begin{aligned} &1. BN(1,0;1,0) = T, \textit{ when } \Delta \textit{ is even;} \\ &2. BN(1,0;1,0) \leq T - [(V_{odd} + 2)/3], \textit{ otherwise,} \end{aligned} \tag{6}$$

where T is the number of cells of Δ, V_{odd} is the number of odd vertices of Δ, and $[x]$ denotes the maximum integer $\leq x$.

Theorem 3. *If Δ is a 2–signs triangulation, and $Max(k_1, k_2) \geq 2$, then*

$$BN(k_1, 0; k_2, 0) = k_1 k_2 T, \tag{7}$$

where T is the number of cells of Δ.

A real algebraic curve γ is called to be set on $\Delta_1, \ldots, \Delta_m \subset \Delta$, provided that

$$\gamma \cap \overline{\Delta_i} \neq \emptyset \quad i = 1, \ldots, m,$$

and

$$\gamma \subset \bigcup_{i=1}^{m} \overline{\Delta_i}.$$

Let a curve

$$L: \quad x = x(t), y = y(t), \quad a \leq t \leq b, \tag{8}$$

be set on the cells

$$\Delta_1, \ldots, \Delta_m \subset \Delta,$$

where $x(t)$ and $y(t)$ are polynomials on the parameter t.

It is clear that the curve L has been become a piecewise curve cut by the cells $\Delta_1, \ldots, \Delta_m$. Denote by

$$L(\Delta): \quad a = t_0 < t_1 < \cdots < t_h < t_{h+1} = b \tag{9}$$

a partition of the interval $[a, b]$ generated by setting L on cells $\Delta_1, \ldots, \Delta_m \subset \Delta$.

For any given $s \in S_k^r(\Delta), s(x(t), y(t))$ is called a parametric spline on t generated by putting $s(x, y)$ on the curve L. The smoothness of $s(t) = s(x(t), y(t))$ is still r, however the degree K of the piecewise polynomial $s(t)$ depends on the degree of $x(t)$ and $y(t)$.

Denote by d the dimension of $S_k^r(t)$, a knot set $\{v_j\}_{j=1}^d \subset [a, b]$ is called (K, r)–suitable, if for any given $\{u_j\}_{j=1}^d$, the following interpolation problem

$$s(v_i) = u_i, \quad j = 1, \ldots, d, \quad s \in S_k^r, \tag{10}$$

has a unique solution.

It is well known that I. J. Schoenberg & A. Whitney ([2]) and S. Karlin & Z. Ziegler ([1]) showed the characterization on (K, r)–suitable knot set.

On the real local branch, we have

Theorem 4. ([8]) *The real curve L given by (8) is a real local branch of Γ : $s(x, y) = 0, s \in S_k^r(\Delta)$, if and only if there exists a (K, r)–suitable knot set included in the intersection Γ and L, where the integer K is fixed by putting $s(x, y)$ on the curve L.*

Of course, $x(t)$ and $y(t)$ given in (8) can be generalized to the spline functions.

Zhi–Xun Su([4]) discussed the piecewise algebraic curve and surface, piecewise algebraic variety, and piecewise algebraic set, etc.

Wen–Ju Wang and Ren–Hong Wang([11]) discussed the irreducibility and isomorphism of the piecewise algebraic sets.

Let D be a domain in k^n, k a fixed algebraically closed field, $\triangle$ be a partition of D consisting of a finite number of hyperplanes. Denote by $S^\mu(\triangle)$ the spline function ring, it is a k-algebra. In fact, $S^\mu(\triangle)$ is a reduced Noetherian ring.

Definition. *Let $\triangle$ be a partition of k^n. A subset Y of k^n is called a piecewise algebraic set of partition $\triangle$ if there exists a subset T of $S^\mu(\triangle)$ such that*

$$Y = Z(T) = \{P \in k^n : s(P) = 0, \textit{for all} \ \ s \in T\}.$$

A piecewise algebraic set Y is called irreducible if $Y = Y_1 \cup Y_2$, with piecewise algebraic sets Y_1 and Y_2, implies $Y = Y_1$ or $Y = Y_2$. An irreducible piecewise algebraic set is called a piecewise algebraic variety.

For any subset Y of k^n, the ideal of Y in $S^\mu(\triangle)$ is defined by

$$I(Y) = \{s \in S^\mu(\triangle) \mid s(P) = 0 \ \text{ for all } \ P \in Y\},$$

and the coordinate ring $S(Y)$ of Y is $S^\mu(\Delta)/I(Y)$.

Let $\overline{k}, k'$ be two extended fields of k, $\xi = (\xi_1, \ldots, \xi_n) \in \overline{k}^n$, $\eta = (\eta_1, \ldots, \eta_n) \in k'^n$, if for any element s of $S^\mu(\triangle), s(\xi) = 0$ implies $s(\eta) = 0$, then we say ξ to be a generic point of η with respect to k or η to be a k-specialization of ξ. Let Y be a piecewise algebraic set, if $I(Y)$ has an extension zero ξ such that all extension zeros of $I(Y)$ are k-specialization of ξ, then ξ is called a generic point of Y.

Let $V \subseteq k^n$ be a piecewise algebraic set of partition $\triangle$, $W \subseteq k^m$ be a piecewise algebraic set of partition $\triangle'$, a mapping $f\colon V \mapsto W$ is called a spline mapping if there are splines $f_i(x_1, \ldots, x_n) \in S^\mu(\triangle), (1 \le i \le m)$ such that

$$f = (f_1(x_1, \ldots, x_n), \ldots, f_m(x_1, \ldots, x_n)).$$

If $f\colon V \mapsto W$ and $g\colon W \mapsto V$ are spline mappings such that $f \circ g = 1 \mid_W, g \circ f = 1 \mid_V$, then we say V and W are isomorphic.

Theorem 5. ([11]) *A piecewise algebraic set Y is irreducible if and only if Y has a generic point.*

Theorem 6. ([11]) *Let $V \subseteq k^n$ be a piecewise algebraic set of $\triangle$, $W \subseteq k^m$ be a piecewise algebraic set of $\triangle'$, $Hom(V, W)$ denotes the set of all spline mappings from V to W, $Hom(S(W), S(V))$ denotes the set of all k-algebraic homomorphisms from $S(W)$ to $S(V)$, then the mapping* $\mathrm{Hom}(V, W) \longmapsto \mathrm{Hom}(S(W), S(V))$

$$f \longmapsto f^*\colon s \mapsto s \circ f$$

is surjective. Moreover $f\colon V \longmapsto W$ is isomorphic if and only if $f^*\colon S(W) \longmapsto S(V)$ is k-isomorphic.

Acknowledgements. Supported by The National Natural Science Foundation of China, The Ph.D project of the State Education Commission of China, and the State Major Key Project *Mechanize proof and applications.*

References

1. Karlin, S., and Z. Ziegler, Tchebycheffian spline functions, SIAM J. Math. Anal. **3** (1966), 514–543.
2. Schoenberg, I. J. and A. Whitney, On Polya frequency functions III, Trans. Amer. Math. Soc. **74** (1953), 246–259.
3. Shi, X. Q. and R. H. Wang, Bezout's number of piecewise algebraic curves, *Intern. Conference Numer. Approx. and Appl., Dalian, Aug. 14–18, 1995.*
4. Su, Z. X., Piecewise algebraic curve and surface and their applications in CAGD, Ph.D thesis, Dalian University of Technology, 1993.
5. Walker, R. J., *Algebraic Curves*, Princeton, New Jersey, 1950.
6. Wang, R. H., The structural characterization and interpolation for multivariate splines, Acta Math. Sinica **18** (1975), 91–106.
7. Wang, R. H., Multivariate spline and its applications in science and technology, Rend. Sem. Mat. Fis. Milano , Vol. LXIII (1993), 213–229.
8. Wang, R. H., Piecewise algebraic curve and its branches, *The Seventh Conf. of Chinese Mathematical Society, Beijing, May 18-21, 1995.*
9. Wang, R. H. et al., *Multivariate Spline Functions and Its Applications* , Science Press, Beijing, 1994.
10. Wang, R. H. and G. H. Zhao, Intersection and local branch of piecewise algebraic curves, *Intern. Conference Numer. Approx. and Appl., Dalian, Aug. 14–18, 1995.*
11. Wang, W. J. and R. H. Wang, The irreducibility and isomorphism of piecewise algebraic sets, *Intern. Conference Numer. Approx. and Appl., Dalian, Aug. 14–18, 1995.*

Ren-Hong Wang
Institute of Mathematical Sciences
Dalian University of Technology
Dalian 116024, P.R.China